HEAVY METALS

Edited by **Hosam El-Din M. Saleh**
and **Refaat F. Aglan**

Heavy Metals
http://dx.doi.org/10.5772/intechopen.71185
Edited by Hosam El-Din M. Saleh and Refaat F. Aglan

Contributors

Olatunde Durowoju, Joshua Edokpayi, Oluseun Popoola, John Odiyo, Prasann Kumar, Shweta Pathak, Hebboul Zoulikha, Eduardo López-Maldonado, Mercedes T. Oropeza-Guzman, Ma. Del Rosario Moreno Virgen, Virginia Hernández Montoya, Rigoberto Tovar Gómez, Omar Francisco González Vázquez, Vanessa Alves, Absolon C. Da Silva Junior, Alessa G. Siqueira, Carolina A. De Sousa E Silva, Jordana De Assis N. Oliveira, Nívia Maria M. Coelho, Vhahangwele Masindi, Khathutshelo Lilith Muedi, Hoduck Kang, Chandra Romika, Ji Jian, Yao Fang, Qianyu Chen, Hengyu Tian, YiHsiu Jen, Chung Shin Yuan, Fei Li, Patricia Mussali-Galante, Efrain Tovar-Sánchez, Isela Hernández-Plata, Miguel Santoyo-Martínez, Leticia Valencia-Cuevas, Fumiyuki Nakajima, Rupak Aryal, Irina Lyanguzova, Vasily Yarmishko, Vadim Gorshkov, Natalie Stavrova, Irina Bakkal, Carlos Escudero-Oñate, Isabel Villaescusa, Randhir Kumar, Tarun Banerjee, Mohammed Abu-Dieyeh, Kamal Usman, Mohammad Al-Ghouti, Hadiya Khalilova, Oupa Ntwampe, Hosam El-Din M. Saleh

First published in London, United Kingdom, 2018 by IntechOpen
IntechOpen is the global imprint of INTECHOPEN LIMITED, registered in England and Wales, registration number: 11086078, The Shard, 25th floor, 32 London Bridge Street
London, SE19SG – United Kingdom
Printed in Croatia

British Library Cataloguing-in-Publication Data
A catalogue record for this book is available from the British Library

Additional hard copies can be obtained from orders@intechopen.com

Heavy Metals, Edited by Hosam El-Din M. Saleh and Refaat F. Aglan
p. cm.
Print ISBN 978-1-78923-360-5
Online ISBN 978-1-78923-361-2

For EU product safety concerns:
IN TECH d.o.o., Prolaz Marije Krucifikse Kozulić 3, 51000 Rijeka, Croatia,
info@intechopen.com or visit our website at intechopen.com.

Meet the editors

Hosam Saleh is a professor of Radioactive Waste Management at the Radioisotope Department, Atomic Energy Authority, Egypt. He received his MSc and PhD degrees in Physical Chemistry from the Cairo University. He is interested in studying innovative economic and environment-friendly techniques for management of hazardous and radioactive wastes. He authored many peer-reviewed scientific papers and chapters. He is an editor of several books related to valuable international publishers.

Refaat F. Aglan is an associate professor of Analytical Chemistry at the Atomic Energy Authority of Egypt. He received his MSc degree from the Cairo University and his PhD degree from the Ain Shams University. He authored more than 25 peer-reviewed scientific papers. He is interested in the development of chemical sensors based on inorganic and organic ligands for determination of toxic and industrially important metals and synthesis of inorganic ion exchanger.

Contents

Preface

Since the prehistoric ages, humankind is exploring heavy metals as persistent, hard, ductile, and technologically applicable materials. It all began with the Copper Age, starting more than 5 millennia BC, when copper was mined as the very first technologically applied metal; this era generated the basic knowledge on metallurgy techniques because copper was not only mined in its solid form but also by reductive conversion of, e.g., malachite. Moreover, this era also experienced the start of exploration of the noble metals such as gold and silver in the solid form, mainly implemented in the manufacturing of metal ornaments. During the third millennium BC, the development of a copper-tin alloy, harder and better to be applied as material than pure copper, resulted in the rise of the Bronze Age in Europe, North Africa, and the Middle East. Around the same time, copper was also blended with zinc, resulting in the until today heavily used alloy brass; this discovery is supposed to have started in the Far East. As we know today, copper possesses another characteristic typical for metals, such as the heavy metal group: they are electrically conductive, which today makes this element of utmost importance in electric coils and wires. Coming back to the prehistoric times, besides bronze and brass, meteoric iron, which contains considerable shares of the heavy metal nickel, was also already known during that time; however, due to its scarcity, it was mainly used for cultic purposes and metal ornament production. Moreover, the use of lead during the Bronze Age is also witnessed by respective historic findings. During the first millennium AD and the first millennium BC, the next technological breakthrough happened in different Eurasian regions, when the reductive exploitation of iron from ores, a chemically not trivial task, was discovered and further developed. From this period, well known as the famous Iron Age, heavy metals were the only technologically used metals at all for many and many centuries.

After the prehistoric times, the heavy metal lead, extracted from its ores, was frequently used especially in the Roman Empire. On the one hand, bivalent lead acted as the cation in the maybe first artificial sweetener: lead acetate was added to wine in order to increase its sweetness, needless to say that this did not contribute to the fact that the wine, which was pressed during that time, would have been distinguished by a special health-promoting effect! Moreover, the Romans used lead, called *plumbum* by them, in its metallic form technologically to make not only coffins and metal seals but also water pipes, which resulted in severe ecological heavy metal pollution and suspected thousands of dead people. Now, it also becomes clear where the English job title "plumber" has its origin.

It lasted until the early nineteenth century, until the first light metals such as magnesium, aluminum, titanium, sodium, potassium, strontium, and many more were detected and later on isolated in their metallic form, which went in parallel to the discovery of various previously unknown heavy metals such as gallium, thallium, or hafnium. Finally, it was in 1817, when the chemistry pioneer Leopold Gmelin raised the terminus "heavy metal" in order to classify the elements known that time into "nonmetals," "light metals," and "heavy metals," five decades before Mendelejev and Meyer established the periodic table of elements, whose gaps still existing that time led to the prediction and subsequent discovery of new chemical elements, among them also additional heavy metals.

Possibilities for the exposure of a heavy metal into different sectors of the ecosphere can exemplarily be visualized for the case of lead; grace to some convenient properties, such as its high ductility and the easy processibility based on its low-melting point, lead to attract early attention for numerous technological applications, such as sealings, lead weights for fishing, batteries, pipes, vessels, alloys for car body repair, antiknock agent for petrol (tetraethyl lead), bullets, and many more. One can easily figure out that these applications offer numerous possibilities to spread lead in soil, water, and air.

We are tremendously optimistic that the exploratory and scientific attempts collected and summarized in the book at hand will encourage researchers all over the globe to deepen their activities in this field and to attract the interest of undergraduates as well as of progressive representatives from relevant various sectors. Primarily, these activities shall boost the impatiently desired breakthrough of "heavy metals" identification and application processes.

Particularly, acknowledgment is given to the publishing process manager, Ms. Dajana Pemac, for her prosperous cooperation, exceptional efforts, and prompt response to my requests. Again, we would like to thank cordially all the contributors to this issue for their supreme work.

Hosam El-Din M. Saleh, PhD
Atomic Energy Authority
Radioisotope Department, Nuclear Research Center
Giza, Egypt

Section 1

Introduction to Heavy Metals

Chapter 1

Introductory Chapter: Introducing Heavy Metals

Martin Koller and Hosam M. Saleh

Additional information is available at the end of the chapter

http://dx.doi.org/10.5772/intechopen.74783

1. Introduction

"Heavy metals" are natural elements characterized by their rather high atomic mass and their high density. Although typically occurring in rather low concentration, they can be found all through the crust of our planet. Commonly, a density of at least 5 g cm^{-3} is used to define a heavy metal and to differentiate it from other, "light" metals. Other, broader definitions for "heavy metals" require an atomic mass higher than 23 or an atomic number exceeding 20; these definitions are highly error prone and confusing. Both alternative definitions cause the inclusion even of nonmetals; resorting to the atomic mass criterion, the maximum number of elements classified as "heavy metals" rockets high to 99 out of the in total 118 building blocks of our universe. Looking at the periodic table of elements, we learn that heavy metals *sensu stricto* (according to the density criterion) occupy the lion's share, namely, columns 3–16, of the periods 4 to 6, encompassing the transition metals, post-transition metals, and lanthanides [1].

Some heavy metals like copper, selenium, or zinc are essential trace elements, with functions indispensible for various biological processes also driving the entire human metabolism [2]. The heavy metal cobalt, acting as the central atom in the vitamin B_{12} complex, is a key player in the reductive branch of the propionic acid fermentation pathway [3]; without this special heavy metal compound, the gourmet would have to do without the unique flavor of Emmentaler cheese. Many heavy metals are of outstanding technological significance, e.g., iron, zinc, tin, lead, copper, tungsten, etc. Recently, different heavy metals act as the central atom of artificially designed "bioinorganic" catalysts for special chemical transformations [4]. Moreover, among them we find precious noble elements like gold, silver, iridium, rhodium, or platinum [5]. On the other hand, many of them, e.g., mercury, cadmium, arsenic, chromium, thallium, lead, and others, classically represent the "dark side of chemistry"; they exert toxic effects already at low concentration [6]. In this context, some heavy metals have gained dubious popularity by being the materials major crimes can be made of [7].

Heavy metals were literally heaven's sent by originating from asteroid impacts. Typically, heavy metals occur in the earth's crust in rather low concentrations between the low ppb ranges (noble metals) and up to 5% (iron); here, heavy metals are mainly found chemically bound in carbonate, sulfate, oxide, or silicate rocks or also occur in their metallic, elemental form. Weathering and erosion resulted in their leaching and mitigation into soil, rivers, and groundwater. About 4–5 billions of years ago, when Earth's mantle was still liquid, heavy metals sank to Earth's center and formed Earth's core, which today predominately consists of the heavy metals iron and nickel [1].

2. Heavy metals and the eco- and biosphere

It is important to emphasize that there are some trace elements among the heavy metal family, which are essential for many biological processes; they are predominantly found in period 4 of the periodic table of elements. For strict aerobic beings as we are, it would not be possible to survive without having cytochromes, which make aerobic life forms breath since the very beginning [8]. Iron also plays a major role in our respiration system as central, oxygen-affine, atom of the blood pigment heme. Copper plays a similar role in the transport of electrons and oxygen, especially as central atom in hemocyanin in mollusks and arthropods [9]. Zinc in turn is pivotal as constituent of zinc finger enzymes [10]. Selenium is described as an antioxidant; further, it is involved in hormone biosynthesis [11]. Cobalt was found to be significant in biosynthesis of complex compounds and different steps in cellular metabolism, especially as central atom in vitamin B_{12}, which is needed for cell division, blood formation, the nervous system, and in propionic acid biosynthesis (*vide supra*) [12]. Moreover, vanadium and manganese are important for regulation and functioning of several enzymes [13], whereas some metabolic functions are also assigned to the typically known-as-toxic elements chromium, arsenic, and nickel. Regarding arsenic, this element was only recently revealed as a natural constituent in herring caviar, where it was shown to substitute phosphorus in phosphatidylcholine-like lipids, the so-called arsenolipids [14]. Further, we should mention the role of molybdenum in some redox reactions [15], in addition to the function of cadmium in the metabolism of some microalgae from the Diatomophyceae class [16]. Moreover, the role of the heavy metal tungsten in the metabolism of prokaryotes is scientifically confirmed [17].

To give an impression on the quantities of heavy metals present in our human body, hence, the "heavy metal load" we are steadily carrying along with us, we can estimate that only about 0.01% of our mass originates from the presence of heavy metals, with iron (about 5 g in a person weighing 70 kg), zinc (2 g), lead, and copper (0.1–0.2 g each) being the top four heavy metals in our body; the rest, from the mass-related perspective, can be considered negligible [18].

Besides the abovementioned leaching and mitigation of heavy metals by erosion and weathering, these elements are mainly mobilized of by the action of humans during their physical (extraction, smelting) or chemical (reductive) release from ores and the subsequent processing for diverse applications. Other processes releasing heavy metals into the ecosphere involve their (agro)industrial, domestic, automotive, medical, electrical, and other technological use, resulting in their extensive distribution in both aquatic and terrestrial environments.

Currently, we witness increasing global worries regarding their possible adverse health effect and their negative enduring impacts on biosystems. Some heavy metals are reported or at least suspected to be carcinogenic (hexavalent chromium, arsenic, cobalt, nickel, antimony, vanadium, mercury), mutagenic (arsenic, vanadium), teratogenic (arsenic), allergenic (nickel), or endocrine-disrupting (silver, copper, zinc, selenium). Others (e.g., thallium) lead to neurological and behavioral alterations, particularly in the case of children, central nervous system damage (mercury, lead, thallium, manganese, and tin), bone marrow damage, and osteoporosis (cadmium); are hepatotoxic and/or nephrotoxic (cadmium, cadmium, mercury, heptavalent manganese); cause heart rhythm disturbances (thallium); or negatively affect the immune system (lead) [19].

3. Determining heavy metals

Classically, quantification of heavy metals involves well-established techniques, such as wet chemical methods (gravimetric, titrimetric, colorimetric, etc.), coupled plasma/atomic emission spectrometry (ICP/AES), inductively coupled plasma with mass spectrometric detection (ICP/MS), or atomic absorption spectroscopy (AAS) [20–22]. Moreover, diverse ion selective electrodes are frequently reported for heavy metal determination [22, 23]. Currently, new, robust, sensitive, selective, inexpensive, and fast optical [24–26], chemical [27, 28], and biological [29–33] sensory systems are currently in status of development. Such advances in analytical chemistry are currently tightly connected to nanotechnology [26, 28, 34]. Moreover, so-called lab-on-paper sensors were also developed for heavy metal determination, as demonstrated for quantification of mercury, silver, copper, cadmium, lead, chromium, and nickel [29]. This sensor, operated by an immobilized enzyme, is an example for so-called biosensors, which synergistically combine the scientific fields of biotechnology and microelectronics; such "biosensors" consist of an immobilized biological component in combination with a transducer [30]. As a very recent technology, so-called genetically encoded fluorescent sensors can be used for monitoring heavy metals inside biological cells and were already assessed for determination of heavy metals like zinc, copper, lead, cadmium, mercury, or arsenic [35].

4. Mitigating heavy metals

To elevate the negative impacts of heavy metals, remediation techniques are increasingly improved in order to address the growing public pressure to reduce prevailing environmental hazards and to bequeath the subsequent generations a future worth living.

Traditional physical, thermal, chelating, and other chemical techniques often display serious shortcomings such as too high cost, excessive expenditure of work, and invasive change of soil properties and microflora [36]. Traditionally, remediation of soils contaminated by heavy metals resorts to simply digging the contaminated soil and subsequently disposing it at landfills. Of course, this disposal strategy merely postpones the eco-problem by shifting it from one location to the next and, moreover, generates hazards connected with transportation of precarious soil and leaching of heavy metals at the ultimate disposal site. In the case of water polluted by heavy metals, alkaline lime precipitation is known as a better advanced and maybe the most efficient traditional technique for treating heavily polluted effluents. Lime precipitation can effectively be used to treat wastewater with metal loads exceeding 1 g L^{-1}. However, the remaining heavy metal-alkali-sludge stills need ultimate disposal [37].

Modern physical and chemical approaches for remediation of heavy metal pollution involve the use of adsorption on new adsorbents such as nano-carriers, ion exchange techniques, removal via advanced membrane filtration techniques, electrodialysis, or photocatalysis. Among these novel physicochemical techniques, new adsorption- and membrane filtration-based methods are most thoroughly investigated and are most commonly applied to treat contaminated wastewater [37, 38]. Among new absorbents, both inorganic (kaolinite, montmorillonite) [39] and organic materials (e.g., agricultural waste or bio-char) [40–43] were studied for heavy metal recovery. In this context, the application of carbon-, metal-, or metal oxide-based nanoparticles as adsorbents benefits from high surfaces susceptible toward metal adsorption and expedient reactivity. Here, the mechanisms of interactions of nanomaterials with, on the one hand, heavy metals and, on the other hand, heavy metal with additional wastewater constituents with metal-binding groups need to be understood in order to optimize the recovery processes [44].

Photocatalysis uses photons from the UV-near vis region of light's electromagnetic spectrum and, when operated in a smart way, is able to degrade toxic organic pollutants in parallel to metal recovery in just one single process step. This technique resorts to photocatalytic semiconductors, e.g., TiO_2, which, when illuminated with UV light, generate highly reductive electrons that in turn reduce heavy metal ions in contaminated wastewater. As an example, photocatalysis was successfully implemented for reduction of the dramatically precarious hexavalent chromium to its about 500 times less toxic trivalent form. In the case of precious noble metals like gold, this process does not only mitigate an environmental pollutant but also mines value-added materials for further use ("photorecovery") [45].

Phytoremediation is an emerging technology to overcome shortcomings of above-discussed methods. During phytoremediation, plants act synergistically with diverse soil microbes, which convert the heavy metals in a form bioavailable for the plants, finally decreasing the concentrations of contaminants in affected environments. Phytoremediation resorts to the ability of many plants for specific and efficient uptake, translocation, and storage of hazardous elements with chemical properties mimicking those of elements essential for plant growth [46]. Being a relatively recent technology, phytoremediation is supposed to be efficient, cost-effective, and ecologically benign; is driven by sunlight as the sole energy source; and enjoys an excellent public acceptance [47]. In the context of heavy metals, new powerful "heavy metal hyperaccumulator" plants are currently assessed for both phytoremediation (getting rid of the unwanted heavy metal) and phytomining (accumulating the precious heavy metals for further use). Such "hyperaccumulators" are characterized by their capacity to take up toxic metal ions at levels of thousands of ppm. In the optimum scenario, the toxic metals are transported from the plant's rhizosphere up to the shoots in the plant's periphery; now, the shoots, enriched with the target contaminants, can easily be harvested and burned for energy generation and, if economically reasonable, recycling the metal from remaining ash [48]. Using aquatic plants, phytoremediation can also be used to cure polluted water bodies. Various plant species have successfully performed in absorbing heavy metals such as arsenic, cadmium, chromium, lead, and even radionuclides from contaminated soil. Among the different phytoremediation categories, phytoextraction can be used to mitigate heavy metals from soil by profiting from its ability to uptake and accumulate those heavy metals, which constitute elements essential for plant development, such as iron, copper, manganese, molybdenum, or nickel. In addition to these essential elements, some chemically similar heavy metals with unknown or not yet confirmed biological function can also be phytoextracted, such as cadmium, chromium, silver, lead, cobalt, selenium, or mercury [49]. Excellent results for phytoremediation of arsenic were reported for the fern *Pteris vittata* L. species [50]; here, the uptake capacity of the plant yielded more than 4 g heavy metal per kg plant material [51]. In the case of lead, several plants, most of all different mustards (*Brassica* ssp.), are described to be able to accumulate between 50 and 100 mg of this heavy metal per gram plant dry mass [46].

Author details

Martin Koller[1] and Hosam M. Saleh[2]*

*Address all correspondence to: hosamsaleh70@yahoo.com

1 Office of Research Management and Service, c/o Institute of Chemistry, NAWI Graz, University of Graz, Graz, Austria

2 Radioisotope Department, Nuclear Research Center, Egyptian Atomic Energy Authority, Cairo, Egypt

References

[1] Duffus JH. Heavy metals' – A meaningless term? International Union of Pure and Applied Chemistry (IUPAC). 2002;**74**:793-807. DOI: 10.1351/pac2002740507

[2] Mertz W. The essential trace elements. Science. 1981;**213**(4514):1332-1338

[3] Stowers CC, Cox BM, Rodriguez BA. Development of an industrializable fermentation process for propionic acid production. Journal of Industrial Microbiology & Biotechnology. 2014;**41**(5):837-852

[4] Terfassa B, Schachner JA, Traar P, Belaj F, Zanetti NCM. Oxorhenium(V) complexes with naphtholate-oxazoline ligands in the catalytic epoxidation of olefins. Polyhedron. 2014;**75**:141-145

[5] Rao CRM, Reddi GS. Platinum group metals (PGM); occurrence, use and recent trends in their determination. TrAC Trends in Analytical Chemistry. 2000;**19**(9):565-586

[6] Duruibe JO, Ogwuegbu MOC, Egwurugwu JN. Heavy metal pollution and human biotoxic effects. International Journal of physical sciences. 2007;**2**(5):112-118

[7] Kaiser J. Toxicologists shed new light on old poisons. Science. 1998;**279**(5358):1850-1851

[8] Groves JT, Haushalter RC, Nakamura M, Nemo TE, Evans BJ. High-valent iron-porphyrin complexes related to peroxidase and cytochrome P-450. Journal of the American Chemical Society. 1981;**103**(10):2884-2886

[9] Redfield AC, Coolidge T, Shotts MA. The respiratory proteins of the blood I. The copper content and the minimal molecular weight of the Hemocyanin of *Limulus polyphemus*. Journal of Biological Chemistry. 1928;**76**(1):185-195

[10] Maret W. Zinc biochemistry: From a single zinc enzyme to a key element of life. Advances in Nutrition. 2013;**4**(1):82-91

[11] Stadtman TC. Selenium biochemistry. Annual Review of Biochemistry. 1990;**59**(1):111-127

[12] Hudson TS, Subramanian S, Allen RJ. Determination of pantothenic acid, biotin, and vitamin B12 in nutritional products. Journal-Association of Official Analytical Chemists. 1984;**67**(5):994-998

[13] Egami F. Origin and early evolution of transition element enzymes. The Journal of Biochemistry. 1975;**77**(6):1165-1169

[14] Viczek SA, Jensen KB, Francesconi KA. Arsenic-containing phosphatidylcholines: A new group of arsenolipids discovered in herring caviar. Angewandte Chemie International Edition. 2016;**55**(17):5259-5262

[15] Mendel RR. Molybdenum: Biological activity and metabolism. Dalton Transactions. 2005;**21**:3404-3409

[16] Lane TW, Saito MA, George GN, Pickering IJ, Prince RC, Morel FM. Biochemistry: A cadmium enzyme from a marine diatom. Nature. 2005;**435**(7038):42

[17] Andreesen JR, Makdessi K. Tungsten, the surprisingly positively acting heavy metal element for prokaryotes. Annals of the New York Academy of Sciences. 2008;**1125**(1):215-229

[18] Iyengar GV. Reevaluation of the trace element content in reference man. Radiation Physics and Chemistry. 1998;**51**(4-6):545-560

[19] Tchounwou PB, Yedjou CG, Patlolla AK, Sutton DJ. Heavy metal toxicity and the environment. In: Molecular, Clinical and Environmental Toxicology. Basel: Springer; 2012. pp. 133-164

[20] Butcher DJ. Advances in inductively coupled plasma optical emission spectrometry for environmental analysis. Instrumentation Science and Technology. 2010;**38**(6):458-469

[21] Feldmann J, Salaün P, Lombi E. Critical review perspective: Elemental speciation analysis methods in environmental chemistry – Moving towards methodological integration. Environmental Chemistry. 2009;**6**(4):275-289

[22] Saar RA, Weber JH. Comparison of spectrofluorometry and ion-selective electrode potentiometry for determination of complexes between fulvic acid and heavy-metal ions. Analytical Chemistry. 1980;**52**(13):2095-2100

[23] Wang J. Stripping analysis at bismuth electrodes: A review. Electroanalysis. 2005; **17**(15-16):1341-1346

[24] Oehme I, Wolfbeis OS. Optical sensors for determination of heavy metal ions. Microchimica Acta. 1997;**126**(3-4):177-192

[25] Rasheed T, Bilal M, Nabeel F, Iqbal HM, Li C, Zhou Y. Fluorescent sensor based models for the detection of environmentally-related toxic heavy metals. Science of the Total Environment. 2018;**615**:476-485

[26] Ullah N, Mansha M, Khan I, Qurashi A. Nanomaterial-based optical chemical sensors for the detection of heavy metals in water: Recent advances and challenges. TrAC Trends in Analytical Chemistry. 2018;**100**:155-166

[27] Vlasov Y, Legin A, Rudnitskaya A. Cross-sensitivity evaluation of chemical sensors for electronic tongue: Determination of heavy metal ions. Sensors and Actuators B: Chemical. 1997;**44**(1-3):532-537

[28] Salimi F, Kiani M, Karami C, Taher MA. Colorimetric sensor of detection of Cr(III) and Fe(II) ions in aqueous solutions using gold nanoparticles modified with methylene blue. Optik-International Journal for Light and Electron Optics. 2018;**158**:813-825

[29] Hossain SZ, Brennan JD. β-Galactosidase-based colorimetric paper sensor for determination of heavy metals. Analytical Chemistry. 2011;**83**(22):8772-8778

[30] Verma N, Singh M. Biosensors for heavy metals. Biometals. 2005;**18**(2):121-129

[31] Ghica ME, Carvalho RC, Amine A, Brett CM. Glucose oxidase enzyme inhibition sensors for heavy metals at carbon film electrodes modified with cobalt or copper hexacyanoferrate. Sensors and Actuators B: Chemical. 2013;**178**:270-278

[32] Ghica ME, Brett CM. Glucose oxidase inhibition in poly(neutral red) mediated enzyme biosensors for heavy metal determination. Microchimica Acta. 2008;**163**(3-4):185-193

[33] Zhylyak GA, Dzyadevich SV, Korpan YI, Soldatkin AP, El'Skaya AV. Application of urease conductometric biosensor for heavy-metal ion determination. Sensors and Actuators B: Chemical. 1995;**24**(1-3):145-148

[34] Aragay G, Merkoçi A. Nanomaterials application in electrochemical detection of heavy metals. Electrochimica Acta. 2012;**84**:49-61

[35] Finney LA, O'halloran TV. Transition metal speciation in the cell: Insights from the chemistry of metal ion receptors. Science. 2003;**300**(5621):931-936

[36] Fu F, Wang Q. Removal of heavy metal ions from wastewaters: A review. Journal of Environmental Management. 2011;**92**(3):407-418

[37] Barakat MA. New trends in removing heavy metals from industrial wastewater. Arabian Journal of Chemistry. 2011;**4**(4):361-377

[38] Premkumar MP, Thiruvengadaravi KV, Kumar PS, Nandagopal J, Sivanesan S. Eco-friendly treatment strategies for wastewater containing dyes and heavy metals. In: Environmental Contaminants. Singapore: Springer; 2018. pp. 317-360

[39] Bhattacharyya KG, Gupta SS. Adsorption of a few heavy metals on natural and modified kaolinite and montmorillonite: A review. Advances in Colloid and Interface Science. 2008;**140**:114-131

[40] Sud D, Mahajan G, Kaur MP. Agricultural waste material as potential adsorbent for sequestering heavy metal ions from aqueous solutions – A review. Bioresource Technology. 2008;**99**:6017-6027

[41] Cimino G, Passerini A, Toscano G. Removal of toxic cations and Cr(VI) from aqueous solution by hazelnut shell. Water Research. 2000;**34**(11):2955-2962

[42] Inyang MI, Gao B, Yao Y, Xue Y, Zimmerman A, Mosa A, Pullammanappallil P, Ok YS, Cao X. A review of biochar as a low-cost adsorbent for aqueous heavy metal removal. Critical Reviews in Environmental Science and Technology. 2016;**46**(4):406-433

[43] Poo KM, Son EB, Chang JS, Ren X, Choi YJ, Chae KJ. Biochars derived from wasted marine macro-algae (*Saccharina japonica* and *Sargassum fusiforme*) and their potential for heavy metal removal in aqueous solution. Journal of Environmental Management. 2018;**206**:364-372

[44] Tang WW, Zeng GM, Gong JL, Liang J, Xu P, Zhang C, Huang BB. Impact of humic/fulvic acid on the removal of heavy metals from aqueous solutions using nanomaterials: A review. Science of the Total Environment. 2014;**468**:1014-1027

[45] Ku Y, Jung IL. Photocatalytic reduction of Cr(VI) in aqueous solutions by UV irradiation with the presence of titanium dioxide. Water Research. 2001;**35**(1):135-142

[46] Ali H, Khan E, Sajad MA. Phytoremediation of heavy metals – Concepts and applications. Chemosphere. 2013;**91**(7):869-881

[47] Ghosh M, Singh SP. A review on phytoremediation of heavy metals and utilization of it's by products. Asian Journal on Energy and Environment. 2005;**6**(4):18

[48] Tangahu BV, Sheikh Abdullah SR, Basri H, Idris M, Anuar N, Mukhlisin M. A review on heavy metals (As, Pb, and Hg) uptake by plants through phytoremediation. International Journal of Chemical Engineering. 2011;**2011**:31. Article ID 939161

[49] Sarwar N, Imran M, Shaheen MR, Ishaque W, Asif M, Kamran MA, Matloob A, Rehim A, Hussain S. Phytoremediation strategies for soils contaminated with heavy metals: Modifications and future perspectives. Chemosphere. 2017;**171**:710-721

[50] Francesconi K, Pornsawan V, Sridokchan W, Goessler W. Arsenic species in an arsenic hyperaccumulating fern, *Pityrogramma calomelanos*: A potential phytoremediator of arsenic-contaminated soils. Science of the Total Environment. 2002;**284**:27-35

[51] Fayiga AO, Ma LQ, Cao X, Rathinasabapathi B. Effects of heavy metals on growth and arsenic accumulation in the arsenic hyperaccumulator *Pteris vittata* L. Environmental Pollution. 2004;**132**(2):289-296

Section 2

Characterization and Removal of Heavy Metals

Chapter 2

Cadmium Iodate Syntheses and Characterization

Zoulikha Hebboul

Additional information is available at the end of the chapter

http://dx.doi.org/10.5772/intechopen.73866

Abstract

Six polymorphs of anhydrous cadmium iodate are characterized, three of which showing second harmonic generation activity (SHG). Single crystals of $Cd(IO_3)2{\cdot}H_2O$ are obtained by slow evaporation of aqueous solutions of $CdCl_2$ and KIO_3. This compound crystallizes in the triclinic space group P1¯, a = 7.119(2), b = 7.952(2), c = 6.646(2)Å, α = 102.17(2)°, β = 114.13(2)°, and γ = 66.78(4)°. Three chemical routes of preparation of chloro cadmium iodate $CdIO_3Cl$ are given. The prepared material was characterized by X-ray diffraction (XRD) and scanning electron microscopy (SEM). The crystal structure has been determined by single-crystal X-ray diffraction methods; the unit cell is orthorhombic with a = 7.270(0) Å, b = 15.995 (0) Å, c = 7.1980 (0) Å V = 837.009(1) Å3, and Z = 8. The space group is CmCa. The cadmium hydroxy-iodate $CdIO_3OH$ is synthesized in the form of transparent platelets in the same way as $CdIO_3Cl$; the unit cell is orthorhombic with a = 11.5245(11) Å, b = 6.7985 (7) Å, c = 4.7303 (4) Å, V = 304.31(1) Å3, and Z = 4.

Keywords: cadmium, iodate, XRD, SHG, polymorphism, mixed ligand

1. Introduction

Cadmium is a chemical element with symbol Cd and atomic number 48. This soft, bluish-white metal is chemically similar to the two other stable metals in group 12, zinc and mercury. Like zinc, it demonstrates oxidation state +2 in most of its compounds, and like mercury, it has a lower melting point than the transition metals in groups 3 through 11. Cadmium and its congeners in group 12 are often not considered transition metals, in that they do not have partly filled *d* or *f* electron shells in the elemental or common oxidation states. The average concentration of cadmium in Earth's crust is between 0.1 and 0.5 parts per million (ppm). It was discovered in 1817 simultaneously by Stromeyer and Hermann, both in Germany, as an impurity in zinc carbonate.

Despite cadmium being a hazardous heavy metal and toxic to almost all living organisms, it is used in batteries, electroplating, nuclear reactors, and semiconductors. In 2009, 86% of cadmium was used in batteries, predominantly in rechargeable nickel-cadmium batteries. Nickel-cadmium cells have a nominal cell potential of 1.2 V. The cell consists of a positive nickel hydroxide electrode and a negative cadmium electrode plate separated by an alkaline electrolyte (potassium hydroxide) [1]. Cadmium electroplating, consuming 6% of the global production, is used in the aircraft industry to reduce corrosion of steel components [2]. It is used too in the control rods of nuclear reactors, acting as a very effective "neutron poison" to control neutron flux in nuclear fission [2]. When cadmium rods are inserted in the core of a nuclear reactor, cadmium absorbs neutrons preventing them from creating additional fission events, thus controlling the amount of reactivity. The pressurized water reactor designed by Westinghouse Electric Company uses an alloy consisting of 80% silver, 15% indium, and 5% cadmium [2]. The cadmium oxide was used in black and white television phosphors and in the blue and green phosphors of color television cathode ray tubes [3]. Cadmium sulfide (CdS) is used as a photoconductive surface coating for photocopier drums [4]. Various cadmium salts are used in paint pigments, with CdS as a yellow pigment being the most common. Cadmium selenide is a red pigment, commonly called cadmium red. To painters who work with the pigment, cadmium provides the most brilliant and durable yellows, oranges, and reds—so much so that during production, these colors are significantly toned down before they are ground with oils and binders or blended into watercolors, gouaches, acrylics, and other paint and pigment formulations. Because these pigments are potentially toxic, users should use a barrier cream on the hands to prevent absorption through the skin [5] even though the amount of cadmium absorbed into the body through the skin is reported to be less than 1% [6]. In PVC, cadmium was used as heat, light, and weathering stabilizers [2, 7]. Currently, cadmium stabilizers have been completely replaced with barium-zinc, calcium-zinc, and organo-tin stabilizers. Cadmium is used in many kinds of solder and bearing alloys, because of a low coefficient of friction and fatigue resistance [2]. It is also found in some of the lowest-melting alloys, such as Wood's metal [8].

The SHG is a nonlinear optical (NLO) process, in which two photons with the same frequency interacting with a nonlinear material are effectively "combined" to generate new photons with twice the energy and therefore twice the frequency and half the wavelength of the initial photons. Second harmonic generation, as an even-order nonlinear optical effect, is only allowed in media without inversion symmetry [9]. It is a special case of sum frequency generation. Half-harmonic generation (a special case of spontaneous parametric down-conversion) is its reverse process where one photon leads to a pair of photons with half the energy and occurs in parallel of the SHG with a lower probability though [10]. The second-order nonlinear susceptibility of a medium is responsible for the creation of SHG, which can convert a small or a large part of the excitation wave depending on some interference conditions that can happen in the medium. It, often called frequency doubling, is also a process in radio communication; it was developed early in the twentieth century and has been used with frequencies in the megahertz range. It is a special case of frequency multiplication.

Metal iodates are of great importance in the investigation of nonlinear optical (NLO) materials. Owing to the presence of stereochemically active lone-pair electrons on IV ions, the IO_3 unit is a favorable NLO-active anionic group with large microscopic second-order NLO susceptibility. The alignment of IO_3 units in polar or non-centrosymmetric (NCS) crystal structures may result in materials with excellent second harmonic generation (SHG) properties [11]. Recently, a few effective routes have been developed for the synthesis of novel NLO iodates, such as the introduction of mixed ligand Cl-, F-, and OH-. These research efforts afforded a large number of metal iodates with excellent SHG properties, including $Bi(IO_3)F_2$ [11] and $Zn(IO_3)OH$ [12].

Cadmium iodate $Cd(IO_3)_2$ is a chemical compound of cadmium and iodate ligand. It is notable for its crystal structure. In 1838, Rammelsberg synthesized anhydrous cadmium iodate by hot mixing concentrated solutions of sodium iodate and cadmium nitrate [13]. A century later in 1940, Oelke and Wagner studied the solubility of this compound in water [14] and thermal behavior [15]; the first diffraction pattern was presented in 1968 [16]. In 1978 the structure of anhydrous cadmium iodate is described as a single orthorhombic phase $P2_12_12_1$ [17, 18], later called δ-$Cd(IO_3)_2$. But in 1980, Liang et al. [19] gave powder diffraction results that do not correspond to this compound [20]. Work carried out in 2005 revealed a polymorphism of anhydrous $Cd(IO_3)_2$ iodate. The $CdCl_2$-IO_3 system has shown that there are no less than five polymorphic phases for cadmium—α-$Cd(IO_3)_2$, β-$Cd(IO_3)_2$, γ-$Cd(IO_3)_2$, δ-$Cd(IO_3)_2$, and ε-Cd $(IO_3)_2$—obtained according to the synthesis conditions [13, 21, 22], and finally in 2017, Hebboul provides existence of a new polymorph ζ-$Cd(IO_3)_2$ [23]. This is a complex chemical system that has a large polymorphism that makes these materials very difficult to develop as crystals for optics, and obtaining large crystals was a problem because of the complexity of the system [20, 24, 25]; all these polymorphic phases are not acentric; those are γ-$Cd(IO_3)_2$, ε-$Cd(IO_3)_2$, and ζ-$Cd(IO_3)_2$; of course δ-$Cd(IO_3)_2$ has a signal of second harmonic generation (SHG) but lower than that of $LiIO_3$ [13]. Ravi Kumar et al. optimized the growth conditions of the δ-Cd $(IO_3)_2$ crystal that was grown from an aqueous solution by employing the slow-cooling technique. The structure was confirmed by single-crystal XRD analysis. $Cd(IO_3)_2$ is transparent from 350 to 2000 nm, thus confirming its wider optical transmission range to extend its applications in the entire visible and NIR region. The optical band gap is found to be 3.85 eV from the Tauc's plot. As an inorganic material, the developed crystal has moderate SHG efficiency, good thermal stability, and mechanical strength, and $Cd(IO_3)_2$ has the promise to be utilized for NLO applications and photonic device fabrications. The results of optical band gap, mechanical, dielectric, and electrical properties of $Cd(IO_3)_2$ are reported for the first time in 2008 [22].

This chapter will investigate the bibliography on synthetic methods and powder spectra for various $Cd(IO_3)_2$ polymorphs and monohydrate cadmium iodate. Moreover, bibliography on the two cadmium mixed ligand compounds, cadmium chloro-iodate $CdIO_3Cl$ and cadmium hydroxy-iodate $CdIO_3OH$.

2. Synthesis methods and powder spectra for various $Cd(IO_3)_2$ polymorphs and monohydrate cadmium iodate

At 20°C and constant concentration of cadmium, it is possible to precipitate in sequence; this fact is used to assign a name—the first polymorph that precipitates is called α-$Cd(IO_3)_2$, the

(1) *Variable cadmium concentration at fixed temperatures and time* (T = 20°C, t = 4 hours)				
Concentration	$C_1 = 94.510^{-3}$ mol/l	$C_2 = 56.810^{-3}$ mol/l	$C_3 = 35.510^{-3}$ mol/l	$C_4 \leq 25.810^{-3}$ mol/l
Polymorphs present in the precipitate	α~β~γ	β~δ > α~γ	γ > β ~α	γ >> α
(2) *Variable reaction time with fixed cadmium concentration* (T = 20°C, $CCd = 94.510^{-3}$ mol/l)				
Reaction time	t = 10 min	t = 1 h	t = 4 h	t = 7 h
Polymorphs present in the precipitate	A	α >> β	α~β~γ	α ≥ γ >> δ
(3) *Variable reaction time with fixed cadmium concentration* (T = 60°C, $C_{Cd} = 94.510^{-3}$ mol/l)				
Reaction time	t = 10 min	t = 1 h	t = 4 h	t = 7 h
Polymorphs present in the precipitate	α ≥ γ	γ > α ≥ δ	γ >> δ >> α	γ >> δ >> α

Table 1. Summary of various results obtained according to synthesis conditions [20].

following β-$Cd(IO_3)_2$, then γ-$Cd(IO_3)_2$, and finally δ-$Cd(IO_3)_2$—heating these four polymorphs or working in nitric acid at 30°C can lead to ε-$Cd(IO_3)_2$ [20]. The kinetic study of the $CdCl_2$-HIO_3 [13] system in aqueous solution is extremely complicated and very sensitive to the procedure. A thorough study of this system by systematically varying the operating conditions was carried out. **Table 1** summarizes the different results obtained according to the four studied items:

1. Variable cadmium concentration at fixed temperature and time
2. Variable reaction time at fixed cadmium concentration and at T = 20°C
3. Variable reaction time with fixed cadmium concentration and at T = 60°C
4. The synthesis in a concentrated nitric acid medium

From acid solutions Abrahams and Nassau have shown that the transition metal iodate d or f could be obtained in the anhydrous state from nitric acid solutions [26–32]. Nitric acid is the medium used by Bach for the polymorph crystallogenesis of the δ-$Cd(IO_3)_2$ [18].

The fifth polymorph ε can be prepared under the same previous conditions but in the 30% nitric acid medium [13, 20].

The comparison of all results shown in the literature shows that the XRD diagram of the polymorph ε has been already published in 1980 without being identified [19].

2.1. Preparations and characterizations of pure phases

2.1.1. *The polymorph α-Cd(IO₃)₂*

This product precipitates in the form of a very fine white powder from a solution of cadmium chloride and the highly concentrated iodic acid, $C_1 = 94.510^{-3}$ mol/l; all very slow evaporation efforts in order to obtain single crystals for structural study proved vain. The powder diagram of this variety is shown in **Figure 1** indicating that the spectrum of polymorph α is characterized by the main peak d = 3.5784 Å [13].

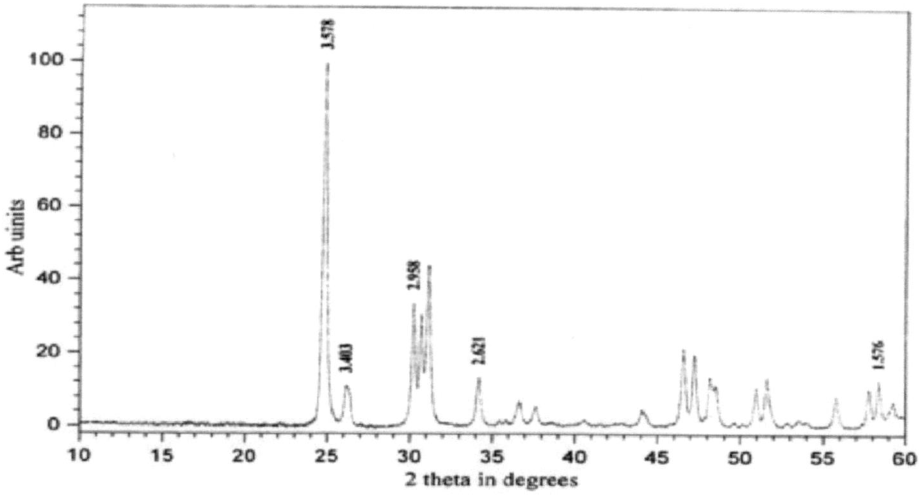

Figure 1. Diffraction diagram of α-$Cd(IO_3)_2$ [20].

2.1.2. *The polymorph β-$Cd(IO_3)_2$*

The polymorph β is obtained in its pure state when heating, at 250°C, the product obtained after a few days of the reaction of cadmium chloride (0.26 g) with lithium iodate (0.5 g) in 70 ml of water. The powder diagram of this variety is shown in **Figure 2** and shows that the spectrum of the polymorphic β is characterized by the main peak d = 3.296 Å [13].

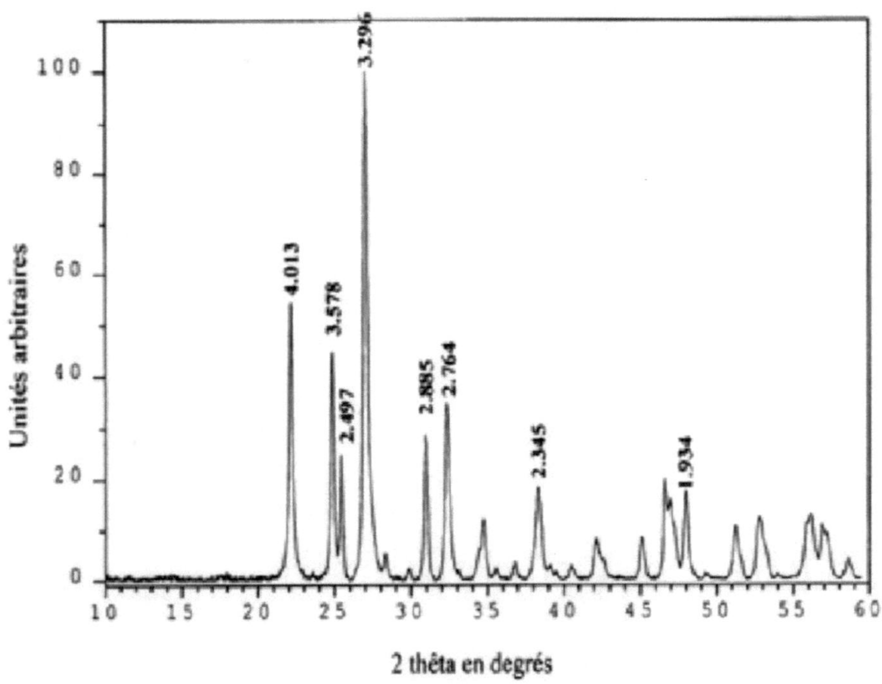

Figure 2. Diffraction diagram of β-$Cd(IO_3)_2$ [20].

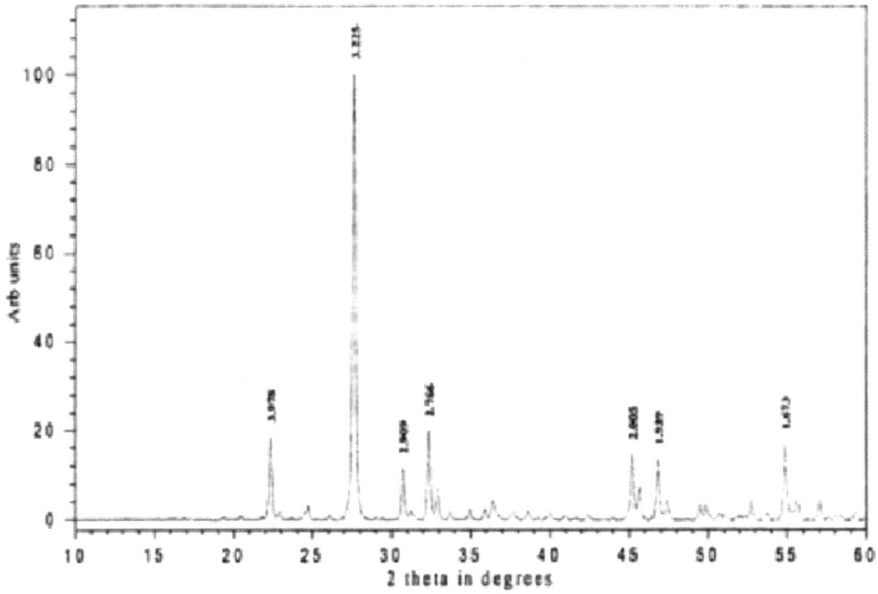

Figure 3. Diffraction diagram of γ-$Cd(IO_3)_2$ [20].

2.1.3. *The polymorph γ-$Cd(IO_3)_2$*

This polymorph can be obtained in its pure state, in the form of a well-crystallized powder, by two different methods: by evaporation at 60°C of a dilute aqueous solution of cadmium chloride and iodic acid ($C_{Cd} < 25\ 10^{-3}$ mol/l) and by evaporation at 70°C of a dilute aqueous solution of cadmium chloride 0.26 and 0.607 g of potassium iodate in 100 ml of water. The crystallogenesis efforts on this compound have not succeeded. It should be noted that this compound remains stable up to 380°C [13, 20, 25] and crystallizes in the non-centro-symmetric space group, because the optical activity test in second harmony generation (Kurtz and Perry test [33]) is positive and visually inferior to that of $LiIO_3$ [13, 20, 24, 25]. The X-ray diffraction pattern shown in **Figure 3** shows a perfectly crystallized product that is not listed in the JCPDS database. It should be noted that in the 2θ = 10°–60° range of its powder spectrum, there are 44 low-intensity peaks compared to the main peak d = 3.22 Å.

2.1.4. *The polymorph δ-$Cd(IO_3)_2$*

A solution containing (0.2 g) of γ-$Cd(IO_3)_2$ and 30 ml (60%)HNO_3 gives after slow evaporation (70°C) a prismatic transparent crystals, slightly yellow [13, 20]. A monocrystal of 11 × 10 × 2 mm^3 shown in **Figure 4** was obtained [22] from a saturated solution of a white powder of $Cd(IO_3)_2$ in a 60% nitric acid medium of HNO_3 at T = 60°C; the total crystallogenesis time is 20–25 days. The white powder used is the product of the reaction of cadmium chloride and iodic acid (1/2) in an aqueous medium.

Unlike other polymorphs, the spectrum of **δ-$Cd(IO_3)_2$** shown in **Figure 5** of 2θ = 10° to 2θ = 60° is characterized by several peaks because it contains more than 60 peaks, six among which are intense (d = 3.93 Å, d = 3.66 Å, d = 3.44 Å, **d = 3.32 Å,** d = 2.49 Å, and d = 1.77 Å).

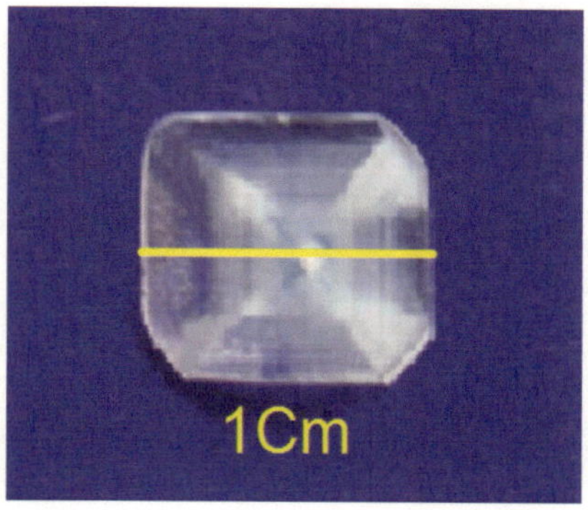

Figure 4. δ-$Cd(IO_3)_2$ morphology [22].

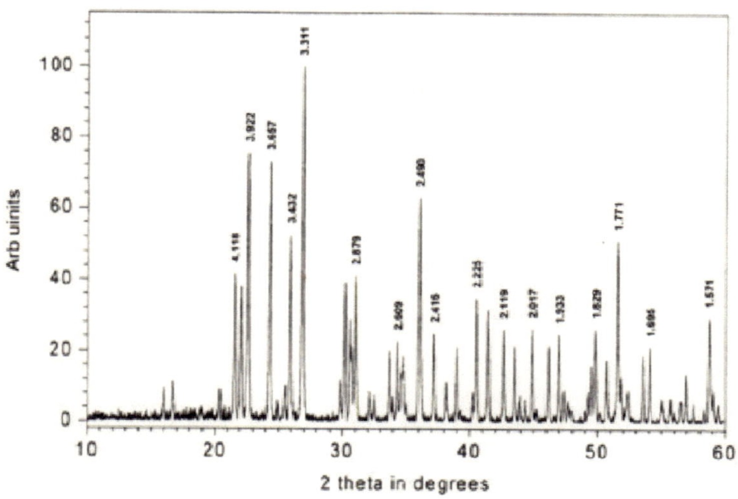

Figure 5. Diffraction diagram of δ-$Cd(IO_3)_2$ [20].

2.1.5. *The polymorph ε-$Cd(IO_3)_2$*

This polymorph constitutes the final product of the thermal evaluation, before the decomposition, of all the polymorphs described previously [20]. This polymorph can be obtained in solution by slow evaporation at 70°C of a solution containing 0.2 g of γ-$Cd(IO_3)_2$ in 30 ml of 30% HNO_3 [13].

Clear colorless needles, about 0.2 mm shown in **Figure 6**, are used to solve the crystalline structure of this polymorph which crystallizes in the centro-symmetric group $Pca2_1$, thus inactive in GSH but thermally stable up to 550°C. The powder diagram is given in **Figure 7** characterized by the main peak d = 3.26 Å.

2.1.6. *The polymorph ζ-$Cd(IO_3)_2$*

2 mmol of anhydrous cadmium chloride and 1 mmol potassium iodate were dissolved in 20 ml of deionized water. The solution was evaporated slowly at room temperature. After 2 days

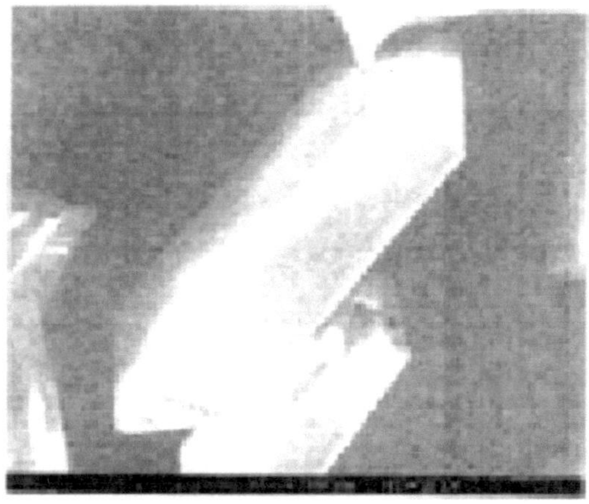

Figure 6. SEM view on the polymorph ε-$Cd(IO_3)_2$ [20].

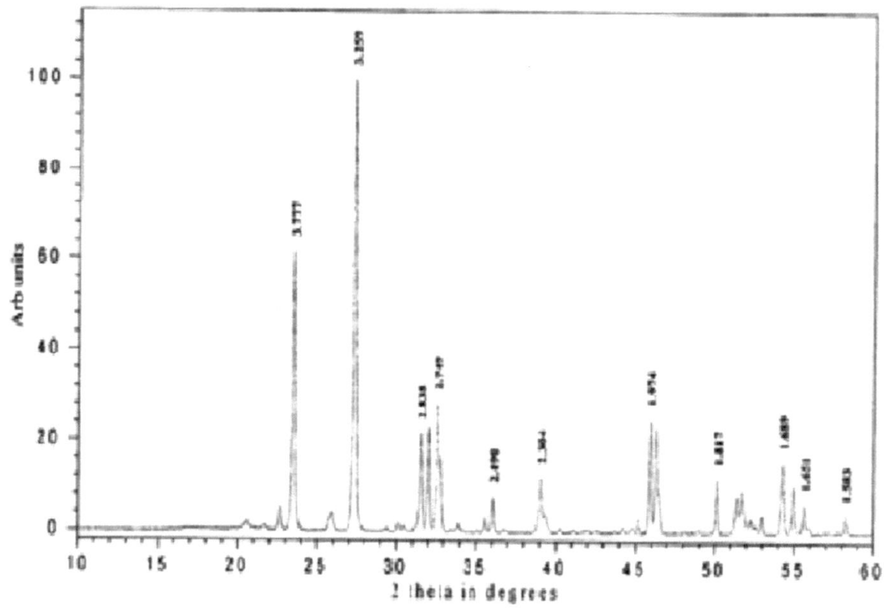

Figure 7. Diffraction diagram of ε-$Cd(IO_3)_2$ [20].

transparent and colorless prismatic of $Cd(IO_3)_2H_2O$ crystals is filtered from the solution. The solution was evaporated slowly at room temperature again; after 4 days, ζ-$Cd(IO_3)_2$ precipitates as colorless crystalline powder. They were filtered, washed with deionized water, and dried at room temperature (yield 30%). The second method consists of preparing the polymorph γ-$Cd(IO_3)_2$ in one step. This compound precipitates at a temperature of 60°C, as colorless crystalline powder from low concentrated aqueous solution of 2 mmol $CdCl_2$ and 1 mmol of KIO_3. The ζ-$Cd(IO_3)_2$ was obtained by heating γ-$Cd(IO_3)_2$ at 450°C for 2 hours [23].

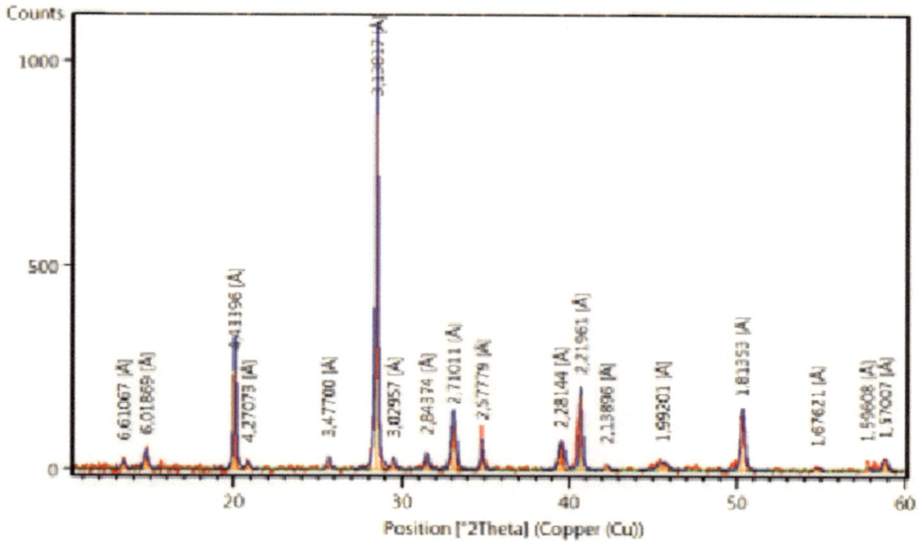

Figure 8. X-ray powder diagram of ζ-$Cd(IO_3)_2$ polymorph [23].

h	k	l	Int	2θ(obs)	2θ(cal)	diff
0	1	3	25.69	20.0196	20.0145	0.0051
*	*	*	1.649	20.7960	******	******
1	3	0	2.22	25.6282	25.6287	−0.0005
−1	3	2	100	28.4564	28.4086	0.0478
2	1	3	1.36	29.4866	29.4287	0.0579
*	*	*	5.40	31.4740	******	*****
1	4	0	10.87	33.0587	33.0709	−0.0122
−3	1	2	7.519	34.8143	34.8034	0.0109
−1	1	6	6.959	39.4989	39.5209	−0.0220
3	2	3	20.99	40.6123	40.5692	0.0431
1	2	6	1.013	42.2563	42.3072	−0.0509
3	0	5	3.915	45.3698	45.3740	−0.0042
−1	4	5	2.662	45.7851	45.7821	0.0030
4	3	1	1.739	49.7340	49.7907	−0.0567
3	0	6	15.11	50.3116	50.3107	0.0009
−1	7	1	1.832	57.8014	57.8082	−0.0068
2	2	8	0.583	58.2564	58.2407	0.0157
−4	3	5	3.258	58.8234	58.8067	0.0167

* Peak wich do not fit with this cell.

Table 2. Indexed reflections, standardized intensities, and 2θ values (°) of ζ-$Cd(IO_3)_2$ polymorph used for cell determination [23].

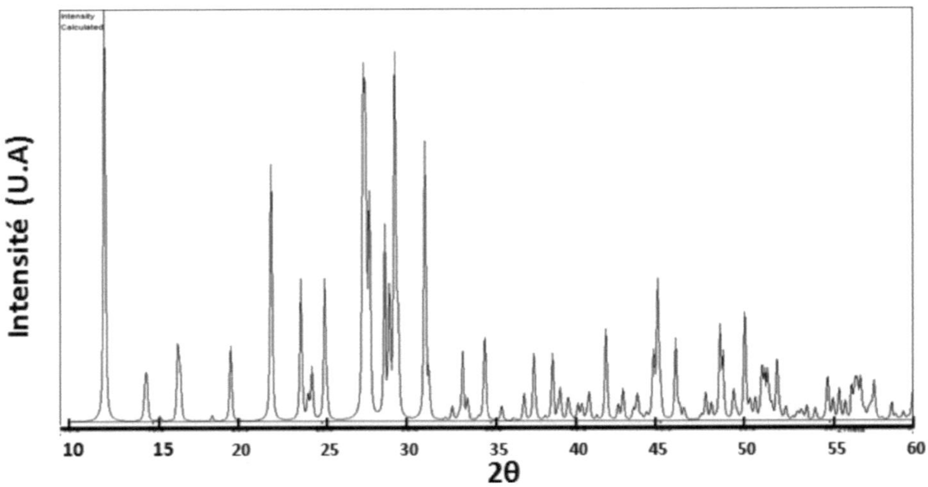

Figure 9. Theoretical powder diagram of $Cd(IO_3)_2H_2O$.

The X-ray powder diagram of this novel compound prepared by the first method is illustrated in **Figure 8**; the analysis shows that it is a polymorph different than α, β, γ, δ, and ε polymorphs; his diagram is characterized by the main peak d = 3.14 Å. We have used the powder diagram for peak indexing (**Table 2**). In the 15°–60° range, 18 peaks have a relative intensity greater than 0.58%; 16 peaks can be indexed using the following monoclinic cell: **a** = 8.4672 Å, **b** = 11.4365 Å, **c** = 14.4432 Å, and β = 91.12° (agreement factor R = 0.057 for all 16 peaks). The two only peaks which do not fit with this cell are positioned at 2θ = 20.796° and 2θ = 31.474° with the relative intensity I = 1.649 and 5.4%, respectively. This ζ-$Cd(IO_3)_2$ polymorph exhibits a noticeable SHG activity which shows that it crystallizes in an acentric space group [23].

2.2. Synthesis method and the powder spectrum for monohydrate $Cd(IO_3)_2H_2O$ iodate

The slow evaporation of a very dilute solution of $CdCl_2$ (0.26 g) and 0.607 g of KIO_3 in 100 ml of distilled water gives, after a few days, clear transparent crystals of prismatic shape and millimetric size of $Cd(IO_3)_2H_2O$. This compound is characterized by its theoretical powder diagram retraced by the Poudrix software [34] (**Figure 9**). This compound crystallizes in the triclinic space group P1¯ a = 7.119(2), b = 7.952(2), c = 6.646(2) Å, α = 102.17(2)°, β = 114.13(2)°, and γ = 66.78(4)° [20].

3. Synthesis method and the powder spectra for the two mixed cadmium-based ligand compounds: cadmium chloro-iodate $CdIO_3Cl$ and cadmium hydroxide-iodate $CdIO_3OH$

In 2014, the cadmium chloro-iodate compound $CdIO_3Cl$ in the form of a prism is synthesized by Yang and Mao. The mixture of 1mmol of $CdCl_2$ and 0.5 mmol of I_2O_5 and 1ml of water sealed in an autoclave aquipped with teflon liner (23 ml) at 200°C for 4 days and than cooled to

30°C at 6°C/h. The crystal structure has been determined by single-crystal X-ray diffraction methods; the unit cell is orthorhombic with a = 7.293(6) Å, b = 16.04(2) Å, c = 7.207(5) Å, V = 843 (1) Å3, and Z = 8. The space group is CmCa [35].

The cadmium hydroxy-iodate $CdIO_3OH$ is synthesized in the form of colorless platelike in the same way as $CdIO_3Cl$ except that they have used the starting reagents: 0.25 mmol of Cd $(CH_3COO)_2 2H_2O$, 2.5 mmol I_2O_5, and 0.625 mmol of K_2CO_3 in a volume of 3 ml of distilled water [35]. The powder diagram for both compounds is shown in **Figure 10**. The crystal structure has been determined by single-crystal X-ray diffraction methods; the unit cell is orthorhombic with a = 11.5245(0) Å, b = 6.7985 (7) Å, c = 4.7303 (4) Å, V = 304.31(1) Å3, and Z = 4. The space group is Pnma [35].

In 2016, two new production methods were published by Hebboul [36] to obtain the $Cd(IO_3)Cl$ compound; the first method (Procedure 1) is a synthesis by double decomposition; it consists of dissolving the KIO_3 in nitric acid 16 N, the same for the $CdCl_2$ whose molar ratio is 2:1; then the reaction mixture is conducted into ambient. In the first minute, the solution becomes opaque, seen by optical microscope after 24 h; it shows the formation of very small needles,

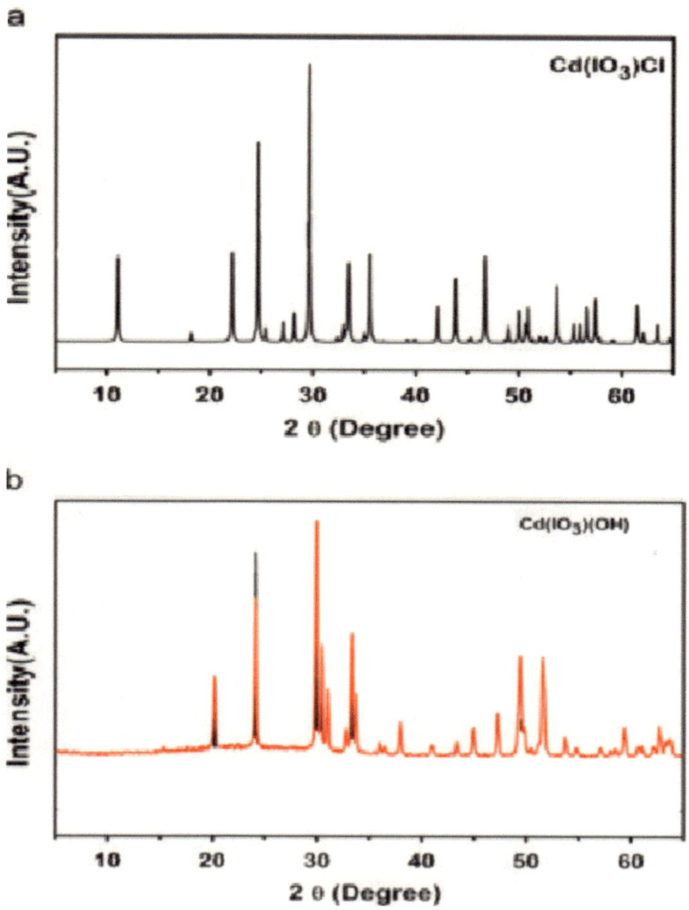

Figure 10. Mixed ligand compound powder diagram [35]. a-$Cd(IO_3)Cl$ and b-$Cd(IO_3)OH$.

but after 7 days, we noticed the increase in the size of the crystals in the form of spherulite needles (**Figure 11**).

The time at which the crystals appear in solution depends strongly on the concentration of nitric acid. Thus, a weakly concentrated solution of nitric acid requires almost dry evaporation, while a more concentrated solution, such as 7 M, allows crystallization when half of the solvent has evaporated. These crystals are filtered, rinsed with distilled water, and then dried at 60 °C in a drying oven. The molar yield of the reaction is around 90%.

The second method (Procedure 2) is a synthesis by substitution of one of the two groups [IO_3^-] of the polymorph γ-$Cd(IO_3)_2$ by a chloride [Cl^-]. Iodic acid HIO_3 (0.148 g) and cadmium chloride $CdCl_2$ (0.48 g) are solubilized in 20 ml of distilled water; then the reaction mixture is brought into ambient; after a few days, we notice the formation of a mixture of two phases, crystals in the form of rods and white powder.

Figure 11. A view taken by the optical microscope after 7 days [37].

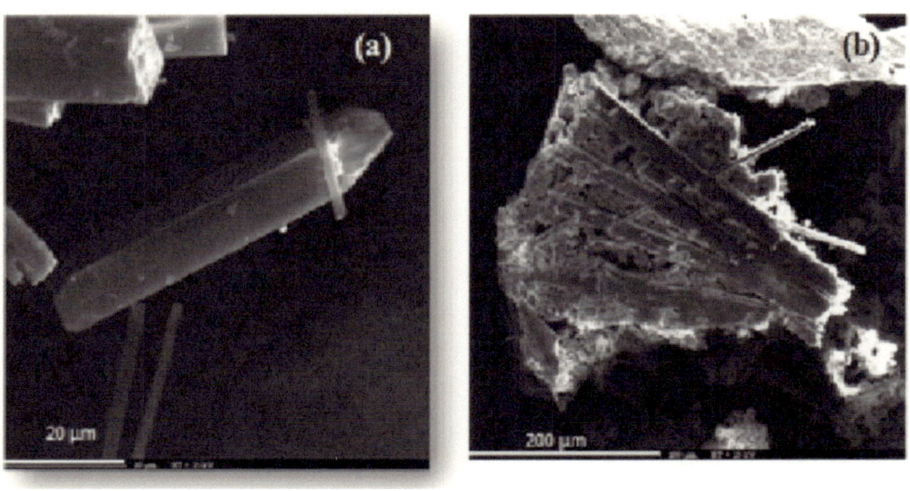

Figure 12. Morphology of the produced samples [36]. a-$CdIO_3Cl$ and b-$CdCl_2H_2O$ et γ-$Cd(IO_3)_2$.

The mixture thus produced is heated at 400°C in a muffle furnace for 2 hours. Finally, the desired compound is obtained in the form of a powder with some trace of residual cadmium chloride which can be removed through washing by using distilled water.

3.1. $CdIO_3Cl$ compound characteristics

The SEM image of **Figure 12a** shows the crystalline appearance of $CdIO_3Cl$ compound obtained by method 1; the crystal has needle morphology. **Figure 12b** shows the morphology of the mixture produced by method 2; a powder of γ-$Cd(IO_3)_2$ is stuck to the $CdCl_2H_2O$ rods

A monocrystalline needle is obtained by Procedure 1 of dimensions 0.09 × 0.03 × 0.03 mm^3 and was chosen and then mounted on a goniometric head. The conditions of the intensity collection and the refinement parameters of the structure are given in **Table 3**. The $CdIO_3Cl$

Formula	$CdIO_3Cl$
Shape color	Colorless needle
Crystal size (mm^3)	0.09 × 0.03 × 0.03
Molecular weight ($g.mol^{-1}$)	322.75
Crystalline system	Orthorhombic
Temperature (K)	293
λ (Ag) (Å)	0.56087
Space group (n°)	CmCa(64)
a (Å) *b* (Å) *c* (Å)	7.270(0) 15.995(0) 7.1980(0)
V (Å3)	837.01
Z	8
Dx ($g.cm^{-3}$)	5.122
Absorption factor (mm^{-1})	6.897
F(000)	1136
Scan area in/θ (°)	3.00 − 21.38
Indication limits (h k l)	$-9 \le 9; -20 \le 20; -9 \le 9;$
Measured reflections	16,006
Independent reflections	522
Number of refined parameters	33
Rint	0.0457
R1	0.0385
ωR2	0.0888
Refinement quality (S) _ρmax/_ρmin ($e.Å^{-3}$)	1.247 1.406/−1.565

Table 3. Crystallographic data and structural refinement of $CdIO_3Cl$ [36].

crystallizes in the orthorhombic system, CmCa space group (No. 64) with the following mesh parameters: a = 7.270(0) Å, b = 15.995 (0) Å, c = 7.1980 (0) Å, V = 837.009 $Å^3$, and Z = 8.

The powder spectrum of **$CdIO_3Cl$ (Figure 13)** was recorded and compared to the spectrum calculated with the Poudrix program [34]. The similarity of the diagrams shows that the phase obtained is pure.

The experimental diagrams (Procedure 2) are refined with the High Score Plus software; the identification of the presented phases is carried out by comparing the obtained experimental diagram with the reference file. The analysis gave us the types of the prepared phases

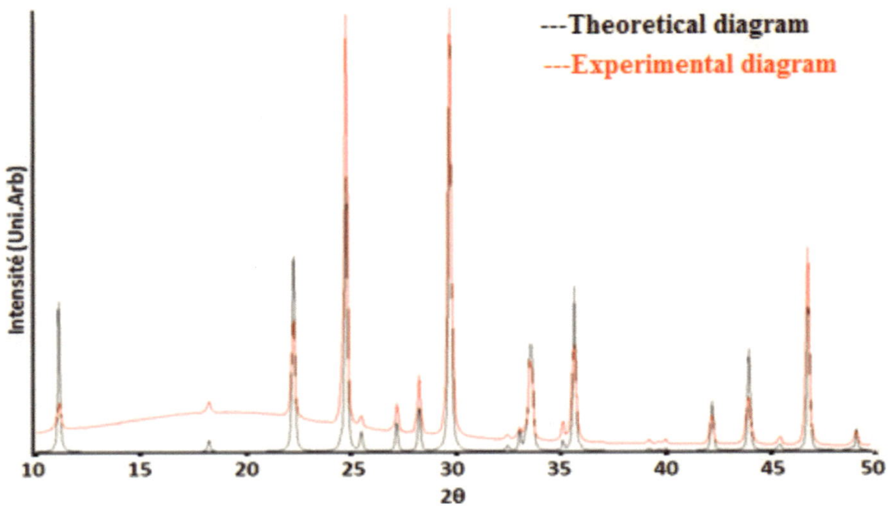

Figure 13. (XRD) Experimental and calculated powder spectra of the $CdIO_3Cl$ compound [36].

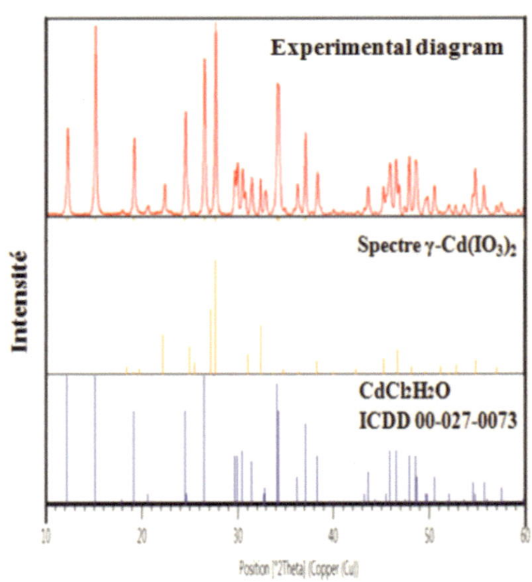

Figure 14. DRX spectrum of the produced mixture [36].

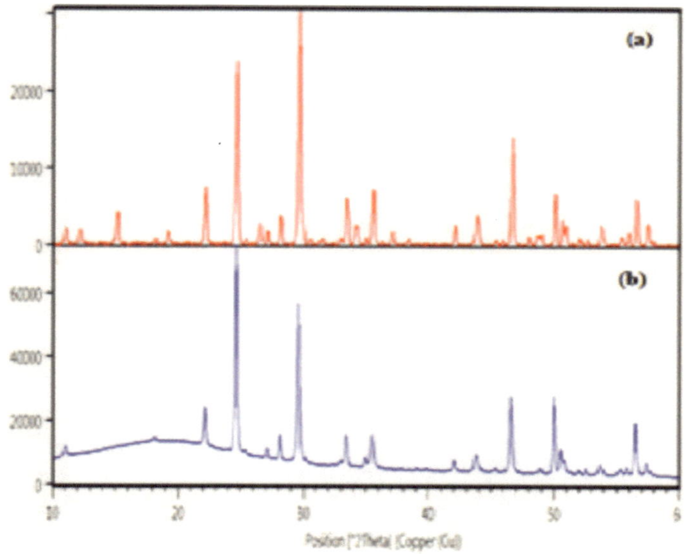

Figure 15. $CdIO_3Cl$ DRX spectrum [36]. a-Before washing and b-After washing.

(**Figure 14**); the comparison with the JCPDS files shows that one of the phases is hydrate cadmium chloride $CdCl_2H_2O$ No.00-027-0073; the other is not listed, but a simple comparison with the powder spectra published on the polymorphs of cadmium iodate [13] confirms the formation of γ-$Cd(IO_3)_2$. After the heat treatment, the powder diagram analysis before and after the washing (**Figure 15a**) and (**Figure 15b**) shows the formation of the $CdIO_3Cl$ good phase.

4. Conclusion

It is still a great challenge to summarize the relationships between structures or chemical compositions of the materials with their NLO properties. It is possible for compounds with the same chemical composition to exhibit several different phases with different structures. In this chapter, we provided the conditions for obtaining six polymorphs of anhydrous cadmium iodate in their pure state, as well as their powder diffractograms, whose three of which showing second harmonic generation activity (SHG), based on the work of Sciences Fondamentales Laboratory, University of Laghouat, in collaboration with Crystallography Laboratory, CNRS Grenoble, France, carried out between 2001 and 2005 and then the work of crystallogenesis by Ravi Kumar et al., in 2008; finally the work was realized in 2017 (**LPCM,** University of Laghouat), and we described the conditions for obtaining the triclinic single crystal of $Cd(IO_3)_2H_2O$. Three chemical routes of preparation of chloro-cadmium iodate $CdIO_3Cl$ are given. The prepared material was characterized by X-ray diffraction (XRD) and scanning electron microscopy (SEM). The crystal structure has been determined by single-crystal X-ray diffraction methods. The hydroxy-cadmium iodate $CdIO_3OH$ is synthesized in the form of colorless platelike. The crystal structure has been determined by single-crystal X-ray diffraction methods.

Author details

Zoulikha Hebboul

Address all correspondence to: z.hebboul@lagh-univ.dz

Physico-Chimie des Matériaux Laboratory (LPCM), University of Amar Telidji, Laghouat, Algeria

References

[1] Krishnamurthy N. Engineering Chemistry (2nd ed). New York: PHI Learning Private Limited; July 2, 2013. pp. 82-83. ISBN 978-81-203-3666-7

[2] Scoullos MJ, Vonkeman GH, Iain T, Zen M. Mercury, Cadmium, Lead: Handbook for Sustainable Heavy Metals Policy and Regulation. Springer; 2001. ISBN 978-1-4020-0224-3

[3] Lee C-H; Hsi CS. Recycling of scrap cathode ray tubes. Environmental Science & Technology. 2002;**36**(1):69-75. Bibcode:2002EnST…36…69L. DOI:10.1021/es010517q. PMID 11811492

[4] Miller LS, Mullin JB. Crystalline Cadmium Sulfide. Electronic materials: From silicon to organics. Springer; 1991. p. 273. ISBN 978-0-306-43655-0

[5] Buxbaum G, Pfaff G. Cadmium Pigments: Industrial inorganic pigments. Wiley-VCH; 2005. pp. 121-123. ISBN 978-3-527-30363-2

[6] Case Studies in Environmental Medicine (CSEM) Cadmium Toxicity. Agency for Toxic Substances and Disease Registry. Archived from the original on 6 June 2011. Retrieved 30 May 2011

[7] Jennings TC. Cadmium Environmental Concerns. PVC handbook. Hanser Verlag; 2005. p. 149. ISBN 978-1-56990-379-7

[8] Brady GS, Brady GS, Clauser HR, Vaccari JA. Materials handbook: An encyclopedia for managers, technical professionals, purchasing and production managers, technicians, and supervisors. McGraw-Hill Professional; 2002. p. 425. ISBN 978-0-07-136076-0

[9] Boyd R. The nonlinear optical susceptibility. Nonlinear Optics (3rd ed); 2007. p. 2. DOI: 10.1016B978-0-12-369470-6.00001-0

[10] Mandel L, Wolf E. Some Quantum Effects in Nonlinear Optics. Optical Coherence and Quantum Optics; 1995. p. 1079. DOI: 10.1017/CBO9781139644105.023

[11] Mao FF, Hu CL, Yang D, Yang BP, Mao JG. Angewandte Chemie International Edition. 2017;**56**:2151-2155. DOI:/10.1002/anie.201611770

[12] Lee DW, Kim SB, Ok KM. Dalton Transactions. 2012;**41**:8348-8353. DOI: 10.1039/C2DT30947E

[13] Bentria B, Benbertal D, Hebboul Z, Bagieu BM, Mosset A. Zeitschrift für Anorganische und Allgemeine Chemie. 2005;**631**(5):894-901

[14] Oelke WC, et Ch. Wagner, pro. Iowa Academy of Science. 1940;**46**:187

[15] Varhelyi C, Kekedy E. Studia Universitatis Babes-Bolyai. Serbian Chemical Society. 1962;**1**:11

[16] Pacesova L, Bohanesova E. Collection of Czechoslovak Chemical Communications. 1968; **33**:1

[17] Svensson C, Abrahams SC, Bernstein JL. Journal of Solid State Chemistry. 1981;**36**:195-204

[18] Bach H, Kuppers H. Acta Crystallographica. 1978;**B34**:263

[19] Liang JK,Yu YD, Ding SL. Acta PhisicaSinica. 1980;**29**(2):252

[20] Bentria B. Ingénierie cristalline pour l'optique non linéaire quadratique - préparation et caractérisation de nouveaux iodates métalliques [Doctoral Thesis]. University of Batna Algeria; 2005

[21] Rammelsberg CF. Annalen der Physik. 1838;**55**:588

[22] Ravi-Kumar SM, Melikechi N, Selvakumar S, Sagayaraj P. Journal of Crystal Growth. 2009;**311**:337-341

[23] Hebboul Z, Benbertal D. Synthesis and characterization of new anhydrous cadmium iodate zeta polymorph ζ-$Cd(IO_3)_2$. Journal of Materials and Environmental Science. 2018;**7**(9):565-569. DOI:org/10.26872/jmes.2018.9.2.62

[24] Hebboul Z. Eaboration et caractérisation de nouveaux iodates bimétalliques pour l'optique non linéaire [Ingeniorate Thesis]. Algeria: University of Laghouat; 2001

[25] Hebboul Z. Elaboration et caractérisation de nouveaux iodates métalliques pour l'optique non linéaire [Magister Thesis]. Algérie: University of Laghouat; 2004

[26] Nassau K, Shiever JW, PrescottBE. Journal of Solid State Chemistry. 1973;**7**:186

[27] Abrahams SC Sherwood RC, Brenstein JL, Nassau K. Journal of Solid State Chemistry. 1973;**7**:205

[28] Nassau K, Cooper AS, Shiever JW, Prescott BE. Journal of Solid State Chemistry. 1973;**8**:260

[29] Abrahams SC, Sherwod RC, Brenstein JL, Nassau K. Journal of Solid State Chemistry. 1973;**8**:274

[30] Nassau K, Shiever JW, Prescott BE, Cooper AS. Journal of Solid State Chemistry. 1974;**11**:314

[31] Nassau K, Shiever JW, Prescott BE. Journal of Solid State Chemistry. 1975;**14**:122

[32] Abrahams SC, Brenstein JL, Nassau K. Journal of Solid State Chemistry. 1976;**16**:173

[33] Kurtz SA, Perry TT. Journal of Applied Physics. 1968;**39**:3798

[34] Laugier J, Bochau B: Poudrix, a suite of programme for interprétation of X-ray experiments. ENSP / LMGP, BP 46, 38042 sain martin D(Héres, France. WWW: http://www.Inpg.fr/LMGp and http://www.ccp14.ac.Uk/tutorial/Impg/

[35] Yang BP, Mao JG. Journal of Solid State Chemistry. 2014;**219**:185-190

[36] Hebboul Z, Taouti MB, Benbertal D. Synthesis and characterization of $CdIO_3Cl$ compound. Journal of Materials and Environmental Science. 2016;**7**(5):1607-1613

[37] Hebboul Z. Elaboration et caractérisation de nouveaux matériaux pour l'optique non linéaire. [Doctoral Thesis]. Algérie: University of Laghouat; 2016

Chapter 3

Removal of Heavy Metals Using Bentonite Clay and Inorganic Coagulants

Oupa I. Ntwampe and Kapil Moothi

Additional information is available at the end of the chapter

http://dx.doi.org/10.5772/intechopen.76380

Abstract

Heavy metals have always been defined as elements with a density higher than 5 g/cm^3. They are regarded as serious wastewater contaminants with detrimental effect to human and environment. Their removal from wastewater poses a serious challenge as they require cost-effective reagent and treatment technique. About 200 mL solution of acid mine drainage (AMD) collected from the Western decant in Krugersdorp, South Africa was poured into five 500 mL glass beakers. Three different sets of experiments (employing mixing, shaking and no mixing) were conducted using a jar test and a shaker with 1.5 g bentonite clay, 20–60 mL of 0.043 M $FeCl_3$ and $Al_2(SO_4)_3$ and a flocculent of bentonite clay and $FeCl_3$ dosage, respectively. The experiments were conducted without pH adjustment. The samples settled for 1 hour after which the pH, conductivity and turbidity were measured. The results show that a combination of bentonite clay and $FeCl_3$ exhibits a better turbidity removal efficiency compared to the samples with bentonite clay, $FeCl_3$ and $AlCl_3$ respectively. The variation of the turbidity removal in the samples with mixing shaking and without mixing is insignificant, showing that destabilization-hydrolysis depends upon the strength of the reagent and the physicochemical properties of the solution. The results also show that hydrolysis occurs at low pH, indicating that it plays an insignificant role in destabilization. The SEM micrographs show that turbidity removal is a physical phenomenon.

Keywords: heavy metals, wastewater, pH, turbidity, destabilization

1. Introduction

Heavy metals have always been defined as elements with a density higher than 5 g/cm^3 [1, 2]. On the contrary, most of the heavy metals have density equal or less than 3 g/cm^3, hence the name has been changed to toxic metals. Heavy metals are geologically occurring substances which react physicochemically to form economic mineral resources such as coal, gold and copper, among others. They have a semi- or non-degradability property and can accumulate in the food chain, causing danger to human [3, 4]. Apart from their toxicity, they play a pivotal role in domestic and industrial activities. Drinking contaminated water containing heavy metals even in very small quantity may be detrimental to human and aquatic life [4]. A wide range of research projects has been exploited relating to new trends of removing heavy metals from industrial wastewater [5]. Some of the metals are both toxic and radioactive, e.g. uranium, which is normally embedded in mineral deposits of other metals such as gold, radium, selenium and thorium. There are some of the heavy metals, i.e. arsenic, cadmium, chromium, lead and mercury, which are classified as priority metals due to high toxicity to public health [6]. Despite their toxic nature, their use is unavoidable as they are necessary in daily activities of economic value, including metal-containing compounds [7–10]. Apart from normal natural causes, heavy metals are also introduced to the environment through natural phenomena such as weathering and volcanic eruptions [7, 11–14]. They are mostly transition elements where some are metalloids, e.g. arsenic, exhibiting toxicity at low level to soil, vegetation, rivers [6, 10, 11]. Environmental degradation is caused by the emissions from power utilities, mining and chemical industries is a serious global concern. The latter is responsible for the high concentration of lead and chromium in several water bodies, predominantly due to lack or ineffective purification systems [15]. On the other hand, some of the heavy metals, i.e. lead and chromium, are toxic and carcinogenic in their oxidation state, e.g. Pb^{2+} and Cr^{3+}, respectively [16]. Coal and gold mining discharges wastes or wastewater (**Figure 1**) are highly polluted with heavy metals. Alternately, they are the source of pyrite (FeS_2), a geologically, chemically or microbially formed mineral derived from the reaction between iron and sulfur under thermal conditions [17].

Apart from the economic value associated with heavy metals (mineral resources), they play a pivotal role in the formation of clay. On the other hand, clay minerals form a larger fraction of coal and also used as catalysts during coal combustion (oxidation). Extensive alteration of rocks to the formation of clay minerals produces relatively pure clay deposits that serve a variety of economic applications such as manufacturing of cooking pots, bricks, porcelain, drainage pipe, floor and wall tiles, tobacco pipes, oil drilling, cat litter, heat resistant tiles, construction of lime mortars, building materials and equipment, among others [18, 19]. Some are fluxing agent and carry some problematic compounds which are responsible for slagging

Figure 1. Geochemical surface feature along the river concentrated with metals (www.earth.illinois.edu).

and clogging of the boilers of the power utilities. Some are emitted to the atmosphere during industrial operation, thus causing a serious environmental catastrophe. Disposal of the chemicals and barren ore deposits which are polluted with heavy metals is the main attribute to the environmental degradation, which is predominant in the dumping sites.

2. Impact of waste materials on the environment

The purpose of this review does not cover the use of the heavy metals in industrial processes, i.e. cyanide and mercury, among others. The impact of heavy metals is mostly apparent in tailing dumps. Tailings contribute towards environmental degradation, and the impact does not take into account the amounts of toxic materials they contain. They host heavy metals, predominantly gangue minerals such as silicates, oxides, hydroxides, carbonates and sulfides. Some of the heavy metals in the tailings exist as iron-sulfide minerals, e.g. pyrite, pyrrhotite, chalcopyrite, arsenopyrite (FeAsS), sphalerite (ZnS) and galena (PbS), whereas others contain sulfur-containing minerals (Pb, Zn, Cd, Se, As) and compounds that are critically harmful. Apart from that, sulfide-bearing tailings can be oxidized during weathering before the formation of the AMD and release metals/metalloids (As, Cd, Co, Cr, Cu, Fe, Hg, Mn, Ni, Pb and Zn) [9]. The toxic metals are distributed throughout the environment by severely leaching out into the surroundings, causing serious health problem due to non-degradable property [20]. The leaching of the heavy metals into the soil may endanger natural population of bacteria, leading to disintegration of bacterial species responsible for nutrient cycling, thus affecting the ecosystem negatively [21]. Under such conditions, they have to devise means of survival under such adverse conditions, i.e. develop and establish detoxifying mechanisms (biotransformation, biosorption and bioaccumulation) [22].

Uranium, the most radioactive element, has a molecular weight of 238.03 mol/g and detrimental to humans through chemical toxicity, radiation and other uranium by-products. It is highly harmful and can lead to kidney, respiratory and neurological dysfunctionalities [23]. It exists in six radioactive isotopes (232–238) as classified from mother element which causes instability of the highly reactive nuclei. Its radioactivity becomes apparent when the nucleus emits minute particles during elemental transformation (radioactive decay), when α and β particles and γ-rays are emitted. **Figure 2** illustrates the spillage of radioactive uranium. There are countries where millions of tons of tailing cover wide areas, considering South Africa (SA) which produced over 43,500 tons of gold and 73,000 tons of uranium within 1953–1995. The basin covers an area of 1600 km^2 leading to 400 km^2 tailings dams which contained 6 billion tons of pyrite tailings and 430,000 tons (600,000 t) low-grade uranium. That resulted in the contamination of about 6000 km^2 of soils gold mining operations [24].

The most detrimental effects of uranium to human and environment include (1) aquatic physiological defects and fatalities, (2) disruption of water bodies, i.e. aquatic physiological-biochemical process, (3) deterrence of aquatic activities due to the presence of Fe, Zn, Cu, Mn and Pb, (4) disturbance of biodiversity of aquatic life, (5) depression of the dissolved oxygen in the water bodies, (6) modification of the nutrient availability which may cause loss of

Figure 2. Spillages and remaining of dumps of uranium in West Rand (SA) [24].

Fraction	Name	Chemical formula
Gauge	Quartz	SiO_2
Mineral	K-feldspar	$KAlSi_3O_8$
	Na-feldspar	$NaKAlSi_3O_8$
	Ca-feldspar	$CaKAl_2Si_2O_8$
	Serisite	KAl (AlSi)(F,OH)
	Chlorite	$(Mg,Fe)_3(Si,Al)_4O_{10}(OH)_2(OH)_6)$
	Calcite	$CaCO_3$
	Dolomite	$Ca,Mg(CO_3)_2$
Sulfide-oxide	Pyrite/pyrrhotite	FeS_2/FeS
	Arsenopyrite	FeAsS
	Galena	PbS
	Sphalerite	ZnS
	Chalcopyrite	$CuFeS_2$
	Magnetite	Fe_3O_4

Table 1. Mineralogy of a tailing dump in South Africa [29].

vegetation, (7) change of the direction of the roots growth and (8) cause of ailments due building up in aerian organs of plants and human.

Lead (Pb) and chromium (Cr) are toxic metals especially to plants and animals and cannot be easily treated due to non-biodegradability and subjection to bioaccumulation in living cells [25, 26]. They can form multiple-oxidation states, where Cr forms Cr^{2+}, Cr^{3+} and Cr^{6+}, the highly toxic and reactive and highly soluble in solutions of varying pH values [27]; however, their toxicity is less compared to Cr^{6+}. Pb is a bluish-white, soft metal, which is highly flexible and ductile and resistant to corrosion [28]. The types of mineral matter (**Table 1**) are complex compounds, which are main constituents of both clay and coal.

Oxidation of the pyrite acidifies the water percolating through the dumps, which then enters the groundwater regime beneath the dumps [30]. This acidic water is believed to be entering

Figure 3. Gold mine tailing dumps in Johannesburg, South Africa (UGS stock photo by George Steinmetz).

streams along the Witwatersrand in SA [30, 31]. Apart from direct metals contamination, tailings are also sources of environmental degradation. Gauteng Province (SA) is highly concentrated with tailings, which contaminate the environment through metals leachate during heavy rainfall [30]. **Figure 3** illustrates tailing dump.

A general characteristic of tailings is the pyrite content of up to 6% pyrite, highly saline, low nutritional value and low organic matter content [32–34]. High sulfide content causes high acidity and high metal concentrations in groundwater around the proximity of the tailings [32]. Rafiei et al. [35] reported a pH value of 7.35 in gold mine tailings in Iran, whereas Mitileni et al. [36] reported pH values of 3.25–6.28 in South Africa. Harish et al. [37] reported pH value of 3.48–8.12 in India. pH is essential to aquatic life as different species behave differently, whereby high soil acidity from mine pollution may destroy microbes responsible for breaking down organic matter into nutrients. In addition, acidity can dissolve aluminum to form free organic materials, which is toxic to plant roots, and also reduce the concentrations of essential nutrient. The characteristic features of gold mine tailings are the elevated concentrations of toxic heavy metals, i.e. As, Cd, Ni, P, Cu, Zn, Co and Hg [38].

3. The effect of chemical treatment to the ecosystem

Industrial and population expansions are the main attributes to increasing concentration of heavy metals, and inevitably, their escalating quantity is uncontrollable as their genesis is natural and anthropogenic [39]. Gold mining has been identified as an operation which results in the accumulation of thousands of voluminous tailings dumps scattered all over the countries, potentially degrading the ecosystem [40]. Nuclear power generation is another source of toxic waste generation as it uses uranium as fuel. The problem with nuclear is its difficulty in handling waste materials. This review study suggests that one of the solutions around the management of toxic minerals is proper mineral processing using mineral liberation-classification, but mineral slipping is inevitable. Although mineral liberation-classification does not produce all the minerals in their pure form, some are in complex compounds which require further separation. There is an enormous amount of waste generated during various processes, especially in gold mining where approximately 99% of the waste disposal consists mainly of ore [41]. The use of bacteria is another technique but the main problem is the rate of production and the timing for their growth as they rely mainly on pH and temperature. The pH-dependent bacteria, e.g. acidophilic, chemolithotrophic iron- and sulfur-oxidizing bacteria, assist in solubilizing the sulfides contained in the gold ore deposits by liberating gold from the deposit [42]. Although biomining poses less danger to the environment than many physico-chemical extraction processes [42], it is perceived to be unsustainable. Bacteria such as *Acidithiobacillus* obtain energy by oxidizing Fe^{2+} to Fe^{3+} ions or elemental sulfur (S^0). Eqs. (1)–(3) illustrate oxidation-reduction of iron and sulfur from pyrite, a mineral exposed during both gold and coal mining. The bacteria (*acidithiobacvillus*) obtain energy by oxidizing ferrous iron (Fe^{2+}) to ferric iron (Fe^{3+}) or reduction of elemental sulphur (S^0) to form sulphuric acid [45, 46], thus resulting in the production of AMD.

$$2\,S^0 + 3\,O_2 + 2\,H_2O \rightarrow 4\,H^+ + 2\,SO_4{}^{2-} \tag{1}$$

$$FeS_2\ (Au) + 2\,Fe^{3+} \rightarrow 3\,Fe^{2+} + 2\,S^0 + (Au) \tag{2}$$

$$FeS_2\ (Au) + 14\,Fe^{3+} + 8\,H_2O \rightarrow 15\,Fe^{2+} + 2\,SO_4{}^{2-} + 16\,H^+ + Au \tag{3}$$

Another distinct characteristic of the bacteria in gold purification includes their ability to excrete ligands that stabilize gold by forming gold-rich complexes and/or colloids [43–45]. This review invokes the functions and integrity of the use of bacteria in gold purification, but residual heavy metals remain a challenge, moreover that they form part of waste which is a danger to the ecosystem.

Mercury, one of the toxic metals used in gold purification, is accomplished by mixing it with mineral ore deposits extracted from the ground or stream beds to form an amalgam. The burning of the latter vaporizes elemental mercury into a toxic plume, resulting the separation of the gold from the ore deposit. The technique is globally regarded as the second largest source of atmospheric mercury pollution after coal combustion [46]. Apart from mercury, cyanide is another toxic metal which is employed in gold purification process (Eq. (4)), a two-stage process which includes extraction and recovery. Eq. (4) shows dissolution of gold:

$$4\,Au(s) + 8\,NaCN(aq) + O_2(g) + 2\,H_2O(l) \rightarrow 4\,NaAu(CN)_2(aq) + 4\,NaOH(aq) \tag{4}$$

Cyanide, one of the toxic metals, is employed in the extraction process of gold. The process also uses another toxic metal, i.e. zinc to cement cyanide solution [46]. Environmental degradation occurs during the spillages of cyanide solution, whereas activated carbon can also be used when the gold content is high. The ground and crushed ore deposits are enclosed in large tanks with agitators to dissolve gold which then adhere to particles of activated carbon. The activated carbon and the gold are separated from the solution, which is discarded together with the leached ore [46]. The residual trace heavy metals from the process act as potential hazards to the ecosystem. It has been noted that sludge from gold and cold operations contains mercury amalgam, cyanide, uranium and mineral matter, among others. Since gold recovery from ore deposit is not 100%, small quantities remained in the tailings, where they will subsequently decompose to form other complexes or react with other toxic substances to form new toxic compounds.

Lime has been used in previous studies to regulate the pH of the AMD solution [47]. The challenge with such a technique is excess sludge generated, which requires further treatment of heavy metals polluted sludge. Instead, the sludge is pumped to large dumps (slimes dumps).

4. The effect of metal salts during the removal of turbid materials

Mine wastewater and AMD are the main carriers of the heavy metals produced from mining operations. According to a trend of various studies to investigate a cost-effective mine wastewater treatment, the use of inorganic coagulants seems to be regarded as a primitive technique. The study conducted in this review study focuses on conventional treatment process using inorganic coagulants. Their advantage is based on their hydrolysis potential to form flocs which are adsorbents. In case of Fe, there are at least four species that co-exist in aqueous solution in the pH range of 1–5, namely Fe^{3+}, $Fe(OH)^{2+}$, $Fe(OH)_2^{+}$ and a very small fraction of dimer $Fe_2(OH)_2^{4+}$ [48]. On the other hand, Al species prefer a slightly higher pH than Fe in a range of 3–8. Apart from positive contributions, naturally formed iron is a source of AMD formation. It reacts with sulfur through a geological phenomenon to form FeS_2, which has negative environmental impact. Pyrite is not toxic as a mineral until it is oxidized by oxygen in an aquatic medium to form sulfuric acid, an environmental degradation agent, Eq. (5).

$$2\,FeS_2 + 7\,O_2 + 2\,H_2O \rightarrow 2\,Fe^{2+} + 4\,SO_4{}^{2-} + 4\,H^+ \tag{5}$$

Fe^{2+} ions may be oxidized to form Fe^{3+} ions (Eq. (5)), which oxidize excess pyrite:

$$FeS_2 + 14\,Fe^{3+} + 8\,H_2O \rightarrow 15\,Fe^{2+} + 2\,SO_4{}^{2-} + 16\,H^+ \tag{6}$$

Eq. (6) yields unstable Fe^{2+} ions species which are further oxidized by excess Fe^{3+} ions that are subjected to hydrolysis when added to wastewater.

The catastrophe associated with the AMD (**Figure 4**) is not only about the surface or ground water pollution and degradation of the quality of the soil, but endanger it poses to aquatic sediments and fauna, allowing heavy metals to seep into the environment.

Figure 4. Diagram of AMD pumped from the underground workings [48].

Long-term exposure to drinking water polluted with AMD may lead to increased rates of ailment such as cancer, decreased cognitive function and appearance of skin lesions. Heavy metals in drinking water may impair the development of the neurons of the fetus, resulting in mental retardation. In addition, AMD is attributed to loss of stream bed, drying of the rivers, limited water supply for growing industrialization and population and limiting the quantity of water for irrigation. Acidified water (AMD) seepage from the mine tailing dumps concentrated with heavy metals causes serious environmental degradation especially stream flow around the vicinity [49, 50]. The condition reduces the pH and adds high metal loads to the water resulting in extreme iron hydroxide precipitation in the stream which endangers aquatic life. Possible evaporation of groundwater from the capillary zone above the water table may contaminate the surface soil layer with heavy metals. Dissolution of the metals on the Earth's crust and surface soil may add to the metals load in the stream. The oxidation of iron introduced into the stream by seepage may behave as redox buffer, controlling the pH of the stream water. Increasing the pH of AMD above 3, either through contact with fresh water or neutralizing minerals, precipitates Fe^{3+} ions to form iron hydroxide species. Other types of iron precipitates are possible, including iron oxides and oxyhydroxides [48]. All such species cause discoloring of the water and smother aquatic plant and animal life, disrupting stream ecosystems.

Literature [47] states that optimal removal of turbid materials occurs in an alkaline medium; however, the findings of this review portray a different view. Despite the findings of this review, the results of the studies on the AMD samples using inorganic coagulants, metal hydroxide and bentonite clay [51, 52] revealed that the removal of turbid materials/heavy metals is dependent upon the physicochemical properties of both the colloidal suspension and the reagents and that which also includes structural morphology of the flocs (SEM micrographs).The hypothesis relating to pH adjustment is not only depend upon the metal hydroxide but also the OH^- ions released from the cleavage of bipolar water molecules. This is based on the high ionic strength of the AMD and the charges in the solution to enhance dehydroxylation of hydrated metal ions, releasing the OH^- ions which react with metal ions to form metal hydroxide precipitates.

5. Factors that affect the removal of heavy metals

A lot of studies have been conducted on the investigation of ideal techniques employable in the removal of heavy metals, but most are costly and complicated in operation. Factors, such as dosage, destabilization-hydrolysis, time, concentration of the heavy metals, type of colloidal suspension, the reagent(s), temperature, retention time, sorption capacity and structural morphology of a substrate, among others, are most influential during treatment. Although pH is considered as a determinant of the removal of turbid materials as alluded later, studies conducted by Ntwampe et al. [53, 54] on paint wastewater and AMD using traditional inorganic coagulants, metal hydroxide or bentonite clay revealed that pH of the colloidal suspension plays an insignificant role in the removal of turbid material/heavy metals. The studies revealed that the optimal removal of turbid materials is a physicochemical phenomenon,

http://dx.doi.org/10.5772/intechopen.76380

which is determined by the rate of destabilization-hydrolysis. Although the reaction consists of two reactions, this review considers them as a single stage because of their co-existence, i.e. destabilization occurs on the colloidal suspension whereas hydrolysis on the metal ions [55].

5.1. Destabilization-hydrolysis reaction

In this reaction, the reagent disturbs the equilibrium state of a colloid due to equal counter forces of van der Waals attractive and electrostatic repulsive force [56, 57]. The reaction which is a physico-chemical phenomenon, determines the effectiveness of the succeeding reactions. The colloid is separated into two ionic regions, namely diffuse and Stern layers to form electrical double layer (EDL) [58]. The efficiency of a reagent during treatment is determined by the ability to compress EDL, resulting in the reduction of the radius which enhances particle-particle collision to form agglomerates, flocs.

5.2. Type of a colloid

Colloids are classified according to hydrophilicity (water-loving) or hydrophobicity (water-hating), where treatment with the former is difficult as turbid materials dissolve completely in the solvent, making flocculation difficult. On the other hand, hydrophobic colloid consists of suspended solids, some of which settle spontaneously due to gravitational force [58].

5.3. Type of a reagent

An effective coagulant is characterized by destabilizing potential attributed to high valence and electronegativity. This is apparent after addition to a colloidal suspension when the metal ions of a salt disperse throughout the colloidal particle to destabilize the system, resulting in flocs formation. Nucleation occurs and small flocs form larger flocs, which remove turbid materials [58].

5.4. Structural morphology

The crystal morphology shown by the scanning electron microscopy (SEM) micrographs is illustrative of the ability of a reagent to adsorb the absorbate. Most of the SEM micrographs which correspond to optimal adsorption consist of dense flocs joined together with limited or without voids [51]. Such a SEM micrograph is indicative of no slipping through of the turbid materials (heavy metals); on the contrary, the SEM micrograph with flocs and voids has less adsorption potential [51].

5.5. pH of the solution

Some studies [59–61] state that the rate of adsorption of turbid materials/heavy metals is more favorable in an acidic colloidal suspension, i.e. low pH. This observation is in agreement with aforementioned studies revealing that the treatment of the AMD using inorganic coagulant(s) without pH adjustment yields optimal adsorption. An acidic solution is highly concentrated with hydronium ions (H^+), which are prone to strengthen the electrostatic forces of attraction between negatively charged heavy metal oxyanions and protonated sorbent, and the reaction occurs according to Eq. (7).

$$Fe^0 + 2\,H^+ + Fe^{2+} + H_2 \tag{7}$$

Fe^{2+} can be further oxidized to form Fe^{3+}.

$$2\,Fe^{2+}{}_{(aq)} + 2\,H^+{}_{(aq)} + 0.5\,O_{2(aq)} \rightarrow 2\,Fe^{3+}{}_{(aq)} + O \tag{8}$$

6. Treatment of the AMD to remove heavy metals

The removal of heavy metals from industrial effluent discharge has been a serious challenge for decades due to the complex compounds formed during chemical reactions. The wastewater consists of metals as pure and trace elements, where the latter always disperse throughout the solution and sorbed onto other compounds or form other compounds during nucleation. The problem associated with trace elements or compounds consisting of trace elements is a plausibility to slip through the adsorbents. The residual heavy metals are caused by the existence of the trace elements in the system [62], a problem caused by low adsorption potential of an adsorbent in a medium with low concentration of absorbate. Technologies such as membrane separation, electrochemical, precipitation, ion exchange and adsorption were investigated to treat wastewater/AMD [63]. The results showed that they are expensive and incur high waste disposal costs, apart from high costs, some studies [64] revealed that some of the new technologies have a tendency of regenerating pollutants back to the environment by producing toxic chemicals such as sodium borohydride ($NaBH_4$). Chemical reaction between iron salt and sodium borohydride to release flammable hydrogen gas [65] is shown in Eq. (9).

$$4Fe^{3+} + 3BH_4^+ + 9H_2O \rightarrow 4F^0\downarrow + 3H_2BO_3^- + 12H^+ + H_2\uparrow \tag{9}$$

Some studies were conducted [51, 54] to reduce turbid materials from AMD using a combination of bentonite clay and inorganic coagulants. Among their studies, a combination of bentonite clay and Na_2CO_3 [53] was employed to remove the heavy metals from the AMD sample. Inorganic coagulants, mainly the metal ions (M^{n+}), have exhibited a significant influence in wastewater and AMD treatment [51–53], including the pivotal role played by the bentonite clay. There are more studies which have been conducted in the treatment of AMD sample using a variety of metal salts without or with a combination of bentonite clay [51–53]. The results show that the turbid materials removal efficiency of inorganic coagulants is optimal when dosed alone or in combination. The observations show that they are all subjected to hydrolysis to form species which are adsorption substrates. On the other hand, bentonite clay also exhibits ionic exchange, sorption and intercalation as the main reactions leading to the removal of turbid materials/heavy metals [52, 53]. The mineralogy of the AMD (**Table 2**), i.e. pH and conductivity of the AMD sample, treated with a combination of bentonite clay and Na_2CO_3 with mixing is shown in **Figure 5**.

Element	Conc. (ppm)
Al	1.171
Ca	182.1
Co	39.7
Cu	0.14
Fe	28.3
K	4.5
Mg	67.3
Mn	35.3
Na	44.5
Ni	27.6
Pb	35.2
As	45.2
Se	0.71
Zn	31.7

Table 2. ICP-OES analyses of untreated AMD sample with clay (5× dilution).

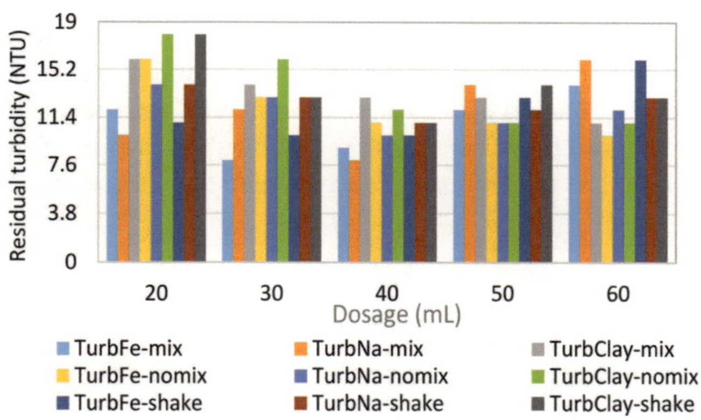

Figure 5. Residual turbidity of AMD with $FeCl_3$, Na_2CO_3 and a combination of bentonite clay and Na_2CO_3 with and without mixing and shaking.

According to **Table 1**, the concentration of the metals, i.e. Ca, Mg and Na, in the AMD is high, 182, 67 and 46 ppm, respectively, and that of non-toxic heavy metals, Fe and Mn, is 28 and 35 ppm, respectively. The concentration toxic metal content, Co, Ni, Pb, As and Zn is also high, the main toxic metals to the ecosystem.

The turbid materials of the AMD (**Figure 5**) are identified as turbidity, which includes the suspended and soluble solids, inorganic material, heavy metals and any trace substances.

The treatment with shaking and without mixing does not form part of this review study and therefore will not be elucidated extensively. Owing to a high turbidity of the AMD sample, residual turbidity below 10 NTU is highly acceptable, 10–15 NTU acceptable, above 15–20 NTU moderately acceptable and above 20 NTU is poor. However, the experimental results of the samples with $FeCl_3$ and Na_2CO_3 mixing are in the range of 8.4–14.0 and 8.2–16.5 NTU, respectively. Notwithstanding the fact that they are both efficient, dosage in a pure form is unaffordable compared to the dosage recommended in the study by Ntwampe et al. [51], which includes the investigation of the efficiency of bentonite clay in AMD treatment. **Table 3** shows the weight percentage of the mineral matter of the bentonite clay.

The weight % mineral content (**Table 3**), where CO_2 and FeO representing siderite ($FeCO_3$) form a larger weight % (33.2 and 45.3 wt. %) of the total complex compounds contained in both coal/gold mineral composition, and the rest are trace elemental compounds. The particle size of the bentonite clay is 180 μm mesh size with intensive mineral liberation (**Table 3**). The mineralogy of the tailing dump (**Table 1**) shows complex compounds of which some are original and others reacted to form new complex compounds; however, the chemical compounds are also representative of the clay minerals. Pulverizing the mineral compound (**Table 1**) results in minerals liberation to form simple compounds (**Table 3**). The elements/metals with high weight percentage include Fe, K and O, an indication of oxidation-reduction of ferric and ferrous ions during chemical reactions. A simple chemical composition of simple compounds (**Table 3**) indicates a high rate of chemical reactions by the colloid or the reagent(s).

The residual turbidity in the samples with bentonite clay mixing is in a range of 9.8–12.0 NTU, without mixing in a range of 10.5–19.2 NTU and shaking in a range of 10.4–13.3 NTU. As indicated, the method of chemical dispersion, i.e. mechanical agitation, is insignificant in this review, and the focus is on the efficiency of the combination of bentonite clay and Na_2CO_3 during AMD treatment. The mineral matter composition and structural configuration of the

Element	Wt.%	Wt.%	Atomic%	Compd%	Formula	No. of ions
C K	9.07	0.46	16.55	33.22	CO_2	2.15
Mg K	0.91	0.08	0.82	1.50	MgO	0.11
Al K	1.56	0.08	1.26	2.94	Al_2O_3	0.16
Si K	5.06	0.13	3.95	10.82	SiO_2	0.51
S K	1.48	0.10	1.01	3.69	SO_3	0.13
Ca K	1.14	0.08	0.62	1.60	CaO	0.08
Ti K	0.53	0.09	0.24	0.89	TiO_2	0.03
Fe K	35.25	0.49	13.84	45.34	FeO	1.79
O	45.02	0.58	61.70			8.00
Total	**100.00**					

Where Wt is weight and Compd is compound.

Table 3. Weight % of the mineral content in the bentonite clay.

bentonite clay play a pivotal role during treatment process. The former identifies the metals and content present in the colloidal suspension, the information which are indicative of inevitable electrochemical reactions. On the other hand, the latter is distinguished by the planar layers, which include octahedral and tetrahedral formation to form T-O-T. According to mineralogy of the bentonite clay (**Table 1**), dehydroxylation and processes are plausible. On the other hand, the structural configuration (T-O-T) is effective in sorption by intercalation of the turbid materials (heavy metals) onto the sheets and porous surface of the bentonite clay. In addition, the mineralogy (**Table 1**) is indicative of a series of physicochemical reactions in the system, especially on the serisite and chlorites, and oxidation of Fe on the iron-bearing compounds (chlorite, pyrite, pyrrhotite, arsenopyrite, chalcopyrite and magnetite). The formation of CO_2 from the decomposition of dolomite is inevitable, including oxidation of metals and other physico-chemical reactions. These form the main chemical reactions occurring in the bentonite clay in the system to form new simple compounds suitable for further reactions or sorption. Na_2CO_3 component in the flocculent also plays a pivotal role which includes increasing of the pH when it solubilizes to $NaCO_3^-$ and CO_3^{2-}. A slight pH rise decreased the solubility of the metals in the colloidal suspension, thus ameliorating the rate of agglomeration. **Table 4** illustrates the ICP-OES results of treated AMD samples.

The results of the treated AMD with a bentonite clay mixing (**Table 4**) show a considerable heavy metals removal of As, Co, Ni, Pb and Zn from 45.5, 39.7, 27.6, 35.2 and 31.7 to 4.1, 3.2, 6.6, 5.4 and 4.8 ppm, respectively. **Figure 6** shows SEM micrograph of the sludge of the AMD samples with $FeCl_3$ (**Figure 6A**) and bentonite clay (**Figure 6B**) mixing, respectively.

Element	Conc. (ppm)
Al	1.023
Ca	142.1
Co	3.2
Cu	0.14
Fe	19.2
K	4.14
Mg	55.7
Mn	23.1
Na	27.4
Ni	6.6
Pb	5.5
As	4.1
Se	0.71
Zn	4.8

Table 4. ICP-OES analyses of treated AMD sample with clay (5× dilution).

Figure 6. SEM of samples the sludge of AMD sample with $FeCl_3$ and bentonite clay respectively (25,000×) [53].

The SEM micrographs show the dry sludge of the AMD (**Figure 6**) with $FeCl_3$ (**Figure 6A**) and bentonite clay (**Figure 6B**) dosage. The former shows flocs which are bound together to form clusters of separate dense and small non-spherical porous structures with some voids in-between surface. The structural morphology is indicative of subjection of shear stress during treatment. On the other hand, the micrograph (**Figure 6B**) exhibits more condensed clusters of flocs with uniform crystal morphology, showing fewer voids compared to those obtained in **Figure 6A**.

7. Conclusion

Heavy metals have exhibited both economic and detrimental implications, a condition which uplifts the economy and degrades the environment, respectively. The removal of heavy metals from the medium (solvent) is not cost-effective as they sometimes appear as trace elements which are not easily removable. On the other hand, the trace elements found in tailings accumulate with time and form new complex and toxic compounds to the environment. Treatment of tailings is a very expensive exercise, which has never been deliberated, thus causing degradation especially after rainfall where the toxic compounds dissolve and disperse throughout the environment. A majority of modern technologies which have been explored are not cost-effective and sustainable, a conclusion emanating from employing sophisticated and foreign equipment. In addition, developing and underdeveloped countries cannot cope with such expenditures.

The only solution to the existing challenge relating to environmental degradation by the mine's wastewater is around a choice of a cost-effective treatment technique, which includes a simple operation and uses the most abundant reagent(s) that are affordable in any way. Natural inorganic coagulants have been used to treat industrial wastewater, but the problem is the damage to plant and equipment when dosed in pure form, i.e. corrosion, scaling or addition to sludge. The present review investigates the efficiencies of a combination of bentonite clay and inorganic coagulants (flocculent) in the treatment of the AMD sample in reduced mass fraction. Bentonite clay was chosen due to chemical composition and crystal morphology, which are indicative of effective physicochemical adsorption. On the other hand, inorganic coagulants also employ a physicochemical reaction to remove turbid materials as exhibited by SEM micrographs, but the bentonite clay is prone to perform better as it applies ion exchange, adsorption and intercalation. The advantage of the mixture of bentonite clay with inorganic coagulants in reduced molarity by half, i.e. 50% molarity ($^{m}/_{m}$%), serves both to reduce their detrimental effect and alleviate over or underdosage, which may cause restabilization and deflocculation, respectively. A flocculent, prepared by a combination of bentonite clay and Na_2CO_3, employed in the present review exhibited an optimal heavy metals removal potential.

Although some of the inorganic coagulants exhibited optimal turbid materials removal, varying atmospheric conditions affect the operating conditions during treatment, thus requiring continuous adjustments. Although it is also plausible that such a phenomenon can still occur with the flocculent used in this study, bentonite clay always behaves as a principal substrate due to multiple adsorption characteristics. The other advantage with this technique is that it does not require pH adjustment as the reagents exhibited a high destabilization-hydrolysis potential, a reaction which is shown by the removal of the turbid materials. The removal of the turbid materials from the AMD indicates that bentonite clay and Na_2CO_3 (flocculent) are complementary.

Author details

Oupa I. Ntwampe* and Kapil Moothi

*Address all correspondence to: ontwampe@gmail.com

Department of Chemical Engineering, School of Chemical Engineering and Built Environment, University of Johannesburg, Doornfontein, Johannesburg, South Africa

References

[1] Barakat MA. New trends in removing heavy metals from industrial wastewater. Arabian Journal of Chemistry. 2011;**4**(4):361-377

[2] Coetzee H et al. An Assessment of Sources, Pathways, Mechanisms and Risks of Current and Potential Future Pollution of Water and Sediments in Gold-Mining Areas of the Wonderfonteinspruit Catchment. Report, WRC, Council for Geosience. 2004. Report No 1214/1/06. 2006

[3] Winde F. Uranium pollution on water – A global perspective on the situation in South Africa. Epidemiological data related to persons living around the gold mine tailings are unavailable. Inaugural Lecture. 22 February 2013. ISBN 978-1-86822-629-0

[4] Bambas-Nolen L, Birn AE, Cairncross E, Kisting S, Liefferink M, Mukhopadhyay B, Shroff FM, van Wyk D. Case study on extractive industries prepared for the lancer commission on global governance: Report from South Africa. The Lancet. 2013;**383**(9917):630-667

[5] Fergusson JE. The Heavy Elements: Chemistry, Environmental Impact and Health Effects. Oxford: Pergamon Press; 1990

[6] Duffus JH. Heavy metals-a meaningless term? Pure and Applied Chemistry. 2002;**74**(5): 793-807

[7] He ZL, Yang XE, Stoffella PJ. Trace elements in agroecosystems and impacts on the environment. Journal Trace Elemental Medical Biology. 2005;**19**(2–3):125-140 [PubMed]

[8] Goyer RA. Toxic effects of metals. In: Klaassen CD, editor. Cassarett and Doull's Toxicology: The Basic Science of Poisons. New York: McGraw-Hill Publisher; 2001

[9] Herawati N, Suzuki S, Hayashi K, Rivai IF, Koyoma H. Cadmium, copper and zinc levels in rice and soil of Japan, Indonesia and China by soil type. Bulletin of Environmental Contamination and Toxicology. 2000;**64**:33-39 [PubMed]

[10] Shallari S, Schwartz C, Hasko A, Morel JL. Heavy metals in soils and plants of serpentine and industrial sites of Albania. Science Total Environment. 1998;**19209**:133-142 [PubMed]

[11] Nriagu JO. A global assessment of natural sources of atmospheric trace metals. Nature. 1989;**338**:47-49

[12] Arruti A, Fernández-Olmo I, Irabien A. Evaluation of the contribution of local sources to trace metals levels in urban PM2.5 and PM10 in the Cantabria region (Northern Spain). Journal Environmental Monitoring. 2010;**12**(7):1451-1414 [PubMed]

[13] Sträter E, Westbeld A, Klemm O. Pollution in coastal fog at Alto Patache, Northern Chile. Environmental Science Pollution Research International. 2010. [Epub ahead of print] [PubMed]

[14] Pacyna JM. Monitoring and assessment of metal contaminants in the air. In: Chang LW, Magos L, Suzuli T, editors. Toxicology of Metals. Boca Raton, FL: CRC Press; 1996

[15] Poguberović SS, Krčmar DM, Dalmacija BD, Maletić SP, Tomašević-Pilipović DD, Kerkez DV, Rončević SD. Removal of Ni (II) and Cu (II) from aqueous solutions using 'green'zero-valent iron nanoparticles produced by oak and mulberry leaf extracts. Water Science and Technology. 2016;**74**(9):2115-2123

[16] Fu F, Wang Q. Removal of heavy metal ions from wastewaters: A review. Journal of Environmental Management. 2011;**92**(3):407-418

[17] Duan J, Gregory J. Coagulation by Hydrolysing Metal Salts. UK: Elsevier B. V; 2002

[18] Swaine DJ. The Significance of Trace Elmer: in Solving Petrogenetic, Problems and Controversies. Athens, Greece: Theophrastus PublicJtii 3s S.A.; 1983. pp. 521-532

[19] Ren D. Mineral matter in coal. In: Dexin H, editor. Coal petrology of China. Xuzhou: 1996. pp. 67-78

[20] Singh R, Gautam N, Mishra A, Gupta R. Heavy metals and living systems: An overview. Indian Journal of Pharmacology. 2011;**43**:246. DOI: 10.4103/0253-7613.81505. [PMC free article] [PubMed]

[21] Piotrowska-Seget Z, Cycoń M, Kozdroj J. Metal-tolerant bacteria occurring in heavily polluted soil and mine spoil. Applied Soil Ecolology. 2005;**28**:237-246. DOI: 10.1016/j.apsoil.2004.08.001

[22] Gadd GM. Metals, minerals and microbes: Geomicrobiology and bioremediation. Microbiology, 2010;**156**:609-643. DOI: 10.1099/mic.0.037143-0. [PubMed]

[23] Golyshina OV, Yakimov MM, Lünsdorf H, Ferrer M, Nimtz M, Timmis KN, Wray V, Tindall BJ, Golyshin PN. *Acidiplasma aeolicum* gen. Nov., sp. Nov., a Euryarchaeon of the family *ferroplasmaceae* isolated from a hydrothermal pool, and transfer of *Ferroplasma cupricumulans* to *Acidiplasma cupricumulans* comb. International Journal of Systematic and Evolutionary Microbiology. 2009;**59**:2815-2823. DOI: 10.1099/ijs.0.009639-0. [PubMed]

[24] Chevrel S, Croukamp L, Bourguignon A, Cottard T. A remote-sensing and GIS-based integrated approach for risk-based prioritization of gold tailings facilities–Witwatersrand, South Africa. In: Fourie AB, Tibbett M, Weiersbye IM, editors. Proceedings of the Third International Seminar on Mine Closure, 14–17 October. Johannesburg, South Africa: Australian Centre for Geomechanics; 2008. pp. 639-650

[25] Ezzatahmadi N, Ayoko GA, Millar GJ, Speight R, Yan C, Li J, Li S, Zhu J, Xi Y. Clay supported nanoscale zero-valent iron composite materials for the remediation of contaminated aqueous solutions: A review. Chemical Engineering Journal. 2016

[26] Fazlzadeh M, Rahmani K, Zarei A, Abdoallahzadeh H, Nasiri F, Khosravi R. A novel green synthesis of zero valent iron nanoparticles (NZVI) using three plant extracts and their efficient application for removal of Cr(VI) from aqueous solutions. Advanced Powder Technology. 2017;**28**(1):122-130

[27] Jiang W, Cai Q, Xu W, Yang M, Cai Y, Dionysiou DD, O'Shea KE. Cr (VI) adsorption and reduction by humic acid coated on magnetite. Environmental Science and Technology. 2014;**48**(14):8078-8085

[28] Aliyah NA. Adsorption of lead using rice husk. [Doctoral dissertation] UMP. 2012

[29] Lottermoser B.Mine Wastes: Characterization, Treatment and Environmental Impacts. New York, NY, USA: Springer; 2007. pp. 1-290

[30] Marsden DD. The current limited impact of Witwatersrand gold mine residues on water pollution in the Vaal River system. Journal of the South African Institute of Mining and Metallurgy. 1986;**86**:481-504

[31] Jones GA, Brierly SE, Geldenhuis SJJ, Howard JR. Research on the Contribution of Mine Dumps to the Mineral Pollution Load at the Vaal Barrage. Water Research Commission Report No. 136/1/89. Pretoria: Water Research Commission; 1988

[32] Vega F, Covelo E, Andrade M, Marcet P. Relationships between heavy metals content and soil properties in minesoils. Analytica Chimica Acta. 2004;**524**:141-150. DOI: 10.1016/j.aca.2004.06.073

[33] Vega FA, Covelo EF, Andrade M. Competitive sorption and desorption of heavy metals in mine soils: Influence of mine soil characteristics. Journal Colloid Interface Science. 2006;**298**:582-592. DOI: 10.1016/j.jcis.2006.01.012

[34] Vega FA, Covelo EF, Andrade M. Competitive sorption and desorption of heavy metals in mine soils: Influence of mine soil characteristics. Journal of Colloid and Interface Science. 2006;**298**:582-592. DOI: 10.1016/j.jcis.2006.01.012. [PubMed]

[35] Rafiei B, Bakhtiari NM, Hashemi M, Khodaei A. Distribution of heavy metals around the Dashkasan Au mine. Internationa Journal of Environmental Research. 2010;**4**:647-654

[36] Mitileni C, Gumbo J, Muzerengi C, Dacosta F. The distribution of toxic metals in sediments: Case study of new union gold mine tailings, Limpopo, South Africa. In: Mine Water – Managing the Challenges. Aachen, Germany: IMWA; 2011

[37] Harish E, David M. Assessment of potentially toxic cyanide from the gold and copper mine ore tailings of Karnataka, India. International Journal of Science and Technology. 2015;**3**:171-178

[38] Da Silva EF, Zhang C, Pinto LSS, Patinha C, Reis P. Hazard assessment on arsenic and lead in soils of Castromil gold mining area, Portugal. Applied Geochemistry. 2004;**19**:887-898. DOI: 10.1016/j.apgeochem.2003.10.010

[39] Akabzaa TM. Boom and Dislocation: A Study of the Social and Environmental Impacts of Mining in the Wassa West District of Ghana. Accra, Ghana: Third World Network, Africa Secretariat; 2000

[40] Dold B. Basic Concepts in Environmental Geochemistry of Sulfidic Mine-Waste Management. Rijeka, Croatia: InTech Open Access Publisher; 2010

[41] Adler R, Rascher JA. Strategy for the Management of Acid Mine Drainage from Gold Mines in Gauteng. Pretoria, South Africa: CSIR; 2007

[42] Acevedo F. The use of reactors in biomining processes. Electronic Journal of Biotechnology. 2000. DOI: 10.2225/vol3-issue3-fulltext-4

[43] Van Hille RP, van Wyk N,Froneman T, Harrison ST. Dynamic Evolution of the Microbial Community in Bio Leaching Tanks. Pfaffikon, Switzerland: Trans Tech Publications; 2013. pp. 331-334 (Advanced Materials Research)

[44] Brehm U, Gorbushina A, Mottershead D. The role of microorganisms and biofilms in the breakdown and dissolution of quartz and glass. Palaeogeography, Palaeoclimatology, Palaeoecology. 2005;**219**:117-129. DOI: 10.1016/j.palaeo.2004.10.017

[45] Baker BJ, Banfield JF. Microbial communities in acid mine drainage. FEMS Microbiological Ecology. 2003;**44**:139-152. DOI: 10.1016/S0168-6496(03)00028-X. [PubMed]

[46] Telmer K, Stapper DA. Practical Guide: Reducing Mercury Use in Artisanal and Small-Scale Gold Mining. Nairobi, Kenya: Geneva, Switzerland: United Nations Environment Programme; 2012

[47] Maree JP. Treatment of industrial effluent for neutralization and sulphate removal. A thesis submitted for PhD at the North West University, RSA. 2004

[48] Flynn CM. Hydrolysis of inorganic iron(III). Journal of American Chemical Society. 1984; **84**:31-41

[49] Coetzee H et al. WRC, Council for Geosience. Report No 1214/1/06. 2004

[50] Tchounwou PB, Yedjou CG, Patlolla AK, Sutton DJ. Heavy Metals Toxicity and the Environment. The publisher's final edited version of this article is available at EXS. 2005

[51] Ntwampe IO, Waanders FB, Bunt JR. Reaction dynamics of iron and aluminium salts dosage in AMD using shaking as an alternative technique in the destabilization-hydrolysis process. International Journal of Science. 2016;**4**:5-23

[52] Ntwampe IO, Waanders FB, Bunt JR. Destabilization dynamics of clay and acid-free polymers of ferric and magnesium salts in AMD without pH adjustment. Water Science and Technology. 2016;**74**(4):861-875

[53] Ntwampe IO, Waanders FB, Bunt JR. Turbidity removal efficiency of clay and a synthetic af-PACl polymer of magnesium hydroxide in AMD treatment. International Journal of Science. 2015;**4**:88-104

[54] Ntwampe IO, Moothi K. Determination of the efficiencies of a flocculant of clay and PFCl/PACl with Na_2CO_3 in the removal of TSS from the AMD. Accepted by Journal of Environmental Science and Technology. 2017

[55] Ntwampe IO, Jewell LL, Hildebrandt D, Glasser D. The effect of mixing on the treatment of paint wastewater with Fe^{3+} and Al^{3+} salts. Journal of Environmental Chemistry and Ecotoxicology. 2013;**5**(1):7-16

[56] Coulson JM, Richardson JF. Chemical Engineering. 6th ed. Vol. 1. London: Elsevier, Butterworth-Heinemann; 1999

[57] Sincero AP, Sincero GA. Physical-Chemical Treatment of Water and Wastewater. Londao, USA: IWA Publishing; 2003

[58] Thomas PR, Allen D, McGregor DL. Evaluation of combined chemical and biological nutrients removal. Water Science Technology. 1996;**34**:285-292

[59] Pratt C, Shilton A, Pratt S, Haverkamp RG, Elmetri I. Effects of redox potential and pH changes on phosphorus retention by melter slag filters treating wastewater. Environmental Science and Technology. 2007;**4**(18):6583-6590

[60] Nozaic DJ, Freeze SD, Thompson P. Water Science and Technology: Water Supply. 2001;**1**: 43-50

[61] Ghaly AE, Snow A, Faber BE. Treatment of grease filter washwater by chemical coagulation. Canadian Biosystem Engineering. 2006;**48**:6.13-6.22

[62] Kabata-Pendia 3rd A, editor. Trace Elements in Soils and Plants. Boca Raton, FL: CRC Press; 2001

[63] Niyogi DK, William Jr ML, McKnight DM. Effects of stress from mine drainage on diversity, biomass, and function of primary producers in mountain streams. Ecosystems. 2002; **5**:554-567

[64] Saif S, Tahir A, Chen Y. Green synthesis of iron nanoparticles and their environmental applications and implications. Nanomaterials. 2016;**6**(11):209

[65] Demirci UB, Akdim O, Andrieux J, Hannauer J, Chamoun R, Miele P. Sodium borohydride hydrolysis as hydrogen generator: Issues, state of the art and applicability upstream from a fuel cell. Fuel Cells. 2010;**10**(3):335-350

Chapter 4

Strategic Design of Heavy Metals Removal Agents through Zeta Potential Measurements

Eduardo Alberto López-Maldonado and
Mercedes Teresita Oropeza-Guzmán

Additional information is available at the end of the chapter

http://dx.doi.org/10.5772/intechopen.74053

Abstract

Industrial wastewater generally contains significant amounts of toxic heavy metals that cause a problem of contamination to the environment. In this chapter, the use of polyelectrolytic waste as new coagulant-flocculating-chelating agents in the separation of Cu, Ni, Zn, Pb, Cd, Cr by a coagulation-flocculation process is discussed. The isoelectric point ($\zeta = 0$) of the residual water was reached with a dose of 2.5 mg chitosan and observed a clarification kinetics = 187.49% T/h, sediment kinetics = 93.96 mm/h and an efficiency of 85% in the removal of heavy metals. With the SEM-EDS analysis and the determination of heavy metals in the treated water, it is shown that the functional groups that chitosan has in its structure have the following order of affinity for the removal of heavy metals from the wastewater model: Cr = 27.64% > Ni = 21.96% > Pb = 21.28% > Zn = 14.68% > Cu = 10.96% > Cd = 3.35% > Ca = 0.12%.

Keywords: heavy metals, zeta potential, wastewater treatment, biopolyelectrolytes, coagulation-flocculation

1. Introduction

The environmental pollution problems increase, with the aim of controlling and reducing the impact of industrial activities that cause damage to the environment, and environmental regulations are increasingly demanding. One of the issues of greatest scientific and technological interest is the care of water quality, derived from the problems of scarcity and water contamination by a variety of chemical elements such as heavy metals [1]. The main sources of heavy metals include mineral processing, pulp and paper industry, printing in the graphic industry, metallurgical, printed board manufacturing, nuclear, mining, battery manufacturing, leather tanning, smelting, petrochemicals, metal finishing and plating, semiconductor manufacturing, textile dyes, ceramic and other industries, which are those that consume significant amounts of water and therefore generate wastewater with a high content of heavy metals (>200 mg/L) that cannot be discharged to water bodies without previously carrying out a treatment process [2, 3].

The most common heavy metals that are often present in industrial wastewater include nickel, zinc, silver, lead, iron, chromium, copper, arsenic, cadmium, and uranium [1].

One of the main reasons why heavy metals should not be discharged into water receiving bodies is that they are nonbiodegradable, toxic, and easy to accumulate at low concentrations in living organisms in general, specifically in humans, and cause serious illnesses such as cancer, nervous system damage, and kidney failures and can be deadly at high concentrations [4].

Currently, there are different technologies to perform the treatment of wastewater containing heavy metals, including adsorption using various adsorbent materials (polymers, carbon, nanomaterials, clays, zeolites), electrodeposition, membrane filtration (ultrafiltration, nanofiltration and reverse osmosis), coagulation-flocculation, chemical precipitation with hydroxides, sulfides and chelating precipitations, electrocoagulation, electrodialysis, ion exchange, biological treatment, photocatalysis, and electroflotation. Each technology has certain advantages and disadvantages. Considering the most demanding environmental legislation, industries need more efficient wastewater treatments to eliminate suspended or dissolved metals [5–20].

In this context, the use of waste from the food industry and renewable sources of biopolyelectrolytes such as shrimp shell waste, nopal mucilage, nejayote, mesquite seeds, coffee, and tule residues as biodegradable coagulating-flocculating-chelating agents is proposed [21, 22]. Due to the chemical nature of the interactions that predominate at the molecular level between the functional groups of the biopolyelectrolytes (BPE) and the heavy metals present in the wastewater, in this work, the zeta potential measurements were used. The zeta potential is a key parameter to determine the surface charge density of the BPE, the isoelectric point (IEP) and define the strategy of adding the BPE to wastewater with heavy metals.

2. Experimental section

2.1. Materials and methods

2.1.1. Materials

The heavy metal wastewater model was prepared with the nitrate salts of each metal ion: $Ca(NO_3)_2$, $Zn(NO_3)_2$, $Cd(NO_3)_2$, $Cu(NO_3)_2$, $Ni(NO_3)_2$, $Pb(NO_3)_2$, and $Cr(NO_3)_3$. Sodium alginate (AG) with viscosity (5–40 cP) were purchased from Sigma Aldrich.

2.2. Methods

2.2.1. Extraction and physicochemical characterization of biopolyelectrolytes

Chitosan is obtained from waste shrimp shells using the adapted method proposed by Goycoolea et al. [23]. Maize gum was obtained by fractional separation, using hexane, ethanol, hydrochloric acid, isopropanol, acetone, and methanol formed by the steps of desalmidonado, deproteinization, delipidation, and delignification, which are adaptations of the methods proposed [24–26]. Mesquite gum was extracted from mesquite seeds using the method reported by Sciarini et al. [27]. For the extraction of nopal mucilage (NM), the methodology was reported by Cárdenas et al. [28]. Extraction of lignin was performed by the delignification method from black liquors of coffee and tule [29]. The process consists of precipitating the lignins by acidification from black liquors obtained from tule and coffee residues. The FTIR spectra were recorded on a FT-IR Spectrometer (Perkin Elmer). Universal Attenuated Total Reflectance (UATR Two) attachment was employed to analyze the biopolyelectrolyte.

2.2.2. $\zeta = f(pH)$ profiles of the alginate, maize gum, mesquite gum, nopal mucilage, coffee and tule lignin and chitosan

The zeta potential of the anionic, cationic, and neutral biopolyelectrolytes was measured in a Stabino Particle Charge Mapping (Microtrac). The measurements were done at ambient temperature in teflon cuvettes. Influence of pH on the zeta potential behavior of each biopolyelectrolytes was studied within a pH range of 2–11 with 0.1 M NaOH and 0.1 M HCl to determine the isoelectric point (IEP) and the highest degree of ionization of the biopolylectrolytes [30].

2.2.3. Evaluation of the chelating performance of the BPE

The performance of the BPE in the separation of heavy metals present in the wastewater model was evaluated using stability analysis to determine the kinetics of coagulation-flocculation, morphology and elemental composition of the formed flocs, content of metals remaining in the treated water, in each test of dosage of BPE to synthetic heavy metals wastewater. Samples of flocs obtained were examined in a SEM ZEISS EVO-MA15, equipped with an EDS (energy dispersive spectroscopy) BRUKER detector for heavy metals-relative estimations. Agilent Technologies 4200 MP-AES were used to determine metals concentration. Stability tests were carried out, performing a transmittance scan every 25 s for 30 min using Turbiscan Lab® Expert.

3. Results and discussion

3.1. Physicochemical characterization of biopolyelectrolytes

Considering that the surface functional groups of the BPE are responsible for the molecular interactions between the biopolyelectrolytes (AG, Ch, Mag, nopal mucilage, Meg, CL, and TL)

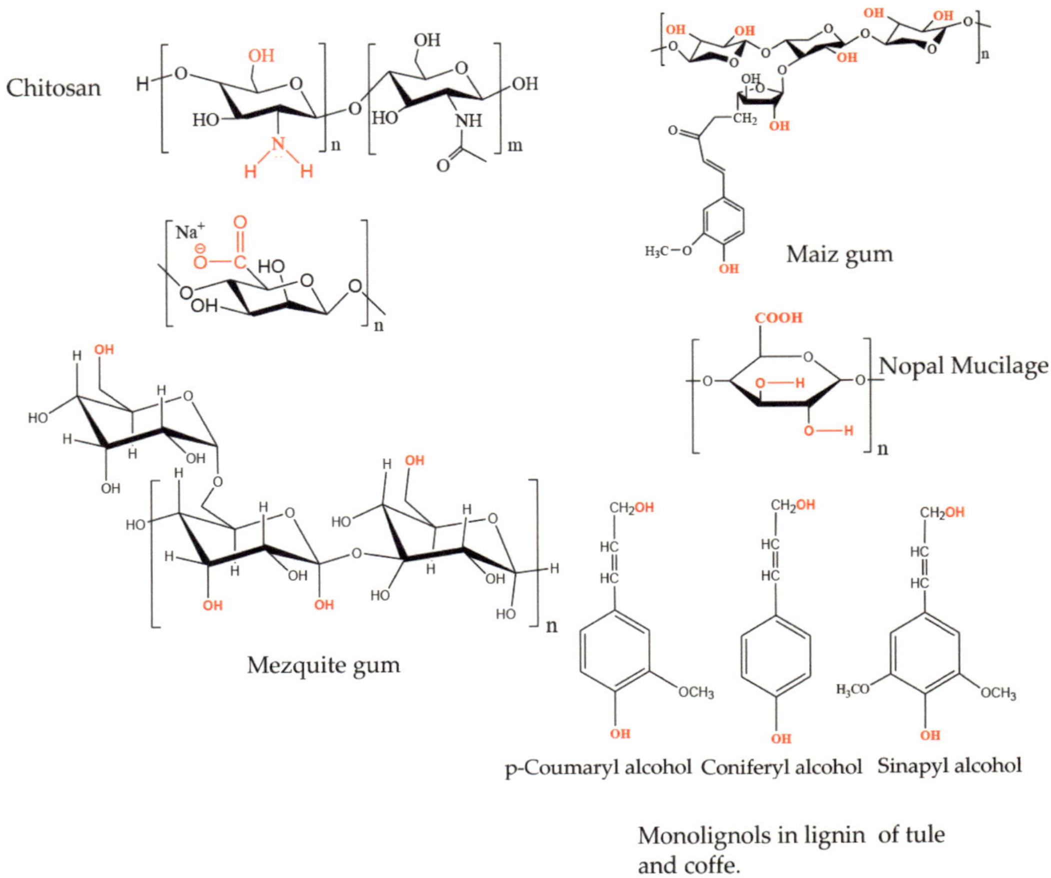

Figure 1. Chemical structure of the BPE.

(see **Figure 1**) and the heavy metals contained in the wastewater, they were studied by FTIR [31]. The most common BPE functional groups are carbonyl, phenolic, acetamido, alcoholic, amido, amino, and sulfhydryl, which have an affinity for heavy metal ions to form complexes or metal chelates [32].

In **Figure 2**, the FTIR spectra corresponding to the anionic biopolyelectrolytes that have higher density of negative surface charge are shown. In **Figure 2a** and **c**, the characteristic signals are shown to the three main compounds present in the coffee and tule lignin: p-coumaryl alcohol, coniferyl alcohol, and sinapyl alcohol. Assignment bands shown at the wavelength 3500–3100 cm^{-1} are attributed to the stretching vibrations of the OH, the stretching vibrations at 2910–2835 cm^{-1} are attributed to the CH group, and the stretching vibrations at 1700–1550 cm^{-1} are attributed to the C—C group. There are also other signals at 1230 cm^{-1} and 1012 cm^{-1}, corresponding to the vibration bands of the C—O.

Figure 2b shows the FTIR spectrum of the anionic biopolyelectrolyte (maize gum) obtained from nejayote. The signal observed at 2923 cm^{-1} corresponds to the vibrations of the CH bonds of the methyl groups and the wide band that appears at 3421 cm^{-1} is due to the stretches of the HO bonds of the hydroxyl groups present in the macromolecule of the maize gum.

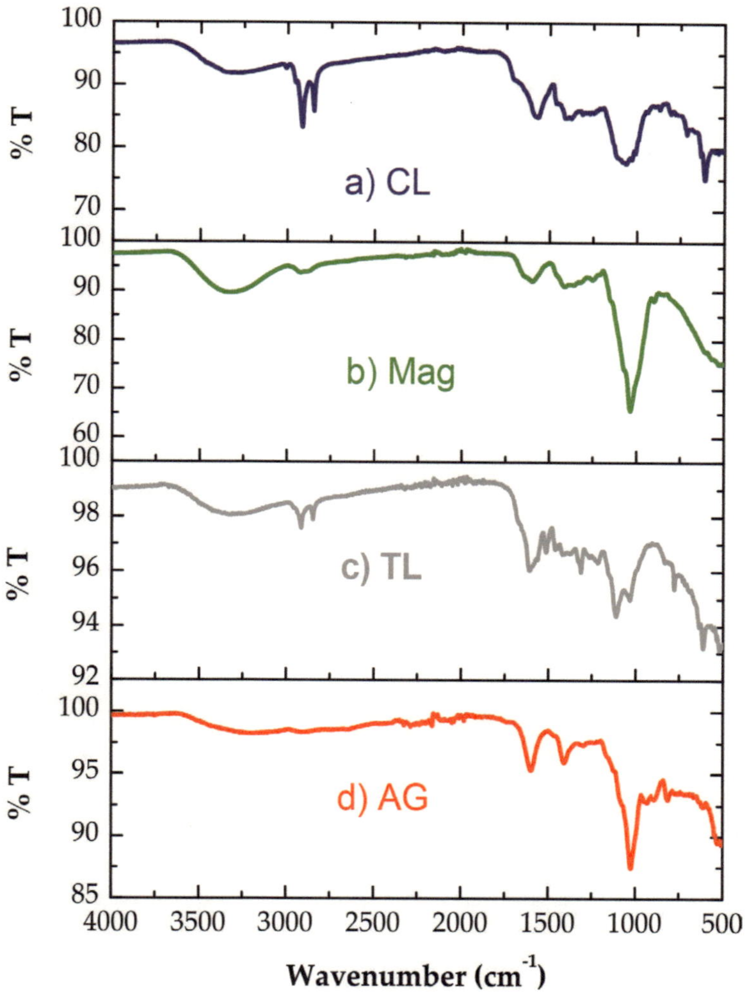

Figure 2. FTIR spectra of biopolyelectrolytes with higher density of negative surface charge: a) Coffe lignin, b) Maize gum, c) Tule lignin, d) Alginate.

The OH groups are those that give anionic character to this biopolyelectrolyte and that favors its interaction with metal ions present in the wastewater.

Figure 2d shows the characteristic signals of the alginate macromolecule, in which the band at the 3500–3200 cm^{-1} region is attributed to the hydroxyl groups and the signal at 1500 cm^{-1} corresponds to the carboxyl groups found distributed along the anionic biopolyelectrolyte chain.

Figure 3 shows the FTIR spectra of the BPE of cationic, non-ionic and slightly anionic character, considering the zeta potential values.

In the spectrum of Ch (**Figure 3a**), the characteristic bands at 3450 cm^{-1} (—OH group), at 3292 cm^{-1} (NH group), at 2919 and 2862 cm^{-1} (CH group), at 1655 cm^{-1} (amide I), at 1580 cm^{-1} (double group $-NH_2$), at 1313 cm^{-1} (amide III), at 1154 cm^{-1} (antisymmetric stress of the COC bridge), at 1082 and 1032 cm^{-1} (skeletal vibrations characteristic of the pyranotic structure) and 896 cm^{-1} (CH tension of the anomeric groups) are shown.

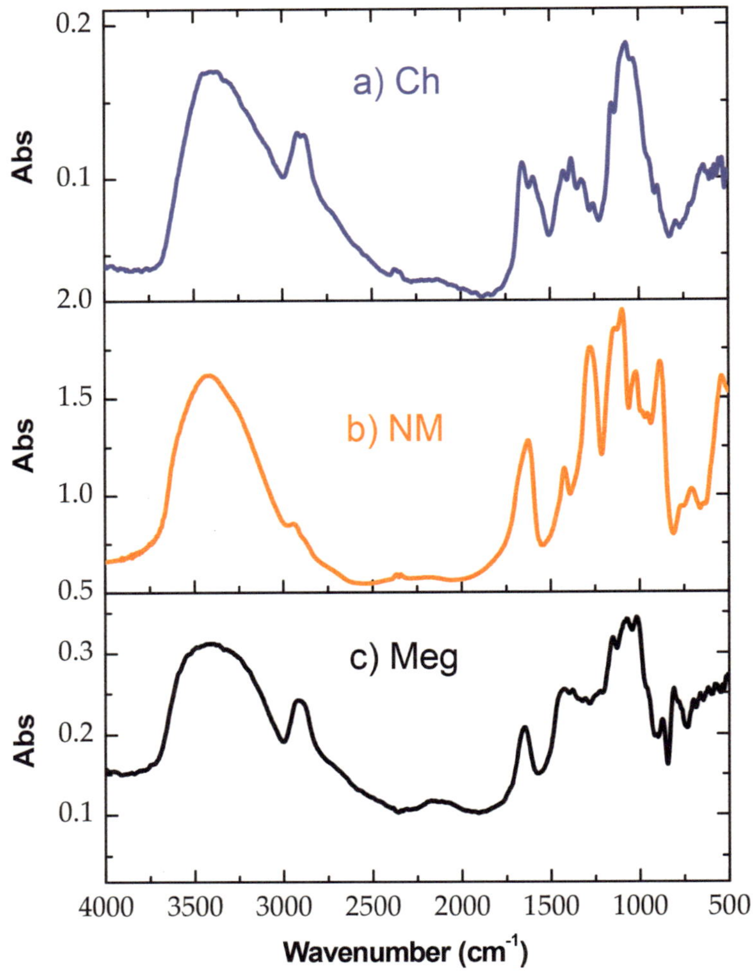

Figure 3. FTIR spectra of the BPE of cationic, non-ionic, and slightly anionic character: a) Chitosan, b) Nopal mucilage, c) Mezquite gum.

The absorption bands that are common for mesquite gum (**Figure 3b**) correspond to the OH-group centered around 3400 cm^{-1} and the C—H found in the band at 2900 cm^{-1}. Other characteristic absorption bands are observed in the bands at 1150–950 cm^{-1}, which are associated with the signals of C—O and carbohydrates (C—O—H). Also, in the bands at 870 cm^{-1} and 813 cm^{-1} indicate the presence of units of β-D mannopyranose and units of α-D galactopyranose.

The nopal mucilage spectrum is shown in **Figure 3c**, and the spectral region at 1350 cm^{-1} to 1750 cm^{-1} assigned to the vibrations of the bands of the nonionized COOH carboxylic groups is observed. Likewise, ionization leads to its disappearance and to the appearance of new bands associated with the symmetric and asymmetric vibrations of the COO— groups, centered approximately between at 1600–1650 cm^{-1} and at 1400–1450 cm^{-1}.

3.2. ζ = f (pH) profiles of the BPE

In **Figure 4**, the ζ = f(pH) plot shows the charge density variation for AG, NM, Meg, CL, TL, Mag and Ch with respect to pH. The change in pH had a distinct effect with each biopolyelectrolyte because of the difference between the functional groups present in the chains of the macromolecules. Sodium alginate is a strong anionic biopolyelectrolyte and maintains negative zeta potential values (ζ = − 480.0 mV) between pH 3.5 and 10, attributed to the carboxylate group (—$COO^{-}Na^{+}$) and was not observed their IEP.

Chitosan has a cationic character at pH < 6.0, reaching a maximum value in zeta potential ζ = 55 mV at pH 3.0. This behavior is attributed to the amine groups that are protonated at pH < pka = 6.3. At pH = 6.5, the isoelectric point of the chitosan was reached and at pH > IEP, the values of zeta potential remained close to neutrality (ζ = 0), and this corresponds to the hydrophobic character acquired by the chitosan macromolecule due to the deprotonation of the amine groups.

The zeta potential values of the NM indicate the acid-base behavior of the carboxyl groups present in the macromolecule. The NM has its isoelectric point at pH = 4.0, observing the formation of NM aggregates at pH < IEP. At pH > IEP, it had an anionic character, and this derived that the carboxyl groups are ionized at pH > pka = 3.5. The behavior zeta potential vs. pH of anionic BPE obtained of nejayote is shown in **Figure 4**, which has a high negative charge density (ζ = −35 mV) in the pH range 6–12, having the isoelectric point close to pH = 2.0.

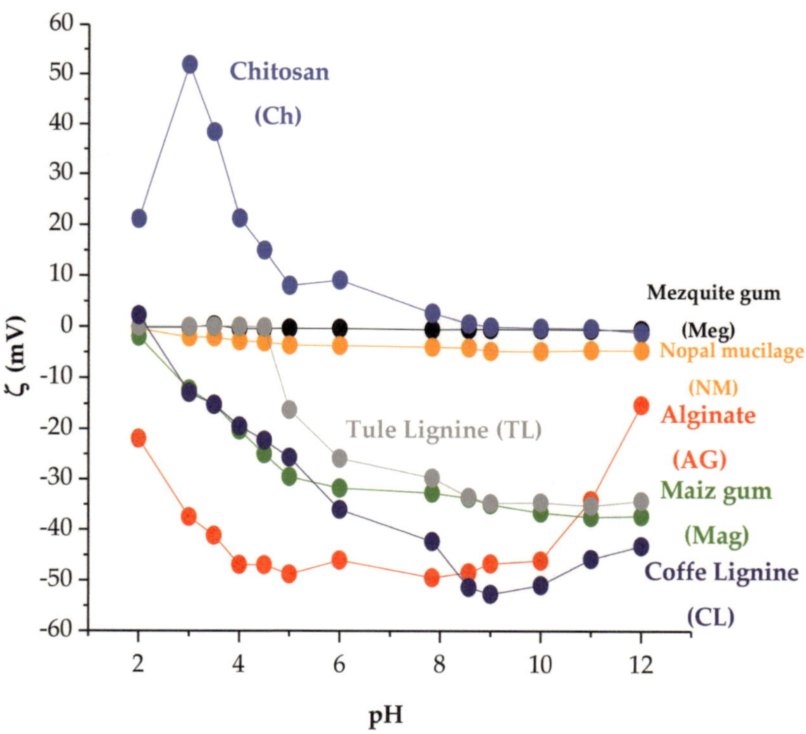

Figure 4. ζ = f (pH) profiles of the BPE.

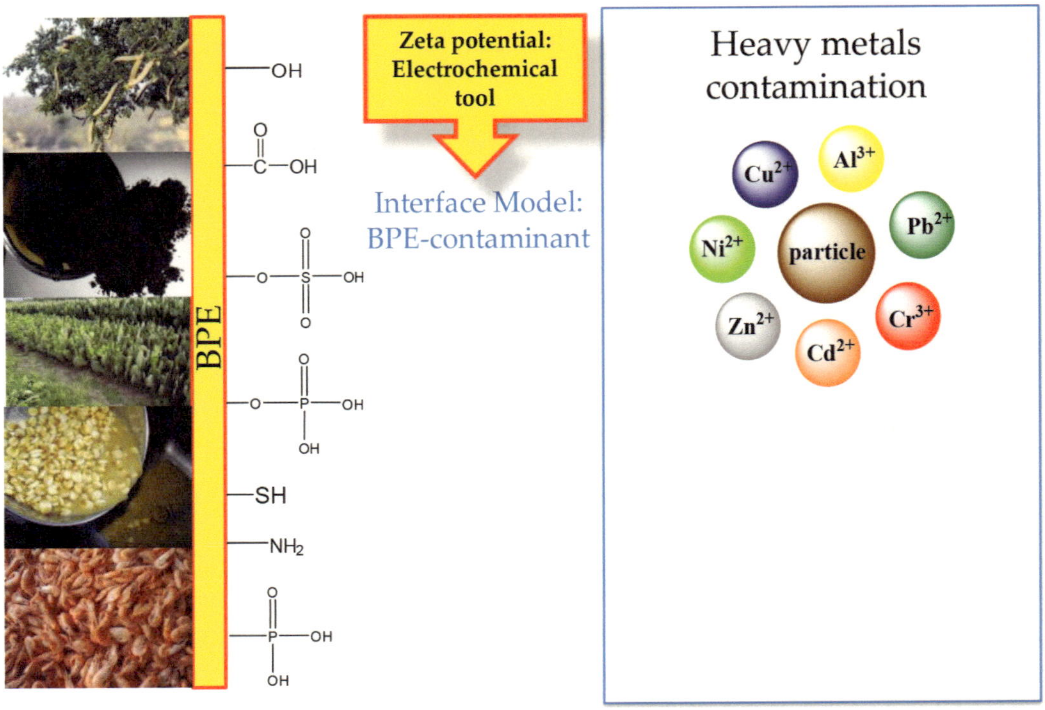

Figure 5. Outline of BPE-contaminant interface model.

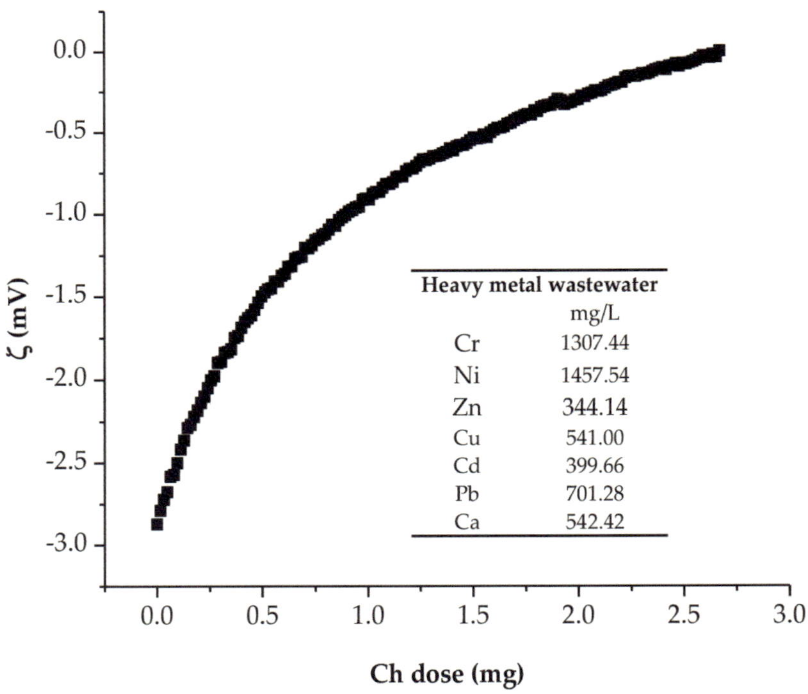

Figure 6. ζ = f (Ch dose) profile with heavy metals wastewater model.

http://dx.doi.org/10.5772/intechopen.74053

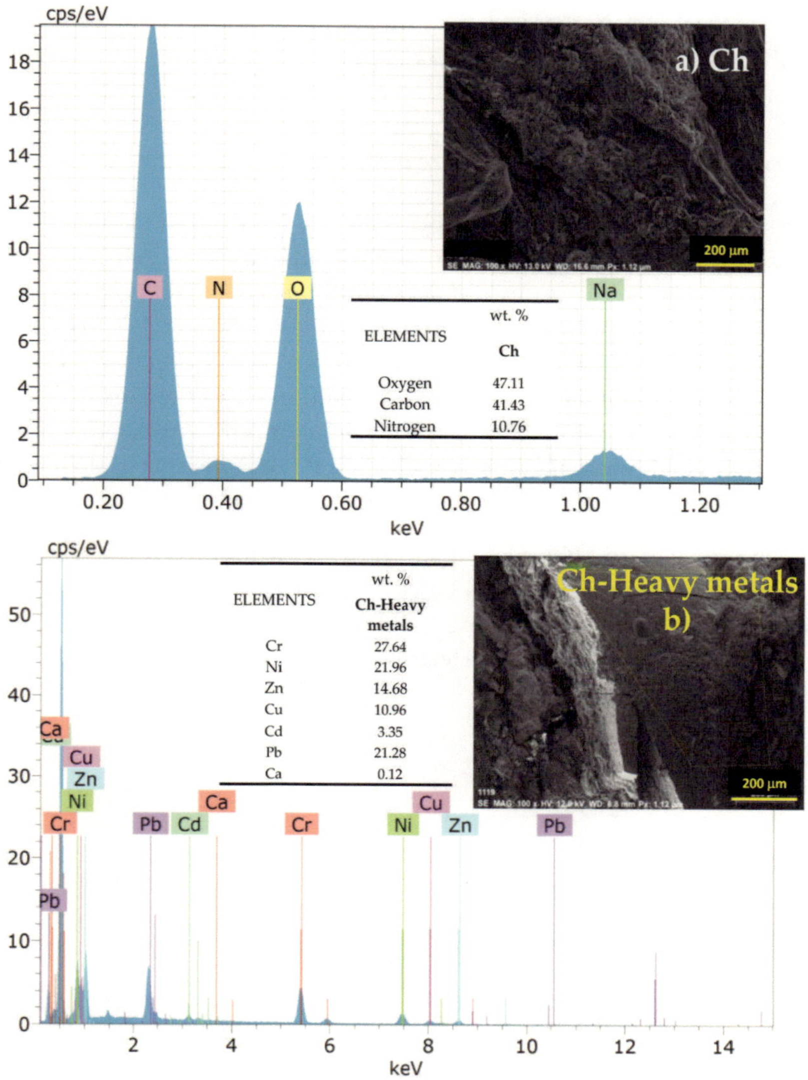

Figure 7. SEM micrograph and EDS spectrum of (a) Ch and (b) flocs of Ch-heavy metals.

The coffee lignin showed an average negative zeta potential value ($\zeta = -40$ mV) at a pH of 6–12, decreasing linearly according to the low pH ($\zeta = -40$ mV to $\zeta = -10$ mV). The lignin chains of coffee present their IEP at pH = 2.0.

Similarly, tule lignin showed this tendency, but with a lower negative charge density ($\zeta = -30$ mV) at pH 2–12, and its IEP at pH = 4.0. This behavior is attributed to the difference in the composition of the three main monomers that contain the lignin and which is a function of its source of origin.

The profiles of $\zeta = f$ (pH) of the BPE allow to determine and predict the physicochemical behavior of the macromolecules in the coagulation processes of flocculation of residual water with heavy metals. Through these profiles, it is possible to determine the pH at which the

BPE have their highest surface charge density (negative or positive), define a strategy to make additions of BPE mixtures sequentially, and orient the dose needed to carry out the separation of heavy metals from waste water.

3.3. Evaluation of the chelating performance of the BPE

The BPE have a great potential to get to replace the synthetic coagulant-flocculant-chelating agents, considering their chemical structure and their surface charge density in aqueous solution. The BPE contain in their structure different functional groups that give rise to the surface charge of the macromolecules and that can interact with the heavy metals frequently found in industrial wastewater. As shown in **Figure 5**, the evaluation of the performance of each BPE can be performed for the separation of metals in a natural way as well as to carry out strategies for modifying the macromolecules to make them more efficient in terms of coagulation-flocculation kinetics and removal of heavy metals. In this chapter, the evaluation of chitosan for the separation of heavy metals in wastewater model is presented.

In **Figure 6**, the profile of $\zeta = f$ (Ch dose) is shown with the residual water model with heavy metals. The model wastewater has a zeta potential value ($\zeta = -2.8$ mV) at pH = 5.0, as the Ch dose increases, the zeta potential of the wastewater rises linearly ($\zeta = -2.8$ mV to $\zeta = -0.5$ mV), in a dose range of Ch 0–1.5 mg. The isoelectric point of the wastewater model was reached with a dose of 2.5 mg of Ch.

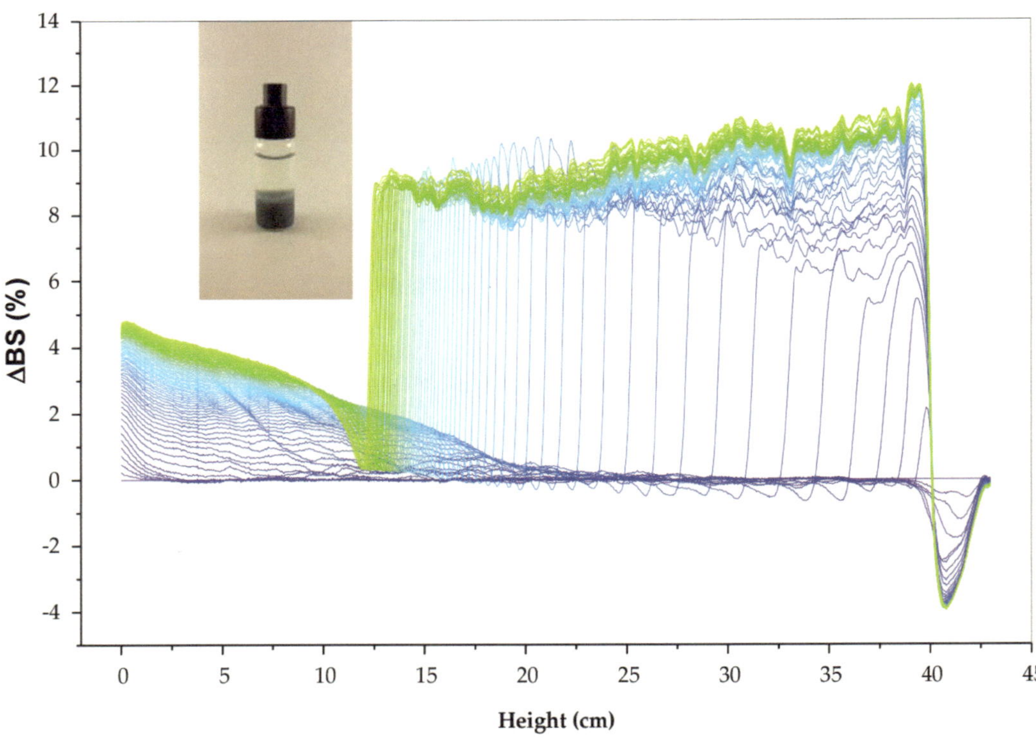

Figure 8. Backscattering profile of NP at optimum dose; data are reported as a function of time (0–120 min).

The superficial chemical analysis of the flocs formed in the coagulation-flocculation process using the best dose of Ch (2.5 mg), at pH = 5.0, **Figure 7a** shows the morphology of the chitosan without interacting with the heavy metals. The EDS spectrum of chitosan shows that it does not contain heavy metals in its natural form. In **Figure 7b**, the morphology of flocs formed with heavy metals and chitosan is shown.

The determination of heavy metals in the treated wastewater demonstrates the effectiveness of chitosan to perform the separation by 85% removal of heavy metals. In **Figure 8**, the backscattering profile (% BS) vs. the height of the vial is shown, with the best dose of chitosan a clarification kinetics = 187.49%T/h and sediment kinetics = 93.96 mm/h.

4. Conclusions

One of the alternatives for the treatment of wastewater with a high content of heavy metals consists of the use of BPE extracted from waste from the food industry and gives it an added value. The zeta potential allows the coagulant-flocculant-chelating agents to be strategically applied for the separation of heavy metals, defining the best physicochemical conditions to make the coagulation-flocculation processes more efficient. At a dose of 2.5 mg chitosan, was reached 85% efficiency of separation of heavy metals and clarification kinetics = 187.49% T/h and sediment kinetics = 93.96 mm/h in wastewater treatment.

Acknowledgements

The authors thank the Autonomous University of Baja California for the financing granted for the development of this project (UABC-PTC-668).

Conflict of interest

The authors state that there is no conflict of interest.

Author details

Eduardo Alberto López-Maldonado[1]* and Mercedes Teresita Oropeza-Guzmán[2]

*Address all correspondence to: elopez92@uabc.edu.mx

1 Faculty of Chemical Sciences and Engineering, Autonomous University of Baja California, Tijuana, Baja California, Mexico

2 Graduate and Research Center of the Technical Institute of Tijuana, Tijuana, Mexico

References

[1] Khan A, Badshah S, Airoldi C. Dithiocarbamated chitosan as a potent biopolymer for toxic cation remediation. Colloids and Surfaces B: Biointerfaces. 2011;**88**:88-95

[2] Kurniawan TA, Chan GYS, Lo WH, Babel S. Physico-chemical treatment techniques for wastewater laden with heavy metals. Chemical Engineering Journal. 2006;**118**:83-98

[3] Babel S, Kurniawan TA. Cr (VI) removal from synthetic wastewater using coconut shell charcoal and commercial activate carbon modified with oxidizing agents and/or chitosan. Chemosphere. 2004;**54**:951-967

[4] Blais JF, Djedidi Z, Cheikh RB, Tyagi RD, Mercier G. Metals precipitation from effluents: Review. Practice Periodical of Hazardous, Toxic, and Radioactive, Waste Management. 2008;**12**:135-149

[5] Chen B, Qu R, Shi J, Li D, Wei Z, Yang X, Wang Z. Heavy metal and phosphorus removal from water by optimizing use of calcium hydroxide and risk assessment. Environmental Pollution. 2012;**1**:38-54

[6] Charerntanyarak L. Heavy metals removal by chemical coagulation and precipitation. Water Science and Technology. 1999;**39**:135-138

[7] Tünay O, Kabdasli NI. Hydroxide precipitation of complexed metals. Water Research. 1994;**28**:2117-2124

[8] Andrus ME. A review of metal precipitation chemicals for metal-finishing applications. Metal Finishing. 2000:20-23

[9] Li YJ, Zeng XP, Liu YF, Yan SS, Hu ZH, Ya M. Study on the treatment of copper-electroplating wastewater by chemical trapping and flocculation. Separation and Purification Technology. 2003;**31**:91-95

[10] Lazaridis NK, Matis KA, Webb M. Flotation of metal-loaded clay anion exchangers. Part I: The case of chromate. Chemosphere. 2001;**42**:373-378

[11] Doyle FM, Liu ZD. The effect of triethylenetraamine (trien) on the ion flotation of Cu2+ and Ni2+. Journal of Colloid and Interface Science. 2003;**258**:396-403

[12] Kongsricharoern N, Polprasert C. Electrochemical precipitation of chromium (Cr6+) from an electroplating wastewater. Water Science Technology. 1995;**31**:109-117

[13] Holt PK, Barton GW, Mitchell CA. The future for electrocoagulation as a localized water treatment technology. Chemosphere. 2005;**59**:355-367

[14] Pansini M, Colella C, De Gennaro B. Chromium removal from water by ion exchange using zeolite. Desalination. 1991;**83**:145-157

[15] Álvarez E, García A. Purification of metal electroplating wastewater using zeolites. Water Research. 2003;**37**:4855-4862

[16] Abdel-Halim SH, Shehata AMA, El-Shahat MF. Removal of lead ions from industrial waste water by different types of natural materials. Water Research. 2003;**37**:1678-1683

[17] Algarra M, Jiménez MV, Rodríguez E, Jiménez A, Jiménez J. Heavy metals removal from electroplating wastewater by aminopropyl-Si MCM-401. Chemosphere. 2005;**59**:779-786

[18] Karabulut S, Karabakan A, Denizli A, Yûrûm Y. Batch removal of copper (II) and zinc (II) from aqueous solutions with low-rank Turkish coals. Separation and Purification Technology. 2000;**18**:177-184

[19] Juang R, Shiau RC. Metal removal from aqueous solutions using chitosan-enhanced membrane filtration. Journal Membrane Science. 2000;**165**:159-167

[20] Ahn KH, Song KG, Cha HY, Yeom IT. Removal of ions in nickel electroplating rinse water using low-pressure nanofiltration. Desalination. 1999;**122**:77-84

[21] Chang Q, Wang G. Study on the macromolecular coagulant PEX which traps heavy metals. Chemical Engineering Science. 2007;**62**:4636-4643

[22] Chitanu GC, Carpov A. Ecologically benign polymers: The case of maleic polyelectrolytes. Environmental Science and Technology. 2002;**36**:1856-1860

[23] Goycoolea F, Agullo E, Mata R. Sources and processes of obtaining. In: Pastor de Abram A, editor. Chitin and Chitosan: Obtaining, Characterization and Applications. 1st ed. Fondo editorial de la Pontificia Universidad Catolica del Perú; 2004. pp. 103-154

[24] Niño-Medina G, Carvajal-Millán E, Lizardi J, Rascon-Chu A, Marquez-Escalante J, Gardea A, Martínez-López A, Guerrero V. Maize processing wastewater arabinoxylans: Gelling capability and cross-linking content. Food Chemistry. 2009;**115**:1286-1290

[25] Zhang Z. Extraction and modification technology of arabinoxylans from cereal byproducts: A critical review. Food Research International. 2014;**65**:423-436

[26] Simpson J. A survey of microorganisms for the production of enzymes that attack the pentosans of wheat flour. Canadian Journal of Microbiology. 1954;**1**:131-139. DOI: 10.1139/m55-017

[27] Sciarini LS, Maldonado F, Ribotta PD, Pérez GT, León AE. Chemical composition and functional properties of *Gleditsia triacanthos* gum. Food Hydrocolloids. 2009;**23**:306-313

[28] Cárdenas A, Goycoolea FM, Rinaudo M. On the gelling behaviour of 'nopal' (*Opuntia ficus indica*) low methoxyl pectin. Carbohydrate Polymers. 2008;**73**:212-222

[29] Sarwar M, Nasima D, Islam M, Iqbal S. Characterization of lignin isolated from some nonwood available in Bangladesh. Bioresource Technology. 2007;**98**:465-469

[30] López-Maldonado EA, Oropeza-Guzman MT, Jurado-Baizaval JL, Ochoa-Terán A. Coagulation-flocculation mechanisms in wastewater treatment plants through zeta potential measurements. Journal of Hazardous Materials. 2014;**279**:1-10

[31] Abiola Oladoja N. Headway on natural polymeric coagulants in water and wastewater treatment operations. Journal of Water Process Engineering. 2015;**6**:174-192

[32] Bhatnagar A, Sillanpää M, Witek-Krowiak A. Agricultural waste peels as versatile biomass for water purification—A review. Chemical Engineering Journal. 2015;**270**:244-271

Section 3

Environmental Effects of Heavy Metals

Chapter 5

Short-Term Response of Plants Grown under Heavy Metal Toxicity

Prasann Kumar and Shweta Pathak

Additional information is available at the end of the chapter

http://dx.doi.org/10.5772/intechopen.75722

Abstract

Sorghum vulgare L. plants when exposed to cadmium nitrate with the concentrations of 70 and 150 ppm per kg of soil for 90 days exhibited phytotoxic responses. The observations of specific responses were dependent on treatment combinations. The significant hazardous effects and oxidative damage of cadmium nitrate (70 and 150 ppm) were evident by increased MDA content and hydrogen peroxide content. However, these responses were reversed by exogenous application of putrescine (2.5 and 5.0 mM) and mycorrhiza (*Glomus*; 150 inoculants per kg of soil), more so, in their combined treatment, at different DAS. But combined treatment of putrescine and mycorrhiza enhanced the stability of sorghum by reducing the ROS production in plant cells. On the basis of the data obtained, it is concluded that plants responded up to 70 ppm cadmium nitrate with stress-induced responses, which were ameliorated by combined application of putrescine and *Glomus* mycorrhiza.

Keywords: agriculture, biotic, cadmium, density, economic, forage

1. Introduction

Cadmium (Cd) is one of the components of the earth's crust and present in several places or in different ecosystems on earth i.e., terrestrial, aquatic and others. Benavides et al. [1] reported that cadmium is one of the most hazardous heavy metals in the atmosphere, soil and aquatic system which is finally going into our food chain and responsible for serious environmental problem leading to the health hazards in the living organisms, for instance, mutagenesis, lung cancer, convulsion and brain damage. The alleviation or inhibition of cadmium in plants has, therefore, caused extensive attention of the whole society [2, 3]. It is must to know about cadmium, its physical and chemical properties, isotopic studies In the atmosphere, Cd can enter by burning of coal, mining of metals as well as refining process which may lead to rise in Cd level in the soil by atmospheric fallout also. If we see the atmospheric fallout of Cd from atmospheric air, it follows in the order Remote area < Rural area < Urban areas. Due to long term effects of cadmium, the countries fixed its tolerance limit. The European Economic Committee proposed the concept of PTE i.e. Potential Toxic Elements. PTE for the Cadmium in soil is 1.0–3.0 mg/kg of dry soil [4]. If we think about the aging of the metal in the soil, then a distinction should be made between persistence of total metals in the soil and the persistence of bioavailable forms of metals. This aging of the metals depends on soil acidity. Evidence of the aging process is provided by studies of metal extractability and liability. There are many terms used to describe and categorize metals, including traces metals, transition metals, micronutrients, toxic metals, heavy metals. Bjerrum's [5] meaning of "heavy metals" depends on the density of the natural type of the metal, and he arranges "heavy metals" as those metals with basic densities over 7 g cm^{-3}. In 1964, the editorial manager of Van Nostrand's Worldwide Reference book of Concoction Science and in 1987, the editors of Allow and Hackh's Compound Lexicon included metals with a thickness more noteworthy than 4 g cm^{-3}. The fortune of various metals viz. Cr, Ni, Cu, Mn, Hg, Compact disc, Pb and metalloids, including As, Sb, and Se, in the encompassing outskirts is of huge misgiving [6], especially close previous mine locales, dumps, following heaps, and impoundments, yet in addition in urban and mechanical focuses. Cause of cadmium in soil is portrayed as agrarian squanders (20%), ooze (38%), manures (2%) and barometrical aftermaths (40%) [7]. Cadmium is one of the most toxic elements with reported carcinogenic effects in humans [8]. Cadmium and cadmium compounds are, compared to other heavy metals, relatively water soluble and mobile compound in most soils, generally more bio- available and tends to bio- accumulate. It induces cell injury and death by interfering with calcium (Ca) regulation in biological system. Cadmium is not essential for plant or animal life (IPCS monographs). Terrestrial plants may accumulate cadmium in the roots where it is found bound to the cell walls [9]. The pH level is one of the most important factors controlling cadmium absorption. The translocation of cadmium is significant in above ground parts with respect to copper and lead. Concentration in roots represents only 2–5 times that in the above ground parts, but cadmium is transferred only with difficulty to reproductive or storage organs of the plant [10]. The polyamine (PA) putrescine (Put) is an important modulator for the mitigation of diverse of stress in the plants [11–18], and it also plays a significant role in the apoptosis and programmed death in both animals and plants [19]. Chemically the PAs are undersized, +ve charged aliphatic amines at cellular pH values and, consequently, bind opposite charged molecules, viz. nucleic acids, acid phospholipids and proteins, consequently [14, 20]. The cellular level of free amino acids in plants depends on the synthesis and degradation of PAs, their bounding with phenolic acids and intracellular transport [14, 15, 21–23]. The inhibitory effects of Cd are manifestations of oxidative stress, which finally reduces crop productivity. Polyamines are involved in abiotic stress tolerance in plants. Increased polyamines level in stressed plants has adaptive significance because of their involvement in the regulation of cellular ionic environment, maintenance of membrane integrity, prevention of chlorophyll loss and stimulation of protein, nucleic acid and protective alkaloids. Interaction of polyamines with membrane phospholipids implicates membrane stability under stress conditions. Polyamines additionally shield the layer from oxidative harm as they go about as free radical foragers. Reaction to abiotic damage and mineral supplement insufficiency is related with the generation of conjugated PAs in plants. Polyamine substances are adjusted in light of the introduction to overwhelming metals. These viably balance out and secure the film frameworks against the dangerous impacts of metal particles. The mycorrhiza *Glomus* is the significant inhibitor of mobility of cadmium ion into the soil solution and defends the plants from cadmium toxicity. Colonization of AM fungal was observed in highly contaminated soil [24, 25]. *Glomus mosseae* was reported in heavy metal contaminated sites [26]. External mycelium of *Glomus mosseae* produce a type of protein called Glycoprotein (Glomalin), which has heavy metal binding sites [27]. Cadmium metals accumulate at these binding sites. The antioxidant level is also increased as a result of association of AM fungi with plants [28–30]. Some fungal strains were isolated in the past which were resistant to heavy metal contamination. Mycorrhizal fungi overcome the stress of heavy metal contamination [31, 32].

We are interested in the study of the sorghum plant because of several reasons, which we find more relevant:

[A] The research of the poisoning effect of the heavy metal Cd on plant mainly focuses on food crops such as rice, wheat and maize, but less on sorghum.

[B] Sorghum frequently used as animal feed so quality assurance related to heavy metals is not understood very well, hence studies on contamination of heavy metal in fodder crop is required.

[C] The sensitivity level of sorghum for cadmium varies. By deciding the sensitivity one can find genes which are responsible for the same. Literature indicates that tolerance level of a heavy metal for sorghum is 1000 ppm onwards. It indicates that it has a wide range of tolerance. The genomic size of the sorghum ranges from 700 to 772 Mbp, and it has been well sequenced. So, search for genes responsible for tolerance is not that difficult now. After finding the gene, it can be transferred into the sensitive crops such as cow pea and other leafy vegetables. Finally, we can get transgenic of the heavy metal tolerance. Da-Lin et al. [33] reported some important changes in growth and physiological characteristics of the sorghum plant under cadmium stress such as height, chlorophyll content, root activities and MDA content. They concluded that Cd effect was the manifestation of oxidative stress. Low concentration of cadmium can promote growth of hard wheat, while under a relatively high concentration, the growth of wheat and tillering were both inhibited and the degree varied among the various species. Liu's [34] research also showed that corn seedling's height under cadmium treatment was reduced significantly as the concentration of cadmium increased with prolonged growth period. These studies showed that lower concentrations of Cd stimulated increase of sorghum height, which may be related to the certain resistance of the sorghum genus plant. The higher concentrations hindered their development; explanations behind the restraining impact of substantial metals to plant development were presumably due to: [A] A progression of physical and chemical responses between the overabundance substantial metal and soil parts which changes soil properties, along these lines influencing soil fruitfulness [35, 36]. For instance, substantial metal contamination can improve the obsession of soil phosphorus, which influenced the plant engrossing phosphorus, consequently influencing the development of plants [37, 38].

[B] Substantial metal caused a lessening in plant photosynthesis, in this way decreasing the plant water and supplement adsorption, which influenced the ordinary development and improvement of sorghum [39].

There are large reports about substantial metal contamination influencing root exercises of Graminaceous plants. For instance, through the hydroponic way, Yang et al. [40] found the impact of sewage straightforwardly flooded on root and seedlings of the wheat. The outcomes demonstrated that the worry of sewage flooded quickened the decrease of wheat seedlings and root, diminishing the root number and the root exercises essentially. Jiang et al. [41] additionally demonstrated that tainted soil made the underlying foundations of the rice seedlings yellow and red, extended the rhizome, root shading was dark colored and yellow, while Huang et al. [42] demonstrated that under framework or soil with cadmium the root exercises altogether diminished.

So, our objective in the present study was to find out the oxidative damage induced by cadmium toxicity in sorghum and to assess the role of putrescine and mycorrhiza in the mitigation of cadmium induced stress responses in sorghum.

2. Material and methods

2.1. Selection of crop sorghum (*Sorghum vulgare* L.)

Sorghum is one of the main staple foods for the world's poorest and most food-unsecured people across the semi-arid tropics. Sorghum belongs to the grass family Graminae. It is a short-day C4 plant. The optimum photoperiod which induces flower formation is between 10 and 11 h. Sorghum is one of the main staple foods for the world's poorest and most food-unsecured people across the semi-arid tropics. Globally, sorghum is cultivated on 41 million hectares to produce 64.20 million tons, with productivity having around 1.60 tonnes per hectare ([96], Directorate of Sorghum Research, Hyderabad). With exception in some regions, it is mainly produced and consumed by poor farmers. India contributes about 16% of the world's sorghum production ([96], DSR Hyderabad). It is the fifth most important cereal crop in the country. In India, this crop was one of the major staple cereals during 1950s and occupied an area of more than 18 million hectares, but has come down to 7.69 million hectares ([96], Directorate of Sorghum Research, Hyderabad). Sorghum requires warm conditions, but it can be grown under a wide range of conditions. It is grown from sea level to as high as 1500 m. It can tolerate high temperature throughout its life cycle better than any crop. It can be grown in the areas having an average annual rainfall 60–100 cm. The minimum temperature for the germination of the seed is 7–10°C. It needs about 26–30°C for optimal growth. Sorghum is characterized by a vastly diverse germplasm in terms of phenotypic and morphological traits. Sorghum can be classified into four main groups depending on their production characteristic: grain sorghum, forage sorghum (FS), high-tonnage sorghum (energy) and sweet sorghum.

Sorghum cultivars are now being considered as candidates in the search for efficient energy crops due to an increased interest in the conversion of biomass to energy.

2.2. Experimental site

The present investigation was carried out in the polyhouse and the laboratory of the Department of Plant Physiology, Institute of Agricultural Sciences, Banaras Hindu University, Varanasi, U.P., India. Its geographical location lies between 25°18′ N latitude to 83°03′E longitudinal and the elevation of the experimental site from the sea level is approximately 75.7 m above the mean sea level.

2.3. Climate condition

Varanasi falls in the Northern India belt of semi-arid to sub humid climate. The normal period of onset of monsoon is the third week of June in this region, which lasts up to the end of September or sometimes third week of October. The normal annual rainfall is about 1100 mm, of which 88% are received from June to September as monsoon season rain, 5–7% in October to December as post monsoon and about 3.3% from January to February as winter season or pre –monsoon rain. The temperature fluctuated in the range of 45–19°C as maximum and 28–7°C as a minimum. The mean relative humidity of the area is about 66%, which rises up to 92% during July to September and falls down to 39% during the end of April to early June.

2.4. Plant and mycorrhizal source

Disease free and healthy, bold seeds of sorghum cv. CSV15 were obtained from the Directorate of Sorghum Research, Hyderabad. The endomycorrhiza *Glomus mosseae* was obtained from the Tata Energical Research Institute, New Delhi.

2.5. Treatments detail

The pot experiment was conducted in the poly house of the Institute of Agricultural Sciences, Banaras Hindu University, Varanasi, with one genotype of sorghum CSV 15. The pot size for the experiment was in the diameter of 30 and 25 cm in height and each of capacity with 10 kg of soil, with a small hole at the bottom. Pots containing soil mix (Soil + FYM in 3:1) are inoculated with seeds of sorghum. Targeted pots were inoculated with Endomycorrhiza *Glomus sp.* (150 inoculants per kg of soil), after that heavy metal stress was created in the plant by the exogenous application of cadmium nitrate in soil. Two best concentrations of heavy metals on the basis of initial screening were selected i.e., 70 and 150 ppm per kg of soil. Putrescine was applied at the rate of 2.5 and 5.0 mM through foliar spray in 7 day interval, starting from seven DAS up to a weak before 90 DAS. The various observations were taken in three stages such as 30, 60 and 90 days after sowing in the concerned pots. The detailed plan of treatments were: Control (T0),Control + Mycorrhiza (T1), Control + 2.5 mM Putrescine (T2), Control +5 mM Putrescine (T3), Control +2.5 mM Putrescine + Mycorrhiza (T4), Control +5 mM Putrescine + Mycorrhiza (T5), 70 ppm $Cd(NO_3)_2$ (T6), 70 ppm $Cd(NO_3)_2$ + Mycorrhiza (T7),

70 ppm $Cd(NO_3)_2$ + 2.5 mM Putrescine (T8), 70 ppm $Cd(NO_3)_2$ + 5 mM Putrescine (T9), 70 ppm $Cd(NO_3)_2$ + 2.5 mM Putrescine + Mycorrhiza (T10), 70 ppm $Cd(NO_3)_2$ + 5 mM Putrescine + Mycorrhiza (T11), 150 ppm $Cd(NO_3)_2$ (T12), 150 ppm $Cd(NO_3)_2$ + Mycorrhiza (T13), 150 ppm $Cd(NO_3)_2$ + 2.5 mM Putrescine (T14), 150 ppm $Cd(NO_3)_2$ + 5 mM Putrescine (T15), 150 ppm $Cd(NO_3)_2$ + 2.5 mM Putrescine + Mycorrhiza (T16), 150 ppm $Cd(NO_3)_2$ + 5 mM Putrescine + Mycorrhiza (T17).

2.6. Design and layout of experiment

The experiment was laid out in completely randomized design (CRD). There were 18 treatments including control. Each treatment was replicated five times.

2.7. Measurement of oxidative damage

Estimation of Lipid Peroxidation (malondialdehyde (MDA) content) in terms of Thiobarbituric Acid Reducing Substances (TBARS) content was estimated by the method given by Heath and Packer [43]. About 0.5 g of leaf tissues from control and treated groups were cut into small pieces and homogenized by the addition of 5 ml of 5% trichloroacetic acid (TCA) solution. The homogenates were then transferred into fresh tubes and centrifuged at 12,000 rpm for 15 min at room temperature. Equal volumes of supernatant and 0.5% thiobarbituric acid (TBA) in 20% TCA solution were added into a new tube and boiled at 96°C for 25 min. The tubes were transferred into an ice bath and then centrifuged at 10,000 rpm for 5 min. The absorbance of the supernatant was measured at 532 nm and corrected for non-specific turbidity by subtracting the absorbance at 600 nm, 0.5% TBA in 20% TCA solution was used as the blank. The amount of MDA-TBA complex (red pigment) was calculated from the extinction coefficient as 155 M^{-1} cm^{-1}. Values of MDA contents were taken from measurements. Results were presented as μmoles MDA g^{-1} FW.

The H_2O_2 content was measured by the method given by Jana and Choudhuri [44]. One hundred mg root samples were extracted using 3 ml of 50 mM sodium phosphate buffer (pH 7.4). The homogenate was centrifuged at 6000 g for 15 min. To determine hydrogen peroxide levels, 3 ml of the extracted solution was mixed with 1 ml of 0.1% titanium sulfate in 20% (w/v) sulfuric acid and then centrifuged at 6000 g for 15 min. The intensity of the yellow color in the supernatant was measured at 410 nm. The hydrogen peroxide level was calculated using extinction coefficient 0.28 μM^{-1} cm^{-1} and was expressed as nmol H_2O_2 g^{-1} tissue FW.

2.8. Statistical analysis

All the numerical data obtained were analyzed through Statistical package of Origin 6.1-advance scientific graphing and data analysis [OriginLab Corporation, One RoundHouse Plaza, Northhampton, MA 01060]. Two-way ANOVA was performed for interaction between mycorrhiza and cadmium treatments. One Way ANOVA was performed for comparing the significance difference among individual means.

http://dx.doi.org/10.5772/intechopen.75722

3. Results

3.1. Malondialdehyde (MDA) [nmole/ml FW] content

Effect of polyamine (putrescine), mycorrhiza and their combination on MDA [nmole/ml FW] was studied in sorghum variety CSV15 under the cadmium stress. Data were recorded at 30, 60 and 90 days after sowing (DAS) (**Figure 1**). It is evident that the average MDA was significantly increased by 35.64, 41.39 and 64.02% when exposed to heavy metal stress (T6) as compared to control (T0) at 30, 60 and 90 DAS of the interval. Similarly, when plants were exposed to higher doses of heavy metal (T12) then MDA was significantly increased by 62.28, 59.44 and 73.67% as compared to control (T0) on the dates of proposed interval. Exogenous application of endomycorrhiza in the soil (T7) showed the mitigation effect by reducing the MDA by 6.05, 2.87 and 7.87% as compared to T6 at 30, 60 and 90 DAS. During treatment, T13 was compared to T12, the MDA reduced significantly by 12.92, 9.85 and 9.63% at proposed DAS. In comparison to T6, the exogenous application of putrescine (T8) showed the mitigating effect by decreasing MDA by 1.84, 6.64 and 14.90% on proposed DAS. The average MDA was significantly reduced as compared to T6 by 11.15, 7.24 and 29.68% when treated with high dose of putrescine (T9) with respect to T8. Similarly, when treatment, T14 was compared with T12, the MDA reduced significantly by 21.47, 17.32 and 18.45% at proposed DAS. The average MDA was significantly reduced as compared to T12 by 33.17, 23.72 and 20.39% when treated

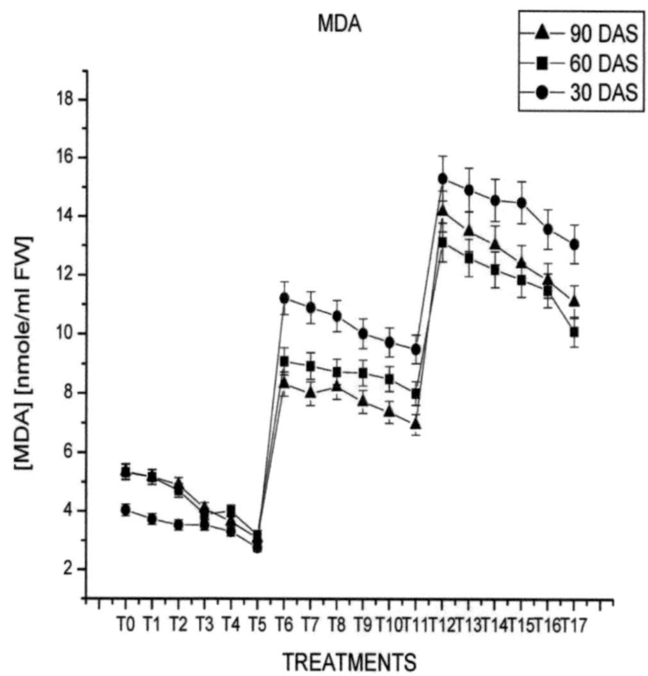

Figure 1. Effects of putrescine and *Glomus* on MDA content in sorghum grown in cadmium contaminated soil.

with high dose of putrescine (T15) with respect to T14. The combination of putrescine and mycorrhiza showed the best mitigation effect in treatment T10 by reducing MDA by 17.65, 11.03 and 36.89% with respect to treatment T6 at proposed DAS. When treatment T11 was compared with treatment T6 then MDA was reduced by 25.51, 20.21 and 42.59%, respectively. A similar effect was seen in the treatment (T16) with respect to treatment T12. In this treatment (T16), the MDA was found to decrease significantly by 43.79, 30.52 and 42.96%, respectively at proposed DAS. The treatment T17 was found to show better results; significant decrease in MDA by 57.32, 56.76 and 55.50% with respect to T12 was observed. So, the combination of putrescine and mycorrhiza showed the best combination for the mitigation of cadmium toxicity for the malondialdehyde.

3.2. H_2O_2 [μ mole/g FW] content

H_2O_2 [μ mole/g FW] content was studied in sorghum variety CSV15 under the cadmium stress. Data were recorded at 30, 60 and 90 days after sowing (DAS) (**Figure 2**). It is evident that the average H_2O_2 was significantly increased by 34.09, 32.02 and 19.37% when exposed to heavy metal stress (T6) as compared to control (T0) at 30, 60 and 90 DAS of the interval. Similarly, when plants were exposed to high doses of heavy metal (T12) then its H_2O_2 was significantly increased by 30.9, 37.55 and 34.66% as compared to control (T0) on the dates of proposed interval. Exogenous application of endomycorrhiza in the soil (T7) showed the mitigation effect by reducing the H_2O_2 by 32.47, 14.21 and 0.54% as compared to T6 at 30, 60 and 90 DAS. During treatment, T13 was compared to T12, the H_2O_2 reduced significantly by 1.13, 2.00 and 1.25% at proposed DAS. In comparison to T6, the exogenous application of

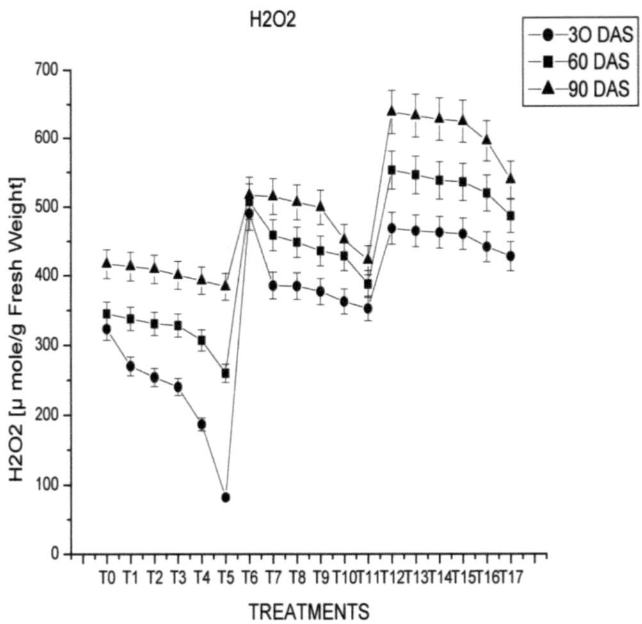

Figure 2. Effects of putrescine and *Glomus* on H_2O_2 content in sorghum grown in cadmium contaminated soil.

putrescine (T8) showed the mitigating effect by decreasing H_2O_2 by 32.74, 17.24 and 2.52% on proposed DAS. The average H_2O_2 was significantly reduced as compared to T6 by 35.10, 20.87 and 4.30% when treated with high dose of putrescine (T9) with respect to T8. Similarly, when treatment, T14 was compared with T12, the H_2O_2 reduced significantly by 1.82, 4.33 and 2.50% at proposed DAS. The average H_2O_2 was significantly reduced as compared to T12 by 2.5, 4.98 and 3.28% when treated with high dose of putrescine (T15) with respect to T14. The combination of putrescine and mycorrhiza showed the best mitigation effect in treatment T10 by reducing H_2O_2 by 17.65, 11.03 and 36.89% with respect to treatment T6 at proposed DAS. When treatment T11 was compared with treatment T6 then significant H_2O_2 was reduced by 42.70, 34.80 and 22.80%, respectively. A similar effect was seen in the treatment (T16) with respect to treatment T12 where H_2O_2 was found to decrease significantly by 8.41, 9.66 and 10.21%, respectively at proposed DAS. The treatment T17 was found to show better results; significant decrease in H_2O_2 by 12.70, 19.24 and 23.89% with respect to T12 was observed. So, the combination of putrescine and mycorrhiza showed the best combination for the mitigation of cadmium toxicity for the hydrogen peroxide.

4. Discussion

The ROS is conceivably unsafe to cell layers, bringing about oxidative corruption of film lipids (lipid peroxidation). Malondialdehyde (MDA) is one of the end breakdown results of lipid peroxidation and can be utilized as a marker of in vivo lipid peroxidation. Cadmium uptake triggered oxidative stress resulting in increased generation of Hydrogen peroxide and lipid peroxidation as reported in earlier studies [45, 46]. To cope with enhanced levels of oxidative stress, plants are equipped with antioxidant system that gets activated under cadmium stress [46, 47]. Ascorbate peroxidase, Peroxidase and Catalase play a crucial role during the degradation of the plant sample grown under cadmium stress. The statistical analysis in the treatment T17 showed significant reduction in MDA content because the most important property of PA conjugates with phenolic acids were their antioxidant characteristic, required by the plants to adapt under stress condition. Bors et al. [48] showed that conjugates with coffeic, cinnamic and ferulic acids displayed higher constraint of binding to reactive oxygen species (ROS) i.e., free polyamines are less efficient radical scavengers than their conjugates. Polyamines conjugation by transglutaminases, especially to Rubisco seems to have a significant role in protecting this protein from protease action, thus protecting its photosynthetic efficiency [49]. Therefore, PAs are likely to play a role in photosynthesis since they are capable of reversing stress-induced damage in the photosynthetic apparatus. Cross-linking of PAs mediated by transglutaminases might play a significant role in polyamine bioactivity for flower development and compatibility in reproduction [50]. Thus, conjugates involved in defense mechanisms against cadmium toxicity and also mediate the regulation of certain growth and developmental events [51, 52]. Similarly, the mycorrhiza *Glomus* acted as the potential inhibitor of mobility of cadmium ion into the soil solution and protected the plants from cadmium toxicity. AM fungal colonization was observed in highly contaminated soil [24, 25]. The presence of *Glomus mosseae* was also reported in heavy metal contaminated sites by some researchers [26]. The external mycelium of certain AM fungi produce a type of protein called Glycoprotein (Glomalin), which has heavy metal binding sites [53]. Heavy metals accumulate at these binding sites. The antioxidant level is also increased as a result of association of AM fungi with plants [28–30]. Some fungal strains were isolated in the past which were resistant to heavy metal contamination. Many mycorrhizal fungi overcome the stress of heavy metal contamination [31, 32].

In the present study, when plants were subjected to cadmium stress, lipid peroxidation was lesser in mycorrhizal treated plant than the non mycorrhizal plants, in severity of cadmium stress. The similar trend was shown when the plant was treated with putrescence. The best results were obtained when mycorrhiza and putrescine were treated together for alleviation of Cd toxicity in sorghum. Mycorrhizal plants along with putrescine displayed lower lipid peroxidation and low hydrogen peroxide production [40, 54–56]. It is widely accepted that diminishing levels are one of the mechanisms by which AM protects plants against diverse stresses [57–59]. It has been suggested that peroxidase could act as an efficient H_2O_2 scavenging system in plant vacuoles in the presence of phenolics and reduced ascorbate [60]. Direct chelation, or binding to polyphenols, was observed with methanol extracts of rhizome polyphenols from Nymphaea for Cr, Pb and Hg [61–96]. Phenolic antioxidants inhibit lipid peroxidation by trapping the lipid alkoxyl radical. This activity depends on the structure of the molecules, and the number and position of the hydroxyl group in the molecules. Santiago et al. [62] hypothesize a cycle where H_2O_2 is scavenged by phenolics through a peroxidase, and phenolics are oxidized to phenoxyl radicals, which can be reduced by amino acids [97–118].

The present perceptions demonstrating a positive connection between metal danger and proline aggregation recommend a defensive part of this amino corrosive against substantial metal poisonous quality. In this specific situation, recommendations have been made that proline gives assurance by keeping up the water adjust, which is frequently exasperates by substantial metals [63, 64], searching hydroxyl radicals [65], chelating overwhelming metals in the cytoplasm [66–68] and lessening metal take-up [69]. Numerous scientists, all things considered, feel that proline gathering is just a side effect of assorted anxieties, and isn't engaged with assurance against metal and different burdens [70–72]. Wu et al. [73] demonstrated that Cu-prompted efflux of K^+ in *Anacystis nidulans* was limited within the sight of proline, proposing that proline conceivably shielded the plasma layer from Cu lethality, or it maybe complexed Cu + in this manner diminishing the centralization of free Cu + in the outer condition. In a consequent paper, Wu et al. [69] exhibited decreased Cu take-up by cells containing high convergences of proline, proposing that proline balanced out the plasma film and furthermore diminished the take-up of Cu by algal cells. The impacts of lead, copper, cadmium and mercury on the substance of proline, were researched in 17-day-old bean seedlings (*Phaseolus vulgaris* L.) developed in Hoagland arrangement spiked with different centralizations of Pb, Cu, Album and Hg. Control and substantial metal-treated plants were developed for 10 days in Hoagland arrangement. A noteworthy increment of proline was identified in essential leaves following ten-day presentation to substantial metals. The most grounded impact on proline, content was found in plants presented to mercury, trailed by the grouping $Cd^{++} > Cu^{++} > Pb^{++}$ [74].

Pioneer studies suggested that peroxidase is not only one of the defense proteins, but an important antioxidant enzyme involved in response to environmental stresses [75]. It constitutes a wide variety of heme containing enzymes involved in many physiological processes in plants. It was reported to be involved in response to biotic and abiotic stresses, auxin catabolism, cell wall lignification and degradation [76]. Peroxidase is involved in the scavenging of Reactive Oxygen Species (ROS). High concentration of cadmium in sorghum resulted in inhibition of several enzymes and an increase in activity of others. Cadmium accumulation in the cellular compartment of the enzyme is a pre-requisite for enzyme inhibition *in vivo* [77, 78]. Peroxidase induction was likely to be related to oxidative reaction at the biomembrane. Metal induced enzymatic changes in plants such as isoperoxidase pattern was used to evaluate the phytotoxicity of metal polluted soils. There is extensive literature indicating the role of peroxidase in developmental processes, possibly due to the involvement of this enzyme in auxin metabolism (auxin oxidation) and in the formation of cross-links between cell wall components. Peroxidase activity was more pronounced in cadmium treated sorghum [79–81]. This high peroxidase activity enables the plants to protect themselves against the oxidative stresses as suggested by Scalet et al. [82]. In fact, in plants, peroxidase protects cell against harmful concentration of hydroperoxides by induction of peroxidases especially anionic peroxidase, which have been found to be involved in response to both abiotic and biotic stresses. Mycorrhiza showed best mitigation against cadmium toxicity with respect to peroxidase activity. In the present study, the activity of peroxidase was significantly reduced due to the presence of *Glomus*, because in soil solution the cadmium ion was trapped by the fungus and its mobility and translocation in the plants was inhibited [83]. The translocation of cadmium depends on several factors. The bioavailability of cadmium to sorghum roots from the soil solution depends on its concentration in soil and it modulated by the presence of organic matter, pH and temperature. It is also affected by the presence of chelating organisms like mycorrhiza. Despite different mobility of metal ions in plants, the cadmium content was generally greater in roots than in the above ground tissue [84–86]. In most environmental conditions, Cd enters first the roots, and consequently they are likely to experience Cd damage first [87]. Similarly, the putrescine is another potential ameliorative agent of heavy metal toxicity in sorghum because of its unique nature. Several studies have shown that PA accumulation occurs under abiotic stresses including drought, salinity, low temperature, heavy metals, and PQ [21, 84, 88–90].

5. Conclusions

Finally, it is concluded that the polyamines like putrescine and mycorrhiza *Glomus* impart significant mitigation of cadmium induced toxicity in sorghum mediated through the defensive role of enzymatic and non enzymatic antioxidants in plants. The MDA and hydrogen peroxide content was found significant increased with cadmium treated plants with respect to plants treated with putrescine and mycorrhiza.

Acknowledgements

Authors are thankful to University Grants Commission for providing Senior Research Fellowship to the first author.

Author details

Prasann Kumar[1]* and Shweta Pathak[2]

*Address all correspondence to: prasann0659@gmail.com

1 Department of Agronomy, School of Agriculture, Lovely Professional University, Jalandhar, India

2 Department of Plant Physiology, Institute of Agricultural Sciences, Banaras Hindu University, Varanasi, UP, India

References

[1] Benavides MP, Gallego SM, Tomaro ML. Cadmium toxicity in plants. Brazilian Journal of Plant Physiology. 2005;**17**:21-34. DOI: 10.1590/S1677-04202005000100003

[2] Uraaguchi S, Mori S, Kuramata M, Kawasaki A, Arora T, Ishikawa S. Root-to shoot Cd translocation via the xylem is the major process determining shoot and grain cadmium accumulation in rice. Journal of Experimental Botany. 2009;**60**:2677-2268. DOI: 10.1093/jxb/erp119

[3] Wang L, Zhou QX, Ding LL, Sun Y. Effect of cadmium toxicity on nitrogen metabolism in leaves of *Solanum nigrum* L. as a newly found cadmium hyperaccumulator. Hazardous Materials. 2008;**154**(1-3):818-825. DOI: 10.1016/j.jhazmat.2007.10.097

[4] Smith CJ, Hopmans P, Cook FJ. Accumulation of Cr, Pb, Cu, Ni, Zn and Cd in soil following irrigation with untreated urban effluents in Australia. Environmental Pollution. 1996;**94**(3):317-323. PMID:15093492

[5] Bjerrum J. Metal amine formation in aqueous solution. In: Elving J, editor. Treatise on Analytical Chemistry. Vol. I. New York: The Interscience Encyclopedia; 1959

[6] Adriano DC. Chromium. In: Trace Elements in the Terrestrial Environment. New York: Springer; 1986. pp. 58-76

[7] Juste C, Mench M. Long term application of sewage sludge and its effect on metal uptake by crops. In: Adriano DC, editor. Biogeochemistry of Trace Metals. Ann. Arbor, London, Tokyo: Leuwis publishers; 1992. pp. 159-193

[8] Goering PL, Waalkes MP, Klaassen CD. Toxicity of cadmium. In: Goyer RAC, Herian MG, editors. Handbook of Experimental Pharmacology: Toxicity of Metals, Biochemical Effects. New York: Springer Verlag; 1994. pp. 189-213

[9] AMAP. AMAP Assessment. Heavy Metals in the Arctic-Pre-Print Files. Oslo, Norway: Arctic Monitoring and Assessment Programme; 2002. p. 870

[10] Mench M, Amans V, Sappin-Didier V, Fargues S, Gomez A, Loffler M, Masson P, Arrouays D. A study of additivies to reduce availability of Pb in soils to plants. In: Iskandar A, Adriano DC, editors. Remediation of Soils Contaminated with Metals. Northwood, UK: Science Reviews; 1997. pp. 202-185

[11] Baron K, Stasolla C. The role of polyamines during in vivo and in vitro development. In Vitro Cellular & Developmental Biology. Plant. 2008;**44**:384-395. DOI: 10.1007/s11627-008-9176-4

[12] Dossantos RW, Schmidt EC, Depmartins R, Latini A, Maraschin M, Horta PA, Bouzon ZL. Effects of cadmium on growth, photosynthetic pigments, photosynthetic performance, biochemical parameters and structure of chloroplasts in the agarophyte *Gracilaria domingensis* (Rhodophyta, Gracilariales). American Journal of Plant Sciences. 2012;**3**:1077-1084. DOI: 10.4236/ajps.2012.38129

[13] Kakkar RK, Sawhney VK. Polyamine research in plants–a changing perspective. Physiologia Plantarum. 2002;**116**:281-292. DOI: 10.1034/j.1399-3054.2002.1160302.x

[14] Kuznetsov V, Radyukina NL, Shevyakova NI. Polyamines and stress: Biological role, metabolism and regulation. Russian Journal of Plant Physiology. 2006;**53**:583-604. DOI: 10.1134/S1021443706050025

[15] Kusano T, Berberich T, Tateda C, Takahashi Y. Polyamines: Essential factors for growth and survival. Planta. 2008;**228**:367-381. DOI: 10.1007/s00425-008-0772-7

[16] Steiner N, Santa-Catarina C, Silveira V, Floh EIS, Guerra MP. Polyamine effects on growth and endogenous hormones levels in *Araucaria angustifolia* embryogenic cultures. Plant Cell, Tissue and Organ Culture. 2007;**89**:55-62. DOI: 10.1007/s11240-007-9216-5

[17] Santa-Catarina C, Silveira V, Scherer GFE, Floh EIS. Polyamine and nitric oxide levels correlate with morphogenetic evolution in somatic embryogenesis of Ocotea catharinensis. Plant Cell, Tissue and Organ Culture. 2007;**90**:93-101. DOI: 10.1007/s11240-007-9259-7

[18] Tun NN, Santa-Catarina C, Beghum T, Silveira V, Handro W, Floh EIS, Scherer GFE. Polyamines induce the rapid biosynthesis of nitric oxide (NO) in *Arabidopsis thaliana* seedlings. Plant & Cell Physiology. 2006;**47**:346-354. DOI: 10.1093/pcp/pci252

[19] Kuhen GD, Phillips GC. Role of polyamines in apoptosis and other recent advances in plant polyamines. Critical Reviews in Plant Sciences. 2005;**24**:123-130. DOI: 10.1080/07352680590953161

[20] Yoda H, Hamaguchi R, Sano H. Induction of hypersensitive cell death by hydrogen peroxide produced through polyamine degradation in tobacco plants. Plant Physiology. 2003;**132**:1973-1981. DOI: 10.1104/pp.103.024737

[21] Bouchereau A, Aziz A, Larher F, Martin-Tanguy J. Polyamines and environmental challenges: Recent development. Plant Science. 1999;**140**:103-125. PMCID: PMC2835953

[22] Bhatnagar P, Minocha R, Minocha SC. Genetic manipulation of the metabolism of polyamines in poplar cells: The regulation of putrescine catabolism. Plant Physiology. 2002;**128**:1455-1469. DOI: 10.1104/pp.010792

[23] Minocha SC, Minocha R. Role of polyamines in somatic embryogenesis. In: Bajaj YPS, editor. Biotechnolgogy in Agriculture and Forestry. Vol. 30. Berlin: Spring-Verlag; 1995. pp. 53-70. DOI: 10.12691/plant-1-2-1

[24] Leung HM, Ye ZH, Wong MH. Survival strategies of plants associated with arbuscular mycorrhizal fungi ontoxic mine tailings. Chemosphere. 2007;**66**:905-915. DOI: 10.1016/j.chemosphere.2006.06.037

[25] Pawlowska TE, Blaszkowski J, Ruhling A. The mycorrhizal status of plants colonizing a calamine spoil mound in southern Poland. Mycorrhiza. 1996;**6**:499-505. DOI: 10.1007/s005720050154

[26] Debiane D, Garcon G, Verdin A, Fontaine J, Durand R, Grandmougin-Ferjani A, Shirali P, Lounces- Hadj Sahraui A. In vitro evaluation of the oxidative stress and genotoxic potenitials of anthracene on mycorrhizal chicory roots. Environmental and Experimental Botany. 2008;**64**:120-127. DOI: 10.1016/j.envexpbot.2008.04.003

[27] Vivas A, Azcon R, Biro B, Barea JM, Ruiz Lozano JM. Influence of bacterial strains isolated from lead polluted soil and their interactions with arbuscular mycorrhizae on the growth of *Trifolium pertense* L.under lead toxicity. Canadian Journal of Microbiology. 2003a;**49**:577-588. DOI: 10.1139/w03-073

[28] Gopi R, Jaleel CA, Sairam R, Lakshamanam GMA, Gomathinayagam M, Paneerselvam R. Differential effects of hexaconazole and paclobutrazol on biomass, electrolyte leakage, lipid peroxidation and antioxidant potential of *Daucus carota* L. Colloids and Surfaces Biointerfaces. 2007;**60**:180-186. DOI: 10.1016/j.colsurfb.2007.06.003

[29] Zhang FQ, Wang YS, Lou ZP, Dong JD. Effect of heavy metal stress on antioxidative enzymes and lipid peroxidation in leaves and roots of two mangrove plant seedlings (*Kandelia candel* and *Bruguiera gymnorrhiza*). Chemosphere. 2007a;**67**:44-50. DOI: 10.1016/j.chemosphere.2006.10.007

[30] Zhang LZ, Wei N, Wu QX, Ping ML. Antioxidant response of *Cucumis sativus* L. to fungicide carbendazin. Pesticide Biochemistry and Physiology. 2007b;**89**:54-59. DOI: 10.1016/j.pestbp.2007.02.007

[31] Hildebrandt U, Regvar M, Bothe H. Arbscular mycorrhiza and heavy metal tolerance. Phytochemistry. 2007;**68**:139-146. DOI: 10.1016/j.phytochem.2006.09.023

[32] Joschim HJ, Makoi R, Ndakidemi PA. The agronomic potential of vesicular-arbuscular mycorrhiza (AM) in cereals-legume mixtures in Africa. African Journal of Microbiology Research. 2009;**11**:664-675

[33] Da-Lin L, Shu-pan Z, Zheng C, Wei-wei Q. Soil cadmium regulates antioxidases in sorghum. Agricultural Sciences in China. 2011;**9**(10):1475-1480

[34] Liu JX. Effects of cadmium and zinc interaction on corn seedlings physiological and biochemical characteristics. Journal of Yichun College. 2004;**26**(6):55-57. DOI: 10.5897/AJB11.848

[35] Cieslinski G, Neilser GH, Hogue EJ. Effect of soil cadmium application and pH on growth and cadmium accumulation in roots, leaves and fruit of strawberry plants. Plant and Soil. 1996;**18**:267-271. DOI: 10.1080/01904169409364791.

[36] Chang ZM, Wu XH. Difference comparison of three alfalfa varities resisitant to cadmium pollution. Pratac. Sci. 2005;**22**(2):20-23

[37] Li J, Gao XH, Guo SR, Zhang RH, Wang X. Effects of exogenous spermidine on photosynthesis of salt-stressed Cuellmis sativus seedlings. Chinese Journal of Ecology. 2007;**26**(10):1595-1599. DOI: 10.1007/s10265-014-0653-z

[38] Zhang EH, Zhang XH, Wang HZ. Adaptation effects of phosphorus stress on different genotypes of faba-bean. Acta Ecologica Sinica. 2004;**24**(8):1589-1593

[39] Qin TC, Ruan J, Wang LJ. Effects of cadmium on plant photosynthesis. Environmental Science and Technology. 2000;**13**:33-35. DOI: 10.5897/AJB11.848

[40] Yang Y, Han X, Liang Y, Ghosh A, Chen J, Tang M. The combined effects of arbuscular mycorrhizal fungi (AMF) and lead (Pb) stress on Pb accumulation, plant growth parameters, photosynthesis, and antioxidant enzymes in *Robinia pseudoacacia* L. PLoS One. 2015;**10**:e0145726. DOI: 10.1371/journal.pone.0145726

[41] Jiang Y, Liang WJ, Zhang YG, Xu YF. Research on effect of sewage irrigation on soil heavy metal environmental capacity and rice growth. Chinese Journal of Eco-Agriculture. 2004;**12**(3):124-127

[42] Huang CC, Chen MW, Hsieh JL, Lin WH, Chen PC, Chien LF. Expression of mercuric reductase from *Bacillus megaterium* MB1 in eukaryotic microalga Chlorella sp. Applied Microbiology and Biotechnology. 2006;**72**:197-205. DOI: 10.1007/s00253-005-0250-0

[43] Heath RL, Packer L. Photoperoxidation in isolated chloroplasts: I. Kinetics and stoichiometry of fatty acid peroxidation. Archives of Biochemistry and Biophysics. 1968;**125**: 189-198. DOI: 10.1016/0003-9861(68)90654-1

[44] Jana S, Choudhuri MA. Glycolate metabolism of three submerged aquatic angiosperms during aging. Aquatic Botany. 1981;**12**:345-354

[45] Liu J, Macarisin D, Wisniewski M, Sui Y, Droby S, Norelli J, et al. Production of hydrogen peroxide and expression of ROS-generating genes in peach flower petals in

response to host and non-host fungal pathogens. Plant Pathology. 2013;**62**:820-828. DOI: 10.1111/j.1365-3059.2012.02683

[46] Singh HP, Batish DR, Kohli RK, Arora K. Arsenic-induced root growth inhibition in mung bean (*Phaseolus aureus* Roxb.) is due to oxidative stress resulting from enhanced lipid peroxidation. Plant Growth Regulation. 2007;**53**:65-73. DOI: 10.1007/s10725-007-9205

[47] Sobrino-Plata J, Meyssen D, Cuypers A, Escobar C, Hernández LE. Glutathione is a key antioxidant metabolite to cope with mercury and cadmium stress. Plant and Soil. 2014;**377**:369-381. DOI: 10.1007/s11104-013-2006-4

[48] Bors W, Langebartels C, Michel C, Sandermann H. Polyamines as radical scavengers and protectants against ozone damage. Phytochemistry. 1989;**28**:1589-1595

[49] Serafini-Fracassini D, Del Duca S, Beninati S. Plant transglutaminases. Phytochemistry. 1995;**40**:355-365

[50] Serafini-Fracassini D, Del Duca S. Transglutaminases: Widespread cross-linking enzymes in plants. Annals of Botany. 2008;**102**:145-152. DOI: 10.1093/aob/mcn075

[51] Bagni N, Tassoni A. Biosynthesis, oxidation and conjugation of aliphatic polyamines in higher plants. Amino Acids. 2001;**20**:301-317. PMID:11354606

[52] Martin-Tanguy J. Metabolism and function of polyamines in plants: Recent development (new approaches). Plant Growth Regulation. 2001;**34**:135-148

[53] Vivas A, Voros I, Biro B, Campos E, Barea JM, Azcon R. Symbiotic efficiency of autochtonous arbuscular *Mycorrhizal fungus* (G. mosseae) and *Brevibacillus brevis* isolated from cadmium polluted soil under increasing cadmium levels. Environmental Pollution. 2003b;**126**:179-189. DOI: 10.1016/S0269-7491(03)00195-7

[54] Evelin H, Kapoor R. Arbuscular mycorrhizal symbiosis modulates antioxidant response in salt-stressed Trigonella foenum-graecum plants. Mycorrhiza. 2014;**24**:197-208. DOI: 10.1007/s00572-0130529-4

[55] Jiang QY, Zhuo F, Long SH, Zhao HD, Yang DJ, Ye ZH, et al. Can arbuscular mycorrhizal fungi reduce Cd uptake and alleviate cd toxicity of *Lonicera japonica* grown in Cd-added soils? Scientific Reports. 2016;**6**:21805. DOI: 10.1038/srep21805

[56] Tan SY, Jiang QY, Zhuo F, Liu H, Wang YT, Li SS, et al. Effect of inoculation with *Glomus versiforme* on cadmium accumulation, antioxidant activities and phytochelatins of Solanum 977 photeinocarpum. PLoS One. 2015;**10**:e0132347. DOI: 10.1371/journal.pone.0132347

[57] Garg N, Singla P. The role of Glomus mosseae on key physiological and biochemical parameters of pea plants grown in arsenic contaminated soil. Scientia Horticulturae. 2012;**143**:92-101. DOI: 10.1016/j.scienta.2012.06.010

[58] Hajiboland R, Aliasgharzadeh N, Laiegh SF, Poschenrieder C. Colonization with arbuscular mycorrhizal fungi improves salinity tolerance of tomato (*Solanum lycopersicum* L.) plants. Plant and Soil. 2010;**331**:313-327. DOI: 10.1007/s11104-009-0255-z

[59] Ruiz-Sánchez M, Aroca R, Muñoz Y, Polón R, Ruiz-Lozano JM. The arbuscular mycorrhizal symbiosis enhances the photosynthetic efficiency and the antioxidative response of rice plants subjected to drought stress. Journal of Plant Physiology. 2010;**167**:862-869. DOI: 10.1016/j.jplph.2010.01.018

[60] Zancani M, Nagy G. Phenol-dependent H_2O_2 breakdown by soybean root plasma membrane-bound peroxidase is regulated by ascorbate and thiols. Journal of Plant Physiology. 2000;**156**:295-299. DOI: 10.1016/S0176-1617(00)80064-4

[61] Lavid N, Schwartz A, Yarden O, TelOr E. The involvement of phenylpropanoids and peroxidase activities in heavy metal accumulation by epidermal glands of the waterlily (Nymphaeaceae). Planta. 2001;**212**:323-331

[62] Santiago LJM, Louro RP, De Oliveira DE. Compartmentation of phenolic compounds and phenylalanine ammonia-lyase in leaves of *Phyllanthus tenellus* Roxb. And their induction by copper sulphate. Annals of Botany. 2000;**86**:1023-1032. DOI: 10.1006/anbo.2000.1271

[63] Costa G, Morel JL. Water relations gas exchange and amino acid content in cd treated lettuca. Plant Physiology and Biochemistry. 1994;**32**:561-570

[64] Schat H, Sharma SS, Vooijst R. Heavy metal-induced accumulation of free proline in a metal-tolerant and nontolerant ecotype of Silene vulgaris. Physiologia Plantarum. 1997;**101**:477-482. DOI: 10.1034/j. 1399-3054.1997.1010304. x

[65] Smirnoff N, Cumber QJ. Hydroxyl radical scavenging activity of compatible solutes. Phytochemistry. 1989;**28**:1057-1060. DOI: 10.1016/0031-9422(89)80182-7

[66] Farago ME, Mullen WA. Plants which accumulate metals. A possible copper-proline complex from the roots of *Armeria maritima*. Inorganica Chimica Acta. 1979;**32**:L93-L94. DOI: 10.1016/0031-9422(89)80182-7

[67] Hemalatha S, Anburaj A, Francis K. Effect of heavy metals on certain biochemical constituents and nitrate reductase activity in *Orzya sativa* L. seedlings. Journal of Environmental Biology. 1997;**18**:313-319. DOI: 10.2478/eko-2014-0012

[68] Pandey P, Tripathi AK. Effect of heavy metals on morphological and biochemical characteristics of *Albizia procera* (Roxb. Benth.) seedling. International Journal of Environmental Sciences. 2011;**1**(5):1009-1018. DOI: 10.2478/eko-2014-0012

[69] Wu JT, Hsieh MT, Kow LC. Role of proline accumulation in response to toxic copper in Chlorella sp. (Chlorophyceae) cells. The Journal of Physiology. 1998;**34**:113-117. DOI: 10.1046/j.1529-8817.1998.340113.x

[70] Bhaskaran S, Smith RH, Newton RJ. Physiological changes in cultured sorghum cells in response to induced water stress. I. Free proline. Plant Physiology. 1985;**79**:266-269. PMCID: PMC1075388

[71] Lutts S, Kinet JM, Bouharmont J. Effects of various salts and of mannitol on ion and proline accumulation in relation to osmotic adjustment in rice (*Oryza sativa* L.) callus cultures. Journal of Plant Physiology. 1996;**149**:186-195. DOI: 10.1016/S0176-1617(96)80193-3

[72] Perez-Alfocea F, Santa-Cruz A, Guerrier G, Bolarin MC. NaCl stress-induced organic solute changes on leaves and calli of *Lycopersicon esculentum*, L. pinnellii and their inter-specific hybrid. Journal of Plant Physiology. 1994;**143**:106-111

[73] Wu JT, Chang SC, Chen KS. Enhancement of intracellular proline level in cells of *Anacystis nidulans* (cyanobacteria) exposed to deleterious concentrations of copper. Journal of Phycology. 1995;**31**:376-379. DOI: 10.1111/j.0022-3646.1995.00376.x

[74] Rodriguez-Serrano M, RomeroPuertas MC, Pazmino DM, Testillano PS, Risueno MC, Del Rio LA, et al. Cellular response of pea plants to cadmium toxicity: Cross talk between reactive oxygen species, nitric oxide, and calcium. Plant Physiology. 2009;**150**:229-243. DOI: 10.1104/pp.108.131524

[75] Flohé L, Ursini F. Peroxidase: A term of many meanings. Antioxidants & Redox Signaling. 2008;**10**:1485-1490. DOI: 10.1089/ars.2008.2059

[76] Wang LK, Hung YT, Shammas NK. Physicochemical Treatment Processes. Vol. 3. New Jersey: Humana Press; 2004, 2004. pp. 141-198

[77] Rabie MH, Eleiwa ME, Aboseoud MA, Khalil KM. Effect of nickel on the content of carbohydrate and some mineral in corn and broad bean plant. Journal of King Saud University – Science. 1992;**4**:37

[78] Sharma SS, Dietz KJ. The significance of amino acids and amino acid-derived molecules in plant responses and adaptation to heavy metal stress. Journal of Experimental Botany. 2006;**57**(4):711-726. DOI: 10.1093/jxb/erj073

[79] Ahmad P, Sharma S, Srivastava PS. In vitro selection of $NaHCO_3$ tolerant cultivars of *Morus alba* (Local and Sujanpuri) in response to morphological and biochemical parameters. Horticultural Science (Prague). 2007;**34**:114-122

[80] Chugh LK, Sawhney SK. Photosynthetic activities of Pisum sativum seedlings grown in the presence of cadmium. Plant Physiology and Biochemistry. 1999;**37**:297-303. DOI: 10.1016/S0981-9428(99)80028-X

[81] Deniz B, Merve H, Sermin E. Evaluation of Lead Removal onto Black Cumin by Using Multi Linear Regression. Ohrid, Republic of Macedonia: BALWOIS 2012; 2012. pp. 1-6

[82] Scalet M, Federico R, Guido MC, Manes F. Perioxidase activity and polyamines changes in response to ozone and simulated acidrain in Aleppo pine needles. Environmental and Experimental Botany. 1995;**35**:417-425. DOI: 10.1016/0098-8472(95)00001-3

[83] Ernst WHO, Verkleij JAC, Schat H. Metal tolerance in plants. Acta Botanica Neerlandica. 1992;**41**:229-248

[84] Jentschke G, Godbold DL. Metal toxicity and ectomycorrhizas. Physiologia Plantarum. 2000;**109**:107-116

[85] Leyval C, Turnau K, Haselwandter K. Effect of heavy metal pollution on mycorrhizal colonization and function: Physiological, ecological and applied aspects. Mycorrhiza. 1997;**7**:139-153. DOI: 10.1007/s005720050174

[86] Schützendübel A, Polle A. Plant responses to abiotic stresses: Heavy metal-induced oxidative stress and protection by mycorrhization. Journal of Experimental Botany. 2002;**53**:1351-1365. DOI: 10.1093/jexbot/53.372.1351

[87] Santia di Toppi L, Gabbrielli R. Response to cadmium in higher plants. Environmental and Experimental Botany. 1999;**41**:105-130 PII: S0098-8472(98)00058-6

[88] Groppa MD, Tomaro ML, Benavides MP. Polyamines as protectors against cadmium or copper induced oxidative damage in sunflower leaf discs. Plant Science. 2001;**161**:481-488. DOI: 10.1007/s00726-006-0343-9

[89] Marschner H. Mineral Nutrition of Higher Plants. 2nd ed. London: Academic Press; 1995. 889 pp

[90] Pang XM, Zhang ZY, Wen XP, Ban Y, Moriguchi T. Polyamine, all-purpose players in response to environment stresses in plants. Plant Stress. 2007;**1**:173-188

[91] Aebi HE. Catalase. In: Bergmeyer HU, editor. Methods of Enzymatic Analysis. 3rd ed. Weinheim, Florida: Verlag Chemie; 1983. pp. 273-286

[92] Asada K. The water –water cycle in chloroplast: Scavenging of active oxygen and dissipation of excess photons. Annual Review of Plant Physiology and Plant Molecular Biology. 1999;**50**:601-639. DOI: 10.1146/annurev.arplant.50.1.601

[93] Abdul Jaleel C, Jayakumar K, Chang-Xing Z, Iqbal M. Low concentration of cobalt increases growth, biochemical constituents, mineral status and yield in *Zea mays*. Journal of Scientific Research. 2009;**1**:128-137. DOI: 10.3329/jsr.vlil.1226

[94] Bates L, Waldren RP, Teare ID. Rapid determination of free proline for water-stress studies. Plant and Soil. 1973;**39**:205-207. DOI: 10.1007/BF00018060

[95] Cho UH, Seo NH. Oxidative stress in Arabidopsis thaliana exposed to cadmium is due to hydrogen peroxide accumulation. Plant Science. 2005;**168**:113-120. DOI: 10.1016/j.plantsci.2004.07.021

[96] DSR Annual Report (2011). Directorate of Sorghum Research, Hyderabad; 2010-11

[97] El-Sayed E, Omran, Afaf A, Abd ER. Mapping and screening risk assessment of heavy metals concentrations in soils of the Bahr El-Baker region, Egypt. Journal of Soil Science and Environmental Management. 2010;**6**(7):182-195. DOI: 10.5897/JSSEM12.010.

[98] Gonzalez-Chavez MC, Carrillo-Gonzalez R, Wright SF, Nichols KA. The role of glomalin, a protein produced by arbuscular mycorrhizal fungi, in sequestering potentially toxic elements. Environmental Pollution. 2004;**130**:317-323. DOI: 10.1016/j.envpol.2004.01.004

[99] Gohre V, Paszkowski U. Contribution of arbuscular mycorrhizal symbiosis to heavy metal phytoremediation. Planta. 2006;**223**:1115-1122. DOI: 10.1007/s00425-006-0225-0

[100] Garg N, Manchanda G. ROS generation in plants: Boon or bane? Plant Biosystems. 2009;**143**(1):81-96. DOI: 10.1080/11263500802633626

[101] Ivanova J, Toncheva-Panova T, Chernev G, Samuneva B. Effect of Ag^{+}, Cu^{2+} and Zn^{2+} containing hybrid nano matrixes on the green algae *Chlorella keissleri*. General and Applied Plant Physiology. 2008;**34**:339-348

[102] Joner EJ, Briones R, Leyval C. Metal-binding capacity of arbuscular mycorrhizal mycelium. Plant and Soil. 2000;**226**:227-234. DOI: 10.1023/A:1026565701391.

[103] Kar M, Mishra D. Catalase, peroxidase, and polyphenoloxidase activities during rice leaf senescense. Plant Physiology. 1976;**57**:315-319. PMID: 16659474

[104] Kaldorf M, Kuhn AJ, Schroder WH, Hildebrandt U, Bothe H. Selective element deposits in maize colonized by a heavy metal tolerance conferring arbuscular mycorrhizal fungus. Journal of Plant Physiology. 1999;**154**:718-728. DOI: 10.1016/S0176-1617(99)80250-8

[105] Kukreja S, Nandval AS, Kumar N, Sharma SK, Unvi V, Sharma PK. Plant water status, H_2O_2 scavenging enzymes, ethylene evolution and membrane integrity of Cicer arietinum roots as affected by salinity. Biologia Plantarum. 2005;**49**:305-308

[106] Mehlhorn H, Lelandais M, Korth HG, Foyer CH. Ascorbate is the natural substarte for plant perioxidase. FEBS Letters. 1996;**378**:203-206. SSDI: 0014-5793(95)01448-9

[107] Mozgawa W, Król M, Bajda T. Application of IR spectra in the studies of heavy metal cations immobilization on natural sorbents. Journal of Molecular Structure. 2009; **924-926**:427-433. DOI: 10.1016/j.molstruc.2008.12.028

[108] Nakano Y, Asada K. Hydrogen peroxide is scavenged by ascorbate specific peroxidase in spinach chloroplasts. Plant & Cell Physiology. 1981;**22**:867-880. DOI: 10.1093/oxfordjournals.pcp.a076232

[109] Pooria G, Tahereh TM, Bijan R. Differential scanning calorimetry techniques: Applications in biology and nanoscience. Journal of Biomolecular Techniques. 2010;**21**(4):167-193. PMCID: PMC2977967

[110] Stoeppler M. Cadmium. In: Merian E, editor. Metals and their Compounds in the Environment: Occurrence, Analyses and Biological Relevance. New York: VCH; 1991. pp. 803-851

[111] Sadasivam S, Manickam A. Biochemical methods for agricultural sciences; 1992. pp. 12-13. Sumner JB, Gjessing EC. Arch. Biochem. 1943;**2**:291

[112] Shah K, Kumar RG, Verma S, Dubey RS. Effect of cadmium on lipid perioxidation, superoxide anion generation and activities of antioxidants enzymes in growing rice seedlings. Plant Science. 2001;**161**:1135-1144. DOI: 10.1016/S0168-9452(01)00517-9

[113] Smeets K, Cuypers A, Lambrechts A, et al. Induction of oxidative stress and antioxidative mechanisms in Phaseolus vulgaris after cd application. Plant Physiology and Biochemistry. 2005;**43**(5):437-444. DOI: 10.1016/j.plaphy.2005.03.007

[114] Sbartai H, Rouabhi R, Sbartai I, Berrebbah H, Djebar RM. Induction of anti-oxidative enzymes by cadmium stress in tomato (*Lycopersicon esculentum*). African Journal of Plant Science. 2008;**2**(8):72-76. Article Number - E4FC5413005

[115] Thomet M, Vogel E, Krahenbuhl U. The uptake of cadmium and zinc by mycelia and their accumulation in mycelia and fruiting bodies of edible mushrooms. European Food Research and Technology. 1999;**209**(5):317-324. DOI: 10.1155/2013/1499120

[116] WHO. Lead Environmental Health Criteria. Geneva: WHO; 1995

[117] Yang X, Baligar VC, Martens DC, Clark RB. Cadmium effects on influx and transport of mineral nutrients in plant species (*Sedum alfredii Hance*). Plant and Soil. 1996;**259**: 181-189. DOI: 10.1080/01904169609365148

[118] Zhang H. Chromium contamination in the soil from an alloy steel factory in Nanjing. China Environmental Science. 1997;**17**(2):80-82

Chapter 6

Impact of Heavy Metals on Forest Ecosystems of the European North of Russia

Irina Lyanguzova, Vasily Yarmishko,
Vadim Gorshkov, Natalie Stavrova and Irina Bakkal

Additional information is available at the end of the chapter

http://dx.doi.org/10.5772/intechopen.73323

Abstract

The article presents the results of long-term monitoring of the state of boreal forest ecosystems (Kola Peninsula, European North of Russia) experiencing industrial pollution. The general regularities and distinctive features of the reaction of various components of pine forest ecosystems to joint action of gaseous and solid pollutants and contamination of albic rustic podzols by heavy metals under field experiment are characterized. A dynamic trend in soil and plant contamination levels has been identified, a vitality structure and growth rates of stands have been compared, as well as the state of undergrowth under airborne and soil pollution, the dynamics of the ground cover under soil pollution by heavy metals is characterized. Based on a comparative analysis of the level of soil contamination and the state of the components of pine forest communities, the limits of their tolerance to the toxic effect of heavy metals have been established.

Keywords: heavy metals, forest ecosystems, boreal forests, albic rustic podzol, stand, undergrowth, ground cover, the European North of Russia

1. Introduction

Ferrous and non-ferrous metallurgy are the most polluting industries; their atmospheric emissions comprise mainly sulfur dioxide and polymetallic dust containing heavy metals (Cu, Pb, Cd, Zn, Fe, Ni, Co, etc.). Companies of Norilsk Nickel Mining and Metallurgical Company (Severonickel, Pechenganickel, Norilsk Nickel) are the most studied point sources of environmental pollution on the territory of Russia. In the zone of their impact, there is a disturbance of terrestrial ecosystems up to their complete degradation with the formation of technogenic wastelands [1–10]. In recent decades, in many economically developed and some developing countries, there has been a reduction in the volume of atmospheric emissions from industrial enterprises as a result of improving production and treatment technologies, as well as reducing production or transferring it to other territories. Maximum volumes of atmospheric emissions (140,000–275,000 tons per year) of non-ferrous metallurgy enterprises occurred from 1970 to the end of the 1980s; by the beginning of the 2000s, their emissions dropped sharply (by 3–5 times); by the end of the 2000s, there was a further reduction in the amount of atmospheric emissions; and at present the gross volume of emissions does not exceed 5000–35,000 tons per year [10, 11]. Due to a decrease in the amount of atmospheric emissions, a decrease in the aerotechnogenic load on ecosystems makes it possible to study dynamic trends in the state of both individual ecosystem components and communities as a whole. In the last decade, such works began to appear in the scientific literature, but there are still a few of them [12–16].

Earlier, we showed that the study of the accumulation of heavy metals that are part of the dusty atmospheric emissions of a non-ferrous metallurgy combine in soils, various types of plants, mosses and lichens adequately reflects the negative effect of the combined effect of sulfur dioxide and polymetallic dust on the forest ecosystems of the North of Russia [1, 4, 8]. In order to detect the toxic effect of polymetallic dust on north-taiga forest ecosystems, field experiments were performed.

2. Material and methods

Laboratory of Ecology of Plant Communities of the Botanical Institute named after V.L. Komarov of the Russian Academy of Sciences for more than 35 years has been monitoring the state of forest ecosystems in the impact and buffer zones of the non-ferrous metallurgy complex Severonickel and conducted experimental studies of the effect of heavy metals on various components of forest ecosystems in the Kola Peninsula (European North of Russia).

Studies of the ecosystems of young (formed after fires and felling with a prescription of 40–50 years) and middle-aged (with the prescription of fire or felling of 80–90 years) pine forests of lichen–green-moss site type were carried out in 1980–2017 on a series of permanent sampling sites, 0.1–0.15 ha in size, located at different distances (8–15, 30–35, 60–80 km) from Severonickel smelter (the Kola Peninsula, Russia) within the impact, buffer and background zones.

On the sampling site, the categories of the vital state of all individuals of the tree layer were determined using two parameters: the relative density of the crown (in % of the density of the crown of healthy individuals in the stand of the background zone) and the degree of damage to the assimilation organs by chlorosis and necrosis. Five categories of vitality for trees and undergrowth were identified: I—healthy, II—weakened, III—severely weakened, IV—dying and V—dead individuals. To calculate the vitality index [17, 18], the percentage of individuals of each category was multiplied by the corresponding coefficient (I—1, II—0.71, III—0.43, IV—0.14 and V—0), and then summarized the values obtained. To study the growth dynamics of trees by diameter, increment cores were selected from 20 trees on the sampling site. The registration and determination of the vitality of small seedling (individuals less than 1.3 m in height) was carried out at 40–200 plots with a size of 1 m^2. The characteristics of the ground cover were determined on regularly located marked plots of 1 m^2 (from 20 to 50 on each sampling site).

Analysis of the content of acid-soluble forms of Ni and Cu (extract 1.0 N HCl) in samples of the organogenic horizon of Al-Fe-humus podzols (albic rustic podzol) was carried out by atomic absorption spectrometry, according to accepted methods [19]. An analysis of the content of Ni and Cu in assimilating organs of plants (*Pinus sylvestris, Vaccinium myrtillus, V. vitis-idaea, V. uliginosum, Empetrum hermaphroditum, Arctostaphylos uva-ursi, Avenella flexuosa* and *Solidago lapponica*) was performed by atomic absorption spectrometry after dry ashing of plant material and transfer of ash to the solution [20].

In 1992, four permanent sampling sites (PSSs) were laid in middle-aged pine forests of the background area of the Kola Peninsula, remote from the Severonickel Smelter beyond 80 km, where no visually observed damages to the plants were registered. One half of the PSS was left as a control. On the surface of the snow cover the other half of PSS was dispersed polymetallic dust identical to emitted into the atmosphere by smelter "Severonickel" in the amount of 352–563 kg/ha. The composition of dust included: Ni—1.3–2.1%, Cu—1.3–1.8% and Co—0.06–0.09% [21]. Disintegration of polymetallic dust led to a spatially very uneven destruction of the ground cover and contamination of the Al-Fe-humus podzol of experimental sites.

The study of the dynamics of the ground cover of the control and experimental plots was performed according to the procedure described above. Chemical analysis of samples of the organogenic horizon of Al-Fe-humus podzol, leaves (needles) of plants (*Pinus sylvestris, Vaccinium myrtillus, V. vitis-idaea, Empetrum hermaphroditum*), living parts of lichen thallus (*Cladonia stellaris, Cl. rangiferina, Cl. mitis* and *Cl. uncialis*) and live parts of moss (*Pleurozium schreberi*) collected in experimental plots were carried out according to the procedure described above.

3. Results and discussion

3.1. Impact of aerial pollution on forest ecosystems around smelter complex "Severonickel"

In the period 1981–1990, the annual volume of atmospheric emissions of SO_2 at the Severonickel Smelter exceeded 220,000 tons on average, 16,000 tons of solids, followed by a gradual decrease in emissions of pollutants, and at present, the annual volume of SO_2 and solid matter emissions at the Severonickel Smelter has decreased by 8 and 5 times, respectively, in comparison with their maximum values (**Figure 1**). Sulfides and metal oxides predominate in the fine-dispersed polymetallic dust and also metallic Ni and Cu [21].

During the study period from 1981 to 2014, in the background area of the Kola Peninsula, the content of acid-soluble forms of Ni and Cu in the organogenic horizon of the soil averaged 10.0 ± 0.9 and 10.5 ± 1.1 mg/kg of soil, respectively, these values were taken as regional background values of heavy metals in forest litter. In the buffer zone, the average concentrations of acid-soluble forms of Ni and Cu in the organogenic horizon of the soil for the entire study period were 78.1 ± 5.9 and 137 ± 21 mg/kg, respectively, which is 7.8 and 13.7 times higher than the regional background values. In the impact zone, the mean Ni and Cu concentrations over the entire observation period were 530 ± 30 and 1035 ± 95 mg/kg soil, respectively, these values are 53 and more than 100 times the regional background values [22].

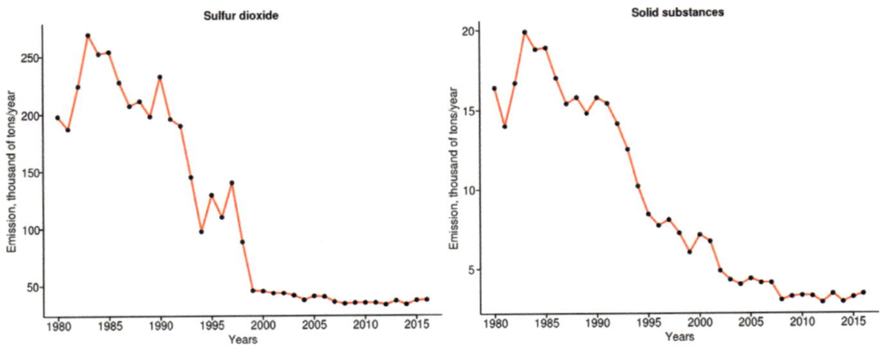

Figure 1. Volume of atmospheric emissions of the Severonickel Smelter (Kola Peninsula, Russia) (after data [22]).

Between the content of acid-soluble forms of Ni and Cu in the organogenic horizon of the soil and the distance from the source of pollution, there is a significant negative correlation ($r = -(0.63–0.70)$, $p < 0.05$), which was ascertained by all specialists who conducted studies in the impact zone of the Severonickel and Pechenganickel Smelters [1, 4–7, 9, 10, 13, 23, 24]. The dependence of the content of acid-soluble forms of Ni and Cu in the organogenic horizon of Al-Fe-podzols from the distance to the contamination source is satisfactorily approximated by the exponential equation (**Figure 2**). It should be noted that the content of acid-soluble forms of Cu in the soil always exceeds that of Ni.

To characterize the level of pollution of habitats, the technogenic load index is used, which is the excess of the content of acid-soluble forms of Ni and Cu in the contaminated organogenic horizon of podzols over their regional background concentrations. During the period of research (1981–2017) within the buffer zone, the technogenic load index averaged 11.2 rel. units, and on the territory of the impact zone – 72 rel. units. Dynamics of the man-caused load index

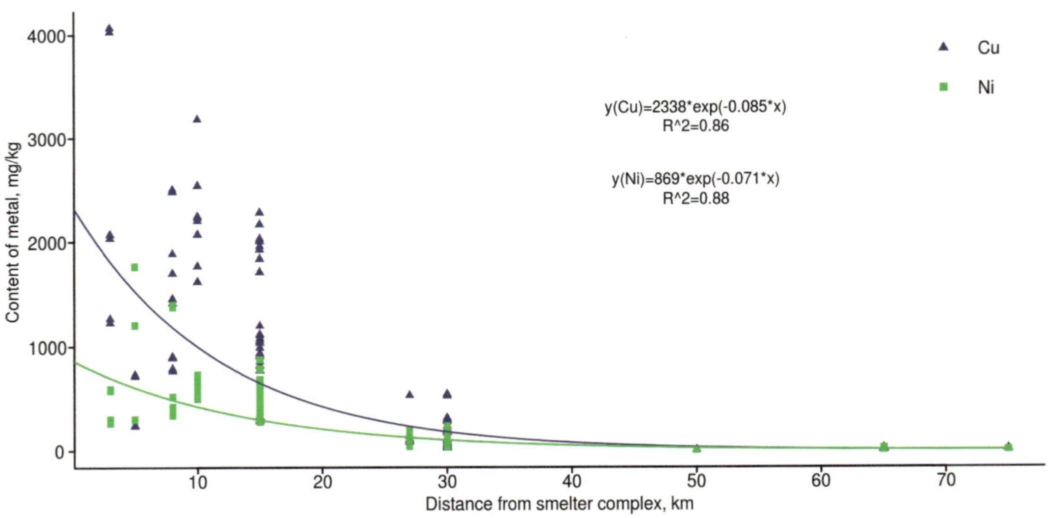

Figure 2. The content of heavy metals in the organogenic horizon of podzols along the profile from the Severonickel Smelter (Kola Peninsula, Russia).

(**Figure 3**) shows that, despite the sharp decrease in atmospheric emissions by the Severonickel Smelter (**Figure 1**), within the buffer zone, level of pollution of the upper soil horizon continues to increase, and within the impact zone remains at a very high level, which we noted earlier [22].

As the source of contamination approaches, the content of heavy metals in the assimilation organs of all plant species under investigation increases (*Pinus sylvestris, Vaccinium myrtillus, V. vitis-idaea, V. uliginosum, Empetrum hermaphroditum, Arctostaphylos uva-ursi, Avenella flexuosa* and *Solidago lapponica*). As an example, **Figure 4** shows the content of Ni and Cu in the needles of *Pinus sylvestris*, which grows at different distances from the Severonickel Smelter. During the period of studies in the background region, the average Ni content in the assimilation organs of all the species studied was 4.9 ± 0.3, Cu—4.8 ± 0.5 mg/kg, in the buffer zone, respectively, Ni—22.3 ± 1.9, Cu—8.8 ± 0.6 mg/kg, which in 2–4.5 times the background values, and within the impact zone their content was Ni—137 ± 10, Cu—42 ± 3 mg/kg, correspondingly 9–27 times their background concentrations [25]. It is necessary to emphasize the specific specificity in the accumulation of heavy metals by different plant species. Within one level of soil contamination (e.g. on the territory of the impact zone), content of the plant species under study varied by a factor of 3–9. It should be noted that, in contrast to the soil, in all plant species, the Ni content is always greater than the Cu concentration, which is due to the faster movement of Ni ions from the contaminated soil to the plant organs above the ground.

In the buffer zone, a fairly high level of accumulation of heavy metals in leaves (needles) persisted in 1981–1997 (**Figure 5**), and subsequently (2008–2014) it was revealed its 1.5–2.5-fold decrease in relation to the maximum value. Within the impact zone, the average level of accumulation of Ni and Cu in the assimilation organs has been consistently decreased from the maximum (300 and 85 mg/kg) to the minimum (47 and 16 mg/kg) values, that is, within the impact zone over the entire observation period, a 5–6-fold decrease in the average content of

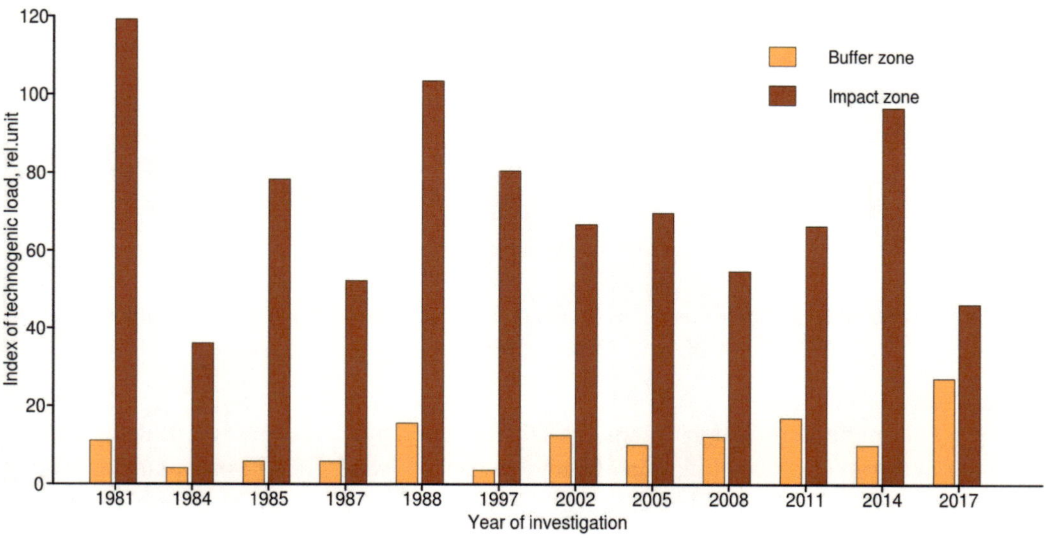

Figure 3. The dynamic trend of the technogenic load index within the buffer and impact zones of the Severonickel Smelter (Kola Peninsula, Russia).

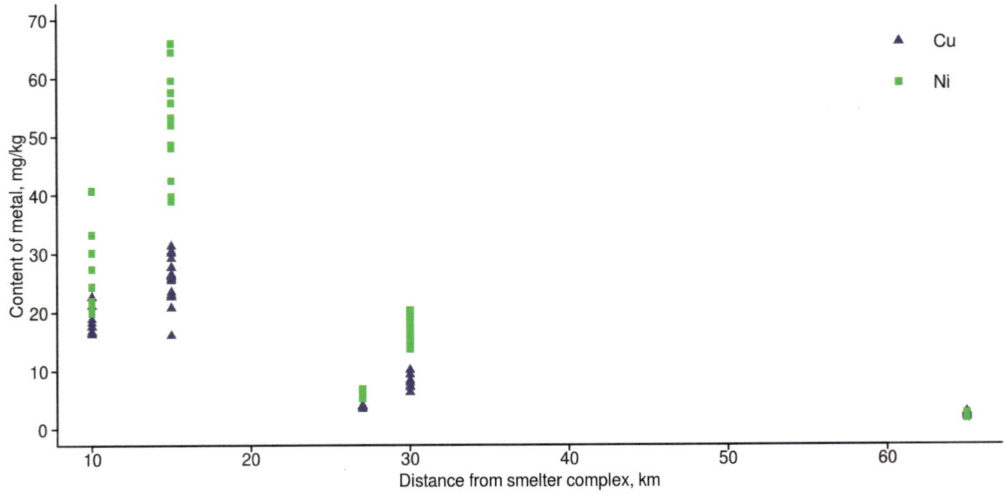

Figure 4. The content of heavy metals in the needles of *Pinus sylvestris* L. along the profile from the Severonickel Smelter (Kola Peninsula, Russia).

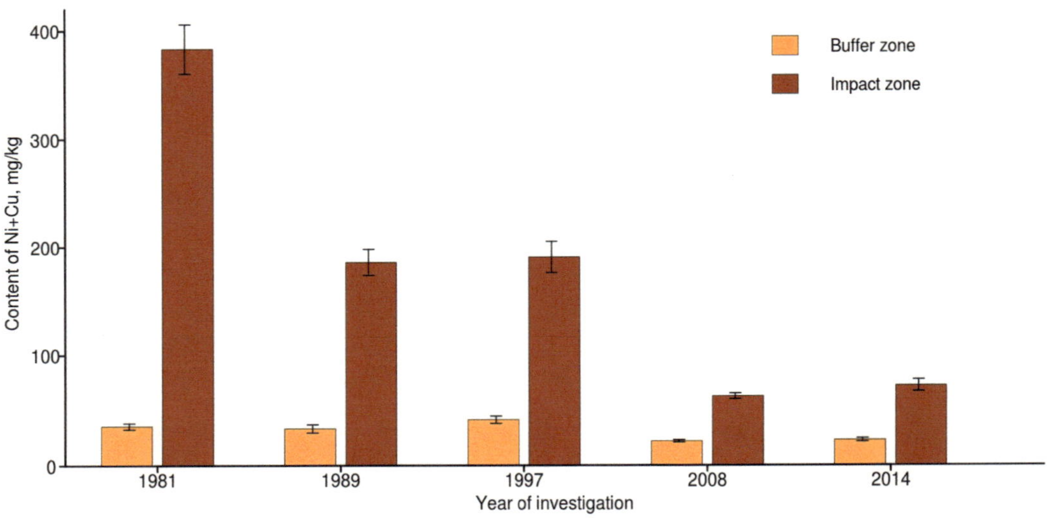

Figure 5. Dynamics of the average content of heavy metals in all species under investigation within the buffer and impact zones of the Severonickel Smelter (Kola Peninsula, Russia).

heavy metals in the assimilation organs of all the plant species under study was detected. The decrease in the level of accumulation of heavy metals by leaves (needles) of plants is largely due to the intensity of the aerial pollution and is significantly less associated with the absorption of heavy metals from contaminated soil. Previously [1, 4, 8, 25] showed that a significant direct correlation ($r = 0.89–0.99$, $p < 0.05$) exists in the gradient of the aerial pollution between the level of contamination by heavy metals of the upper soil horizon and their accumulation in the leaves (needles) of plants, but up to 80% metals enter the plant from contaminated air, possibly in the form of dust deposits on the surface of the leaf blade.

Thus, the direction of dynamic trends in the level of contamination by heavy metals of the organogenic horizon of Al-Fe-humus podzols and the content of Ni and Cu in plant assimilation organs differs significantly. Within the buffer zone, the technogenic load index continues to increase, and in the impact zone the content of heavy metals in the soil remains at a very high level. Thanks to a 5–8-fold reduction in atmospheric emissions of the Severonickel Smelter experiences a 2–16-fold decrease in the content of Ni and Cu in the leaves (needles) of plants, which is caused by a decrease in the amount of solids coming from the polluted air to the sheet surface.

Long-term studies on permanent sample plots allowed to assess the general patterns of the dynamics of the state of trees and stands of *Pinus sylvestris* in the pine forests of the Kola Peninsula in space and time.

One of the main criteria for assessing the vital state of trees and stands is the nature of the development of their assimilation apparatus, because the degree of tree crowning, the vitality of needles and the duration of its life determine the productivity of forest communities, their resistance to stressful environmental factors. During the period of studies in the background pine forests, the life expectancy of pine needles varied from 5.7 to 6.7 years and did not change significantly during the study period (**Figure 6**). In these habitats, only a small part (no more than 5%) of pine needles had chlorosis and/or necrosis, which occupied an area of less than 5% of the total surface, only the old needles (5 years and older), the area of chlorosis and necrosis sometimes reached 20% surface, which is associated with age-related changes in assimilation organs [4, 26].

In the buffer zone in the period of high aerial emission (1982–1990), the life expectancy of pine needles was 3.8–4.4 years, which is on average 2–3 years less than in the background area

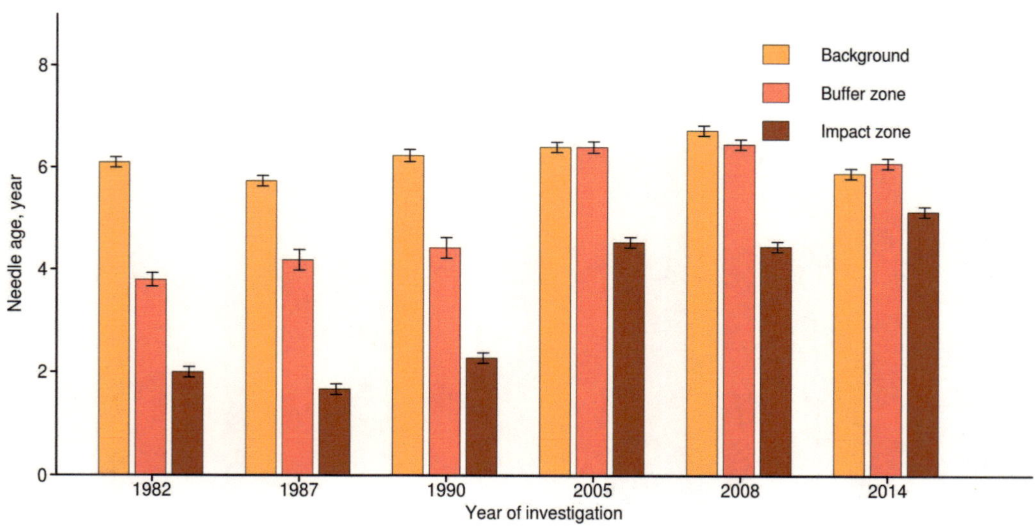

Figure 6. Dynamics of the life expectancy of *Pinus sylvestris* needles in the background, buffer and impact zones of the Severonickel Smelter (Kola Peninsula, Russia).

(**Figure 6**). Later (2005–2014), there was a significant increase in the life expectancy of the needles to 6.1–6.4 years, and the duration of its life ceased to differ significantly from its background value. At the beginning of the research (1988), only 66% of the 1-year-old needles were classified as healthy, more than 30% of the needles of this age had traces of chlorosis and necrosis, with increasing needle age, proportion of healthy needles decreased and the area of injuries increased, 2–3% of needles had necrosis of red-brown color. Studies in 2008 showed that the bulk (93–98%) of *P. sylvestris* needles of 1-year-old age had no traces of damage. Spots of chlorosis and pinpoint necrosis were found only in a small part (6%) of needles of 1–3 years of age [4, 26].

On the territory of the impact zone during the initial period of research (1982–1987), the average life expectancy of pine needles did not exceed 2 years. Against the backdrop of a sharp decrease in the volume of atmospheric emissions of the Severonickel Smelter experienced a twofold increase in the life expectancy of needles, and now its value is approaching the background value of this parameter (**Figure 6**). In 1988, only 25% of 1-year-old needles were classified as healthy in this area, the remaining 75% of 1-year-old needles were covered with chlorosis and necrosis, all needles of 2- and 3-year-old age were damaged. In addition to spotted chlorosis and point necrosis, spotted, girdle and marginal necrosis was noted on the needles, as well as apical necrosis of needles. A re-examination of 2008 showed that the needles of the current year had no visible traces of damage, the proportion of 1-year-old healthy needles increased threefold and was 74%, with the increase in the age of the needles, the degree of its damage increased, but the area of the damaged needle surface was significantly smaller compared with the data of 1988 [4, 26].

An analysis of the vitality patterns of the tree stand reveals not only the natural processes of its formation, development and self-sustainment, but also reflects the impact of stress factors such as fires, felling and airborne pollution. During the period of research, the vital state index very slowly decreased from 0.96 to 0.83 (**Figure 7**), which is due to the natural process of differentiation of individuals in the categories of vitality in middle-aged stands [4, 18, 26].

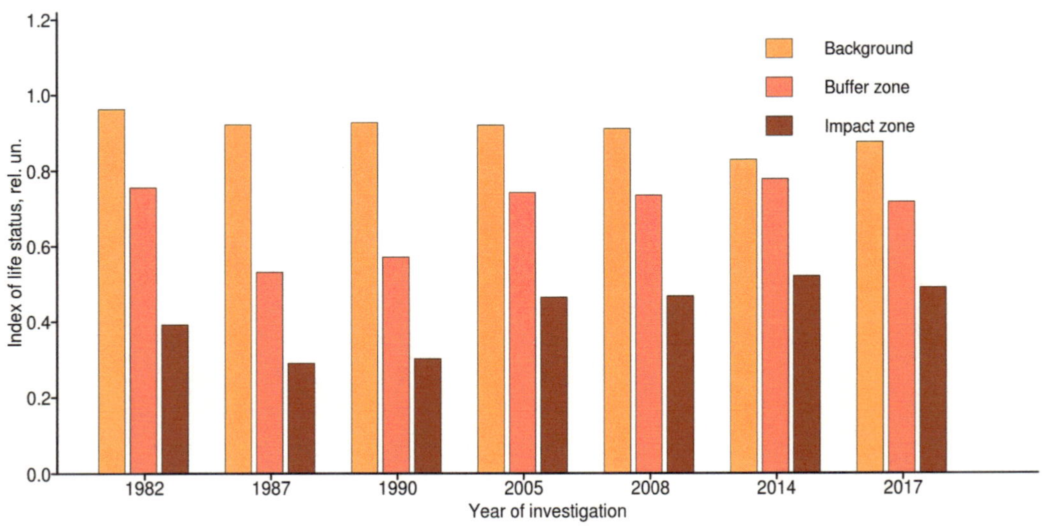

Figure 7. Dynamic trend of the vitality index of *Pinus sylvestris* stands in the background, buffer and impact zones of the Severonickel Smelter (Kola Peninsula, Russia).

In the buffer zone in 1982, the proportion of healthy individuals did not exceed 50%, the proportion of the weakened ones was 31% and the severely weakened and dying – 19%, then the state of the *Pinus sylvestris* stands worsened. The vitality index of pine stands in 1982 was 0.76, in 1987–1990s it decreased on average to 0.55. Then, with a reduction of the degree of needle damage, an increase in the proportion of healthy individuals up to 60% occurred, which led to an increase in the vitality index of the stand to 0.72–0.78 (**Figure 7**).

In pine stands of the impact zone, in 1982, severely weakened and dying individuals predominated, vital state index was 0.39, which is 2.5 times less than its background value (**Figure 7**). In 1987–1990 the state of the stands in this zone continued to deteriorate, the index of the vital state decreased to its minimum value of 0.30. Since 2005, the vitality spectrum has become fuller due to the appearance of healthy individuals of Scots pine, whose share was almost a quarter (23%) of all individuals in the stand, the proportion of weakened individuals somewhat decreased to 22–23%. The index of the vital state has significantly increased in relation to the beginning of the period of research, and in the period 2005–2017 its value was in the range 0.47–0.52, however, this is more than 1.5 times less than the background values.

A comparison of the dynamics of the annual increment of *Pinus sylvestris* tree trunks revealed a different pattern of changes in this parameter in the background and in the impact zone (**Figure 8**). In the background pine forests, the area of annual rings was practically on the same level for the entire observation period, varying from 0.7 to 2.0 cm^2. In the impact zone until the middle of the 1990s the area of annual rings varied within 0.2–0.7 cm^2 and since 2000 it has increased sharply.

Comparative analysis of *Pinus sylvestris* annual rings in the background and in the impact zone showed that the direction of changes in this parameter was fundamentally different during two periods of observations: the period of high aerial emission (1985–1999) and the period of 5–8-fold decrease of atmospheric emission (2000–2014). In the second observation period, a reliable increase in the area of annual rings ($z = 2.47–4.67$, $p = 0.000–0.01$) is recorded with respect to the first observation period, both in the background and in the impact zone. However, in a background the area of the ring increased, on average, just 1.3 times, and in the impact zone—an

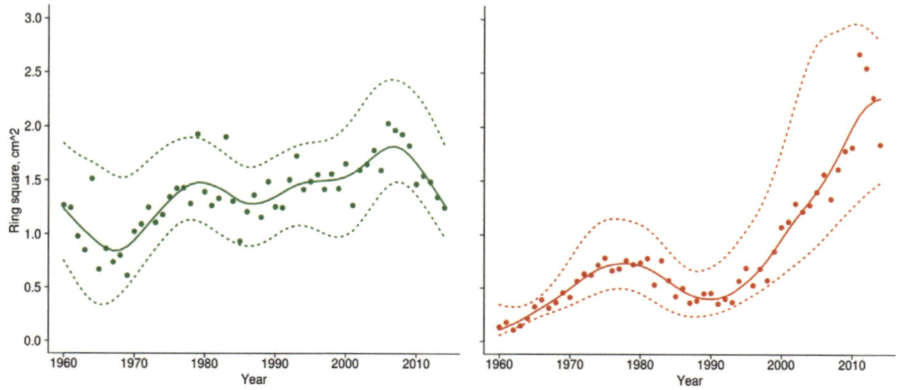

Figure 8. The annual increment of *Pinus sylvestris* trunks in the background area and impact zone of the Severonickel Smelter (Kola Peninsula, Russia).

average of 3.4 times. In 2000–2014, the area of annual rings did not differ significantly in the background and in the impact zone ($z = 0.28$, $p = 0.78$). Thus, in response to a 5–8-fold decrease in the volume of atmospheric emissions by the Severonickel Smelter both in the buffer and impact zones, a positive reaction of the *Pinus sylvestris* stands was revealed: the life expectancy of needles was prolonged, area of needles damage by chlorosis and necrosis was reduced. All this led to an improvement of the stands vitality and the increase in the area of annual increments.

An important role in assessing the stability of pine forests and the possibility of their restoration in disturbed areas is played by studying the process of renewal of the main forest-forming species. For this purpose, a comparative analysis of the abundance and vitality structure of the *Pinus sylvestris* seedlings was carried out. At the beginning of the study, the total number of seedlings in the background area and the buffer zone was 10,000–12,000 individuals/ha and did not differ significantly in both points, and after 5 years it decreased to 2500–3000 individuals/ha. On the territory of the impact zone, the number of seedlings with a similar dynamics was significantly (approximately 3 times) smaller (**Figure 9a**).

The main reasons for the decrease in the number of pine seedlings in the impact zone, as compared to the background, is the decrease in the production of viable seeds and the toxic effect of heavy metals accumulating in the soil. In the pine forests of the impact zone, the

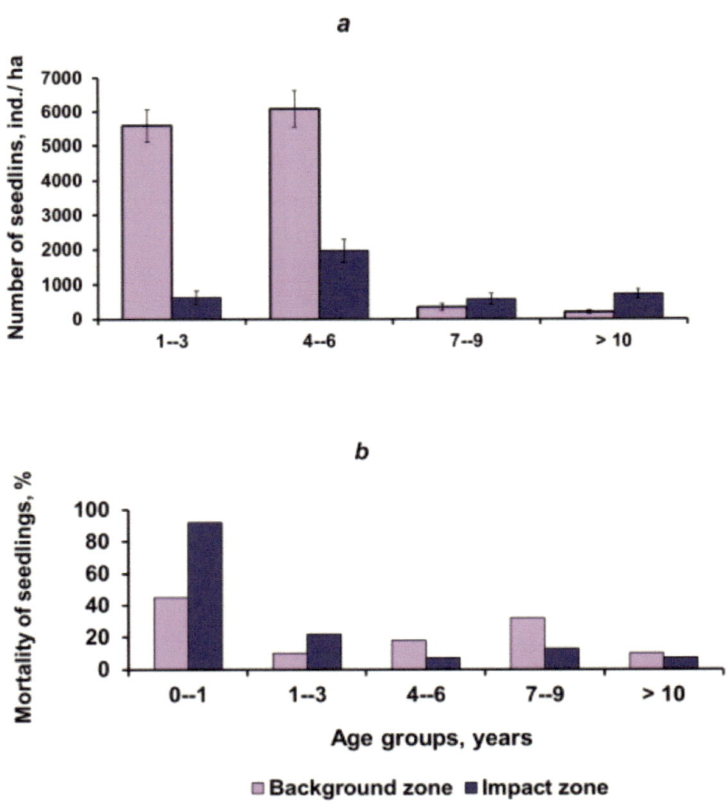

Figure 9. Number (*a*) and average annual mortality (*b*) of different age groups of *Pinus sylvestris* seedlings in the background area and impact zone of the Severonickel Smelter (Kola Peninsula, Russia) at the beginning of the research.

number of female cones on the tree decreases 8–10 times, the number of normally developed seeds in the cone – 3 times, the laboratory germination of seeds, formed in favorable climatic conditions, 2 times. According to the results of a field experiment specially conducted in the background area of the Kola Peninsula [27], in which seeds of *Pinus sylvestris* were germinated on soils from the background (control) and impact zones, field germination of seeds on soils with a high level of contamination (index of technogenic load 120 rel. units) was 1.5–2 times less than in control (10.7 and 19.0%, respectively). Analysis of the mortality dynamics of sprouts and seedlings in the course of their development showed that, both in the background and contaminated soil, the maximum mortality of individuals was observed in the first year of life (**Figure 9b**), but in the second case it was 2 times higher (45 and 92%, respectively). An analogous relationship was maintained during the next 3 years of seedlings life with significantly lower mean annual mortality. The reason for the intensive dying off of pine sprouts on soils with a high level of contamination by heavy metals is a disruption in the development of root systems: the average length of the main root decreases 7.5–8 times (respectively 12 ± 7 and 93 ± 18 mm) compared to the control, the number of lateral roots first order – 10 times (3 ± 4 and 37 ± 13 pieces). A similar reaction of root system to a high level of contamination of soils by heavy metals is typical for many other species of coniferous woody plants [9, 28–30].

In the age groups of pine seedlings from 4–6 to 10–15 years, the rate of extinction of individuals on background soils becomes 1.5–2 times higher than in contaminated ones (**Figure 9a**), but by the end of the experiment the density of 15-year-olds in the background and contaminated soils is leveled and averages 1–2 ind./m^2. This allows us to conclude that if in the impact zone for pine seedlings under the age of 3 years the soil contamination is the leading factor determining survival, then in the subsequent period of seedlings development its role is largely leveled.

In the background area in the vitality spectrum of the 5-year-old seedlings, severely weakened individuals predominate (**Figure 10a**), the total proportion of healthy and weakened individuals is about 30%. In the impact zone, healthy individuals are absent, 5-year-old pine individuals refer mainly to categories severely weakened and dying. The vital state of the 10-year-old pine seedlings in the background area is much better, and in the impact zone it is worse than the 5-year-old (**Figure 10**). The total proportion of healthy and weakened individuals in the background increases in the 10-year-old pine seedlings to 55%, and the proportion of severely weakened ones decreases. In the impact zone, despite the increased decay of the most weakened specimens, the proportion of dying increases almost twofold, and the proportion of weakened and severely weakened individuals decreases by 3–4 times.

The decrease in the amount of atmospheric emissions and the improvement in the vital state of pine stands in the impact zone contribute to an increase in the production and viability of pine seeds, which in turn, even with a persistent high level of soil contamination by heavy metals, leads to an increase in the number of emerging pine sprouts. During the last decade, the number of sprouts and 1-year-old pine seedlings in the impact zone was comparable to the background values in some of the most favorable for germination years (e.g. 2006, 2007). However, the mortality of seedlings in the period from 1 to 3–5 years in the impact zone continues to be very high.

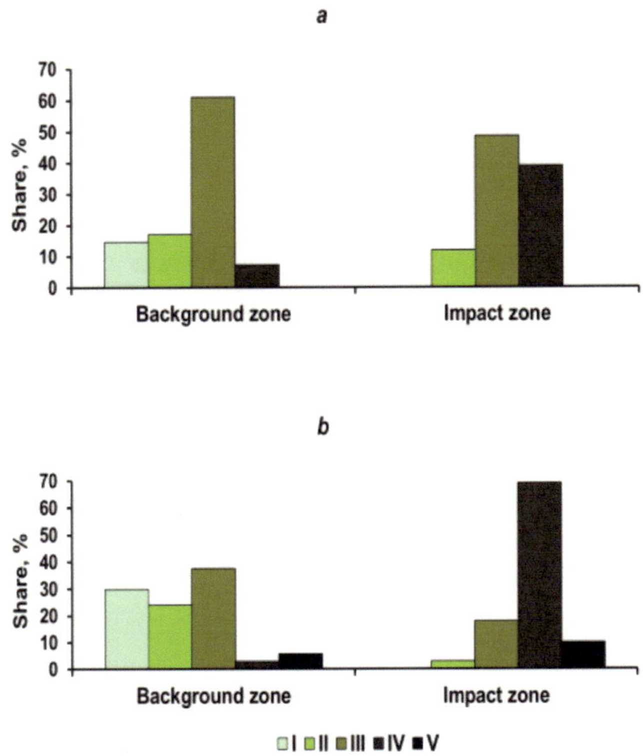

Figure 10. The vitality structure of the 5-year (*a*) and 10-year (*b*) *Pinus sylvestris* seedlings in the background area and the impact zone of the Severonickel Smelter (Kola Peninsula, Russia). Status categories: I – healthy; II – weakened; III – severely weakened; IV – dying; V – dead.

Thus, despite a significant decrease in the intensity of aerotechnogenic pollution, there was no significant improvement in the process of renewal of Scots pine in conditions of the impact zone.

Ground cover is one of the most important components of boreal forests. The moss-lichen layer, fall off and litter align daily fluctuations in humidity and temperature. Aerial pollution leads to the destruction of the lower tiers and the loss of their environment-forming functions [31, 32].

In the background pine forests where the fire was 60 years ago, an increase in the total projective coverage of the dwarf-shrub and herb layer from 17% (1984–1992) to 24% (2006–2017) was recorded during the observation period (**Figure 11**). This increase was due to increase in the coverage of all the main species: *Empetrum hermaphroditum, Vaccinium vitis-idaea* and *Vaccinium myrtillus.*

The total coverage of the moss-lichen layer during the observation period did not change and averaged 70–75% (**Figure 12**). Substantial restructuring is recorded in the structure of the moss-lichen layer. At the beginning of the period under review (1984, the age of fire of 60 years), the main dominants of the stage were fruticose lichens of the genus *Cladonia,* whose coverage was ~50–60%. The projective covering of bryophytes in total amounted to 10%, of which the covering of the dominant of the moss cover of the later stages of succession – *Pleurozium schreberi* – was 4%. With the prescription of a fire of 80–90 years (2006–2017), the projective coverage of fruticose lichens decreased 1.5–2 times; the projective coverage of

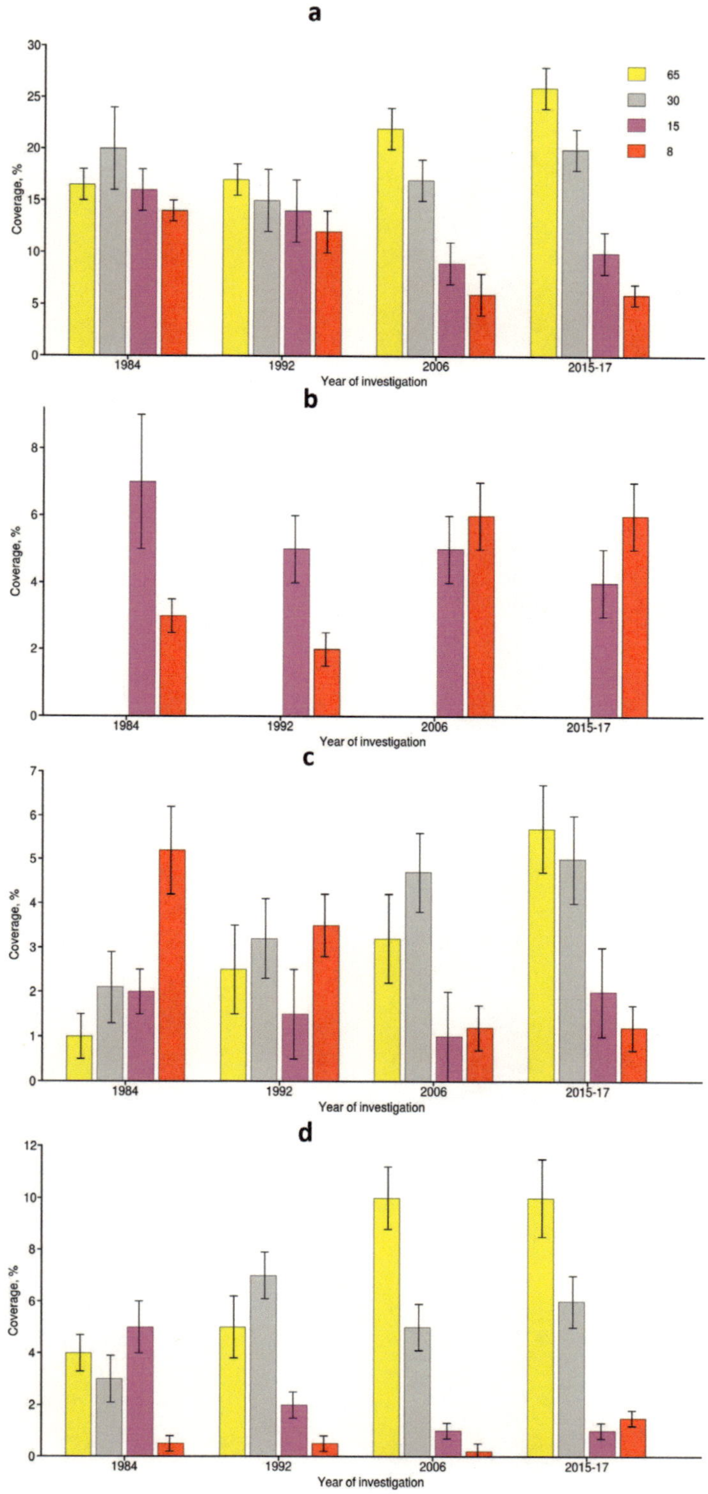

Figure 11. The 33-year dynamics of the total coverage and coverage of some species of the dwarf-shrub and herb layer at various distances from the Severonickel Smelter (Kola Peninsula, Russia): *a* – total coverage of the layer; *b* – coverage of *Arctostahyllos uva-ursi*; *c* – coverage of *Empetrum hermaphroditum*; *d* – coverage of *Vaccinium vitis-idaea.*

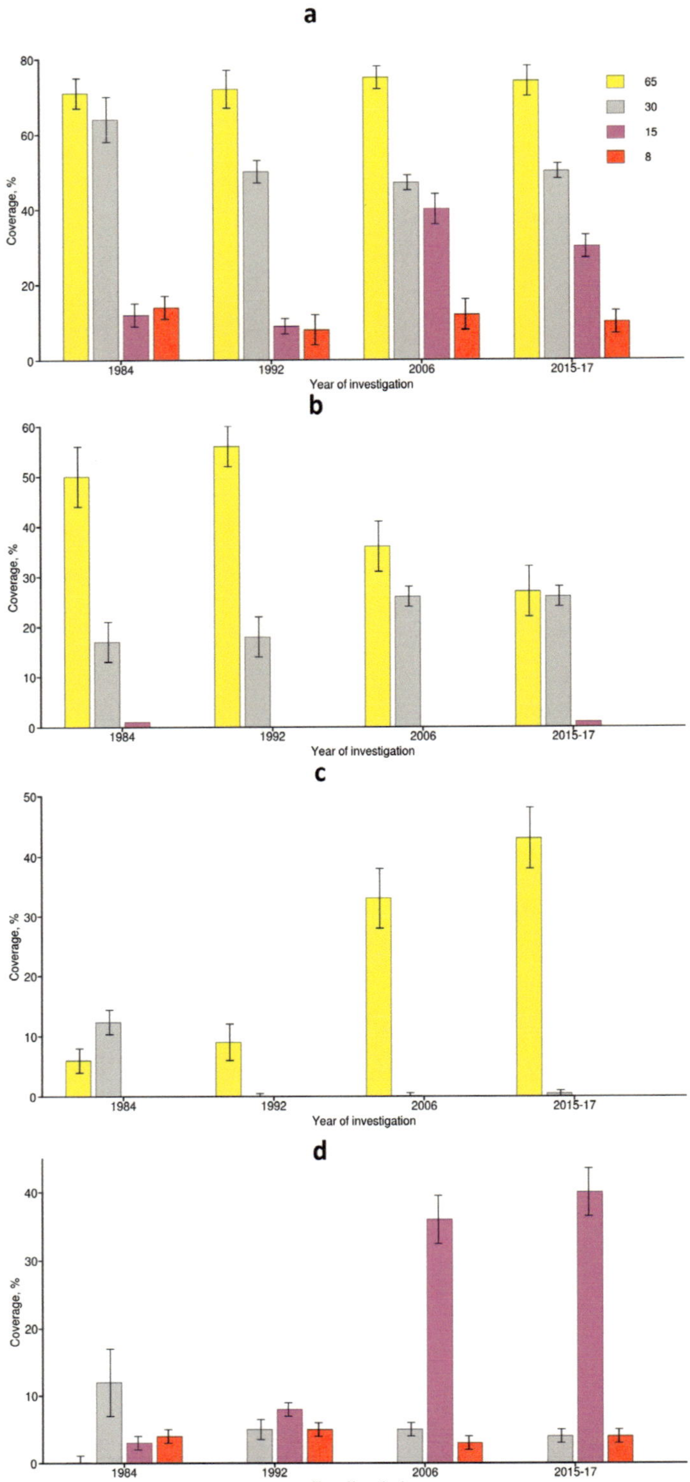

Figure 12. The 33-year of characteristics of the moss-lichen layer at various distances from the Severonickel Smelter (Kola Peninsula, Russia): *a* – total coverage of the layer; *b* – coverage of fruticose lichens; *c* – coverage of *Pleurozium schreberi*; *d* – total coverage of the crustaceous lichens and primary thallus of *Cladonia* sp.

Pleurozium schreberi increased to 35–40% (**Figure 12**). The observed changes in the state of the ground cover are due to the growth of the thickness of the forest litter [33] and the change in the wind regime under the canopy of the stand as its height increases.

In the background pine forests burned 140 ago during the period under review, all parameters of dwarf-shrub and herb layer as well as of moss-lichen layer remained stable [4]. The covering of the dwarf-shrub and herb layer was 25–26%, in the cover co-dominated *Empetrum hermaphroditum, Vaccinium vitis-idaea* and *V. myrtillus*. Covering of the moss-lichen layer was ~85%, the basis of the lichen cover were fruticose lichens of the genus *Cladonia,* moss – *Pleurozium schreberi.*

In the buffer zone in 1984, the projective coverage of the dwarf-shrub and herb layer was 18% and did not change significantly from 1984 to 2017 (**Figure 11**), dwarf-shrubs of the genus *Vaccinium* and *Empetrum hermaphroditum* predominated. In general, for the total coverage, the composition of the dominant species and the character of the long-term dynamics, the dwarf-shrub and herb layer of pine forests in the buffer zone and in the background area does not differ significantly.

In 1984, the total coverage of the moss-lichen layer of pine forests of the buffer zone with a fire age of 50 years did not differ from the coverage in the background communities and was 64% (**Figure 12**), but the cover was dominated by early succession species of lichens of genus *Cladonia* (25%) and *Trapeliopsis granulosa* (10%). In 1992, a decrease in the projective coverage of the moss-lichen layer was registered to 50%, a virtually complete disappearance of the dominant of the moss cover – *Pleurozium schreberi,* a decrease in the projective coverage of crustose forms of lichens (from 14 to 5%), and an increase in the coverage of fruticose lichens of the genus *Cladonia* (from 18 to 26%). All the registered changes in the structure of the moss-lichen layer remain to this day.

In pine forests with the age of fire of 140 years, the total projective coverage of the moss-lichen layer from 1984 to 2006 decreased from 54 to 37% due to the continued decrease in the coverage of fruticose lichens of the genus *Cladonia* and the elimination of *Pleurozium schreberi.* During the period from 2006 to 2015, there was a certain positive trend: an increase in the overall projective cover of the layer to 50%, the appearance of a dominant species of the final stage of succession – *Pleurozium schreberi* (0.1–0.2%) and an increase in the coverage of fruticose lichens from (10 to 18%). However, throughout the study period, crustose lichens (covering an average of ~7%), as well as *Cladonia* species with unbranched podecia (7%) took a significant part in the composition of the ground cover, which is completely untypical for such a prescription of the fire and indicates prolonged oppression of the moss -lichen layer. A feature of the communities of the buffer zone with the age of fire over 140 years is participation in the cover of liverworts, covering an average of 10–15%. This phenomenon is also untypical for the background.

The data presented in **Figures 11** and **12** characterized the pine communities of the impact zone with the age of fire at the beginning of the experiment (1984) – 130 years and ~50 years. Projective covering of the dwarf-shrub and herb layer in 1984 and 1992 in the impact zone did not differ from the background and averaged ~16%. At the final stages of the study (2006 and 2017) it was found the decrease in the total coverage of the layer to 10 and 6%, respectively. The observed decrease in coverage was due to a decrease in the coverage of species of the genus *Vaccinium* (130 years after the fire), as well as *Empetrum hermaphroditum* and *Calluna* *vulgaris* (50 years after the fire), whose coverage decreases from 4–6 to 1%. It should be noted that, unlike the background and buffer zones, the dwarf-shrub and herb layer in the impact zone is formed mainly by the dwarf-shrub *Arctostaphylos uva-ursi*, whose coverage is 2–6%. This phenomenon is due, on the one hand, to the very high content of heavy metals in the organogenic horizon of the soil, where the main part of roots and subterranean shoots of dwarf-shrubs is located, on the other hand, the destruction of the moss-lichen layer and the degradation of forest litter, which plays an important role in maintaining the moisture capacity of the upper soil horizons.

The total coverage of the moss-lichen layer in the impact zone was significantly different from the background at all stages of the study and was 10–40% (**Figure 12**). In this case, the dominant species typical for background areas – fruticose lichens of the genus *Cladonia* and moss *Pleurozium schreberi* were almost completely absent. The layer was composed only of cruticose lichens and primary thalli of different lichen species. Significant differences were found in dynamics of the projective coverage of lichen mat in communities with different fire ages. For the period from 1984 to 2017 in the community with a fire age of 50 years, the total projective coverage did not change significantly and was ~10%. The absence of changes is due to the location of the community in the upper part of the slope, where the processes of soil erosion are actively developing and a continuous flushing of the forest litter (which in 1984 was 1–2 cm) was carried out by rain and meltwater. This led to the exposure of the mineral horizons of the soil and large boulders, preventing the development of lichen cover, which occurred only in restricted areas. In a community with a fire age of 130 years, the projective coverage of lichens was ~10% and increased sharply to 40% in 2006–2017, which was due to a change in the state of the substrate on which the lichen cover was formed. The prolonged decrease in the fall off incoming (due to the destruction of the stand by the early 1990s) and the formation of forest litter made it possible to increase the coverage of lichen species resistant to pollution. Despite this, nowadays, the state of the moss-lichen layer of pine forests of the impact zone is estimated as completely destroyed: the total projective coverage is 2–8 times less than in the background, all the main dominant species typical to communities with fire age of more than 70 years are missing.

Comparison of dynamics of ground vegetation in pine forests of the impact zone of and rate of its recovery in field experiment with the complete removal of ground vegetation conducted in the background area [27], allows to separate the effects of soil pollution and comprehensive aerial pollution by sulfur dioxide together with polymetallic dust. Under the conditions of the field experiment (unpolluted air – polluted soils taken from the impact zone), a significant slowing of the restoration of the dwarf-shrub and herb layer as well as of moss-lichen layer was recorded. This explains the lack of positive dynamics of the state of the ground cover of pine forests with a significant decrease in the level of aerial pollution.

Thereby, in the airborne polluted environment, the natural succession dynamics of the ground cover is disrupted. Regardless of the fire age, the state of the ground cover of communities in the buffer zone roughly corresponds to that in forests burned~30–40 years ago, in the impact zone ~10 years ago. In the impact zone, in the situation when forest litter is available, a cover is formed by the crustose lichens and the primary thallus of lichens of the genus *Cladonia*. The moss cover is absent. Reduction of atmospheric emissions did not affect the state of the ground cover (both in buffer and impact zones) due to the persistent high level of contamination of the upper horizon of the soil with heavy metals.

3.2. Experimental study of impact of soil contamination by heavy metals on forest ecosystems

In the control sampling sites the total content of acid-soluble forms of Ni and Cu in the organogenic horizon of Al-Fe-humus podzols did not exceed 20 mg/kg on average. In the experimental plots, the content of these forms of heavy metals varied within the limits of: Ni – 9.4–260, Cu – 22–615 mg/kg, average concentrations of acid-soluble forms of heavy metals in the forest litter of experimental sites are shown in **Figure 13**. The index of technogenic load varied in the range from 2.0 to 45.5 rel. units, on average, it was 15.0 rel. units and did not differ significantly between the experimental sites. There is no correlation between the amount of polymetallic dust introduced and the technogenic load index. No correlation was found between the amount of deposited polymetallic dust and the index of technogenic load.

The content of heavy metals in the needles of *Pinus sylvestris* trees and dwarf-shrub (*Vaccinium myrtillus, V. vitis-idaea, Empetrum hermaphroditum*) leaves in the experimental plots was significantly higher than their background values in only 1.3–3 times. In the living parts of thalli of lichens (*Cladonia stellaris, Cl. rangiferina, Cl. mitis* and *Cl. uncialis*) and live parts of moss (*Pleurozium schreberi*) collected in control and experimental sites, the content of heavy metals either did not differ significantly, or an insignificant increase in Ni and Cu with respect to their background values. This insignificant increase in the content of Ni and Cu in the plant material of the experimental sites indicates a weak migration of heavy metals from contaminated soil to the ground parts of plant organisms, and their concentrations in tissues do not exceed the toxicity threshold of heavy metals for plants, mosses and lichens.

Thus, under the conditions of the field experiment, the increased concentrations of heavy metals in plant assimilation organs and living parts of lichens and mosses do not reach the lower limit of tolerance and are not lethal for their vital activity.

The dynamics of the projective covering of the dwarf-shrub and herb and moss-lichen layers of the control and experimental sites is shown in **Figure 14**. At the initial stage (1992), both

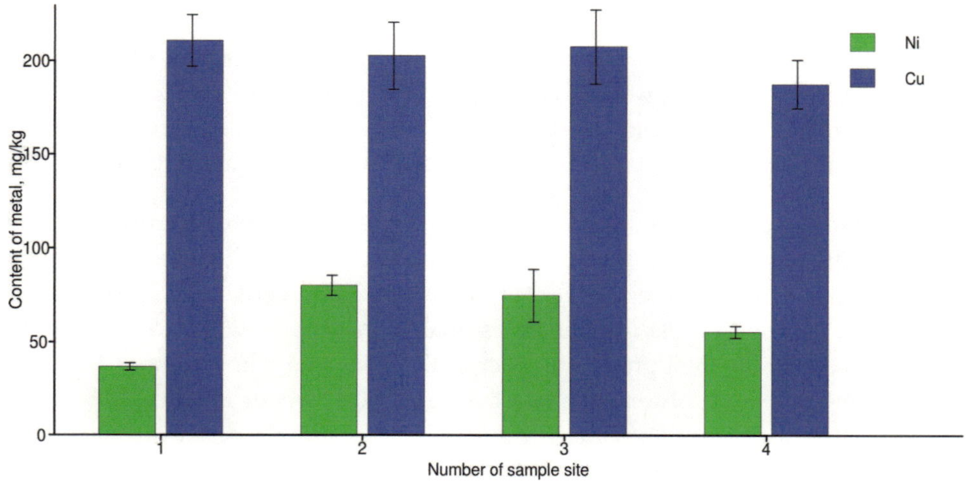

Figure 13. The average content of acid-soluble forms of Ni and Cu in the organogenic horizon of Al-Fe-humus podzol in the experimental sites.

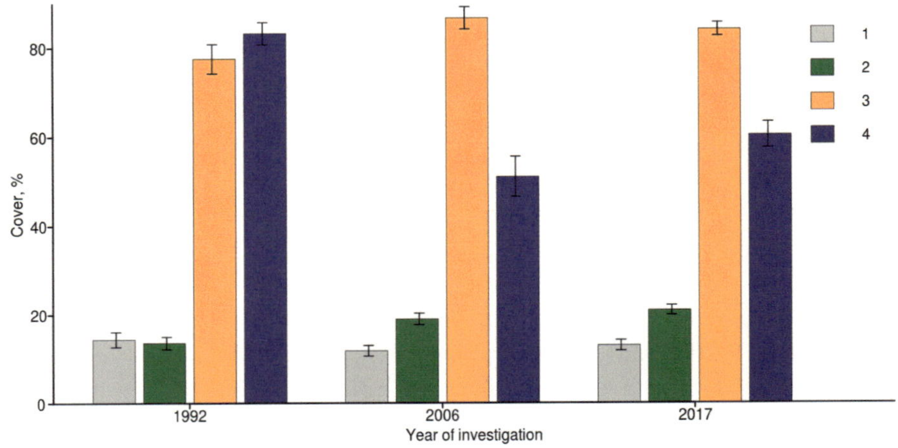

Figure 14. Dynamics of covearages (%) of dwarf-shrub and herb (1, 2) and moss-lichen (3, 4) layers in control (1, 3) and experiment (2, 4) sites.

sites had the same parameters of the dwarf-shrub and herb layer and slightly differed in the overall projective covering of the moss-lichen layer.

In the control community, the dynamics of the ground cover proceeded in accordance with the laws of post-fire succession. A significant change in the total projective coverage of dwarf-shrub and herb layer for the period 1992–2017 was not recorded (**Figure 14**), which was due to the multidirectional dynamics of individual species: an increase in the coverage of *Vaccinium vitis-idaea, V. myrtillus* and a decrease in the coverage of *Arctostaphyllos uva-ursi* and *Calluna vulgaris* (**Table 1**). The total projective coverage of the moss-lichen layer in 2017 was significantly higher by 7% compared with that in 1992, while the total coverage of lichens did not change significantly, and the bryophytes increased threefold. In 2017, the main dominant of the lichen cover was *Cladonia stellaris*, a moss cover – *Pleurozium schreberi*, the projective covering of which increased more than 12 times relative to the beginning of the experiment.

At the experimental site the disturbance has been revealed in succession dynamics of the dwarf-shrub and herb layer. It was the absence of a decrease in the coverage of *Calluna vulgaris*, a more significant (twofold) increase in the *Vaccinium vitis-idaea* coverage and, as a result, an increase in the overall projective cover of the layer in 1.5 times (**Figure 14**, **Table 1**).

The most significant differences in the dynamics of species cover were recorded in the moss-lichen layer. Fifteen years after the start of the experiment (2006), in the experimental site, the decrease in the projective coverage of *Cladonia mitis* was 2 times greater, and the increase in *Cladonia rangiferina* and *Cl. stellaris* – 2 times less than in the control, which led to a significant (by 40%) decrease in the total projective cover of the layer as a whole (**Figure 14**, **Table 1**). During this time in the control areas, the total coverage of the layer increased by 10%. A partial restoration of the moss-lichen layer was registered 25 years after the dust dispersion. The difference with the control was reduced to 30%. The direction of the changes in the coverage of two dominants of the lichen mat – *Cladonia mitis* and *Cl. stellaris* in the control and experiment

Species and characteristics	Control site		Experimental site			
	Year of investigation					
	1992	2006	2017	1992	2006	2017
	a	b	c	d	e	f
Dwarf-shrubs, total cover	**14.4**	**11.8^{e}**	**13.0^{f}**	**13.6^{e,f}**	**19.0^{b,d}**	**21.0^{c,d}**
Vaccinium vitis-idaea L.	3.2^{c}	3.4^{c}	5.8^{a,b}	1.8^{e,f}	4.2^{d,f}	7.5^{d,e}
Vaccinium myrtillus L	0.2^{c}	1.1	1.8^{a}	0.2^{f}	1.0	2.3^{d}
Empetrum hermaphroditum Hagerup	0.3	0.2	0.9	0.9	0.3	0.6
Arctostaphylos uva-ursi (L.) Spreng.	1.0^{b,c}	<0.1^{a}	0.1^{a}	0.5^{f}	0.4	0.1^{d}
Calluna vulgaris (L.) Hull	9.7^{c}	7.1^{e}	4.3^{a,f}	10.0	13.0^{b}	10.4^{c}
Lichens, total cover	**73.4**	**81.3**	**70.7**	**80.2**	**50.0**	**58.3**
Cladonia mitis Sandst.	40.0^{b,c,d}	21.5^{a,c}	3.4^{a,b,f}	52.0^{a,e,f}	16.4^{d,f}	7.3^{c,d,e}
Cladonia rangiferina (L.) Weber ex F.H. Wigg.	8.8^{b}	25.5^{a,c,e}	16.5^{b,f}	9.5^{f}	14.4^{b,f}	23.7^{c,d,e}
Cladonia stellaris (Opiz) Pouzar & Vězda	9.2^{b,c}	26.6^{a,c,e}	48.4^{a,b,f}	7.1^{f}	10.6^{b,f}	19.5^{c,d,e}
Cladonia uncialis (L.)Weber ex F.H. Wigg.	8.4^{b,c}	4.8^{a,c}	1.4^{a,b}	6.6^{e,f}	3.1^{d,f}	1.6^{d,e}
Cladonia deformis (L.) Hoffm.	0.1	<0.1	<0.1^{f}	0.1	0.9	0.2^{c}
Cladonia gracilis (L.) Willd.	0.3^{d}	0.2^{e}	0.1	0.5^{a}	0.5^{b}	0.3
Cladonia crispata (Ach.) Flot.	1.8^{c}	0.9	0.5^{a,f}	1.1	1.5	1.2^{c}
Cladonia cornuta (L.) Hoffm.	1.0^{b,c}	0.1^{a,e}	0.1^{a,f}	0.8	0.5^{b}	1.1^{c}
Stereocaulon paschale (L.) Hoffm.	3.2^{c}	1.4	0.3^{a}	1.6	1.3	1.5
Mosses, total cover	**4.1^{c}**	**5.3^{c,e}**	**13.3^{a,b,f}**	**3.0**	**1.1^{b}**	**2.0^{c}**
Pleurozium schreberi (Willd. ex Brid.) Mitt.	0.9^{c}	4.7^{c,e}	12.2^{a,b,f}	0.1^{f}	0.1^{b,f}	1.4^{c,d,e}
Polytrichum spp.	2.4^{b,c}	0.4^{a}	0.7^{a}	2.4^{e,f}	0.7^{d}	0.3^{d}

Notes: The names of species are given in accordance with the latest (on November 25, 2017) data on the systematics of plants [34] and fungi [35]. The letters indicate significant (at $p \leq 0.05$) differences between the characteristics of the ground cover.

Table 1. Dynamics of the most active species in the control and experimental sites.

was the same. But in the first case, those changes were more contrasting (the differences in the coverings of these species reached 7–11 times) in comparison with the dynamics of covering of the same species in the experimental area (the differences did not exceed 3–7 times). The nature of the dynamics of the cover of one more dominant of the lichen mat *Cl. rangiferina* differed in the control and experimental areas: in control after 15 years (2006) the maximum value of the projective cover of this species was recorded, and then it decreased; a gradual increase in the *Cl rangiferina* coverage was observed in the experimental site.

In the experimental site, soil contamination with heavy metals led to inhibition of the restoration of both the total coverage of mosses and its dominant – *Pleurozium schreberi*. Twenty-five years after the introduction of polymetallic dust, projective covering of bryophytes was seven times less than in the control (**Table 1**).

Thus, soil contamination with heavy metals led to a disruption of the succession processes in the ground cover of pine forests in the intermediate stages of recovery after a fire. In general, the slowing down of the succession processes can be estimated in 15–20 years.

4. Conclusions

Analysis and generalization of the results of long-term (1980–2017) monitoring of the state of the north taiga pine forests in the zone affected by atmospheric emissions from the non-ferrous metallurgy Smelter, as well as the results of field experiments in the Kola Peninsula (Russia) allow us to characterize the spatial and temporal changes in the state of forest ecosystems of pine forests and their components.

1. Against the backdrop of a 5–8-fold reduction in atmospheric emissions from the Severonickel Smelter (the Kola Peninsula, Russia), the level of pollution of the upper soil horizon in the buffer zone continues to increase, and within the impact zone remains very high. During the study period, the range of variation of the technogenic load index of the organogenic horizon of Al-Fe-humus podzols is 4–30 in the buffer zone, and 35–120 rel. units. in the impact zone.

2. The nature of the dynamics of the content of heavy metals in plant assimilation organs differs significantly from the level of pollution of the upper soil horizon. In response to a 5–8-fold decrease in the atmospheric emissions of Severonickel Smelter, there was a 2–16-fold decrease in Ni and Cu in the plant assimilation organs, which was due to a decrease in the amount of dust particles coming from the polluted air. The level of accumulation of heavy metals by assimilation organs is associated with physiological features of plant species: the maximum content of Ni and Cu is recorded in the leaves of *Empetrum hermaphroditum*, the minimum – in the leaves of *Arctostaphylos uva-ursi*.

3. In response to a sharp reduction in the atmospheric emissions of the Severonickel Smelter both in the buffer and impact zones, a positive trend is observed in the state of *Pinus sylvestris* stands. The vital state of the stands improved, the degree of pine needles damage by chlorosis and necrosis decreased. This led to a 1.5–2.5-fold increase in the duration of its life, and 3–4 times the area of annual tree trunk increment. The restoration of natural renewal process of *Pinus sylvestris* in the impact zone is possible only if the level of pollution of the upper horizon of soils by heavy metals is reduced.

4. Ander aerial pollution, the natural succession dynamics of the ground cover is disturbed: parameters of pine forests of the buffer zone, regardless of their successional age, correspond to forests with a fire age of ~30–40 years, in the impact zone ~10 years. Decrease in

the volume of atmospheric emissions by the Severonickel Smelter did not lead to an improvement in the state of the ground cover both in the buffer zone and in the impact zones due to the persistent high level of soil contamination with heavy metals.

5. Experimental soil contamination (the technogenic load index 2–45 rel. units), led to a decrease in the total cover and to a change in the species structure of both the dwarf-shrub and herb and moss-lichen layers. With an index of technogenic load, an average of 15 rel. Units, the total coverage of the dwarf-shrub and herb layer increased 1.5 times, and the moss-lichen layer decreased by 1.5 times. These changes are caused by a change in the proportion of dominant species of dwarf-shrubs (*Calluna vulgaris, Vaccinium vitis-idaea*) and lichens (*Cladonia mitis, Cl. rangiferina, Cl. stellaris*). As a result, in the ground cover of pine forests, which are in the intermediate stages of recovery after a fire, a 15–20-year delay in the succession processes is recorded.

6. Soil contamination with polymetallic dust (in the range of the technogenic load index 2–45 rel. units) caused a 1.5–3-fold increase in the content of Ni and Cu in the plant material of the experimental sites. Such increase in the content of heavy metals indicates their weak migration from contaminated soil to the above ground parts of plant. Concentrations of Ni and Cu in the range of 1.5–15 mg/kg in plant assimilation organs and living parts of lichens and mosses do not exceed the toxicity threshold of heavy metals for plants.

Author details

Irina Lyanguzova*, Vasily Yarmishko, Vadim Gorshkov, Natalie Stavrova and Irina Bakkal

*Address all correspondence to: ilyanguzova@binran.ru

Komarov Botanical Institute RAS, St. Petersburg, Russia

References

[1] Norin B, Yarmishko V, editors. Influence of Industrial Air Pollution on Pine Forests of the Kola Peninsula. Leningrad: Botanical Institute of Academy of Sciences of Soviet Union; 1990. 195 p. (in Russian)

[2] Vodyanitskii Y. Contamination of soils with heavy metals and metalloids and its ecological hazard (analytic review). Eurasian Soil Science. 2013;**46**(7):793-801. DOI: 10.1134/S1064229313050153

[3] Vodyanitskii Y, Plekhanova I, Prokopovich E, Savichev A. Soil contamination with emissions of nonferrous metallurgical plants. Eurasian Soil Science. 2011;**44**(2):217-226. DOI: 10.1134/S1064229311020177

[4] Yarmishko V, editor. Dynamics of Forest Communities of the Northwestern Russia. St. Petersburg: VVM; 2009. 276 p. (in Russian)

[5] Evdokimova G, Kalabin G, Mozgova N. Contents and toxicity of heavy metals in soils of the zone affected by aerial emissions from the Severonickel Enterprise. Eurasian Soil Science. 2011;**44**(2):237-244. DOI: 10.1134/S1064229311020037

[6] Lukina N, Nikonov V. Biogeochemical Cycles in Northern Forests Affected by Technogenic Air Pollution. Apatity: Kola Scientific Center of Academy of Sciences of Soviet Union. 1996. Vol. 1, 213 p. Vol. 2, 192 p. (in Russian)

[7] Lukina N, Nikonov V. Nutrient Balance of the Forests of Northern Taiga: Natural and Technogenic Aspects. Apatity: Kola Scientific Center of Academy of Sciences of Soviet Union; 1998. 316 p. (in Russian)

[8] Yarmishko V, editor. Environmental Problems of the Northern Plant Communities. St. Petersburg: VVM; 2005; 450 p. (in Russian)

[9] Kozlov M, Zvereva E, Zverev V. Impacts of Point Polluters on Terrestrial Biota. New York: Springer-Verlag, 2009; 466 p

[10] Kozlov M, Zvereva E. Industrial barrens: Extreme habitats created by non-ferrous metallurgy. Reviews in Environmental Science and Biotechnology. 2007;**6**:231-259

[11] Environmental Monitoring of the Zone Affected by Emissions of the Kola Smelter [Internet]. Available from: http://www.kolagmk.ru

[12] Lyanguzova I. Dynamics of nickel and copper concentrations in plants of pine forests of the Kola Peninsula under conditions of industrial air pollution. Rastit Resursy. 2008;**44**(4):91-98 (in Russian)

[13] Evdokimova G, Mozgova N. Dynamics of the state of forest ecosystems exposed to emissions from a copper-nickel smelter over 30 years. In: Proceedings of the Conference on Ecological Problems in Northern Regions and Ways to Solve Them; 14-16 October 2008; Apatity: Kola Scientific Center of RAS; 2008. Part 1. pp. 68-73 (in Russian)

[14] Zverev V. Mortality and recruitment of mountain birch (*Betula pubescens* ssp. *czerepanovii*) in the impact zone of a copper-nickel smelter in the period of significant reduction of emissions: The results of 15-year monitoring. Russian Journal of Ecology. 2009; **40**(4):254-260

[15] Sukhareva T, Lukina N. Mineral composition of assimilative organs of conifers after reduction of atmospheric pollution in the Kola peninsula. Russian Journal of Ecology. 2014;**45**(2):95-102. DOI: 10.1134/S1067413614020088

[16] Chernen'kova T, Basova E, Kabirov R. Regeneration successions of northern taiga spruce forests under reduction of aerothechnogenic impact. Contemporary Problems of Ecology. 2011;**4**(7):742-757. DOI: 10.1134/S199542551107006X

[17] Alexeyev V. Impact of air pollution on far North forest vegetation. Science of the Total Environment. 1995;**160/161**:605-617

[18] Yarmishko V, Gorshkov V, Stavrova N. *Pinus sylvestris* L. vital state structure in the tree layer of pine forest with different degree and type of anthropogenic disturbance (Kola peninsula). Rastit Resursy. 2003;**39**(4):1-19 (in Russian)

[19] Fomin G, Fomin A. Control of Quality and Environmental Safety According to International Standards: A Handbook. Moscow: Protektor; 2001. 304 p. (in Russian)

[20] Yarmishko V, Lyanguzova I, editors. Methods of the Investigation of forest communities. St. Petersburg: SRI of Chemistry; 2002. 240 p. (in Russian)

[21] Barcan V. Nature and origin of multicomponent aerial emissions of the copper-nickel smelter complex. Environmental International. 2002;**28**:451-456

[22] Lyanguzova I, Goldvirt D, Fadeeva I. Spatiotemporal dynamics of the pollution of Al-Fe-humus podzols in the impact zone of a nonferrous metallurgical plant. Eurasian Soil Science. 2016;**49**(10):1189-1203. DOI: 10.1134/S1064229316100094

[23] Kashulina G, Pereverzev V, Litvinova T. Transformation of the soil organic matter under the extreme pollution by emissions of the Severonikel smelter. Eurasian Soil Science. 2010;**43**(10):1174-1183

[24] Barcan V. Leaching of nickel and copper from a soil contaminated by metallurgical dust. Environmental International. 2002;**28**(1-2):63-68

[25] Lyanguzova I. Dynamic trend of heavy metal contents in plants and soil under different industrial air pollution regimes. Russian Journal of Ecology. 2017;**48**(4):311-320. DOI: 10.1134/S1067413617040117

[26] Yarmishko V. Scots Pine and Aerial Pollution in the European North. St. Petersburg: Research Institute of Chemistry of St. Petersburg State University; 1997. 210 p. (in Russian)

[27] Stavrova N, Lyanguzova I, Gorshkov V, Bakkal I. Dynamics of forest ecosystems components recovery in field experiment of soil pollution by heavy metals. Rastit Resursy. 2007;**43**(3):48-65 (in Russian)

[28] Wotton D, Jones D, Phillips S. The effect of nickel and copper deposition from a mining and smelting complex on coniferous regeneration in the boreal forest of Northern Manitoba. Water, Air and Soil Pollution. 1986;**31**:349-358

[29] Arduini I, Godbold D, Onnis A. Infuence of copper on root growth and morphology of *Pinus pinea* L. and *Pinus pinaster* Ait. seedlings. Tree Physiology. 1995;**15**:411-415

[30] Veselkin D. Reduction of absorbing root length in Siberian fir and Siberian spruce under heavy metal pollution. Lesovedenie. 2003;**3**:65-68 (in Russian)

[31] Kershaw K. Studies of lichen-dominated systems. XV. The temperature and humidity profiles in a *Cladina alpestris* mat. Canadian Journal of Botany. 1975;**53**(22):2614-2620

[32] Likens G. Biogeochemistry of a Forested Ecosystem. 3rd ed. NY, Heidelberg, Dordrecht, London: Springer, 2013. 208 p

[33] Gorshkov V, Bakkal I, Stavrova N. Postfire recovery of forest litter in scots pine forests in two different regions of boreal zone. Silva Fennica. 1996;**30**(2-3):209-2019. DOI: 10.14214/sf.a9233

[34] The Plant List. Version 1.1. [Internet]. Available from: http://www.theplantlist.org (Accessed: Nov 25, 2017)

[35] Index Fungorum. Version 1. [Internet]. Available from: http://www.indexfungorum.org/Names/names.asp (Accessed: Nov 25, 2017)

Chapter 7

Environmental Contamination by Heavy Metals

Vhahangwele Masindi and Khathutshelo L. Muedi

Additional information is available at the end of the chapter

http://dx.doi.org/10.5772/intechopen.76082

Abstract

The environment and its compartments have been severely polluted by heavy metals. This has compromised the ability of the environment to foster life and render its intrinsic values. Heavy metals are known to be naturally occurring compounds, but anthropogenic activities introduce them in large quantities in different environmental compartments. This leads to the environment's ability to foster life being reduced as human, animal, and plant health become threatened. This occurs due to bioaccumulation in the food chains as a result of the nondegradable state of the heavy metals. Remediation of heavy metals requires special attention to protect soil quality, air quality, water quality, human health, animal health, and all spheres as a collection. Developed physical and chemical heavy metal remediation technologies are demanding costs which are not feasible, time-consuming, and release additional waste to the environment. This chapter summarises the problems related to heavy metal pollution and various remediation technologies. A case study in South Africa mines were also used.

Keywords: heavy metals, environment, contamination, legal requirements, pollution

1. Environment

Environment can be referred to the surroundings within which humans exist. These are made up of: the land, the water and the atmosphere of the earth; microorganisms, plant and animal life; any part or combination of the first two items on this list and the interrelationships among and between them and the physical, chemical, aesthetic and cultural properties and conditions of the foregoing that influence human health and well-being. It is also characterised by a number of spheres that influence its behaviour and intrinsic value. The most important sphere of the environment is the biosphere because it harbours the living organisms. This is the sphere where you find living organisms (plants and animals) interacting with each and their nonliving environment (soil, air and water). In the late centuries, industrialisation and globalisation have impaired pristine environments and their ability to foster life. This has introduced components that compromise the holistic functioning of the environment and its intrinsic values [1].

1.1. Environmental contamination

An environment can be polluted or contaminated. Pollution differs from contamination; however, contaminants can be pollutants, and pose detrimental impact on the environment. From literature, pollution is defined as the introduction by man, directly or indirectly, of substances or energy into the environment resulting in such deleterious effects as harm to living resources, hazards to human health, hindrance to environmental activities and impairment of quality for use of the environment and reduction of amenities. Contamination on the other hand is the presence of elevated concentrations of substances in the environment above the natural background level for the area and for the organism. Environmental pollution can be referred to undesirable and unwanted change in physical, chemical and biological characteristics of air, water and soil which is harmful for living organisms—both animal and plants. Pollution can take the form of chemical substances or energy, such as noise, heat or light [2].

Pollutants, the elements of pollution, can either be foreign substances/energies or naturally occurring contaminants.

1.1.1. Types of pollutants

Environmental pollutants continue to be a world concern and one of the great challenges faced by the global society. Pollutants can be naturally occurring compounds or foreign matter which when in contact with the environment cause adverse changes. There are different types of pollutants, namely inorganic, organic and biological. Irrespective of pollutants falling under different categories, they all receive considerable attention due to the impacts they introduce to the environment. The relationship between environmental pollution and world population has become an inarguable directly proportional relationship as it can be seen that the amount of potentially toxic substances released into the environment is increasing with the alarming growth in global population. This issue has led to pollution being a significant problem facing the environment.

1.1.1.1. Inorganic pollutants

Industrial, agricultural and domestic wastes contribute to environmental pollution, which cause adverse harm to human and animal health. From such sources, inorganic pollutants are released. Inorganic pollutants are usually substances of mineral origin, with metals, salts and minerals being examples [2]. Studies have reported inorganic pollutants as material found naturally but have been altered by human production to increase their number in the environment. Inorganic substances enter the environment through different anthropogenic activities such as mine drainage, smelting, metallurgical and chemical processes, as well as natural processes. These pollutants are toxic due to the accumulation in the food chains [3].

1.1.1.2. *Organic pollutants*

Organic pollution can be briefly defined as biodegradable contaminants in an environment. These sources of pollution are naturally found and caused by the environment, but anthropogenic activity has also been contributing to their intensive production to meet the human needs. Some of the common organic pollutants which have been noted to be of special concern are human waste, food waste, polychlorinated biphenyls (PCBs), polybrominated diphenyl ethers (PBDEs), polycyclic aromatic hydrocarbons (PAHs), pesticides, petroleum and organochlorine pesticides (OCPs) [4].

Organic pollutants have gained attention as they have become a major problem in the environment. Properties of organic pollutants, amongst others, such as high lipid solubility, stability, lipophilicity and hydrophobicity have recently made organic pollutants termed persistent. These properties give organic pollutants the ability to easily bioaccumulate in the different spheres of the environment, thus causing toxicological effects [5, 6].

1.1.1.3. *Biological pollutants*

Biological pollutants are described as pollutants which exist as a result of humanity's actions and impact on the quality of aquatic and terrestrial environment. This type of pollutants include bacteria, viruses, moulds, mildew, animal dander and cat saliva, house dust, mites, cockroaches and pollen. Studies have documented different sources of these pollutants, including pollens originating from plants; viruses transmitted by people and animals; bacteria carried by people, animals, and soil and plant debris [7].

2. Heavy metals

Although there is no specific definition of a heavy metal, literature has defined it as a naturally occurring element having a high atomic weight and high density which is five times greater than that of water [8]. Among all the pollutants, heavy metals have received a paramount attention to environmental chemists due to their toxic nature. Heavy metals are usually present in trace amounts in natural waters but many of them are toxic even at very low concentrations [9]. Metals such as arsenic, lead, cadmium, nickel, mercury, chromium, cobalt, zinc and selenium are highly toxic even in minor quantity. Increasing quantity of heavy metals in our resources is currently an area of greater concern, especially since a large number of industries are discharging their metal containing effluents into fresh water without any adequate treatment [3].

Heavy metals become toxic when they are not metabolised by the body and accumulate in the soft tissues. They may enter the human body through food, water, air or absorption through the skin when they come in contact with humans in agriculture, manufacturing, pharmaceutical, industrial or residential settings. Industrial exposure accounts for a common route of exposure for adults. Ingestion is the most common route of exposure in children. Natural and human activities are contaminating the environment and its resources, they are discharging more than what the environment can handle [9, 10] (**Figure 1**).

2.1. Sources of heavy metals

Heavy metals can emanate from both natural and anthropogenic processes and end up in different environmental compartments (soil, water, air and their interface) (**Figure 2**).

2.1.1. Natural processes

Many studies have documented different natural sources of heavy metals. Under different and certain environmental conditions, natural emissions of heavy metals occur. Such emissions include volcanic eruptions, sea-salt sprays, forest fires, rock weathering, biogenic sources and wind-borne soil particles. Natural weathering processes can lead to the release of metals from their endemic spheres to different environment compartments. Heavy metals can be found in the form of hydroxides, oxides, sulphides, sulphates, phosphates, silicates and organic compounds. The most common heavy metals are lead (Pb), nickel (Ni), chromium (Cr), cadmium (Cd), arsenic (As), mercury (Hg), zinc (Zn) and copper (Cu). Although the aforementioned heavy metals can be found in traces, they still cause serious health problems to human and other mammals [9].

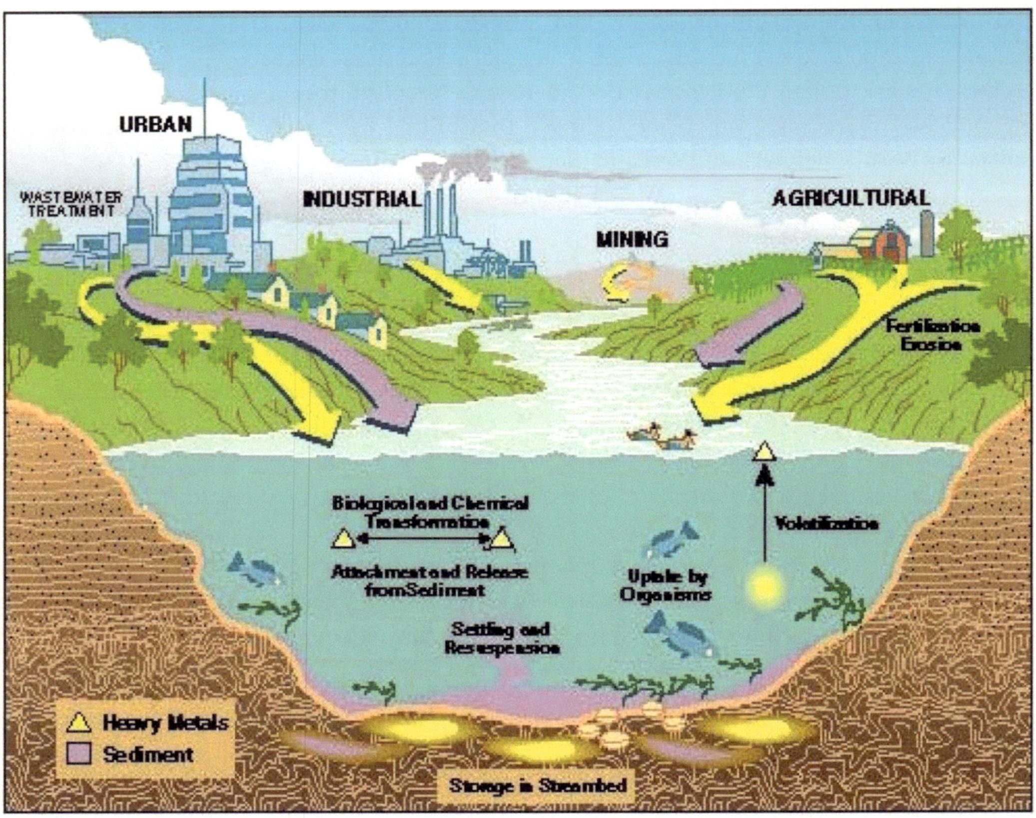

Figure 1. Sources and sinks of heavy metals [11].

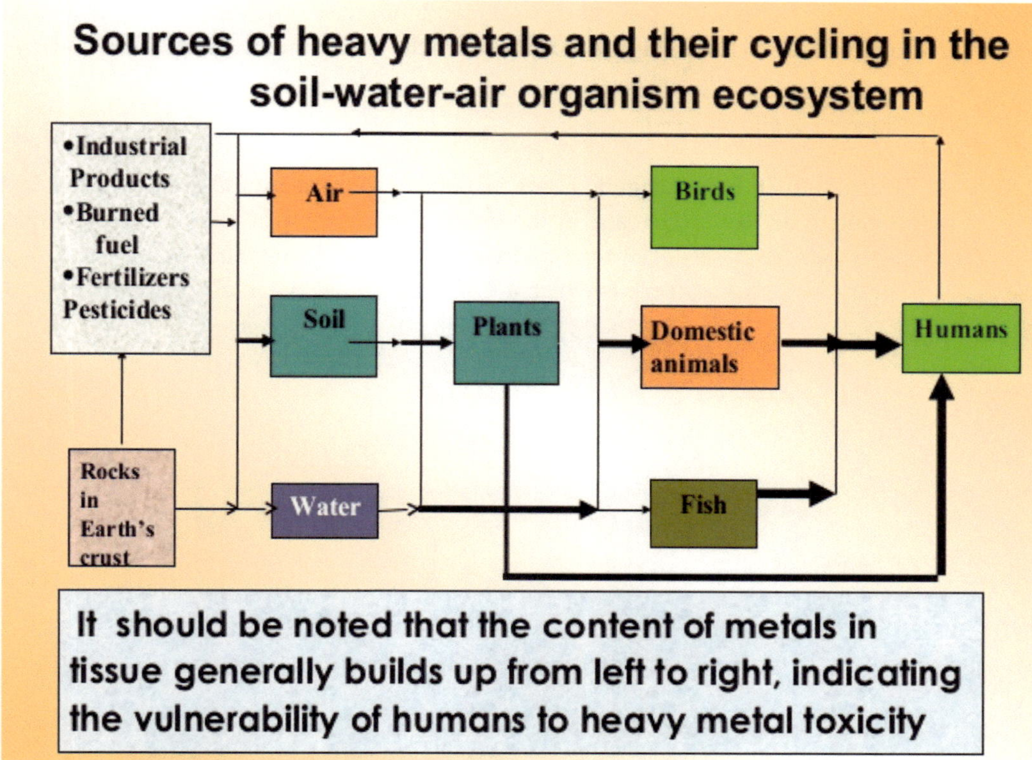

Figure 2. Sources of heavy metals and their cycle in the environment [12].

2.1.2. Anthropogenic processes

Industries, agriculture, wastewater, mining and metallurgical processes, and runoffs also lead to the release of pollutants to different environmental compartments. Anthropogenic processes of heavy metals have been noted to go beyond the natural fluxes for some metals. Metals naturally emitted in wind-blown dusts are mostly from industrial areas. Some important anthropogenic sources which significantly contribute to the heavy metal contamination in the environment include automobile exhaust which releases lead; smelting which releases arsenic, copper and zinc; insecticides which release arsenic and burning of fossil fuels which release nickel, vanadium, mercury, selenium and tin. Human activities have been found to contribute more to environmental pollution due to the everyday manufacturing of goods to meet the demands of the large population [10].

2.2. Environmental impacts of heavy metals

The presence of heavy metals in the environment leads to a number of adverse impacts. Such impacts affect all spheres of the environment, that is, hydrosphere, lithosphere, biosphere and atmosphere. Until the impacts are dealt with, health and mortality problems break out, as well as the disturbance of food chains. **Figure 3** summarises the health impacts of heavy metals.

2.3. Effect of heavy metals contamination

Heavy metals contamination is becoming a serious issue of concern around the world as it has gained momentum due to the increase in the use and processing of heavy metals during various activities to meet the needs of the rapidly growing population. Soil, water and air are the major environmental compartments which are affected by heavy metals pollution.

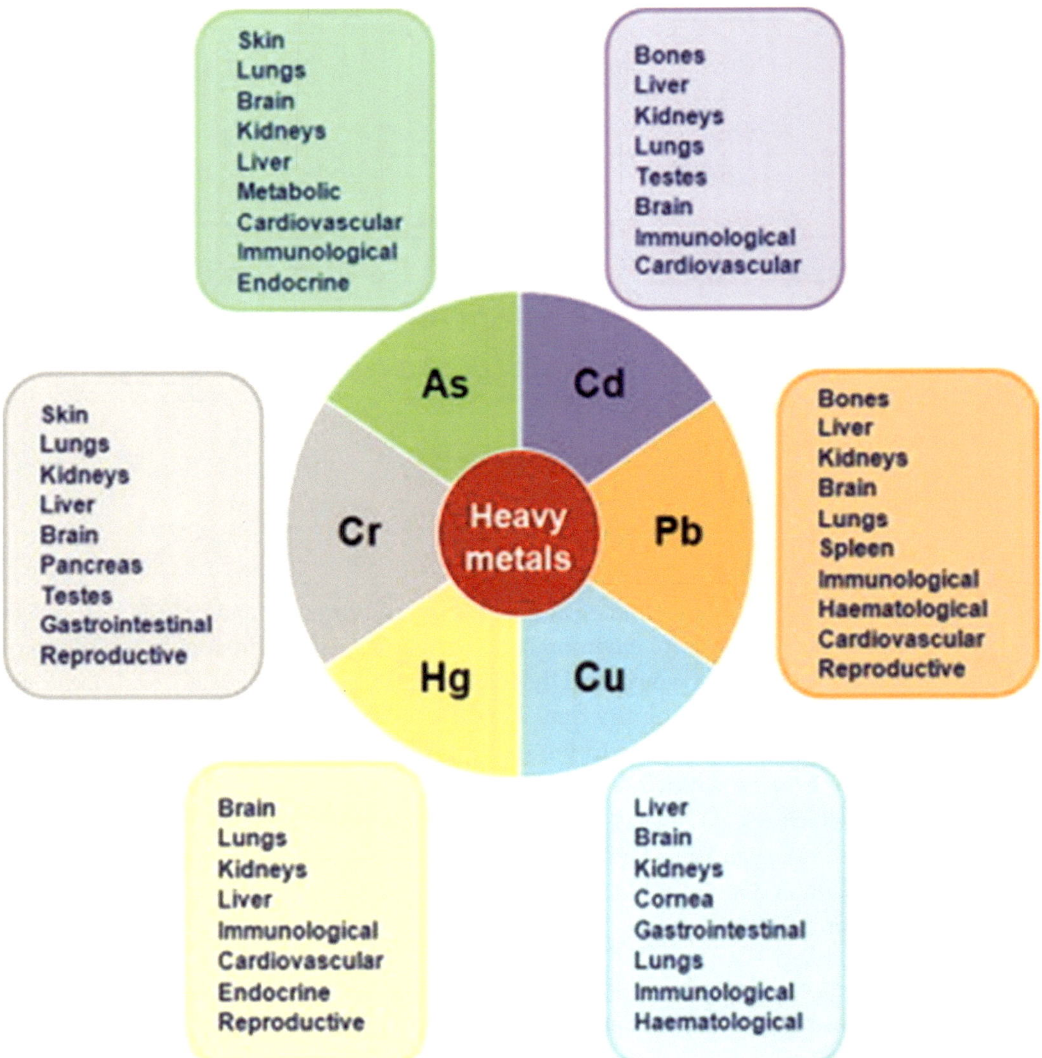

Figure 3. Impacts of heavy metals on the environment [13].

2.3.1. *Effect on soil*

Emissions from activities and sources such as industrial activities, mine tailings, disposal of high metal wastes, leaded gasoline and paints, land application of fertilisers, animal manures, sewage sludge, pesticides, wastewater irrigation, coal combustion residues and spillage of petrochemicals lead to soil contamination by heavy metals. Soils have been noted to be the major sinks for heavy metals released into the environment by aforementioned anthropogenic activities. Most heavy metals do not undergo microbial or chemical degradation because they are nondegradable, and consequently their total concentrations last for a long time after being released to the environment [5, 14].

The presence of heavy metals in soils is a serious issue due to its residence in food chains, thus destroying the entire ecosystem. As much as organic pollutants can be biodegradable, their biodegradation rate, however, is decreased by the presence of heavy metals in the environment, and this in turn doubles the environmental pollution, that is, organic pollutants and heavy metals thus present. There are various ways through which heavy metals present risks to humans, animals, plants and ecosystems as a whole. Such ways include direct ingestion, absorption by plants, food chains, consumption of contaminated water and alteration of soil pH, porosity, colour and its natural chemistry which in turn impact on the soil quality [15].

2.3.2. *Effects on water*

Although there are many sources of water contamination, industrialisation and urbanisation are two of the culprits for the increased level of heavy metal water contamination. Heavy metals are transported by runoff from industries, municipalities and urban areas. Most of these metals end up accumulating in the soil and sediments of water bodies [15].

Heavy metals can be found in traces in water sources and still be very toxic and impose serious health problems to humans and other ecosystems. This is because the toxicity level of a metal depends on factors such as the organisms which are exposed to it, its nature, its biological role and the period at which the organisms are exposed to the metal. Food chains and food webs symbolise the relationships amongst organisms. Therefore, the contamination of water by heavy metals actually affects all organisms. Humans, an example of organisms feeding at the highest level, are more prone to serious health problems because the concentrations of heavy metals increase in the food chain [16].

2.3.3. *Effects on air*

Industrialisation and urbanisation, due to rapid world population growth, have recently made air pollution as a major environmental problem around the world. The air pollution was reported to have been accelerated by dust and particulate matters (PMs) particularly fine particles such as $PM_{2.5}$ and PM_{10} which are released through natural and anthropogenic processes. Natural processes which release particulate matters into air include dust storms, soil erosion, volcanic eruptions and rock weathering, while anthropogenic activities are more industrial and transportation related [17].

Particulate matters are important and require special attention as they can lead to serious health problems such as skin and eyes irritation, respiratory infections, premature mortality and cardiovascular diseases. These pollutants also cause deterioration of infrastructure,

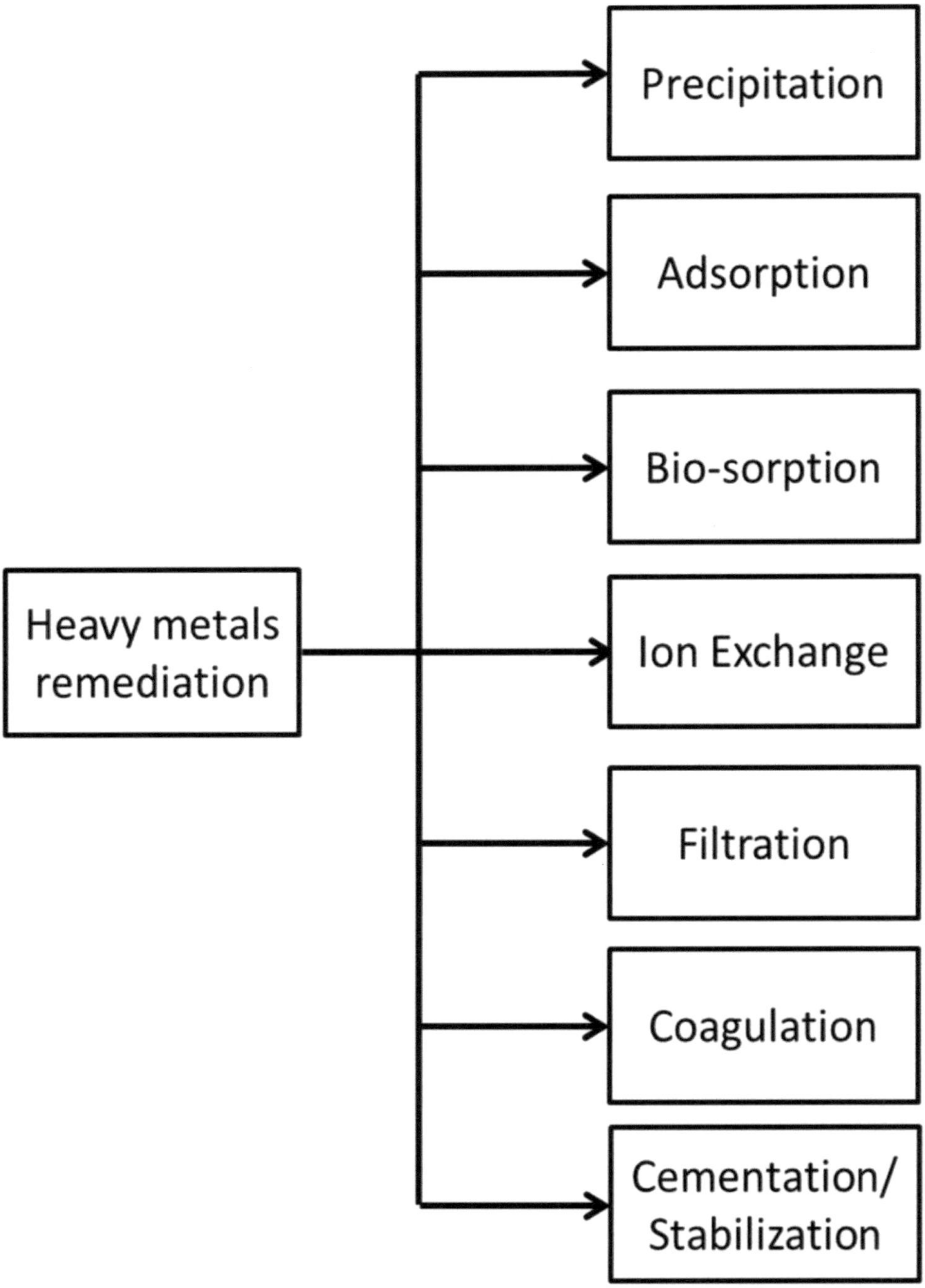

Figure 4. Mechanisms for the removal of heavy metals [20].

corrosion, formation of acid rain, eutrophication and haze [9]. Amongst others, heavy metals such as group 1 metals (Cu, Cd, Pb), group 2 metals (Cr, Mn, Ni, V and Zn) and group 3 metals (Na, K, Ca, Ti, Al, Mg, Fe) originate from industrial areas, traffic and natural sources, respectively [17, 18].

2.4. Mechanisms of remediating heavy metals

Treatment processes for acid mine water typically generate high-density sludge that is heterogeneous due to variety of metals, metalloids and anionic components, and this makes it difficult to dispose the sludge [19]. Recent researches have therefore focused on the recovery of chemical species from acid mine drainage (AMD) and secondary sludge. This is aimed at recovering valuable resources and also enabling easier and safer disposal of the treated sludge, hence reducing their environmental footprints. Disposal of metal ladened waste to landfills and waste retention ponds/heaps lead to secondary pollution of surface and subsurface water resources. It may also lead to soil contamination, hence affecting their productivity [19].

In order to protect the human health, plants, animals, soil and all the compartments of the environment, proper and careful attention should be given to remediation technologies of heavy metals. Most physical and chemical heavy metal remediation technologies require handling of large amounts of sludge, destroy surrounding ecosystems and are very expensive [19] (**Figure 4**).

2.4.1. Precipitation

A variety of alkaline chemical reagents have been used over the years for neutralisation of acid mine drainage (AMD) in order to increase the pH and consequently precipitate and recover the metals. The most common alkaline reagents used for sequential recovery of minerals resources from AMD are limestone ($CaCO_3$), caustic soda (NaOH), soda ash (Na_2CO_3), quicklime (CaO), slaked lime ($Ca(OH)_2$) and magnesium hydroxide ($Mg(OH)_2$) [21]. Some processes have recovered metals at varying pH regimes (**Table 1**) and synthesised commercially valuable materials such as pigments and magnetite [22]. Some minerals are recovered and sold to metallurgical industries, hence off-setting the treatment costs [19].

2.4.2. Adsorption

Adsorption occurs when an adsorbate adheres to the surface of an adsorbent. Due to reversibility and desorption capabilities, adsorption is regarded the most effective and economically

Metal ion	pH	Metal ion	pH	Metal ion	pH
Al^{3+}	4.1	Hg^{2+}	7.3	Cd^{2+}	6.7
Fe^{3+}	3.5	Na^{+}	6.7	Fe^{2+}	5.5
Mn^{2+}	8.5	Pb^{2+}	6.0	Cu^{2+}	5.3
Cr^{3+}	5.3	Zn^{2+}	7.0		

Table 1. pH values at which metals in AMD precipitate [21].

viable option for the removal of metals from aqueous solution. Although efficient, adsorption is not effective with very concentrated solution as the adsorbent easily gets saturated with the adsorbate. It is only feasible for very dilute solutions, is labour intensive because it requires frequent regeneration and it is not selective in terms of metal attenuation [21]. Adsorption is therefore not applied in a large scale of metal remediation.

2.4.3. Ion exchange

Ion exchange is the exchange of ions between two or more electrolyte solutions. It can also refer to exchange of ions on a solid substrate to soil solution. High cation exchange capacity clay and resins are commonly used for the uptake of metals from aqueous solutions. However, this method requires high labour and is limited to certain concentration of metals in the solution. This system also operates under specific temperature and pH. Natural and synthetic clays, zeolites and synthetic resins have been used for removal and attenuation of metals from wastewater [19, 23].

2.4.4. Biosorption

Biosorption refers to the removal of pollutants from water systems using biological materials, and it entails the absorption, adsorption, ion exchange, surface complexation and precipitation. Biosorbents have an advantage of accessibility, efficiency and capacity. This process is readily and easily available. Regeneration is easy, hence making it very favourable. However, when the concentration of the feed solution is very high, the process easily reaches a breakthrough, thus limiting further pollutant removal [24].

2.4.5. Membrane technologies

The use of membrane technologies for the recovery of acid mine drainage is very effective for water that has high concentration of pollutants. It uses the concentration gradients phenomenon or the opposite which is reverse osmosis. There are different types of membranes that are used for mine water treatment including: ultrafiltration, nano-filtration, reverse osmosis, microfiltration and particle filtration [19, 25, 26].

3. Case study of South Africa acid mine drainage

South Africa is well endowed by mineral reserves and this has triggered its immense dependence on mineral resources for gross domestic product and economy. However, the legacy of coal and gold mining has left in its wake serious environmental problems. The major problem is acid mine drainage. Acid mine drainage (AMD) is formed from the hydro-geochemical weathering of sulphide-bearing rocks (pyrite, arsenopyrite and marcasite) in contact with water and oxygen [23, 27]. This reaction is also catalysed by iron (Fe) and sulphur-oxidising microorganisms [28, 29]. In a nutshell, the formation of AMD can be summarised as follows [19, 23, 30, 31]:

$$2FeS_{2(s)} + 7O_{2(g)} + 2H_2O_{(l)} \rightarrow 2Fe^{2+}_{(aq)} + 4SO^{2-}_{4(aq)} + 4H^{+}_{(aq)} \quad (1)$$

The oxidation of sulphide to sulphate solubilises the ferrous iron (Fe(II)), which is subsequently oxidised to ferric iron (Fe(III)):

$$4Fe^{2+}_{(aq)} + O_{2(g)} + 4H^{+}_{(aq)} \rightarrow 4Fe^{3+}_{(aq)} + 2H_2O_{(l)} \quad (2)$$

Either these reactions can occur spontaneously or can be catalysed by microorganisms (sulphur- and iron-oxidising bacteria) that derive energy from the oxidation reaction [26]. The ferric cations produced can also oxidise additional pyrite into ferrous ions:

$$FeS_{(s)} + 14Fe^{3+}_{(aq)} + 8H_2O_{(l)} \rightarrow 15Fe^{2+}_{(aq)} + 2SO^{2-}_{4(aq)} + 16H^{+}_{(aq)} \quad (3)$$

The net effect of these reactions is to produce H^+ and maintain the solubility of the ferric iron [32]. Because of the high acidity and elevated concentration of toxic and hazardous metals, AMD has been a prime issue of environmental concern that has globally raised public concern [33].

The discharge of metalliferous drainage from mining activities has rendered the environment unfit to foster life [22]. Pragmatic approaches need to be developed to counter for this mining legacy that is perpetually degrading the environment and its precious resources [21]. Researches and piloted studies have indicated that active and passive approaches can be successfully adopted to treat acid mine drainage and remove potentially toxic chemical species [23, 31]. The presence of Al, Fe, Mn and sulphates is a prime concern in addition to the trace of Cu, Ni, Pb and Zn [29]. Metalloids of As and earth alkali metal (Ca and Mg) are also present in significant levels [33]. Several studies have shown the feasibility of treating acid mine drainage to acceptable levels as prescribed by different water quality guidelines, but the resultant sludge has been an issue of public concern due to its heterogeneous and complex nature loaded with metal species [23, 34].

Based on that evidence, research studies have been firmly embedded on the recovery of valuable minerals from AMD [19, 23]. There are several mechanisms used for the recovery of chemical components from AMD including: precipitation [35], adsorption [36], biosorption [24], ion exchange [19, 25, 26], desalination [37] and membrane filtration [38, 39]. Out of those techniques, precipitation has been the promising technology due to the ability to handle large volumes of water with very little dosage [35]. Adsorption and ion exchange have a challenge of poor efficiency at elevated concentrations and quick rate of saturation. Membrane technologies have the problem of generating brine that creates another environmental liability. Desalination has a problem of producing salts that has impurities, hence making them unsuitable for utilisation. Freeze desalination has been the promising technology, but it has never been tried in a large scale [19, 23, 34].

3.1. Impacts of heavy metals in South Africa

South Africa's geology is rich in coal and mineral reserves which contain key metals such as gold, platinum and copper. The significant volume of mineral and coal reserves has made mining serve as a backbone in the development and growth of the country's economy. This is evident from the massive number of mines found around the country. However, mining has been noted to cause inimical impacts to the human health, organisms and environment as a whole, with water resources being the most common victim of the pollution [40].

The mining of coal and gold for multilateral uses exposes pyrite to oxidising agents. Iron hydroxide and sulphuric acid are toxic chemical species to living organisms when introduced into water resources (both surface and underground). This deteriorates the natural form of the water bodies and its ability to foster life. Acid mine drainage has very low pH of about <1.4 to >3 [41, 42]; high TDS, EC and other metals in toxic concentration. Previous studies documented the following concentrations in AMD: < 75 ppm to >47,800 acidity; <3560 to >41,700 SO42- ppm; <460 to >12,270 ppm total Fe; <17,400 to 37,700 μg/L Zn; <270 to >13,000 μg/L Cu; <520 to >1500 μg/L Co; <75 to >360 μg/L Ni; <8 to >30 μg/L Pb and 6 to 30 μg/L Cd [41–44].

However, the above-mentioned concentrations depend on the pH of the AMD—concentrations decrease when pH increases. When exposed to such conditions, mortality and diseases are most likely to occur in organisms, as well as other health [45]. In addition, AMD destroys ecosystems of organisms and also negatively impacts on the economy of the country. Heavy metals in active and abandoned mines in South Africa have impacted both surface and underground water.

Parameter	Gold AMD*	Coal AMD**	Neutral drainage†	DWS industrial	DWS irrigation
pH	2.3	2.5	6.5	5.0–10.0	6.5–8.4
EC	22,713	13,980	500	0–250	>540
Na	248.4	70.5	20.1	—	430–460
K	21.6	34.2	29.1	—	—
Mg	2.3	398.9	861.8	—	—
Ca	710.8	598.7	537.5	—	—
Al	134.4	473.9	0.01	—	5.0–20
Fe	1243	8158.2	0.07	0.0–10	5.0–20
Mn	91.5	88.2	25.0	0.0–10.0	0.02–10.0
Cu	7.8	—	—	—	0.2–5.0
Zn	7.9	8.36	0.16	—	1.0–5.0
Pb	6.3	—	—	—	0.2–2.0
Co	41.3	1.89	0.29	—	0.05–5.0
Ni	16.6	2.97	0.21	—	0.2–2.0
SO_4^{2-}	4635	42,862	4603	0–500	—

*Gold mining AMD [44].
**Coalmining AMD.
†Neutral drainage water [40, 42, 45, 47–51].

Table 2. The relevant criteria for discharge of acidic and sulphate-rich water as compared to DWS water quality guidelines.

3.2. Legal requirements of water quality

The National Environmental Management Act (NEMA) 108 of 1998, stipulates that everyone has the right to live in an environment which is safe and unlikely to pose any deleterious effects to their health. The legislative requirements for industrial effluents are primarily governed by the Department of Water Affairs DWS Water Quality Guidelines [46]. This purpose requires that any person who uses water for industrial purposes shall purify or otherwise treat such water in accordance with requirements of DWA [41, 46–48]. The relevant criteria for discharge of acidic and sulphate-rich water are given in **Table 2**.

As shown in **Table 2**, mine effluents in South Africa are dominated by dissolved Fe, Al, Mn, Ca, Na, Mg and traces of Cu, Co, Zn, Pb and Ni. These concentrations are far above the legal requirements.

4. Deleterious effects of acid mine drainage on terrestrial and aquatic ecosystems

The introduction of effluents from mining activities into receiving streams can severely impact aquatic ecosystems through habitat destruction and impairment of water quality. This will eventually lead to reduction in biodiversity of a given aquatic ecosystem and its ability to sustain life. The severity and extent of damage depends on a variety of factors including the frequency of influx, volume and chemistry of the drainage and the buffering capacity of the receiving stream [22, 52–58].

4.1. Acidity

When metals in AMD are hydrolysed, they lower the pH of the water making it unsuitable for aquatic organisms to thrive [52]. AMD is highly acidic (pH 2–4), and this promotes the dissolution of toxic metals [44]. Those toxic species exert hazardous effects on terrestrial and aquatic organisms [23]. Also, if the water is highly acidic, only acidophile microorganisms will thrive on such water with the rest of aquatic organisms migrating to other regions which are conducive to their survival. Many streams contaminated with AMD are largely devoid of life for a long way downstream. To some aquatic organisms, if the pH range falls below the tolerance range, probability of death is very high due to respiratory and osmoregulation failure. Acidic conditions are dominated by H^+ which is adsorbed and pumps out Na from the body which is important in regulating body fluids [23, 52, 53, 56–65].

4.2. Toxic chemical species

Exposure of aquatic and terrestrial organisms to potentially toxic metals and metalloids can have devastating impacts to living organisms [44, 66, 67]. Toxic chemical species present in AMD have been reported to be toxic to aquatic and terrestrial organisms. They are associated with numerous diseases including cancers. Some of these chemical species may accumulate

Element	DWA limit	Ecological impacts of AMD
Al	<0.5 mg/L	Prolonged exposure to aluminium has been implicated in chronic neurological disorders such as *dialysis dementia* and Alzheimer's disease. Severe aesthetic effects (discolouration) occur in the presence of iron or manganese
Fe	<1 mg/L	Severe aesthetic effects (taste) and effects on plumbing (slimy coatings). Slight iron overload possible in some individuals. Chronic health effects in young children and sensitive individuals in the range of 10–20 mg/L, and occasional acute effects towards the upper end of this range
Mn	<0.2 mg/L	Very severe, aesthetically unacceptable staining. Domestic use unlikely due to adverse aesthetic effects. Some chance of manganese toxicity under unusual conditions
Cu	<1 mg/L	Gastrointestinal irritation, nausea and vomiting. Severe taste and staining problems. Severe poisoning with possible fatalities. Severe taste and staining problems
Mg	<200 mg/L	Water aesthetically unacceptable because of bitter taste users if sulphate present. Increased scaling problems. Diarrhoea in most new consumers
Zn	<5 mg/L	Bitter taste; milky appearance. Acute toxicity with gastrointestinal irritation, nausea and vomiting. Severe, acute toxicity with electrolyte disturbances and possible renal damage

Table 3. Effects of selected AMD metals on the health of living organisms.

and be biomagnified in living organisms, hence threatening the life of higher trophic organisms such as birds [68]. Lead causes blood disorders, kidney damage, miscarriages and reproductive disorders and is linked to various cancers. The exposure of living organisms to toxic chemical species in AMD can also lead to nausea, diarrhoea, liver and kidney damage, dermatitis, internal haemorrhage and respiratory problems. Epidemiological studies have shown a significant increase in the risk of lung, bladder, skin, liver and other cancers on exposure to these chemical species. Effects of Al, Fe, Mn, Cu, Mg and Zn on the health of living organisms are summarised in **Table 3** [44, 56, 67].

Author details

Vhahangwele Masindi[1,2]* and Khathutshelo L. Muedi[1]

*Address all correspondence to: masindivhahangwele@gmail.com

1 Council for Scientific and Industrial Research (CSIR), Built Environment (BE), Hydraulic Infrastructure Engineering (HIE), Pretoria, South Africa

2 Department of Environmental Sciences, School of Agriculture and Environmental Sciences, University of South Africa (UNISA), Florida, South Africa

References

[1] Sands P. Principles of International Environmental Law. 2nd ed. London: Cambridge; 2003

[2] Wong MH. Environmental Contamination: Health Risks and Ecological Restoration. United States of America: Taylor & Francis Group; 2012

[3] Salomons W, Forstner U, Mader P. Heavy Metals: Problems and Solutions. Berlin, Germany: Springer-Verlag; 1995

[4] El-Shahawi MS, Hamza A, Bashammakhb AS, Al-Saggaf WT. An overview on the accumulation, distribution, transformations, toxicity and analytical methods for the monitoring of persistent organic pollutants. Talanta. 2010;**80**:1587-1597

[5] Lepp NW. Effect of heavy metal pollution on plants. Metals in the Environment, Pollution Monitoring Series, Applied Science Publishers. Department of Biology. Liverpool, United Kingdom: Liverpool Polytechnic; 2012;**2**

[6] Van Ael E, Covaci A, Blust R, Bervoets L. Persistent organic pollutants in the Scheldt estuary: Environmental distribution and bioaccumulation. Environmental International. 2012;**48**:17-27

[7] Elliott M. Biological pollutants and biological pollution—an increasing cause for concern. Marine Pollution Bulletin. 2003;**46**:275-280

[8] Banfalvi G. Cellular Effects of Heavy Metals. Netherlands, London, New York: Springer; 2011

[9] Herawati N, Suzuki S, Hayashi K, Rivai IF, Koyoma H. Cadmium, copper and zinc levels in rice and soil of Japan, Indonesia and China by soil type. Bulletin of Environmental Contamination and Toxicology. 2000;**64**:33-39

[10] He ZL, Yang XE, Stoffella PJ. Trace elements in agroecosystems and impacts on the environment. Journal of Trace Elements in Medicine and Biology. 2005;**19**(2–3):125-140

[11] Garbarino JR, Hayes HC, Roth DA, Antweiler RC, Brinton TI, Taylor HE. Contaminants in the Mississippi River: Heavy metals in the Mississippi River. Reston, Virginia: U.S. GEOLOGICAL SURVEY CIRCULAR; 1995. pp. 1133

[12] Brady D, Stoll AD, Starke L, Duncan JR. Bioaccumulation of metal cations by *Saccharomyces cerevisiae*. Applied Microbiology and Biotechnology. 1994;**41**:149-154

[13] García-Niño WR, Pedraza-Chaverrí J. Protective effect of curcumin against heavy metals-induced liver damage. Food and Chemical Toxicology. 2014;**69**:182-201

[14] Athar M, Vohora SB. Heavy Metals and Environment. New Delhi: New Age International (P) Limited; 2001

[15] Musilova J, Arvay J, Vollmannova A, Toth T, Tomas J. Environmental contamination by heavy metals in region with previous mining activity. Bulletin of Environmental Contamination and Toxicology. 2016;**97**:569-575

[16] Lee G, Bigham JM, Faure G. Removal of trace metals by coprecipitation with Fe, Al and Mn from natural waters contaminated with acid mine drainage in the Ducktown Mining District, Tennessee. Applied Geochemistry. 2002;**17**(5):569-581

[17] Soleimani M, Amini N, Sadeghian B, Wang D, Fang L. Heavy metals and their source identification in particulate matter (PM2.5) in Isfahan City, Iran. Journal of Environmental Sciences. 2018. In press

[18] Ventura LMB, Mateus VL, de Almeida ACSL, Wanderley KB, Taira FT, Saint'Pierre TD, Gioda A. Chemical composition of fine particles (PM2.5): Water-soluble organic fraction and trace metals. Air Quality, Atmosphere and Health. 2017;**10**:845-852

[19] Nleya Y, Simate GS, Ndlovu S. Sustainability assessment of the recovery and utilisation of acid from acid mine drainage. Journal of Cleaner Production. 2016;**113**:17-27

[20] Yusuf M, Elfghi FM, Zaidi SA, Abdullah EC, Khan MA. Applications of graphene and its derivatives as an adsorbent for heavy metal and dye removal: A systematic and comprehensive overview. RSC Advances. 2015;**5**:50392-50420

[21] Masindi V, Gitari MW, Tutu H. Passive Remediation of Acid Mine Drainage. LAP Lambert Academic Publishing; 2016

[22] Masindi V. A novel technology for neutralizing acidity and attenuating toxic chemical species from acid mine drainage using cryptocrystalline magnesite tailings. Journal of Water Process Engineering. 2016;**10**:67-77

[23] Simate GS, Ndlovu S. Acid mine drainage: Challenges and opportunities. Journal of Environmental Chemical Engineering. 2014;**2**:1785-1803

[24] Silvas FPC, Buzzi DC, Espinosa DCR, Tenório JAS. Biosorption of AMD metals using *Rhodococcus opacus*. Revista Escola de Minas. 2011;**64**:487-492

[25] Buzzi DC, Viegas LS, Rodrigues MAS, Bernardes AM, Tenório JAS. Water recovery from acid mine drainage by electrodialysis. Minerals Engineering. 2013;**40**:82-89

[26] Park S-M, Shin S-Y, Yang J-S, Ji S-W, Baek K. Selective recovery of dissolved metals from mine drainage using electrochemical reactions. Electrochimica Acta. 2015;**181**:248-254

[27] Nordstrom DK, Blowes DW, Ptacek CJ. Hydrogeochemistry and microbiology of mine drainage: An update. Applied Geochemistry. 2015;**57**:3-16

[28] Baker BJ, Banfield JF. Microbial communities in acid mine drainage. FEMS Microbiology Ecology. 2003;**44**:139-152

[29] Hallberg KB. New perspectives in acid mine drainage microbiology. Hydrometallurgy. 2010;**104**:448-453

[30] Amos RT, Blowes DW, Bailey BL, Sego DC, Smith L, Ritchie AIM. Waste-rock hydrogeology and geochemistry. Applied Geochemistry. 2015;**57**:140-156

[31] Johnson DB, Hallberg KB. Acid mine drainage remediation options: A review. Science of the Total Environment. 2005;**338**:3-14

[32] Candeias C, Ávila PF, Ferreira da Silva E, Ferreira A, Salgueiro AR, Teixeira JP. Acid mine drainage from the Panasqueira mine and its influence on Zêzere river (Central Portugal), Journal of African Earth Sciences, 99, Part 2. 2014. pp. 705-712

[33] Akinwekomi V, Maree JP, Zvinowanda C, Masindi V. Synthesis of magnetite from iron-rich mine water using sodium carbonate, Journal of Environmental Chemical Engineering. 2017

[34] Kefeni KK, Msagati TAM, Mamba BB. Acid mine drainage: Prevention, treatment options, and resource recovery: A review. Journal of Cleaner Production. 2017;**151**:475-493

[35] Seo EY, Cheong YW, Yim GJ, Min KW, Geroni JN. Recovery of Fe, Al and Mn in acid coal mine drainage by sequential selective precipitation with control of pH, Catena, 148, Part 1. 2017. pp. 11-16

[36] Masindi V, Gitari MW, Tutu H, DeBeer M. Efficiency of ball milled south African bentonite clay for remediation of acid mine drainage. Journal of Water Process Engineering. 2015;**8**: 227-240

[37] Mulopo J. Continuous pilot scale assessment of the alkaline barium calcium desalination process for acid mine drainage treatment. Journal of Environmental Chemical Engineering. 2015;**3**:1295-1302

[38] Meschke K, Herdegen V, Aubel T, Janneck E, Repke J-U. Treatment of opencast lignite mining induced acid mine drainage (AMD) using a rotating microfiltration system. Journal of Environmental Chemical Engineering. 2015;**3**:2848-2856

[39] Mhamdi M, Elaloui E, Trabelsi-Ayadi M. Adsorption of zinc by a Tunisian Smectite through a filtration membrane. Industrial Crops and Products. 2013;**47**:204-211

[40] Gitari M, Petrik L, Etchebers O, Key D, Iwuoha E, Okujeni C. Treatment of acid mine drainage with fly ash: Removal of major contaminants and trace elements. Journal of Environmental Science and Health - Part A Toxic/Hazardous Substances and Environmental Engineering. 2006;**41**:1729-1747

[41] Van der Linde M, Feris L. Compendium of South African Environmental Legislation. Pretoria, South Africa: Pretoria University Law Press; 2010

[42] Madzivire G, Gitari WM, Vadapalli VRK, Ojumu TV, Petrik LF. Fate of sulphate removed during the treatment of circumneutral mine water and acid mine drainage with coal fly ash: Modelling and experimental approach. Minerals Engineering. 2011;**24**:1467-1477

[43] Africa S, Van der Linde M, Feris L. Compendium of South African Environmental Legislation. Pretoria University Law Press; 2010

[44] Tutu H, McCarthy TS, Cukrowska E. The chemical characteristics of acid mine drainage with particular reference to sources, distribution and remediation: The Witwatersrand Basin, South Africa as a case study. Applied Geochemistry. 2008;**23**:3666-3684

[45] Madzivire G, Petrik LF, Gitari WM, Ojumu TV, Balfour G. Application of coal fly ash to circumneutral mine waters for the removal of sulphates as gypsum and ettringite. Minerals Engineering. 2010;**23**:252-257

[46] Biswas AK, Tortajada C, Izquierdo R. Water Quality Management: Present Situations, Challenges and Future Perspectives. New York, USA: Taylor & Francis; 2014

[47] Madzivire G, Maleka P, Lindsay R, Petrik LF. Radioactivity of mine water from a gold mine in South Africa. WIT Transactions on Ecology and the Environment. 2013;**178**:147-158

[48] Madzivire G, Maleka PP, Vadapalli VRK, Gitari WM, Lindsay R, Petrik LF. Fate of the naturally occurring radioactive materials during treatment of acid mine drainage with coal fly ash and aluminium hydroxide. Journal of Environmental Management. 2014;**133**:12-17

[49] Masindi V, Gitari MW, Tutu H, De Beer M. Application of magnesite–bentonite clay composite as an alternative technology for removal of arsenic from industrial effluents. Toxicological & Environmental Chemistry. 2014:1-17

[50] Gitari WM, Petrik LF, Etchebers O, Key DL, Iwuoha E, Okujeni C. Passive neutralisation of acid mine drainage by fly ash and its derivatives: A column leaching study. Fuel. 2008; **87**:1637-1650

[51] Masindi V, Gitari MW, Tutu H, De Beer M. Application of magnesite–bentonite clay composite as an alternative technology for removal of arsenic from industrial effluents. Toxicological & Environmental Chemistry. 2014;**96**:1435-1451

[52] Torres E, Ayora C, Jiménez-Arias JL, García-Robledo E, Papaspyrou S, Corzo A. Benthic metal fluxes and sediment diagenesis in a water reservoir affected by acid mine drainage: A laboratory experiment and reactive transport modeling. Geochimica et Cosmochimica Acta. 2014;**139**:344-361

[53] Šucha V, Dubiková M, Cambier P, Elsass F, Pernes M. Effect of acid mine drainage on the mineralogy of a dystric cambisol. Geoderma. 2002;**110**:151-167

[54] Peretyazhko T, Zachara JM, Boily JF, Xia Y, Gassman PL, Arey BW, Burgos WD. Mineralogical transformations controlling acid mine drainage chemistry. Chemical Geology. 2009; **262**:169-178

[55] Netto E, Madeira RA, Silveira FZ, Fiori MA, Angioleto E, Pich CT, Geremias R. Evaluation of the toxic and genotoxic potential of acid mine drainage using physicochemical parameters and bioassays. Environmental Toxicology and Pharmacology. 2013;**35**:511-516

[56] Mohapatra BR, Douglas Gould W, Dinardo O, Koren DW. Tracking the prokaryotic diversity in acid mine drainage-contaminated environments: A review of molecular methods. Minerals Engineering. 2011;**24**:709-718

[57] Martins M, Santos ES, Faleiro ML, Chaves S, Tenreiro R, Barros RJ, Barreiros A, Costa MC. Performance and bacterial community shifts during bioremediation of acid mine drainage from two Portuguese mines. International Biodeterioration & Biodegradation. 2011;**65**:972-981

[58] Levings CD, Varela DE, Mehlenbacher NM, Barry KL, Piercey GE, Guo M, Harrison PJ. Effect of an acid mine drainage effluent on phytoplankton biomass and primary production at Britannia Beach, Howe Sound, British Columbia. Marine Pollution Bulletin. 2005; **50**:1585-1594

[59] Sun M, Ru X-R, Zhai L-F. In-situ fabrication of supported iron oxides from synthetic acid mine drainage: High catalytic activities and good stabilities towards electro-Fenton reaction. Applied Catalysis B: Environmental. 2015;**165**:103-110

[60] Strosnider WH, Nairn RW. Effective passive treatment of high-strength acid mine drainage and raw municipal wastewater in Potosí, Bolivia using simple mutual incubations and limestone. Journal of Geochemical Exploration. 2010;**105**:34-42

[61] Mitsch WJ, Wise KM. Water quality, fate of metals, and predictive model validation of a constructed wetland treating acid mine drainage. Water Research. 1998;**32**:1888-1900

[62] Mapanda F, Nyamadzawo G, Nyamangara J, Wuta M. Effects of discharging acid-mine drainage into evaporation ponds lined with clay on chemical quality of the surrounding soil and water. Physics and Chemistry of the Earth, Parts A/B/C. 2007;**32**:1366-1375

[63] Mahmoud KK, Leduc LG, Ferroni GD. Detection of Acidithiobacillus ferrooxidans in acid mine drainage environments using fluorescent in situ hybridization (FISH). Journal of Microbiological Methods. 2005;**61**:33-45

[64] Gray NF, Delaney E. Measuring community response of bentic macroinvertebrates in an erosional river impacted by acid mine drainage by use of a simple model. Ecological Indicators. 2010;**10**:668-675

[65] Gray JB, Vis ML. Reference diatom assemblage response to restoration of an acid mine drainage stream. Ecological Indicators. 2013;**29**:234-245

[66] Skoczyńska-Gajda S, Labus K. Acid mine drainage within the abandoned lignite mining area-Muskau Arch. Biuletyn Państwowego Instytutu Geologicznego. 2011:643-650

[67] Rubin H, Rubin K, Siodlak A, Skuza P. Assessment of contamination of the bottom sediments of the Stola River with selected metals and metalloids within the urban-industrial area of Tarnowskie Góry. Biuletyn Państwowego Instytutu Geologicznego. 2011:615-624

[68] Jooste S, Thirion C. An ecological risk assessment for a South African acid mine drainage. Water Science and Technology. 1999;**39**:297-303

Chapter 8

Estimate of Heavy Metals in Soil Using Combined Geochemistry and Field Spectroscopy in Miyi Mining Area

Jian Ji, Fang Yao, Chen Qian-Yu and Tian Heng-Yu

Additional information is available at the end of the chapter

http://dx.doi.org/10.5772/intechopen.73663

Abstract

Heavy metal-contaminated soil and water is a major environmental issue in the mining areas. However, as the heavy metals migrate frequently, the traditional method of estimating the soil's heavy metal content by field sampling and laboratory chemical analysis followed by interpolation is time-consuming and expensive. This chapter intends to use field hyperspectra to estimate the heavy metals in the soil in Bai-ma, De-sheng and YuanBaoshan mining areas, Miyi County, Sichuan Province. By analyzing the spectra of soil, the spectral features derived from the spectra of the soils can be found to build the models between these features and the contents of Mn and Co in the soil by using the linear regression method. The spectral features of Mn are 2142 and 2296 nm. The spectral features of Co are 1918, 1922 and 2205 nm. With these feature spectra, the best models to estimate the heavy metals in the study area can be built according to the maximal determination coefficients (R^2). The determination coefficients (R^2) of the models of retrieving Mn and Co in the soil are 0.645 and 0.8, respectively. The model significant indexes of Mn and Co are 2.04507E-05 and 7.73E-06. These results show that it is feasible to predict contaminated heavy metals in the soils during mining activities for soil remediation and ecological restoration by using the rapid and cost-effective field spectroscopy.

Keywords: contaminated heavy metals in the soils, spectral measured, spectral analysis

1. Introduction

Due to the development of industries in recent decades, the demand for mineral resources is also growing. However, the mining and post-processing of mineral resources will increase the heavy metals that permeate and accumulate in the soil. These heavy metals are in abundance, in terms of persistence and toxicity, which can inhibit soil functions and increase the soil pollution [1–3]. And when accumulated to harmful levels, the heavy metals in soil may pose an environmental risk and threaten human health through contaminating the food chain, water and air [4–6]. Thus, the pollution of soil by heavy metals is considered as one of the major environmental problems, and the monitoring of heavy metal content is very important for environment management in mining areas.

The traditional method for estimating the heavy metal contents involves the field data measure and laboratory analysis of soil samples. Although direct sampling can provide an accurate measurement of both the intensity and diversity of soil contaminants at specific sites, however, these procedures are often time-consuming and costly, and it can only analyze the point samples [7–9]. So rapid, periodic monitoring of heavy metals in the areas vulnerable to pollution is important. The development of remote-sensing technology, especially hyperspectral, provides a possibility for the rapid and large-scale distribution of heavy metals, which can acquire the continuous spectrum of the target. Many studies have shown that the spectral curves of heavy metal-contaminated soil and the spectral curves of uncontaminated soil have a difference [10, 11]. Although soil heavy metals are spectrally inactive, their relationships with spectrally active soil properties, such as clay and Fe oxides, may allow for their visible and near-infrared diffuse reflectance spectroscopy (VNIRS) prediction [12]. Thus, the relationship between heavy metal content and soil spectrum was used to establish heavy metal retrieval model to predict and map the heavy metal content in the relevant areas [11, 13–15].

In this chapter, the spectral sampling of soil samples was obtained by ASD Fieldspec III spectroradiometer and the contents of Mn and Co were measured by chemical analysis. Then, the feature spectra can be obtained from the results of spectroscopic analysis to establish the heavy metals' retrieval models. Then, the parameters of these models can used to explore the feasibility of using hyperspectral data to retrieve soil heavy metals for soil remediation and ecological restoration.

2. Materials and methods

2.1. Study area

Miyi Country (26°42′–27°10′ N, 101°44′–102°15′ E) is located in the north of Panzhihua City, the southwest of Sichuan Province. The terrain is high in the south and low in the north. In this chapter, the Bai-ma, De-sheng and Yuan Bao-shan mining areas in Miyi County are selected as the study areas, as shown in **Figure 1**. The study area is located in the east of the Yalong River and northwest of the Anning River. The environment in the mining area and surroundings have been contaminated by heavy metals which can enter into the soil through discharge and infiltration and which exist in the waste residue and waste liquor generated in the mining process, especially Mn and Co.

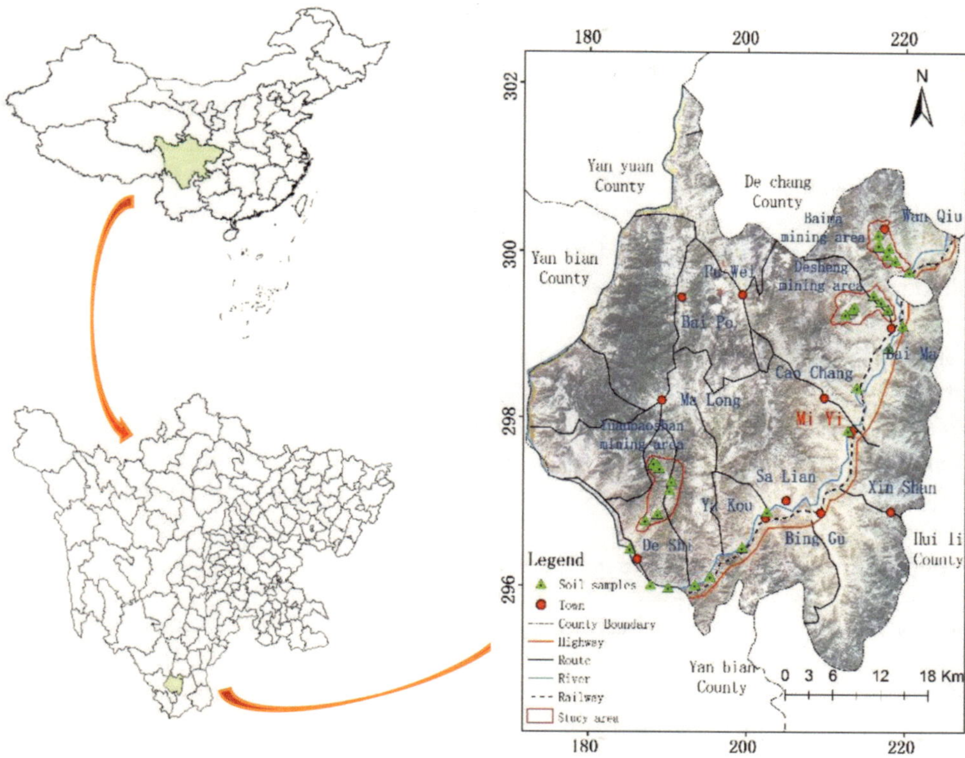

Figure 1. The study area locations of sampling points.

2.2. Data collection

The spectra of 55 soil samples were collected in June 2015; the locations of the sample points are shown in **Figure 1**. Meanwhile, 32 out of the 55 soil samples had been chemically analyzed for the Mn and Co contents by conventional digestion methods using inductively coupled plasma mass spectrometer (ICP-MS). The ICP-MS is the most popular ion source in analytical chemistry for elemental mass spectrometry. In ICP-MS, a mass spectrometer is coupled to an ICP torch by an interface including sampler and skimmer cones so that representative samples of the plasma can be transmitted through its orifices to the mass analyzer [16].

The soil sample's spectrum was obtained from a high spectral resolution ASD Fieldspec III spectroradiometer, which covers the visible and near-infrared (350–2500 nm) region and offers a spectral resolution between 3 and 10 nm, interpolated to 1 nm. Illumination was provided by an ASD high-reflectance probe when collecting soil spectra in the field, while a halogen bulb was used as the light source while collecting water spectra in the laboratory. Each sample was measured three times and the average value was calculated afterwards [11].

2.3. Spectral pre-processing

The soil spectra may contain noise or error that was introduced by operating in situ measurement instruments improperly or using an in situ measurement instrument that is not calibrated properly [17]. And, heavy metals are spectrally featureless in the visible and near-infrared parts of the electromagnetic spectrum. Thus, a serious predicament is observed while dealing with the analysis of overlapping bands of the analytes and interferences which make the extraction of qualitative and quantitative data difficult [18]. Pre-processing of the spectrum is often required to reduce the effect of noise and enhance the spectral signature. Meanwhile, Savitzky–Golay differentiation is a commonly used spectral pre-treatment method, and in practice the first and second derivatives eliminate the interference of the baseline or background, improve sensitivity and detect and enhance minor or subtle spectral features [18, 19]. Obtained spectra were continuum removed and normalized to enhance the spectral absorption features. The continuum that is a convex hull of straight-line segments is fitted over a reflectance spectrum and subsequently removed by division or ratioing relative to the complete reflectance spectrum [8].

2.4. Spectral analysis and model development

In view of the weak relationship between soil spectroscopy and heavy metals, the logarithmic treatment of feature bands can be used to enhance their relationship [8]. In this chapter, the reflectivity of all bands is extracted to create a single-band reflectivity matrix. The band reflectivity from the second band to the last band was selected as the outer loop and the first band to the penultimate band was selected as the inner loop. Then, the band ratio matrix can be obtained when the outer loop is divided by the inner loop, and the band normalization matrix can be obtained by the difference of reflectivity divided by the summation between the outer loop band reflectance and the inner loop, and the multivariate analysis matrix of two bands and three bands can be obtained by combining two or three output feature bands randomly. Then, these matrices were used for Pearson correlation analysis with the soil heavy metal content matrices. The Pearson correlation coefficient is a measure of the linear dependence (correlation) between two variables X and Y, the greater the absolute value of the correlation coefficient, the greater the correlation between the two variables [20, 21]. Thus, it is possible to predict heavy metal contents in the soil with the high correlation between heavy metal content and soil spectrum [22].

In this chapter, the methods of smoothed, the first derivative, the second derivative and the continuum removal of the spectral data were performed by the View Spec Pro and ENVI to eliminate background noise and enhance the spectral feature. The methods of ratio, normalization and multivariate analysis are used to enhance the correlation between heavy metal content and feature spectra. Finally, the IBM SPSS software was used to establish retrieval model.

3. Results

3.1. Geochemical analysis

Mn and Co are the predominant heavy metals in mining waste in the study area, so they were selected as indicators of the environmental impacts from mining activity. The chemical analysis

http://dx.doi.org/10.5772/intechopen.73663

results of soil samples are shown in **Table 1**. From **Table 1**, we can see that the contents of Mn made a great difference with the contents of Co in soil. The standard deviation (SD) of Mn in the soil was relatively high (773.6), which indicates that its concentrations are of little great difference, while the contents of Co in the soil is of small difference.

3.2. Spectral analysis

The visual inspection of the measured soil spectra with different pre-treatment methods showed a significant difference. The original spectral curve and the curves pre-processed by the first derivative, the second derivative and the continuum removal method of soil samplings in the study area are shown in **Figure 2**. The feature spectra can be obtained from these spectrum curves, as shown in **Table 2**. From **Figure 2**, we can see that only 12 feature spectra can be selected as the feature bands from the original soil spectral curve at 473, 791, 1395, 1413, 1926 nm and so on; however, more feature spectra can be selected as feature bands from the soil spectral curves after pre-processing at 584, 1382, 1396, 1403, 1421, 1452, 1890, 1906 nm and so on.

	Max	Min	Mean	SD
Mn	4114.0	531.3	1536.1	773.6
Co	111.9	6.00	43.5	25.87

Table 1. Content of heavy metals in soil statistics (unit: mg/kg).

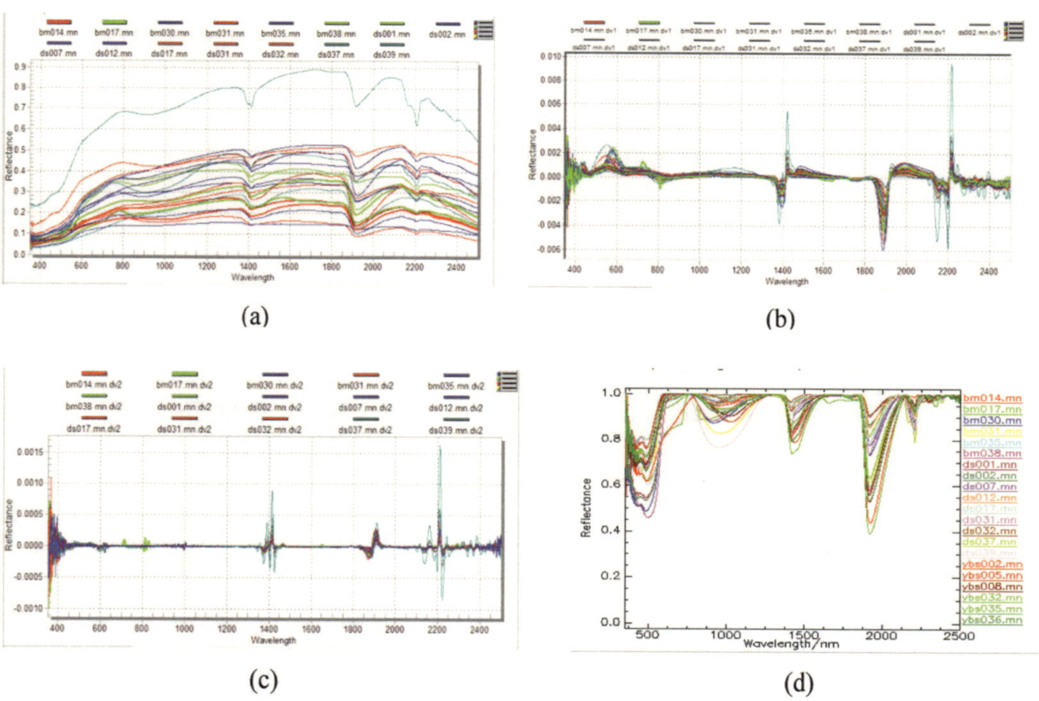

Figure 2. Reflectance spectra of soil samples. (a) Raw reflectance spectral, (b) first-derivative reflectance spectral, (c) second-derivative reflectance spectral, (d) continuum removed reflectance spectral.

	Original spectra	1st spectra	2nd spectra	Continuum removed spectra
Feature spectra	473, 791, 1395, 1413, 1854, 1926, 2136, 2170, 2208, 2243, 2259, 2320,	439, 465, 549, 584, 1382, 1396, 1403, 1421, 1452, 1482, 1766, 1890, 1926, 1993, 2121, 2151, 2172, 2200, 2215, 2234, 2280, 2361, 2375, 2398	550, 590, 625, 995, 1006, 1374, 1393, 1403, 1411, 1425, 1466, 1883, 1906, 2138, 2163, 2196, 2209, 2220, 2240, 2291, 2368, 2387	452, 486, 627, 765, 810, 962, 1029, 1285, 1414, 1698, 1786, 1835, 1918, 1922, 2142, 2205, 2236, 2267, 2296, 2342, 2371, 2386, 2411

Table 2. Feature spectra of four kinds of curves (unit: nm).

3.3. Model development

The spectral features selected from the reflectance spectra are used for spectral analysis and the band combination with the maximum Pearson correlation coefficient (R) is selected as the feature band to build the inversion models of heavy metals. And the regression equations of heavy metals' concentrations in the soil are presented in **Table 3**. For the heavy metal in the soil, the determination coefficients (R^2) of the regression equations are Mn: 0.645 and Co: 0.8. And the determination coefficient (R^2) of the regression equations indicates that the measured heavy metals have a strong relationship with spectral features. Specifically, the ratio of bands at 2124 and 2296 nm has a strong relationship with the contents of Mn, and the bands at 1918, 1922 and 2205 nm have strong relationships with the contents of Co. And the significance indexes of these regression equations are less than 0.05.

The relationship of the measured and predicted concentrations of Mn and Co in soil is shown in **Figure 3**. From the scatter diagrams, we can see that there is a good linear relationship between the measured and predicted concentrations of Mn and Co.

The contents of 12 test soil samples of Mn and Co can be calculated by these regression equations. Then the F-test was carried out to validate the feasibility of these regression models for predicting heavy metal contents, as shown in **Table 4**. From the statistics, we can see that the difference of the mean values of Mn and Co is smaller and the P-value of the F-test between measured and predicted values of Co and Mn in the soil is less than 0.05. This indicates that the models can be used to predict the heavy metal contents in the study area.

	Feature bands (nm)	R^2	Regression equations	Significance F
Mn	2142, 2296	0.645	$Y = -33749.8 + 34703.04X$	2.04507E − 05
Co	1918, 1922, 2205	0.800	$Y = -235.03-7507.17X1 + 7452.81X2 + 333.65X3$	7.73E − 06

Note: X corresponding to R2142/R2296; X1, X2, X3 corresponding to R2121, R2234, R2398, respectively.

Table 3. Spectra parameters and regression equation of Mn and Co.

http://dx.doi.org/10.5772/intechopen.73663

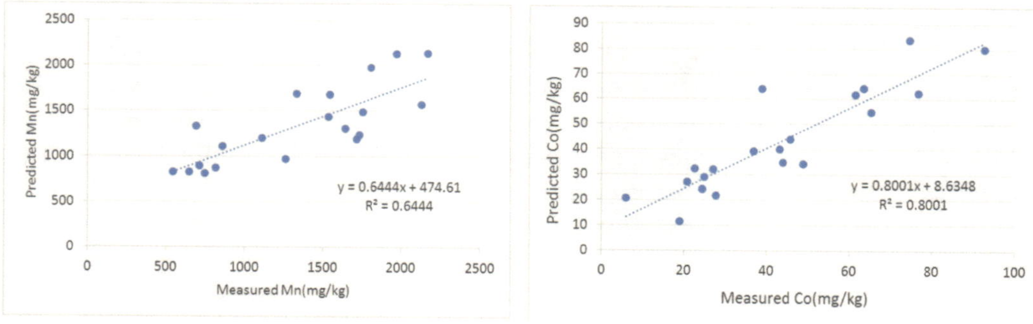

(a) (b)

Figure 3. Scatter plots of the measured values and predicted values: (a) Mn and (b) Co.

	Measurement			Prediction			P(F ≤ f)
	Max	Min	Mean	Max	Min	Mean	
Mn	4114	531.3	1871.6	2290.1	924.9	1474.7	0.003
Co	111.9	13.1	44.0	82.2	22.1	48.4	0.008

Table 4. Summary statistics of measured and predicted heavy metal concentration and F-test (concentration unit: mg/kg).

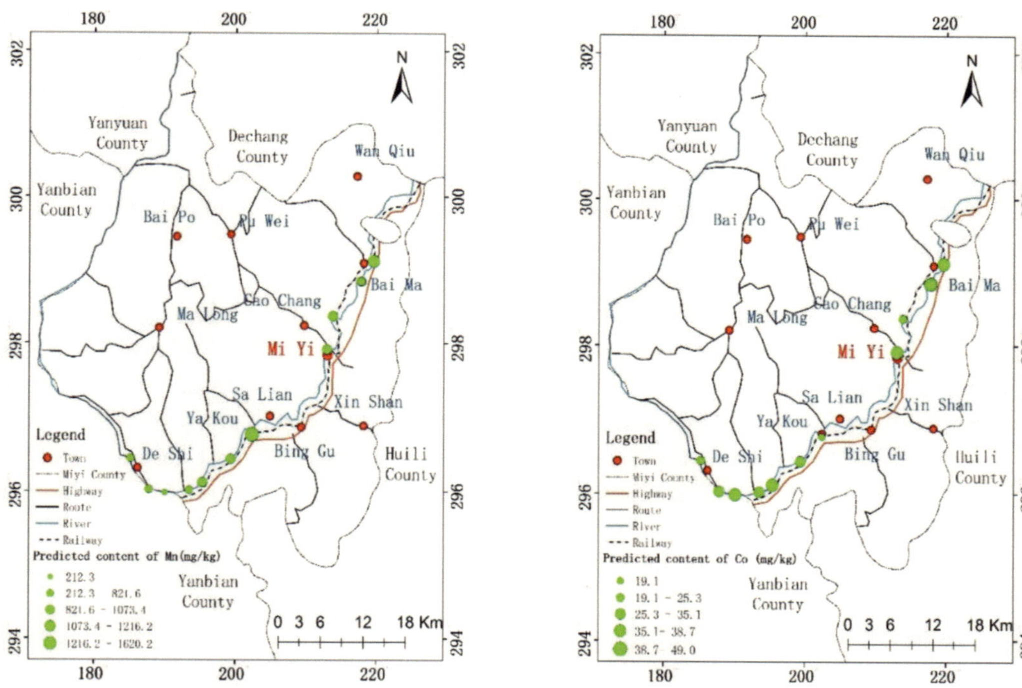

(a) (b)

Figure 4. Predicted content of Mn and Co. (a) The predicted content of Mn and (b) the predicted content of Co.

The results of the predicted content of Mn and Co in soil are shown in **Figure 4**. From the content distribution of the test samplings, we can see that the sampling sites are mainly distributed along the Anning River area. From **Figure 4(a)**, we can see that the content of Mn in the vicinity of Yakou Town is the largest, and it is decreased with the water that flows southward. This indicates that the water has a great influence on the distribution of the content of Mn. From **Figure 4(b)**, we can see that the contents of Co are high in Yuanbaoshan county and Desheng mining area. This indicates that the water has a poor influence on the distribution of the content of Co. The reasons of the high content of Co are (1) the leakage of minerals and slag during transportation and (2) the combustion of Co which is not complete.

4. Discussion

The roughness of soil surface, molecular vibration and electron transition can be changed by the particles of heavy metals adsorbed by the soil organic matter, which makes it is possible to use the soil spectrum to invert the soil heavy metals. The quantitative relationship between the spectrum and the heavy metal content was established by using spectral analysis and chemical analysis of the soil samples. Then the heavy metal content of Co and Mn can be obtained to provide prediction data for regional soil quality research and treatment. From **Figure 4**, we can see that the content distribution of each heavy metal is irregular in the study area for the influences of the river, vegetation and mineral transportation. The flow of river could help transform gradients to the location and extent of heavy metal pollution. The absorption of vegetation could reduce the heavy metal pollution in the soil and the mineral transportation could lead to the jumping change of heavy metal pollution.

Compared to traditional methods, the field of hyperspectral method has many advantages such as fast, efficient, wide coverage and nondestructive to estimate the heavy metals' contents. It can provide predictive data for mine environment monitoring to improve the efficiency of monitoring and management of mine area and protect the surrounding residents' normal life quality. But it still needs much time to collect spectral data and build models, and the monitoring area is limited, and the field sample collection and spectral measurements may contain errors. Therefore, the following research work is to (1) consider the influence of temperature, altitude, weather and other factors on the spectrum to improve the accuracy of the model, (2) obtain the spectral data at the same time when the field data collection is obtained and (3) acquire the Hyperion or AVIRIS imagery of the study area but considering the influence of vegetation, rock and atmosphere on soil spectrum.

5. Conclusions

In this chapter, a fast and convenient method to get the heavy metals' contents in the soils of the study area is described, which can provide a prediction for the eco-remediation of heavy metals in the mining area, leading to assign the human and other resources properly. So the time of remediating the heavy metals contaminated soils can be shortened and prevent the further spread of heavy metals in the soils.

Acknowledgements

The authors would like to thank Wang Jun-wei from Space Star Technology Co., LTD., and Yang Wen-bing from China Nonferrous Metals in Changsha Survey and Design Institute Co., LTD., for the help in the collection of spectra data. The authors would also like to thank Song Lian from Nanjing University and Gao Xiao-jie from the National Geographic Information Bureau of Surveying and Mapping for their help in the chapter's literal correction.

Author details

Jian Ji*, Fang Yao, Chen Qian-Yu and Tian Heng-Yu

*Address all correspondence to: jianji@21cn.com

Key Laboratory of Geoscience Spatial Information Technology, Ministry of Land and Resources of the PR China, Chengdu University of Technology, Chengdu, China

References

[1] Ding Q et al. Effects of natural factors on the spatial distribution of heavy metals in soils surrounding mining regions. Science of the Total Environment. 2017;**578**:577-585. DOI: 10.1016/j.scitotenv.2016.11.001

[2] Li N et al. Effects of double harvesting on heavy metal uptake by six forage species and the potential for phytoextraction in field. Pedosphere. 2016;**26**(5):717-724. DOI: 10.1016/S1002-0160(15)60082-0

[3] Yang Y et al. Risk assessment of heavy metal pollution in sediments of the Fenghe River by the fuzzy synthetic evaluation model and multivariate statistical methods. Pedosphere. 2016;**26**(3):326-334. DOI: 10.1016/S1002-0160(15)60046-7

[4] Cui X et al. Concentrations of heavy metals in suburban horticultural soils and their uptake by *Artemisia selengensis*. Pedosphere. 2015;**25**(6):878-887. DOI: 10.1016/S1002-0160(15)30068-0

[5] Liu M et al. Monitoring stress levels on rice with heavy metal pollution from hyperspectral reflectance data using wavelet-fractal analysis. International Journal of Applied Earth Observation and Geoinformation. 2011;**13**(2):246-255. DOI: 10.1016/j.jag.2010.12.006

[6] Zheng R et al. Land use effects on the distribution and speciation of heavy metals and arsenic in coastal soils on Chongming Island in the Yangtze River Estuary, China. Pedosphere. 2016;**26**(1):74-84. DOI: 10.1016/S1002-0160(15)60024-8

[7] Schuerger AC et al. Comparison of two hyperspectral imaging and two laser-induced fluorescence instruments for the detection of zinc stress and chlorophyll concentration in bahia grass (*Paspalum notatum* Flugge.). Remote Sensing of Environment. 2003;**84**(4): 572-588. DOI: 10.1016/S0034-4257(02)00181-5

[8] Choe E et al. Mapping of heavy metal pollution in stream sediments using combined geochemistry, field spectroscopy, and hyperspectral remote sensing: A case study of the Rodalquilar mining area, SE Spain. Remote Sensing of Environment. 2008;**112**(7):3222-3233. DOI: 10.1016/j.rse.2008.03.017

[9] Zhuang P et al. Heavy metal contamination in soil and soybean near the Dabaoshan Mine, South China. Pedosphere. 2013;**23**(3):298-304. DOI: 10.1016/S1002-0160(13)60019-3

[10] Trevors J. Water, air, and soil pollution. An International Journal of Environmental Pollution. 2010;(Suppl 1):1

[11] Song L et al. Estimate of heavy metals in soil and streams using combined geochemistry and field spectroscopy in Wan-sheng mining area, Chongqing, China. International Journal of Applied Earth Observation and Geoinformation. 2015;**34**:1-9. DOI: 10.1016/j.jag.2014.06.013

[12] St. Luce M et al. Visible near infrared reflectance spectroscopy prediction of soil heavy metal concentrations in paper mill biosolid- and liming by-product-amended agricultural soils. Geoderma. 2017;**288**:23-36. DOI: 10.1016/j.geoderma.2016.10.037

[13] Liu M et al. Estimating regional heavy metal concentrations in rice by scaling up a field-scale heavy metal assessment model. International Journal of Applied Earth Observation and Geoinformation. 2012;**19**:12-23. DOI: 10.1016/j.jag.2012.04.014

[14] Liu Y et al. Heavy metal contamination of agricultural soils in Taiyuan, China. Pedosphere. 2015;**25**(6):901-909. DOI: 10.1016/S1002-0160(15)30070-9

[15] Liu M et al. Regional heavy metal pollution in crops by integrating physiological function variability with spatio-temporal stability using multi-temporal thermal remote sensing. International Journal of Applied Earth Observation and Geoinformation. 2016;**51**:91-102. DOI: 10.1016/j.jag.2016.05.003

[16] Aghaei M, Lindner H, Bogaerts A. Optimization of operating parameters for inductively coupled plasma mass spectrometry: A computational study. Spectrochimica Acta Part B: Atomic Spectroscopy. 2012;**76**:56-64. DOI: 10.1016/j.sab.2012.06.006

[17] John RJ. Introductory Digital Image Processing: A Remote Sensing Perspective. 3rd ed. United States: Pearson Education; 2007

[18] Parmar A, Sharma S. Derivative UV-vis absorption spectra as an invigorated spectrophotometric method for spectral resolution and quantitative analysis: Theoretical aspects and

analytical applications: A review. TrAC Trends in Analytical Chemistry. 2016;**77**:44-53. DOI: 10.1016/j.trac.2015.12.004

[19] Zheng KY et al. Pretreating near infrared spectra with fractional order Savitzky-Golay differentiation (FOSGD). Chinese Chemical Letters. 2015;**26**(3):293-296. DOI: 10.1016/j.cclet.2014.10.023

[20] Erdeljić V et al. Distributed lags time series analysis versus linear correlation analysis (Pearson's R) in identifying the relationship between antipseudomonal antibiotic consumption and the susceptibility of Pseudomonas aeruginosa isolates in a single Intensive Care Unit of a tertiary hospital. International Journal of Antimicrobial Agents. 2011;**37**(5):467-471. DOI: 10.1016/j.ijantimicag.2010.11.030

[21] Susan Prion ERNC, Katie Anne Haerling PR. Making sense of methods and measurement: Pearson product-moment correlation coefficient. Clinical Simulation in Nursing. 2014;(11):587-588

[22] Chen T et al. Identification of soil heavy metal sources and improvement in spatial mapping based on soil spectral information: A case study in northwest China. Science of the Total Environment. 2016;**565**:155-164. DOI: 10.1016/j.scitotenv.2016.04.163

Chapter 9

Tempospatial Distribution, Gas: Solid Partition, and Long-Range Transportation of Atmospheric Mercury at an Industrial City and Offshore Islands

Yi-Hsiu Jen and Chung-Shin Yuan

Additional information is available at the end of the chapter

http://dx.doi.org/10.5772/intechopen.74051

Abstract

This chapter measured atmospheric mercury from two cases of small-scale regions to large-scale regions, and further investigated the tempospatial variation of atmospheric mercury, gas-particulate partition, the transportation routes of mercury, and the comparison of mercury concentration in urban areas and stationary sources. In a heavily polluted industrial city, Kaohsiung, field measurement results showed that total gaseous mercury (TGM) and Hg_p concentrations were in the range of 2.38–9.41 and 0.02–0.59 ng/m^3 with the highest concentrations of 9.41 and 0.59 ng/m^3, respectively. Moreover, the partition of atmospheric mercury was apportioned as 92.71–99.17% TGM and 0.83–7.29% Hg_p. The hot spots of atmospheric mercury were allocated at two regions in Kaohsiung City, including a steel industrial complex in the south and a petrochemical industrial complex in the north. In a coastal site of the Penghu Islands, the field measurement results showed that the average TGM concentration during the monitoring periods was 3.17 ± 1.17 ng/m^3 with the range of 1.17–8.63 ng/m^3, as the highest concentration being observed in spring, while the average TGM concentrations in the daytime were typically higher than that at nighttime. Therefore, prevailing wind direction and air mass transportation routes potentially played critical roles on the variation of TGM concentration at the Penghu Islands.

Keywords: atmospheric mercury, tempospatial variation, gas-particulate partition, backward trajectory simulation, long-range transportation

1. Introduction

The United States had already listed 189 hazardous air pollutants (HAPs) for control in Title III of the Clean Air Act Amendments (CAAA) of 1990 [1, 2]. Among them, 11 are the toxic heavy metals As, Be, Cd, Co, Cr, Hg, Ni, Mn, Pb, Sb, and Se. In 2005, the U.S.A. Environmental Protection Agency (USEPA) became the first agency to amend the Maximum Achievable Control Technology (MACT) standards into growth control quotas for mercury emissions in order to increase the original goal of a 30% reduction in mercury emissions to 70% by 2018 [2]. This was also done to encourage countries around the world to revise their own reduction goals for mercury emissions. Till now, the pollutants of mercury still cannot be completely removed by using chemical method, and continue to endanger the health of humans and other organisms.

Mercury (Hg) is a persistent, toxic, and bio-accumulative heavy metal, and is currently regulated by the USEPA and the United Nations Environment Programme (UNEP) [3–10], and atmospheric mercury has been claimed by UNEP as another global environmental issue followed greenhouse gases (GHGs) [1, 2]. Because of its unique physicochemical properties and potential for long-range transportation via the atmospheric dispersion pathway, it could be deposited worldwide [4, 5]. Due to its characteristics of persistence and bioaccumulation through food chain, mercury could damage the brain and nervous systems in human body. Thus, many countries are becoming increasingly concerned about atmospheric mercury pollution recently.

Accordingly, this chapter aimed to investigate the tempospatial variation and partition of atmospheric mercury at an industrial metropolitan area, and to explore the long-range transportation path at the marine boundary layer (MBL), and to compare the atmospheric mercury concentration level with other major cities all over the world.

2. Tempospatial variation and partition of atmospheric mercury in ambient air of an industrial city

2.1. Background

The mercury emission was mainly generated from combustion sources and metallurgy smelting processes in Taiwan. Among them, the combustion sources include coal-fired boilers, municipal and industrial waste incinerators, petrochemical refineries, cremators, etc., while the metallurgy smelting processes include integrated steel plants, electric arc furnace plants, secondary metal smelters, etc. The economic development of metro Kaohsiung mainly relied on heavy industries located adjacent to the metropolitan area. Coupled with heavy traffics, it causes poor ambient air quality of Kao-Ping Air Quality Zone in Taiwan. Moreover, metro Kaohsiung is the largest and most intensive industrial city in Taiwan, accounting for about 70% of the stationary emission sources, where has two large-scale coal-fired power plants, an integrated steel plant, several metallurgy smelters, and three petrochemical refineries. It would cause quite high emissions of mercury in metro Kaohsiung, Taiwan. In this chapter, the field measurement of atmospheric mercury was designated to investigate the tempospatial variation and partition of gaseous and particulate mercury during the wet and dry seasons, and further correlated atmospheric mercury with meteorological parameters and criteria air pollutants measured in Kaohsiung City, located at the coastal region of southern Taiwan.

2.2. Selection and description of sampling sites

Field measurement of atmospheric mercury speciation and concentration was conducted at a coastal background site and six sensitivity sites in Kaohsiung City, including Nan-zhi (22°44′00″ North latitude, 120°19′41″ East longitude, S1), Ren-wu (22°41′20″ North latitude, 120°19′57″ East longitude, S2), Zuoying (22°40′29″ North latitude, 120°17′34″ East longitude, S3), Cia-jin (23°37′56″ North latitude, 120°17′16″ East longitude, S4), Cianjhen (22°36′18″ North latitude, 120°18′30″ East longitude, S5), and Hsiao-kang (22°33′57″ North latitude, 120°20′15″ East longitude, S6). The coastal background site along coastline far away from the emission sources is located at the campus of National Sun Yat-Sen University (22°37′38″ North latitude, 120°16′01″ East longitude). The location of the coastal background site and six sensitivity sites for sampling total gaseous mercury (TGM) and particulate mercury (Hg_p) in Kaohsiung City is shown in **Figure 1**.

These sensitivity sites were mainly located at the ambient air quality monitoring stations in Kaohsiung City. Among them, sites S5–S6 were nearby a steel industrial complex in southern Kaohsiung, sites S1 and S2 were close to a petrochemical industrial complex in northern Kaohsiung. Active sampling of TGM and Hg_p were conducted consecutively for 24 h at each site in the wet and dry seasons from June to December of 2010 in Kaohsiung City. Among them, wet season started from June to September, while dry season started from October to December. This study intended to investigate the seasonal variation, the spatial distribution, and the partition of TGM and Hg_p at a coastal background site and six sensitivity sites in an industrial city.

2.3. Seasonal variation and partition of TGM and Hg_p

Figure 2 illustrates the seasonal variation of TGM and Hg_p concentrations at the northern and southern Kaohsiung as well as the coastal background site during the wet and dry

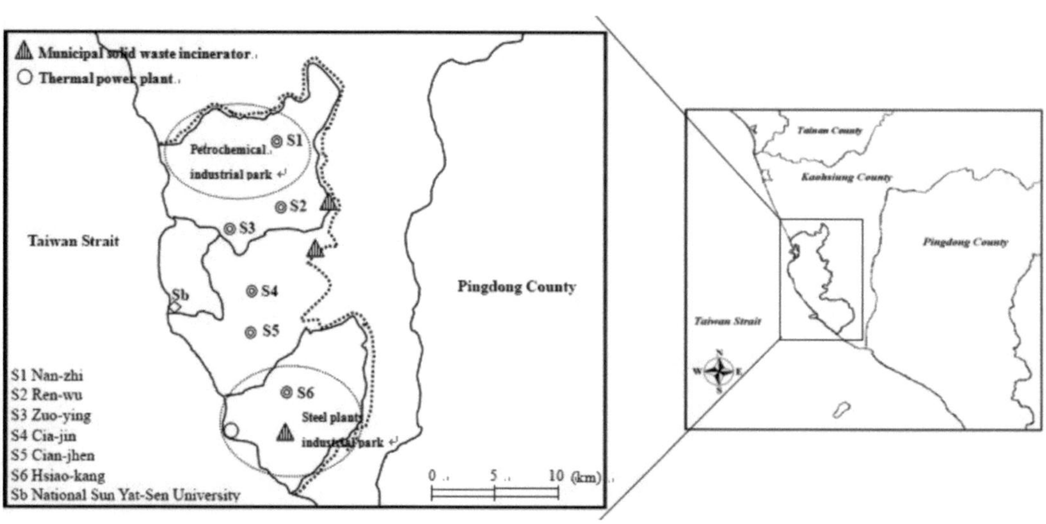

Figure 1. The location of six sensitivity sites and the coastal background site for TGM and Hg_p sampling in Kaohsiung City.

seasons. The field measured meteorological data are listed in **Table 1**, and the field measurement data with mean, standard deviation (SD), and partition of TGM and Hg_p are summarized in **Table 2**.

During the wet season, ambient temperature, relative humidity, wind speed, and UV_B were 28.9 ± 0.7°C, 78.3 ± 1.7%, 2.2 ± 0.2 m/s, and 3.4 ± 0.5 UVI, respectively, which were mostly higher than those during the dry season (**Table 2**). The prevailing wind blew from southwest and northeast during the dry season, and the prevailing wind blew from northwest and northeast during the wet season were reported. It was mainly attributed to the fact that

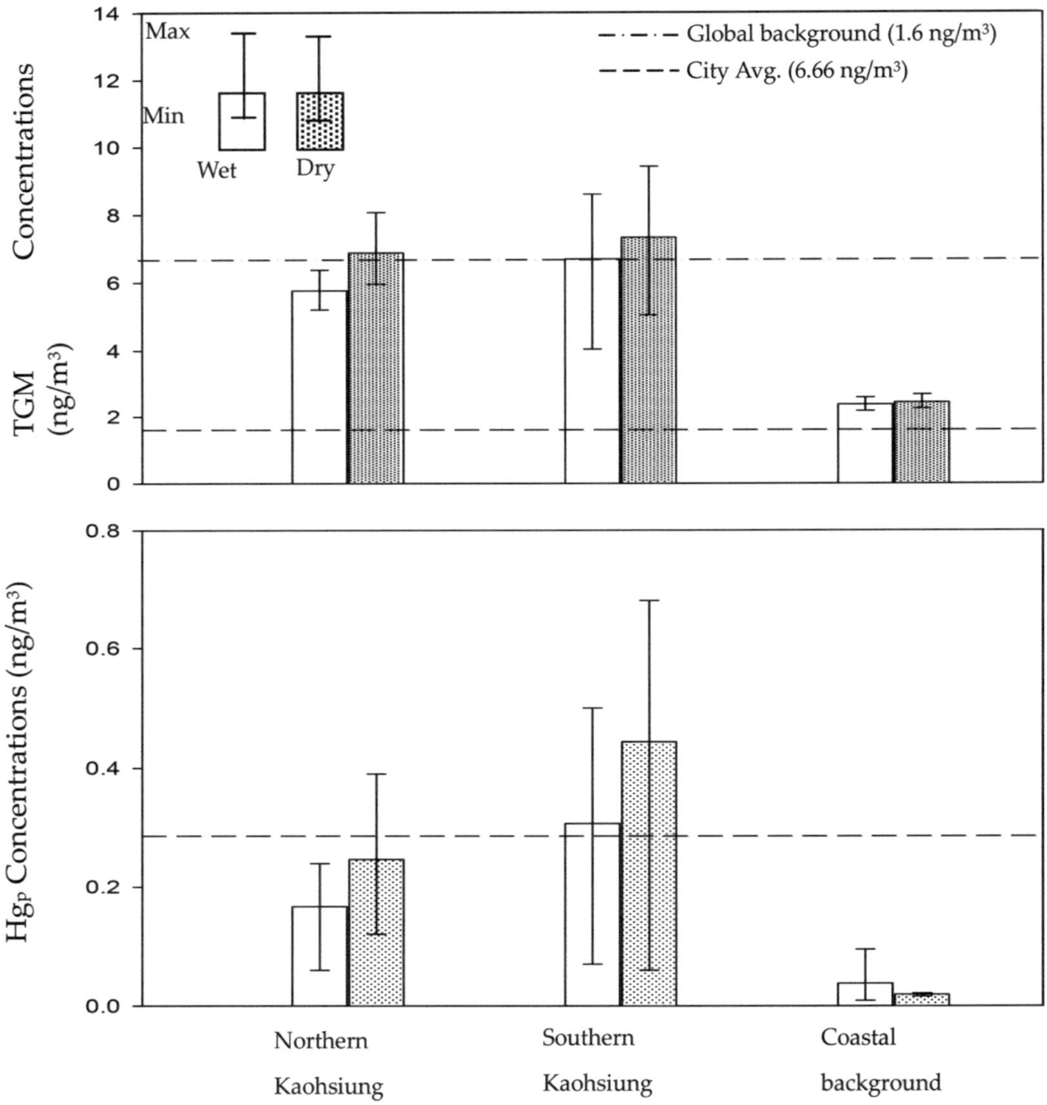

Figure 2. Tempospatial variation of TGM and Hg_p concentrations during the wet and dry seasons in the northern and southern Kaohsiung sites and the coastal site.

Seasons	Temp. (°C)	RH (%)	Rainfall (mm)	Rainy days	UV_B (UVI)	WS (m/s)	WD
Wet	28.9 ± 0.7	78.3 ± 1.7	427.9 ± 305.1	50	3.4 ± 0.5	2.2 ± 0.2	SW, NE
Dry	23.7 ± 3.4	73.3 ± 3.1	63.0 ± 97.6	12	2.1 ± 0.3	1.9 ± 0.1	NW, NE

Temp.: ambient temperature; RH: relative humidity; Rainy days: days with rainfall ≧ 0.01 mm; WS: wind speed; WD: wind direction; UV_B: ultra-violet radiation.

Table 1. Meteorological parameters monitored in Kaohsiung City during the atmospheric mercury sampling periods for wet and dry seasons.

sea-land breezes blew frequently in Kaohsiung City. However, the wet season is the heat convection season (i.e., hurricane season), in which the rainfall was about 427.9 ± 305.1 mm and the rainy days (rainfall ≧ 0.01 mm) were about 50 days. The dry season commonly blew the northeast monsoon, in which the rainfall and rainy days were much less than those during the wet season. Therefore, the aforementioned meteorological condition is considered as the differential between the dry and wet seasons in Kaohsiung City.

Field measurement data showed that the concentrations of TGM and Hg_p were 6.23 ± 1.62 ng/m³ and 0.23 ± 0.17 ng/m³, respectively, during the wet season, while the concentrations of TGM and Hg_p were 7.09 ± 1.57 ng/m³ and 0.35 ± 0.25 ng/m³, respectively, during the dry season in Kaohsiung City. The concentrations of atmospheric mercury during the dry season was obviously higher than those during the wet season. It showed that meteorological condition and atmospheric dispersion played a critical role on the seasonal variation of atmospheric mercury concentration. However, the seasonal concentrations of TGM and Hg_p did not vary

Seasons	Types of Hg	Northern Kaohsiung sites		Southern Kaohsiung sites				Coastal background site
		S1	S2	S3	S4	S5	S6	Sb
Wet season (n = 12)	TGM (ng/m³)	5.72	6.37	5.20	4.04	7.42	8.60	2.38
	Hg_p (ng/m³)	0.20	0.24	0.06	0.05	0.35	0.50	0.02
	Mean ± SD (ng/m³)	TGM		6.23 ± 1.62				
		Hg_p		0.23 ± 0.17				
Dry season (n = 12)	TGM (ng/m³)	5.95	8.06	6.59	5.03	7.50	9.41	2.44
	Hg_p (ng/m³)	0.23	0.39	0.12	0.08	0.59	0.68	0.03
	Mean ± SD (ng/m³)	TGM		7.09 ± 1.57				
		Hg_p		0.35 ± 0.25				
Mean ± SD of atmospheric mercury (ng/m³)		TGM		6.66 ± 1.42				2.41 ± 0.04
		Hg_p		0.29 ± 0.21				0.03 ± 0.01

S1: Nan-zhi site; S2: Ren-wu site; S3: Zuo-ying site; S4: Cia-jin site; S5: Cian-jhen site; S6: Hsiao-kang site; Sb: coastal site.

Table 2. Seasonal variation of TGM and Hg_p concentrations at six sensitivity sites and the coastal background site in Kaohsiung City.

much at the coastal background site, thus the seasonal variation has insignificant influences on regions where atmospheric mercury concentrations were high.

The atmospheric mercury concentrations at southern Kaohsiung were mostly higher than those at northern Kaohsiung during the wet and dry seasons, and their average concentrations were respectively 1.12 and 1.79 times of those at northern Kaohsiung. As a whole, the average concentrations of TGM and Hg_p in Kaohsiung City were about 2.94 and 11.7 times, respectively, higher than those at the coastal background site during the wet season, and were about 2.6 and 11.5 times, respectively, during the dry season. Overall, the average TGM and Hg_p concentrations were 6.66 ± 1.42 and 0.29 ± 0.21 ng/m^3, respectively, in Kaohsiung City. The TGM concentration in Kaohsiung City was about 4.2 times and 2.8 times higher than the background TGM concentration of the North Hemisphere (1.6 ng/m^3) and at the coastal background site in Kaohsiung City (2.4 ng/m^3), respectively. It showed that Kaohsiung City as a heavy industrial city was highly polluted by atmospheric mercury.

Figure 3 illustrates the partition of Hg_p during the wet and dry seasons in Kaohsiung City. The results showed that TGM was the main mercury species, accounting for 94.56–99.59% of atmospheric mercury during the wet season, and 92.71–99.37% of atmospheric mercury during the dry season. Furthermore, Hg_p concentration had a tendency to increase with the distance from the emission sources. The maximum partition of Hg_p were up to 20–40% of total atmospheric mercury (TAM) in the ambient air [11–13]. The partition of Hg_p was generally lower than 1% of TAM in the rural areas, about 1–3% in the metropolitan areas, and beyond 5% in the industrial areas [14–16]. **Figure 1** shows that site Sb is in the rural area, sites S3 and S4 are in the metropolitan areas, and other four sites are in the industrial areas in Kaohsiung City. Moreover, the partition of Hg_p during the dry season was higher than that during the wet season. It implied that the amount of rainfall, the number of rainy days, and relative humidity might correlate to the partition of Hg_p.

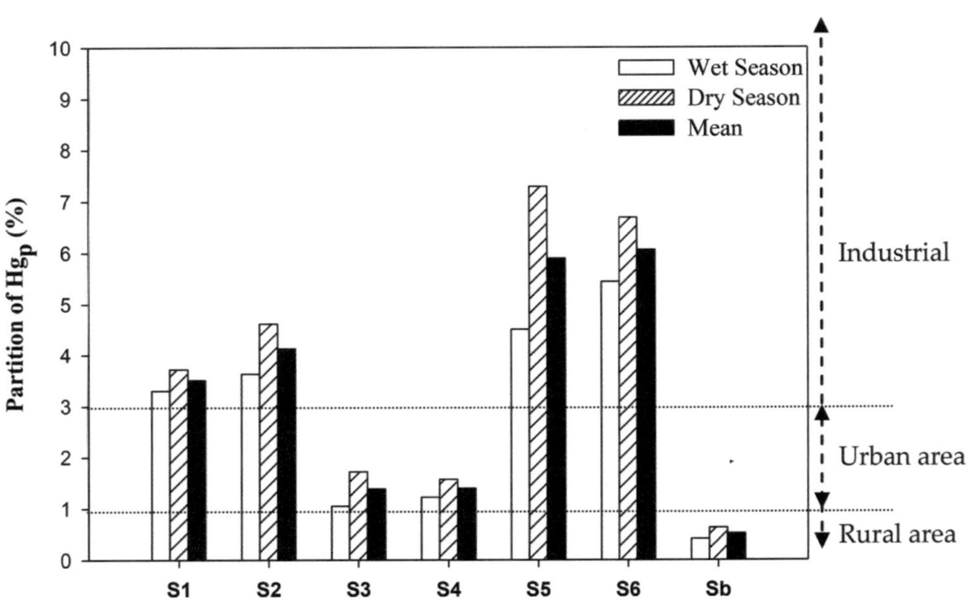

Figure 3. Partition of Hg_p during the wet and dry seasons at six sensitivity sites (S1–S6) and the coastal site (Sb).

2.4. Spatial distribution of TGM and Hg_p

The concentrations of TGM and Hg_p measured at each site are summarized in **Table 2**. As far as the spatial distribution of atmospheric mercury during the wet and dry seasons in Kaohsiung City was concerned, the atmospheric mercury concentrations (TGM of 6.37 ng/m^3; Hg_p of 0.24 ng/m^3) at site S2 was the highest and followed by sites S1 and S3 in northern Kaohsiung, while those (TGM of 8.60 ng/m^3; Hg_p of 0.50 ng/m^3) at site S6 was the highest and followed by sites S5 and S4 in southern Kaohsiung. During the dry season, the atmospheric mercury concentrations (TGM of 8.06 ng/m^3; Hg_p of 0.39 ng/m^3) at site S2 was also the highest and followed by sites S3 and S1 in northern Kaohsiung, while those (TGM of 9.41 ng/m^3; Hg_p of 0.50 ng/m^3) at site S6 was the highest and followed by sites S5 and S4 in the southern Kaohsiung.

The mapping software (SURFER) was further used for plotting the concentration contour of atmospheric mercury in Kaohsiung City. This software uses the grid difference as the calculation basis, and interpolarates the data of atmospheric mercury concentration obtained from each sampling site into the map of Kaohsiung City. This study aimed to explore the spatial distribution of atmospheric mercury concentration in Kaohsiung City. As shown in **Figure 4**, the atmospheric mercury concentrations of northern and southern Kaohsiung were obviously

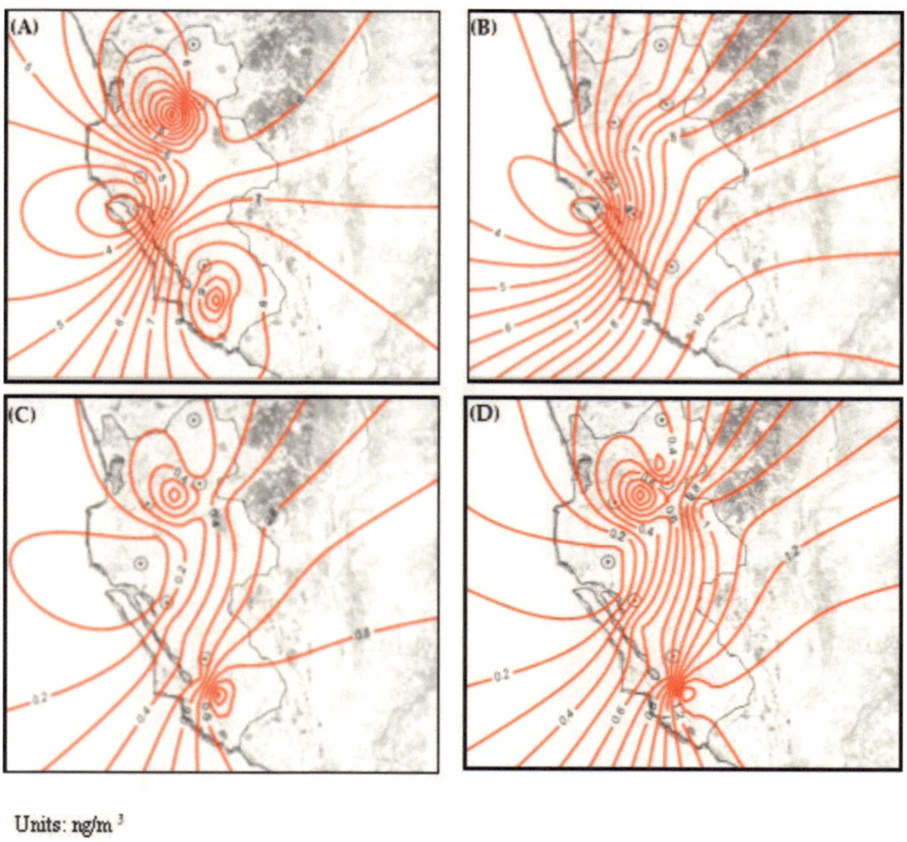

Figure 4. Spatial distribution of atmospheric mercury in Kaohsiung City. (A) TGM concentration during the dry season, (B) TGM concentration during the wet season, (C) Hg_p concentration during the dry season, and (D) Hg_p concentration during the wet season.

affected by the mercury emission sources. Two major high mercury concentration regions concurred with the petrochemical industrial district in northern Kaohsiung and the steel manufacturing complex in southern Kaohsiung. The main emission sources in northern Kaohsiung were petrochemical manufacturing complex, municipal and industrial waste incinerators, cremators, etc. In southern Kaohsiung, mercury was mainly emitted from steel smelters, electric arc furnaces, cement plants, petroleum refineries, coal-fired power plants, municipal waste incinerators, etc. Consequently, the atmospheric mercury concentration in southern Kaohsiung was higher than that in northern Kaohsiung, which was attributed to higher mercury emission in southern Kaohsiung than that in northern Kaohsiung. As a whole, the ambient air quality of Kaohsiung City was highly affected by the heavily densed industries, thus the concentration of atmospheric mercury in the metropolitan area was much higher than the background level during the wet and dry seasons. Both numbers of the emission sources and the consumption of fossil fuels in southern Kaohsiung were higher than those in northern Kaohsiung.

2.5. Comparison of TGM and Hg_p levels with other stationary sources

Figure 5 compares the concentrations of TGM and Hg_p at different mercury emission sources in southern Taiwan [12, 13]. The highest TGM concentrations were observed at a steel plant, which was approximately 2.6 times higher than those at Tainan Scientific Complex, even up to 15 times for Hg_p concentration. Similarly, other stationary sources were also higher than those observed at Tainan Scientific Complex for both TGM and Hg_p concentrations. **Figure 6** illustrates the partition of TGM and Hg_p for various mercury emission sources in southern Kaohsiung.

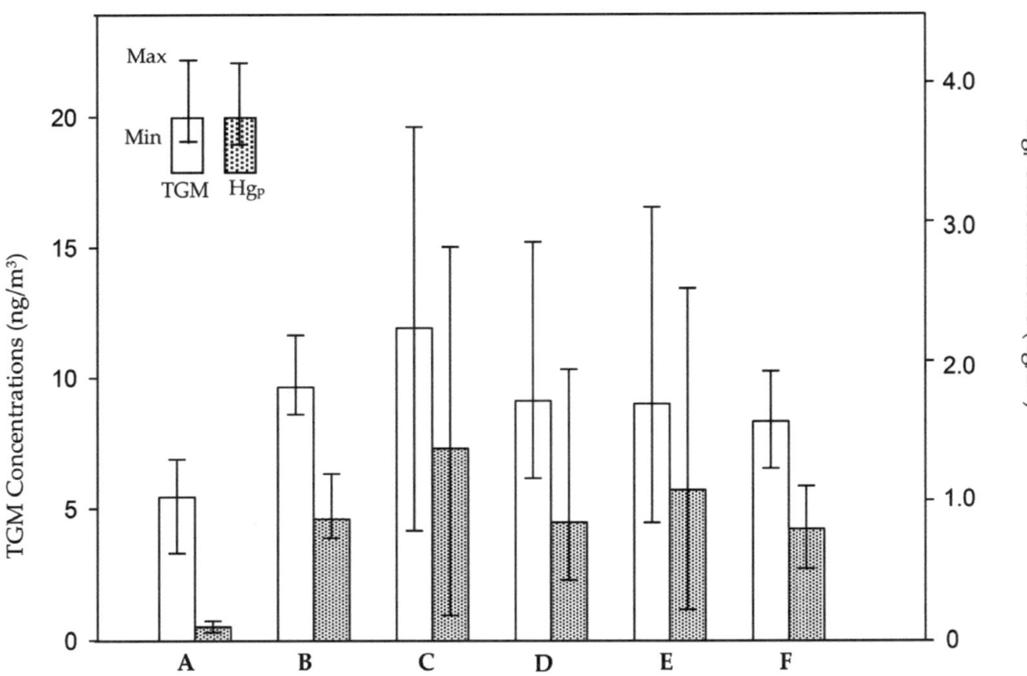

Figure 5. Comparison of TGM and Hg_p concentrations with various mercury emission sources in southern Kaohsiung (A: the semi-conductor complex, B: the petroleum refinery, C: the steel plant, D: the coal-fired power plant, E: the electric arc furnace, F: the municipal waste incinerator).

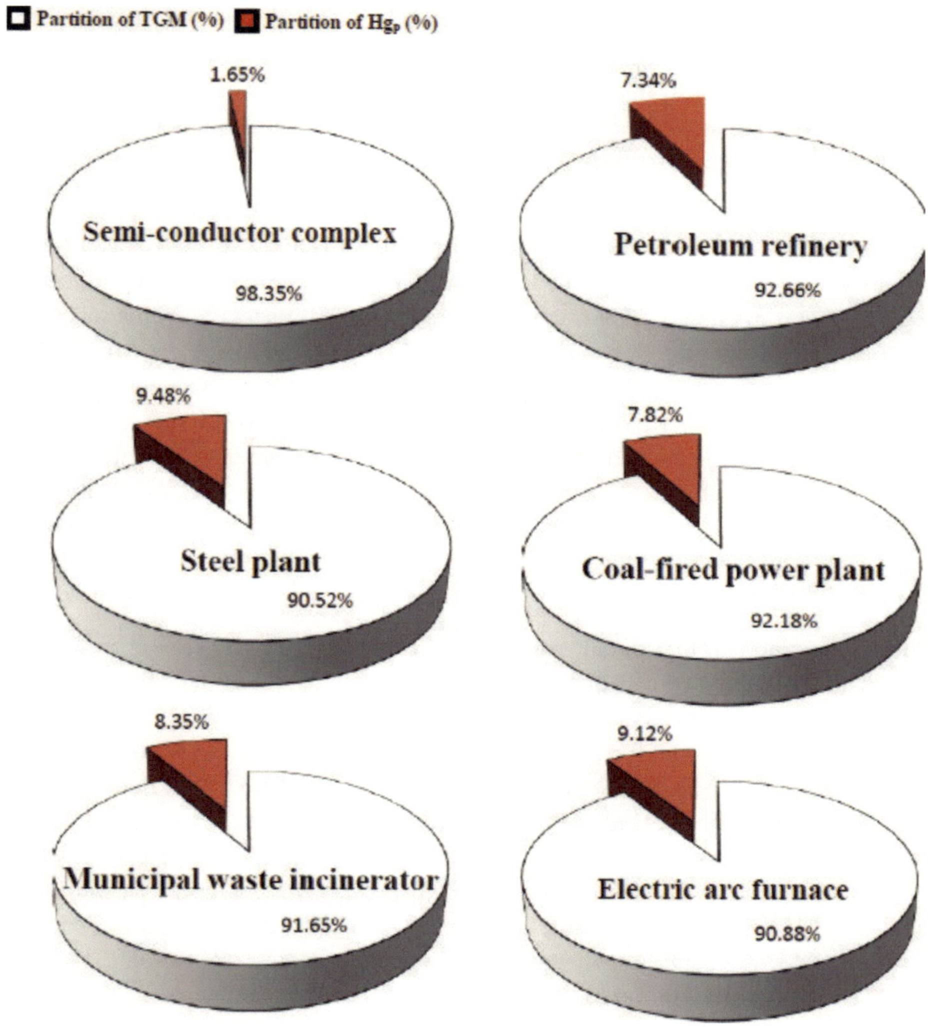

Figure 6. Location of the TGM monitoring site (★) at the Penghu Islands.

The partition of Hg_p at Tainan Scientific Complex was only 1.7%, which was similar to those observed at most urban areas (1–3%), and was much lower than other stationary sources (e.g., coal fired power plants, waste incinerators, and steel plants). Their partition of Hg_p ranged from 7.3 to 9.5%, which contributed much more Hg_p to the ambient air [6, 15, 17, 18]. Overall speaking, TGM was the dominant species of atmospheric mercury in southern Taiwan.

3. Total gaseous mercury concentration at marine boundary layer and associated long-range transportation

3.1. Background

Penghu Islands are located at the middle of the Taiwan Strait between the southeastern China and Taiwan. The islands are dominated by subtropical weather and mainly influenced by East

Asian monsoons. Previous studies reported that, during the seasons of spring and winter, the biomass and fossil fuel burning frequently occurring in Southeast Asia and Southwest China increase the levels of atmospheric mercury in Taiwan owning to long-range transportation [11, 18]. Since there are no significant mercury sources at the Penghu Islands, it can be treated as the background site in the region.

In this chapter, a one-year field monitoring protocol was conducted to investigate the seasonal and daily variations of TAM concentration at the Penghu Islands, as the correlation of TGM concentration with the meteorological parameters and several criteria air pollutants being further examined and discussed. More importantly, a backward trajectory simulation model was further applied to explore the transportation of TGM to the Penghu Islands with respect to the transportation routes for those observed at 500 m above the sea level during the monitoring seasons. While the TGM has been a continuing issue of great concern worldwide, the results of this study would help development more effective management strategies to control the adverse influences associated with the effect of TGM on the environment and human health as well in the Penghu Islands and the areas possibly affected by the long-range transportation of TGM.

3.2. Monitoring sites

According to the meteorological data obtained from 1985 to 2011, dry season with the rainfall of about 800 mm started from April to September. **Table 3** summarizes the meteorological data measured at the Penghu Islands during the TGM monitoring seasons, indicating that the prevailing winds were blown from the east, northeast, northwest, and south, with the ambient temperatures of 13.4–31.2°C, the relative humidity of 53.2–91.3%, and the wind speeds of 0.5–9.7 m/s. The TGM monitoring site was located on the roof of a four-floor building, which was approximately 12 m above the ground and 50 m and 500 m far from the coastline and the major roads, respectively. Overall, the TGM monitoring site was located at the marine boundary layer (MBL), reducing the possible interferences resulted by Hg emission sources. The atmospheric TGM was continuously monitored at the Hsiaomen site (23°38′471″ North latitude, 119°30′316″ East longitude) located at the northwestern coastline of the Penghu Islands (see **Figure 7**) for 15 continuous days in each season from March of 2011 to February of 2012.

Seasons	n	Temp. (°C)	RH (%)	WS (m/s)	WD
Spring	360	19.9 ± 2.1	72.5 ± 7.5	4.5 ± 2.2	ENE
Summer	360	29.8 ± 0.6	79.3 ± 2.7	2.2 ± 0.6	S
Fall	360	24.8 ± 0.6	75.1 ± 4.9	6.1 ± 1.5	ENE
Winter	360	16.1 ± 1.5	82.2 ± 4.7	6.2 ± 1.3	ENE
Mean ± SD		22.7 ± 5.9	77.3 ± 4.3	4.8 ± 1.9	—
Max		33.2	91.3	9.7	—
Min		13.4	53.2	0.5	—

Temp.: ambient temperature; RH: relative humidity; WS: wind speed; WD: wind direction; SD: standard deviation.

Table 3. Meteorological data recorded at the Penghu Islands during the TGM sampling periods.

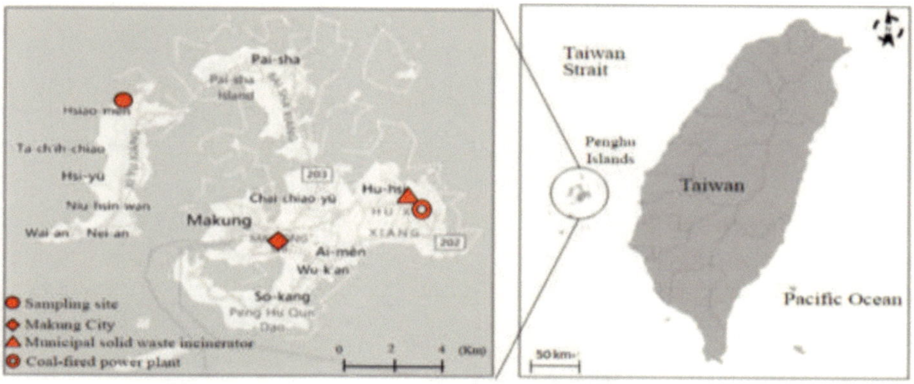

Figure 7. Variation of TGM concentration during four monitoring seasons at the Penghu Islands.

This chapter collected the meteorological data and the criteria air pollutant concentrations from the Makung Air Quality Station during the TGM monitoring periods, and discussed the correlation coefficients derived from the TGM concentrations with the meteorological parameters and the criteria air pollutants. Among them, the correlation coefficient (R) was used to describe the correlation between TGM concentration, meteorological data, and criteria air pollutant concentrations. The meteorological parameters included the ambient temperature (Temp.), relative humidity (RH), wind speed (WS), and wind direction (WD), while the criteria air pollutants of interest included SO_2, NO_X, CO, O_3, PM_{10}, and $PM_{2.5}$. Four seasons are defined as March to May (spring), June to August (summer), September to November (fall), and December to February (winter), respectively. 72-h backward trajectories were simulated by using a NOAA HYSPLIT model with the National Centers for Environmental Prediction's Global Data Assimilation System (NCEP-GDAS) meteorological dataset used as the model input in this study. All backward trajectories started with an arrival height of 500 m above the sea level. By using the NOAA-HYSPLIT model, the dates of the highest TGM concentration at the Penghu Islands in different seasons were determined and the transportation routes of air masses toward the Penghu Islands during the monitoring periods were then simulated.

This information was further applied to examine the possible transportation routes conveying TGM from the upwind sources to the Penghu Islands. Additionally, we used the local fire map to identify the possible sources of TGM drawn by the FIRMS web (http://firms.modaps.eosdis.nasa.gov/firemap/) with the moderate resolution imaging spectroradiometer (MODIS) data from National Aeronautics and Space Administration (NASA), which could illustrate the region and status of biomass burning during the monitoring periods.

3.3. Seasonal and daily variation of TGM concentration

Table 4 summarizes the average and standard deviation of TGM and criteria air pollutant concentrations measured at the Penghu Islands. The highest concentrations of PM_{10} and $PM_{2.5}$ observed in spring were 54.60 ± 15.56 and 31.07 ± 10.59 μg/m^3, respectively, while the lowest concentrations of PM_{10} and $PM_{2.5}$ occurred in summer were 36.63 ± 8.69 and 17.70 ± 6.48 μg/m^3, respectively. Spring and winter are two major seasons frequently blowing Asian dusts from

Seasons	n	TGM (ng/m^3)	PM_{10} ($\mu g/m^3$)	$PM_{2.5}$ ($\mu g/m^3$)	SO_2 (ppb)	NOx (ppb)	O_3 (ppb)	CO (ppm)
Spring	360	4.31 ± 1.87	54.60 ± 15.56	31.07 ± 10.59	1.70 ± 0.72	5.61 ± 0.08	61.58 ± 5.00	0.27 ± 0.06
Summer	360	1.81 ± 0.15	36.63 ± 8.69	17.70 ± 6.48	2.21 ± 0.63	6.15 ± 1.23	33.17 ± 7.70	0.15 ± 0.03
Fall	360	3.03 ± 0.40	45.24 ± 18.49	21.40 ± 6.08	1.68 ± 0.52	3.48 ± 0.46	51.86 ± 5.99	0.18 ± 0.04
Winter	360	3.51 ± 0.67	48.65 ± 9.79	23.81 ± 12.18	2.00 ± 0.08	5.48 ± 1.63	38.59 ± 5.87	0.28 ± 0.08
P-value		3.32×10^{-55}	6.94×10^{-3}	1.93×10^{-3}	1.91×10^{-1}	1.36×10^{-7}	4.00×10^{-18}	3.04×10^{-10}
Mean ± SD		3.17 ± 1.06	46.28 ± 7.51	23.50 ± 5.64	1.90 ± 0.25	5.18 ± 1.17	46.30 ± 12.86	0.22 ± 0.06
Range		1.17–8.63	17.18–77.68	1.32–63.07	1.02–8.63	2.43–14.09	11.35–88.08	0.16–0.57

The concentrations observed among the seasons were significantly different by the analysis of ANOVA (Analysis of Variance) at the confidence level of 95%.

Table 4. Continuous monitoring data of TGM concentrations and their mean, standard deviation (SD), range, and the concentrations of criteria air pollutants.

the northern China to the Penghu Islands [19], which significantly increased the concentrations of PM_{10}. Overall speaking, the concentrations of SO_2 and NO_x measured at the Penghu Islands were much lower than other cities in Taiwan, indicating no significant local sources in the Penghu Islands. Moreover, O_3 concentration was relatively higher compared to other criteria air pollutants at the Penghu Islands.

Field monitoring of TGM at the Penghu Islands showed that the average concentration of TGM was 3.17 ± 1.06 ng/m^3 with the range of 1.17–8.63 ng/m^3. The concentration of TGM in four monitoring seasons were ordered as: spring (4.34 ± 1.87 ng/m^3) > winter (3.51 ± 0.67 ng/m^3) > fall (3.03 ± 0.40 ng/m^3) > summer (1.81 ± 0.15 ng/m^3). Moreover, the average TGM concentration (3.17 ng/m^3) at the Penghu Islands was approximately two times of the background TGM concentration of North Hemisphere (1.6 ng/m^3). The Penghu Islands are likely to be influenced by long-range transportation of TGM in spring and winter, causing higher TGM concentrations in spring and winter than those in summer and fall. The lowest seasonal average TGM concentration of 1.81 ± 0.15 ng/m^3 was observed in summer, which was slightly close to the background TGM concentration of North Hemisphere, suggesting that air masses blown from the South China Sea were relatively clean.

Figure 8 illustrates the variation of TGM concentrations during the four monitoring seasons at the Penghu Islands. The daily concentration of TGM in spring increased significantly from April 2nd to the peak TGM concentration (8.63 ng/m^3) observed on April 10th. The lowest TGM concentration ranging from 1.50 to 2.70 ng/m^3 was compatibly observed in summer. Moreover, the concentrations of criteria air pollutants were also the lowest in summer compared to other seasons (**Table 4**), suggesting that the ambient air quality in summer was relatively better than other seasons at the Penghu Islands, and the TGM concentrations was 1.7–2.8 times lower than other seasons.

Unlike other seasons, the TGM concentrations fluctuated frequently in fall and winter. In winter, it increased rapidly from 3.74 to 5.08 ng/m^3 on December 28th, and then decreased

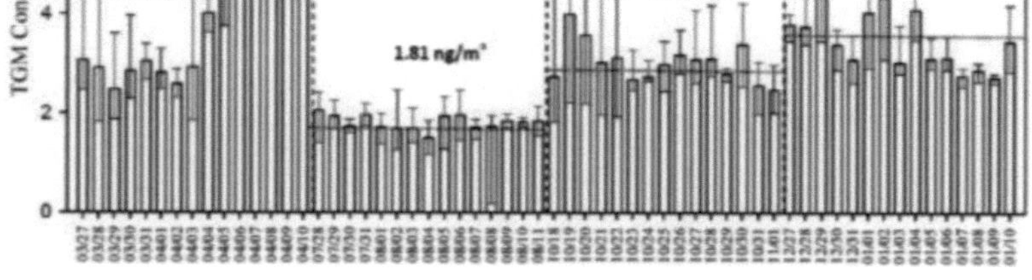

Figure 8. Hourly variation of TGM concentrations during four monitoring seasons at the Penghu Islands.

to 3.38 ng/m^3 on December 30th. The highest peak concentration (4.35 ng/m^3) occurred on December 29th and then leveled off on January 5th. Long-term TGM monitoring at Mt. Lulin background air quality monitoring station showed that Taiwan was highly influenced by atmospheric mercury and gaseous pollutants from China and Southeast Asia in spring and winter [11]. Backward trajectory simulation results indicated that the atmospheric mercury detected in Taiwan significantly increased due to frequent biomass burning originated from the Southeast Asia and industrial emissions from the North China in spring and winter. It suggested that the TGM concentrations at the Penghu Islands might be also influenced by the atmospheric mercury emitting and transporting from these regions. While, air masses blown from the Pacific Ocean had relatively low contribution to the TGM concentration.

3.4. Hourly variation of TGM concentration

Figure 9 illustrates the hourly variation of TGM concentrations during four seasons at the Penghu Islands. It showed that the variation of hourly TGM concentrations were relatively steady in summer. The TGM concentration increased from 6:00 am, gradually reached its concentration peak at 11:00 am, and then decreased after noontime. The increase of TGM concentration resulted probably from the following two processes: (a) UV radiation could temporally transform Hg^{+}, Hg^{2+}, and Hg_p to volatile Hg^0, and subsequently entered to the atmosphere [6]); (b) the downward mixing from the enhanced TGM aloft may increase the levels of TGM concentration as the destroying of nocturnal inversion layer [20]. Except

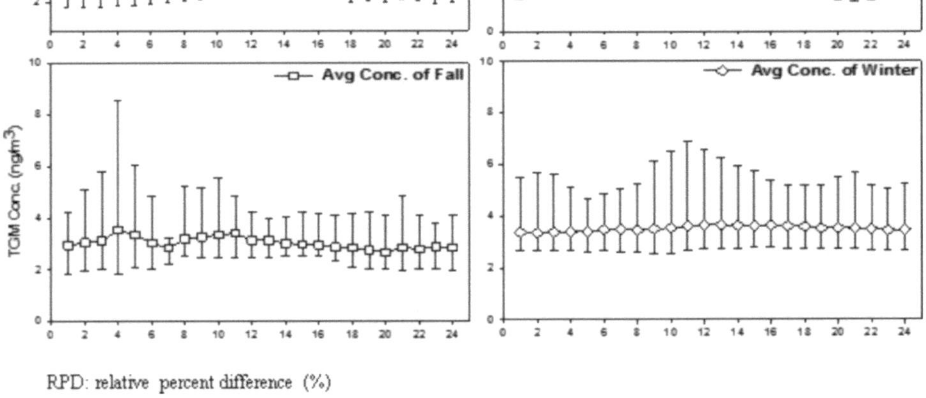

Figure 9. Frequency distribution of TGM concentration covering the four monitoring seasons.

for the morning time, the TGM concentration was less variable through the whole day. The average TGM concentrations in the daytime (6:00 am–6:00 pm) were typically higher than those at nighttime (6:00 pm–next 6:00 am). These findings are attributed to the effects from the variation of the height of atmospheric boundary layer. Additionally, it might be influenced by local anthropogenic activities, such as open burning, mobile sources of fishing boats, vehicles, etc.

The magnitude of hourly variation (differentia between the maximum and the minimum TGM concentrations) was lower in summer and winter with the relative percent differences (RPD) of 61.9 and 88.1%, respectively, and higher in spring and fall with the RPD of 130.7 and 107.8%, respectively. However, previous studies showed that the monitored TGM concentration varies at low altitudes in the typical rural areas [11, 21, 22]. This study revealed that the concentration of TGM monitored at the Penghu Islands was mainly influenced by both local biomass burning and long-range transportation.

Figure 10 illustrates the frequency distribution of TGM detected during four monitoring seasons. It showed that the TGM concentrations followed a lognormal distribution pattern in the range of 2.0–4.5 and 6.0–7.5 ng/m^3, accounting for approximately 80.0% of total frequency in spring. The values ranging from 1.5 to 2.5 ng/m^3 accounted for approximately 89.4% of total frequency in summer. Similarly, the values in the range of 2.0–4.5 and 3.5–5.5 ng/m^3 accounted for 93.3 and 61.7% of total frequency in fall and winter, respectively. However, the episodes with high TGM concentrations (>9.0 ng/m^3) were frequently observed in spring, fall, and winter. It is worth noting that the frequency distribution of TGM appeared to follow two different trends, as shown in **Figure 10**. The frequency distributions of TGM levels in summer and fall were unimodal distribution, while those in spring and winter followed a multi-modal distribution.

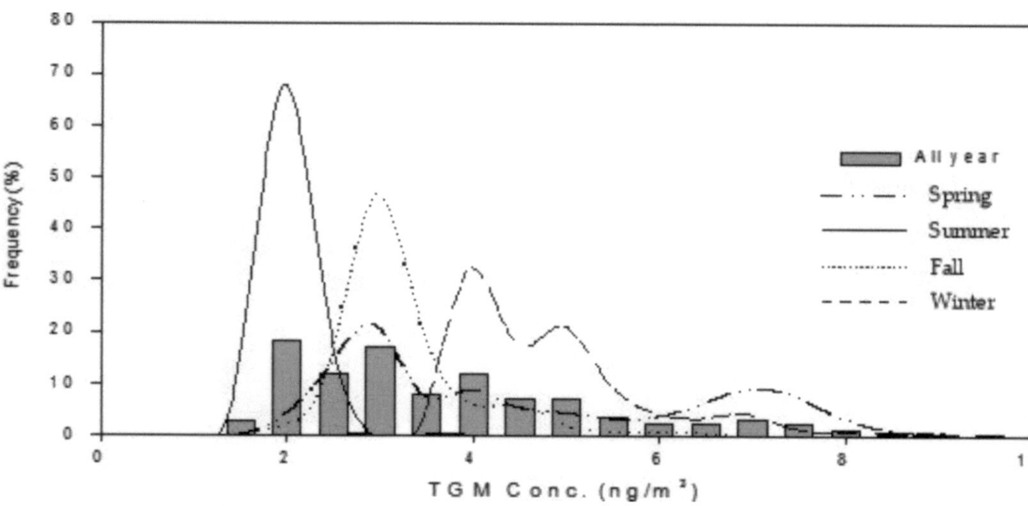

Figure 10. Pollution rose of TGM in four seasons at the Penghu Islands.

3.5. Transportation routes of TGM toward the Penghu Islands

Figure 11 illustrates the pollution rose of TGM for four monitoring seasons at the Penghu Islands. High TGM concentrations were observed mainly in the wind directions of 0–90° and followed by 270–360° in spring, and 60–120° in fall and winter. There were no significant mercury emission sources at the northwestern or southwestern Penghu Islands, suggesting that the high concentrations of TGM were transported remotely from China. An increasing number of studies have shown that biomass burning, industrial combustion, and ocean evaporation are three major emission sources of TGM in the regional scale [10, 23–25]. **Figure 12** illustrates the backward trajectories and the TGM concentration percentage of air masses toward the Penghu Islands in four seasons.

In spring, the TGM concentrations for air masses transported conveying from routes (1) and (2) ranged from 3.55–7.12 ng/m^3, accounting for approximately 89% of TGM data, which was possibly dominated by those transported from local stationary sources and open burning with air masses toward the Penghu Islands. Thus, the southern China were another possible sources contributing to the TGM levels at the Penghu Islands.

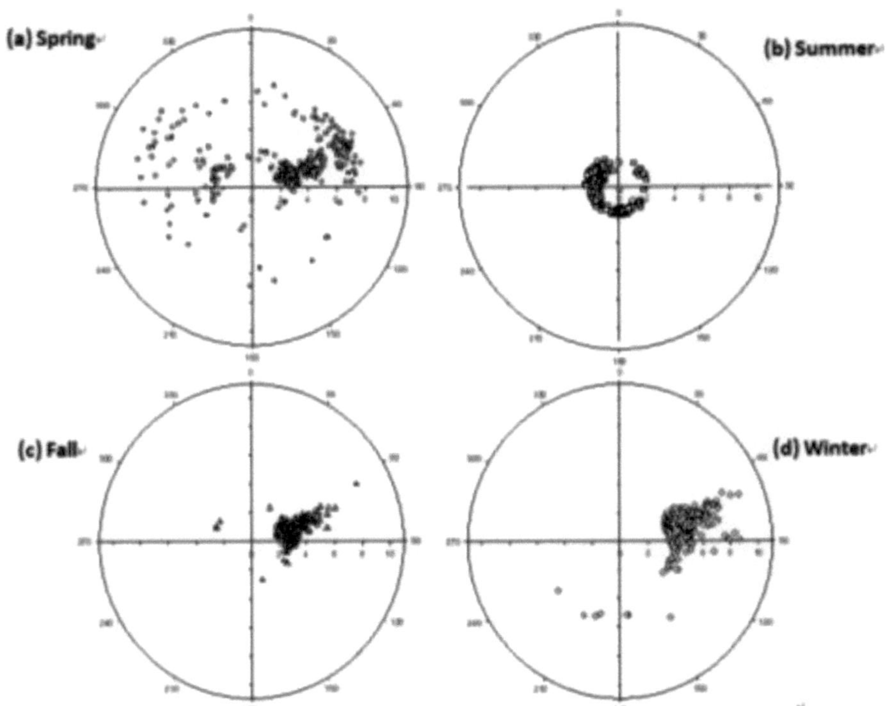

Figure 11. Backward trajectories of air masses and TGM concentration percentage for different pathways during the monitoring seasons at the Penghu Islands.

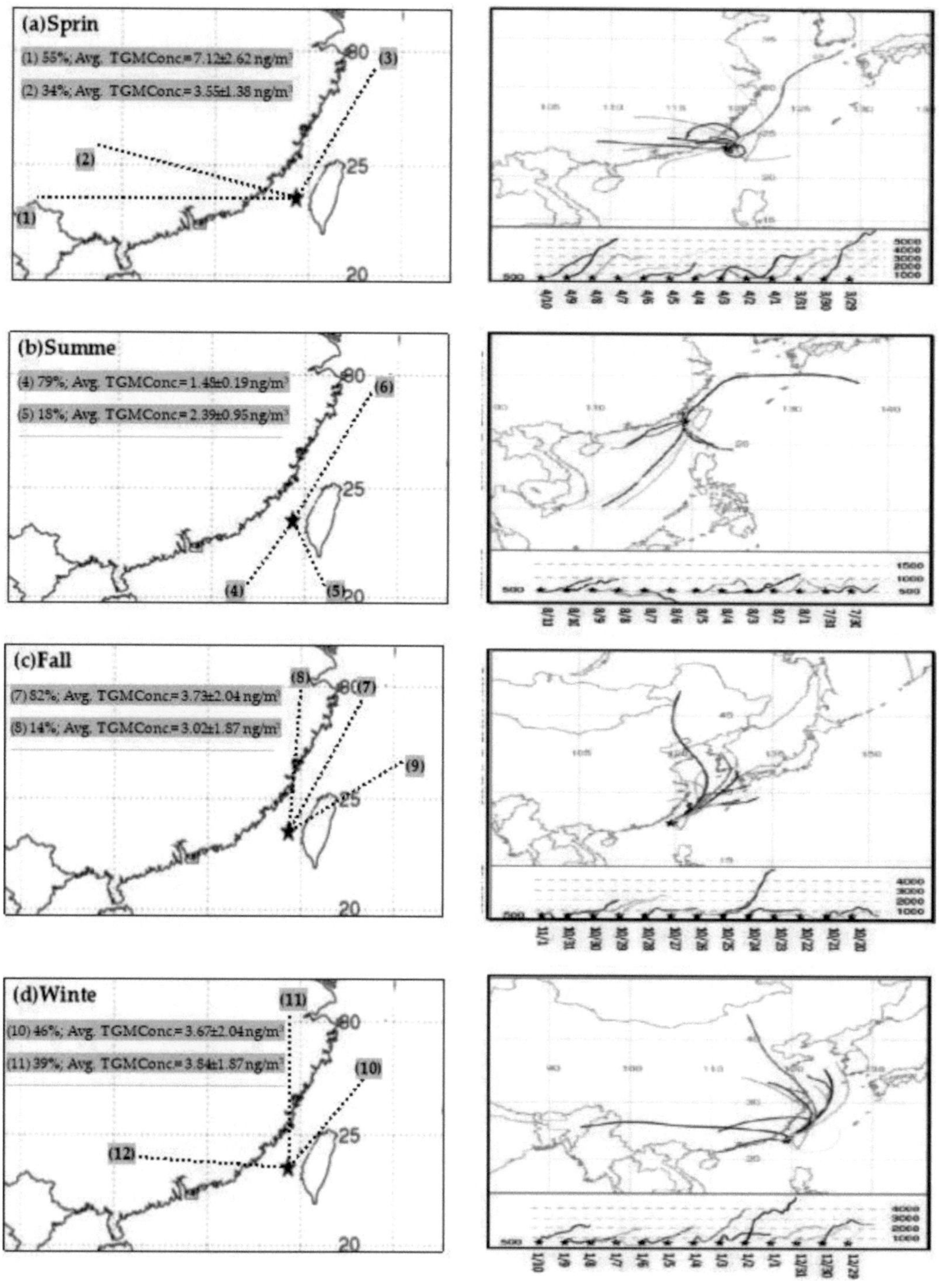

Figure 12. The percentage of the TGM concentration attributed from long-range transportation at the Penghu Islands.

In summer, the TGM concentrations for air masses conveying from routes (4) and (5) ranged from 1.48–2.39 ng/m^3, accounting for approximately 97% of TGM data, which were transported from the South and the East China Sea to the Penghu Islands, dramatically increasing the TGM concentrations at the Penghu Islands during the summer monitoring period. Consequently, the TGM concentrations in summer were relatively lower than other seasons, and the average concentration of TGM was close to the background TGM concentration of North Hemisphere (approximately 1.6 ng/m^3).

In fall, the TGM concentrations for the air masses conveying from routes (7) and (8) ranged from 3.02 to 3.73 ng/m^3, accounting for approximately 96% of TGM data, which seemed to be mainly transported from the northern China, Korea, and Japan with air masses toward the Penghu Islands, resulting in approximately 1.67 times higher TGM concentration in fall than those in summer.

In winter, the TGM concentrations for the air masses conveying from routes (11) and (12) ranged from 3.67 to 3.84 ng/m^3, accounting for approximately 85% of TGM data. The potential sources conveying the TGM toward the Penghu Islands in winter were from the southern Asia, northern China, and Mongolia, increasing the TGM concentrations up to 1.94 times of those in summer.

In this study, the minimum TGM value of 1.5 ng/m^3 in the summer monitoring period, can be treated as the background concentration of TGM at the Penghu Islands. Therefore, the TGM concentration attributed from long-range transportation can be obtained by Eq. (1), and illustrated in **Figure 13**:

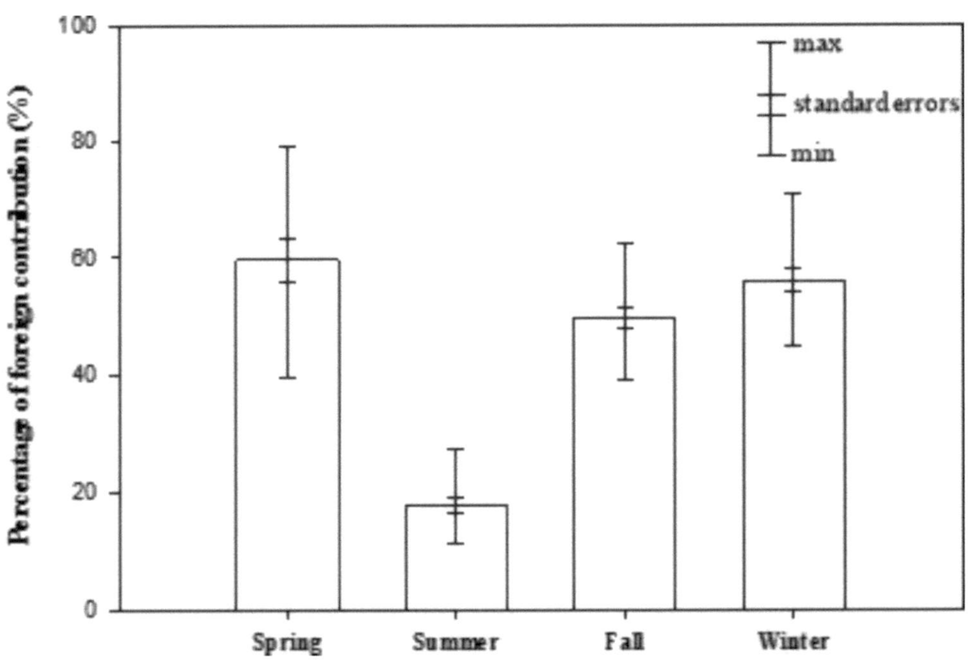

Figure 13. Fire maps air mass during four monitoring seasons in the East Asia in (a) spring, (b) summer, (c) fall, and (d) winter.

$$\text{Contribution percentage} = (\text{VMH} - \text{BC})/\text{VMH} \times 100\% \tag{1}$$

where contribution percentage is the TGM concentration attributed from cross-boundary transportation (%); VMH is the monitored TGM concentrations (ng/m^3); BC is the background concentration of Hg (i.e., 1.5 ng/m^3). The contribution percentage of TGM concentration attributed from cross-boundary transportation at the Penghu Islands were further compared. The percentages of cross-boundary transportation were ordered as: spring (59.6 ± 15.0%) > winter (55.9 ± 7.6%) > fall (49.7 ± 6.2%) > summer (16.8 ± 6.9%). In addition, the maximum percentage of 79.1% was observed in spring. It showed that the TGM concentrations in spring at the Penghu Islands were highly influential, which were mainly affected by the long-range transportation.

High humidity and rainfall frequency at the Penghu Islands could also explain relatively lower TGM concentrations measured in winter. Previous study reported that the occurrence of biomass burning such as forest fires from February to April in the Southeast Asia and the Indochina Peninsula emitting a large amount of mercury-containing pollutants to the atmosphere [26]. **Figure 14** illustrates the fire maps obtained from the FIRMS web fire maps and air mass transportation routes in four seasons. These fire maps explained why high TGM concentration occurred in spring at the Penghu Islands. During the monitoring periods, the hot spot of fires occurred densely in spring than other seasons could emit a large amount of TGM to the atmosphere, and then transported toward the Penghu Islands.

As illustrated in **Figure 15**, the TGM concentrations monitored at the Penghu Islands were compared to those observed at islands and seas in East Asia. The TGM concentrations in the ambient air were ordered as: An-Myun (4.61 ± 2.21 ng/m^3) > Jeju (3.85 ± 1.68 ng/m^3) > Penghu (3.17 ± 1.06 ng/m^3) > Yellow Sea (2.61 ± 0.50 ng/m^3) > South China Sea (2.32 ± 2.62 ng/m^3) > Okinawa Islands (2.04 ± 0.38 ng/m^3) [10, 22, 25–27]. The results showed that the TGM

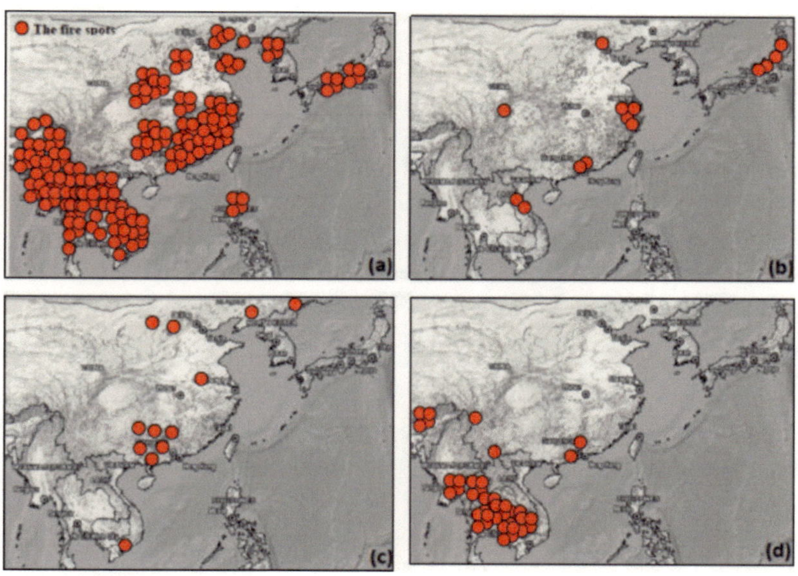

Figure 14. FThe concentration map of TGM at islands and seas in East Asia. (a) An-Myun Island (Nguyen et al., 2007); (b) Yellow Sea (Ci et al., 2011); (c) Jeju Island (Nguyen et al., 2010); (d) Okinawa Island (Chand et al., 2008); (e) Penghu Islands (This study); and (f) South China Sea (Fu et al., 2010).

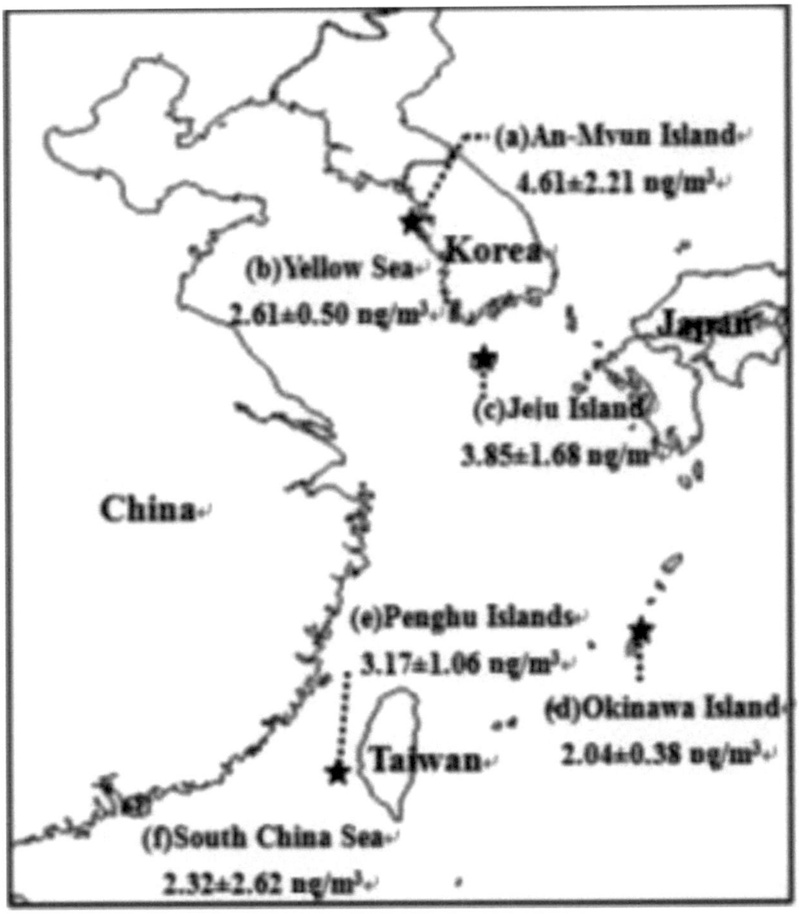

Figure 15. The concentration map of TGM at islands and seas in East Asia. (a) An-Myun Island; (b) Yellow Sea; (c) Jeju Island; (d) Okinawa Island; (e) Penghu Islands; and (f) South China Sea.

concentrations decreased continuously from the northern islands to the southern islands. The TGM concentrations observed at the offshore islands were generally lower than those close to the continent or main island (except Yellow Sea). It further explained why the high levels of TGM concentration at the offshore islands were much easily influenced by Hg emission sources from the anthropogenic sources in the continent or main island.

In summary, in addition to the local sources and open burning, the concentration of TGM at the Penghu Islands was mainly influenced by the long-range transportation of air masses, as the prevailing wind direction and air mass transportation routes potentially playing the critical roles on the variation of TGM concentration in the atmosphere.

4. Conclusions

This chapter investigated the atmospheric mercury by using the modified sampling and analytical methods from two cases of small-scale regions to large-scale regions, respectively,

http://dx.doi.org/10.5772/intechopen.74051

which further investigated the tempospatial variation of atmospheric mercury, gas-particulate partition, transportation routes of mercury, and comparison of mercury concentration in urban areas and stationary sources. According to the results of this field study, several major conclusions are summarized as follows.

The tempospatial variation and the partition of TGM and Hg_p in Kaohsiung City were 6.66 ± 1.42 ng/m^3 and 0.29 ± 0.21 ng/m^3, respectively. The TGM concentration was approximately 4.1 times higher than the background concentration of 1.6 ng/m^3 in North Hemisphere. Two high mercury concentration regions in Kaohsiung City concurred with the petrochemical complex in Northern Kaohsiung, and the steel manufacturing complex in Southern Kaohsiung.

The TGM and Hg_p concentrations in Kaohsiung City were generally higher than those of other Taiwanese cities during the wet and dry seasons. The concentrations of TGM measured in the cities of Taiwan were generally higher than Tokyo and Seoul, however, lower than other cities in China. The burning of coal for space heating in wintertime makes China the main mercury emission source in the world. Moreover, Japan, Korea, and Taiwan are under the leeward of China, which are also in the major atmospheric mercury transportation routes. When the northeastern monsoon prevails, it resulted in the increase of TGM and Hg_p concentrations in the cities of East Asia.

TGM concentration monitored at the Penghu Islands was 3.17 ± 1.06 ng/m^3 with the range of 1.17–8.63 ng/m^3, and were ordered as: spring > winter > fall > summer. Summer is the only season close to the background TGM concentration of Northern Hemisphere at the Penghu Islands. While the hourly variation, TGM concentration typically increased in the morning (8:00 am–1:00 pm), reached its peak concentration, and then decreased in the late afternoon (after 2:00 pm).

Air masses transported from the southern and northern China, the southern Asia, Korea, Japan, and Mongolia might affect the Hg levels at the Penghu Islands during the monitoring seasons. The concentrations of TGM might be influenced by the mercury-polluted air masses to be transported remotely from areas or local stationary combustion and mobile sources. While air masses transported toward the Penghu Islands was dominated by that transported from South China Sea in summer, the TGM concentration levels at the Penghu Islands appeared to be lower than other seasons.

High TGM concentration observed at the Penghu Islands in spring might be attributed to the following three reasons: (a) local emissions from field open burning, local stationary combustion, and mobile sources; (b) long-range transportation from biomass burning in Southeast Asia or neighboring Chinese coastal cities, and (c) long-range transportation through Asian dusts from North China.

Acknowledgements

The authors gratefully acknowledge the financial support and kind assistance from National Sun Yat-Sen University of Air Pollution Control Laboratory (APCL). The authors would like to express their sincere appreciation for its financial support to accomplish this study.

Author details

Yi-Hsiu Jen and Chung-Shin Yuan*

*Address all correspondence to: ycsngi@mail.nsysu.edu.tw

Institute of Environmental Engineering, National Sun Yat-Sen University, Kaohsiung City, Taiwan, ROC

References

[1] UNEP. Global Mercury Assessment, December. 2002. http://www.chem.unep.ch/mercury/report/gma-report-toc.htm

[2] UNEP. Part a: Global Emissions of Mercury to the Atmosphere, UNEP/AMAP 2012 Technical Report, July. 2012

[3] Nater EA, Grigal DF. Regional trends in mercury distribution across the Great Lakes states, north central USA. Nature. 1992;**358**:139-141

[4] Mason RP, Fitzgerald WF, Morel MM. The biogeochemical cycling of elemental mercury: Anthropogenic influences. Geochimica et Cosmochimica Acta. 1994;**58**:3191-3198

[5] Mason RP, Sheu GR. Role of the ocean in the global mercury cycle. Global Biogeochemical Cycles. 2002;**16**(4):1093-1107

[6] Schroeder WH, Munthe J. Atmospheric mercury—An overview. Atmospheric Environment. 1998;**32**:809-822

[7] Lin CJ, Pehkonen SO. The chemistry of atmospheric mercury: A review. Atmospheric Environment. 1999;**33**:2067-2079

[8] Boening DW. Ecological effects, transport, and fate of mercury: A general review. Chemos. 2000;**40**:1335-1351

[9] Clarkson TW, Magos L. The toxicity of mercury and its compounds. Critical Reviews in Toxicology. 2006;**36**:609-662

[10] Fu XW, Feng XB, Dong ZQ, Yin RS, Wang JX, Tang ZR, Zhang H. Atmospheric gaseous elemental mercury (GEM) concentrations and mercury depositions at a high-altitude mountain peak in south China. Atmospheric Chemistry and Physics Discussions. 2010;**9**:23465-23504

[11] Sheu GR, Lin NH, Wang JL, Lee CT, Ou CF, Yang CFO, Wang SH. Temporal distribution and potential sources of atmospheric mercury measured at a high-elevation background station in Taiwan. Atmospheric Environment. 2010;**44**:2393-2400

[12] Jen YH, Yuan CS, Lin YC, Lee CG, Hung CH, Tsai CM, Tsai HH, Ie IR. Partition and tempospatial variation of gaseous and particulate mercury at a unique mercury-contaminated remediation site. Journal of the Air & Waste Management Association. 2011;**61**:1115-1123

[13] Jen YH, Yuan CS, Hung CH, Ie IR, Tsai CM. Tempospatial variation and partition of atmospheric mercury during wet and dry seasons at sensitivity sites within a heavily polluted industrial city. Aerosol and Air Quality Research. 2013;**13**:13-23

[14] Poissant L, Pilote M, Beauvais C, Constant P, Zhang HH. A year of continuous measurements of three atmospheric mercury species (GEM, RGM, and Hg_p) in Southern Quebec, Canada. Atmospheric Environment. 2005;**39**:1275-1287

[15] Feng XB, Tang SL, Shang LH, Yan HY, Sommar J, Lindqvist O. Total gaseous mercury in the atmosphere of Guiyang, PR China. Science of the Total Environment. 2003;**304**:61-72

[16] Fu X, Feng X, Zhu W, Zheng W, Wang S. Total particulate and reactive gaseous mercury in ambient air in the eastern slope of Mt. Gongga area. Applied Geochemistry. 2008;**23**(3):408-418

[17] Sheu GR, Mason RP. An examination of methods for the measurements of reactive gaseous mercury in the atmosphere. Environmental Science & Technology. 2001;**35**(6):1209-1216

[18] Sheu GR, Lin NH, Wang JL, Lee CT. Monitoring of atmospheric mercury concentration at surface sites in Taiwan. Atmospheric Chemistry and Physics. 2009;**30**:1-10

[19] Yuan CS, Lin HY, Wu CH, Liu MH. Preparing of sulfurized powdered activated carbon from waste tires using an innovative compositive impregnation process. Journal of the Air & Waste Management Association. 2004;**54**:862-870

[20] Stamenkovic J, Lyman S, Gustin MS. Seasonal and diel variation of atmospheric mercury concentrations in the Reno (Nevada, USA) Airshed. Atmospheric Environment. 2007;**41**:6662-6672

[21] Mao H, Talbot R. Speciated mercury at marine, coastal, and inland sites in New England—Part 1: Temporal variability. Atmospheric Chemistry and Physics Discussions. 2011;**11**:32301-32336

[22] Nguyen HT, Kim MY, Kim KH. The influence of long-range transport on atmospheric mercury on Jeju Island, Korea. Science of the Total Environment. 2010;**408**(6):1295-1307

[23] Blanchard P, Froude FA, Martin JB, Clark HD, Woods JT. Four years of continuous total gaseous mercury (TGM) measurements at sites in Ontario, Canada. Atmospheric Environment. 2002;**36**:3735-3743

[24] Chand D, Jaffe D, Prestbo E, Swartzendruber PC, Hafner W, Penzias PW, Kato S, Takami A, Hatakeyama S, Kajii Y. Reactive and particulate mercury in the Asian marine boundary layer. Atmospheric Environment. 2008;**42**:7988-7996

[25] Ci ZJ, Zang XS, Wang ZW, Niu ZC, Diao XY, Wang SW. Distribution and air-sea exchange of mercury (Hg) in the Yellow Sea. Atmospheric Chemistry and Physics Discussion. 2011;**11**:2881-2892

[26] Sigler JM, Lee X, Munger W. Emission and long-range transport of gaseous mercury from a large-scale Canadian boreal forest fire. Environmental Science & Technology. 2003;**37**:4343-4347

[27] Nguyen H, Kim KH, Kim MY, Hong S, Youn YH, Shon ZH, Lee J. Monitoring of atmospheric mercury at a global atmospheric watch (GAW) site on An-Myun Island, Korea. Water, Air, & Soil Pollution. 2007;**185**:149-164

Section 4

Phytoremediation of Heavy Metals

Chapter 10

Phytoremediation of Arsenic Contaminated Water

Randhir Kumar and Tarun Kumar Banerjee

Additional information is available at the end of the chapter

http://dx.doi.org/10.5772/intechopen.72238

Abstract

The present investigation deals with the detoxification of arsenic contaminated water using phytoremediation technique. Three macrophytes *Azolla pinnata, Lemna minor,* and *Hydrilla verticillata* were exposed to 1.0 ppm of an arsenic salt (sodium arsenite) separately as well as in combination (*ALH*) for 10 days. The concentration of arsenic in control (wild) macrophytes was below detectable limit. Following exposure, the concentration of arsenic increased steadily in all the plants, and after 10 days, the efficacy of arsenic depletion in phytoremediated media was in the order: *A. pinnata* (88.06%) > *L. minor* (82.56%) > *H. verticillata* (77.53%) and 85.50% when applied in combination (*ALH*). It was found that *A. pinnata* can detoxify the arsenic contaminated water most efficiently.

Keywords: arsenic contamination, bioaccumulation, macromolecular depletion macrophytes, phytoremediation

1. Introduction

In recent years, the areas having arsenic contamination in the ground as well as surface waters are enlarging rapidly in India and its neighboring countries [1–3]. Arsenic is for the most part distributed into the nature in form of either metalloids or chemical compounds, which causes a variety of pathogenic conditions as well as cutaneous and visceral malignancies [4]. Arsenic shows toxicity even at low exposures [5] and causes black foot disease [6]. It is posing great challenge to environmental biologists as well as toxicologist to negotiate the problem. A cost effective technologies are needed to eliminate it from the contaminated water. Phytoremediation is a novel, cost effective and eco-friendly bioremediation technology for environmental cleanup. Bioremediation using macrophytes has been a successful tool to detoxify metal contaminations from variously polluted effluents [7, 8]. Under present evaluation, the arsenic removal competencies from arsenic contaminated water were assessed using three widely distributed aquatic macrophytes (*Azolla pinnata, Lemna minor,* and *Hydrilla verticillata*).

2. Materials and methods

The experimental aquatic plants (*A. pinnata, L. minor,* and *H. verticillata*) were collected from the Agrofarm pond of the Banaras Hindu University, Varanasi, India. For experimentation, monoculture of individual plant was prepared at ambient laboratory temperature under natural photoperiods. Prior to phytoremediation, they were rinsed gently with the tap water (having dissolved O_2 6.3 mg L^{-1}, pH 7.2, water hardness 23.2 mg L^{-1} and room temperature 28 ± 3°C, arsenic concentration below detectable limit) to remove debris. They were subsequently acclimated in tap water for a total period of 2 weeks prior to their application in decontamination activity. The test solution was synthesized by adding 1.0 mg of arsenic in 1.0 L of tap water. This concentration (5%) of the LC_{50} value of *Clarias batrachus* [9] was used for toxicity analysis of the arsenic contamination using fish bioassay technique. Ten grams (fresh weight) of each of the macrophyte was transferred to 10 L of the media. For control, same quantities of each of the plants were put into separate aquaria bearing plain tap water. Growths of each of the macrophytes were also evaluated. For each sampling period, separate experimental and control setups were established.

The percentage of removal efficiency of arsenic by aquatic macrophytes was calculated (**Table 1**) by using the following formula:

$$\% \text{ Removal efficiency} = \frac{C_1 - C_2}{C_1} \times 100 \quad (1)$$

where C_1 is the initial concentration of arsenic in the media, and C_2 is the final concentration of arsenic in the media.

Macrophytes	Initial as concentration in media (ppm)	Residual as concentration in media (ppm)	Efficiency (%)
A. pinnata	1.0	0.119 (11.9%)	88.06
L. minor	1.0	0.174 (17.4%)	82.56
H. verticillata	1.0	0.224 (22.4%)	77.53
Mixture of three macrophytes (*ALH*)	1.0	0.145 (14.5%)	85.50

()Denotes percentage of arsenic residue in media after 10 days of treatment.

Table 1. Arsenic removal efficiency of experimental macrophytes from media.

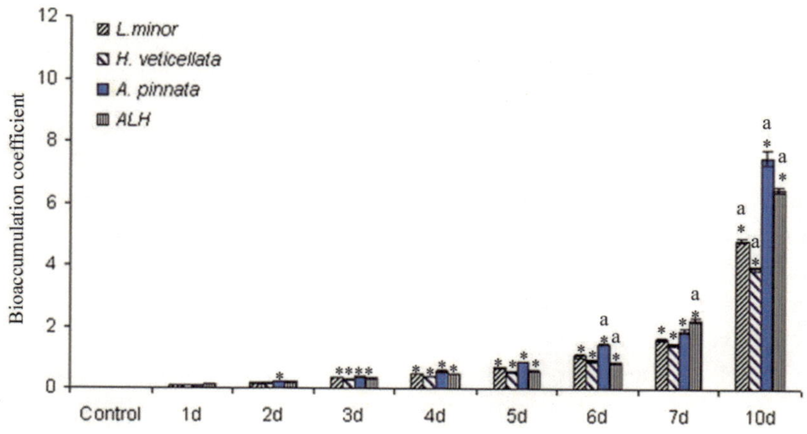

Figure 1. Bioaccumulation coefficient of arsenic at different exposure day. Data are shown as mean ± SEM. The criterion for significant differences set at $p < 0.05$. *Indicates significant differences when compared with 1 day exposed controls. [a]When exposed groups were compared with just previous exposed group.

The bioaccumulation coefficient is defined as the ratio of the concentration of arsenic in the plant and the concentration of residual arsenic in the medium where the plants are growing [10]. The bioaccumulation coefficient was calculated as follows:

$$\text{Bioaccumulation coefficient} = \frac{\text{Arsenic concentration in plant}}{\text{Arsenic concentration in media}} \tag{2}$$

The average bioaccumulation coefficients for the aquatic plants tested here are illustrated in **Figure 1**.

Magnesium ions were analyzed by flame photometer. For chlorophyll estimation, the total chlorophyll was extracted in 80% chilled acetone using Arnon's method [11], prior to observation using a spectrophotometer. The protein content was estimated following Lowry et al. method [12]. Test water was bioremediated with the macrophytes for 10 days. About 1.0 g of each of the plants was collected from experimental setups on different periods (0, 1st, 2nd, 3rd, 4th, 5th, 6th, 7th, and 10th days). For the arsenic analysis, the harvested plants were dried at 80°C. The dried plant materials were digested with acid mixture of HNO_3:$HClO_4$ (5:1 v/v) on a hot plate till a clear solution was obtained. Double distilled water was added to make up the volume to 5 mL. A graphite furnace atomic absorption spectrophotometer (Perkin-Elmer 2380) was used to analyze the arsenic accumulation. All results related to each plant setups were expressed as means followed by standard error of mean. Treatment effects were determined by analyses of variance using the general linear model procedure of the standard statistical analysis system followed by Dunnett's *t*-test at a 5% probability used for post hoc comparisons to separate treatment differences.

3. Results and discussion

Following exposure to the arsenic solution, the macrophytes did not exhibit significant morphological alteration up to 4 days. Then marked deterioration in the physical appearance of all the three plants was noted in the four experimental setups. The deterioration of the plant health became more extensive after 10 days. However, condition of *A. pinnata* deteriorated early and after 7 days it became obvious. The natures of deterioration in these plants were more or less identical whether treated separately or in combination. The concentration of magnesium ions (Mg^{+2}) in these plants did not show much alteration in all the three plants excepting after 10 days when the alteration was statistically significant. The chlorophyll content of all the three plants showed marked decline following phytoremediation applied singly or in combination of the three macrophytes (**Table 2**). Reduction in the chlorophyll content has been observed in variously phytoremediated plants [13]. However, it has been confirmed that the absorption of heavy metals produces phytotoxic effects on plants resulting in inhibition of chlorophyll synthesis and biomass production that often leads to death [14, 15]. The other reason for depletion of chlorophyll content may be due to impaired uptake of essential elements, damaged photosynthetic components or due to increased chlorophyll activity causing chlorophyll regeneration [16]. The protein concentration of these plants showed progressive depletion, which was statistically significant after 10 days in all the experimental setups. Other toxic metals have also caused reduction in plant protein contents [17].

During phytoremediation, the concentration of arsenic in the media decreased progressively, and the residual arsenic concentrations were 11.93, 17.49, and 22.46 % in *A. pinnata, L. minor,* and *H. verticillata,* respectively. The arsenic concentration remained 14.5% in the medium when it was phytoremediated jointly by all the three macrophytes (**Table 1**).

Biomass of all the three macrophytes (*A. pinnata, L. minor,* and *H. verticillata*) increased with exposure period in the order of *A. pinnata* (11.55 ± 0.551) > *H. verticillata* (11.25 ± 0.635) > *L. minor* (11.00 ± 0.550) after 7 days (**Table 3**).

This increase perhaps may partially be due to progressive accumulation of arsenic by these plants. The increase was insignificant after 7 days and onward of exposure. All these plants continued to exhibit detectable arsenic concentrations (**Table 4**). However, after 14 days, they decayed extensively. The bioaccumulation coefficients for the aquatic plants tested in this experiment are shown in **Figure 1**. The results display the bioaccumulation coefficient and illustrate the difference in arsenic accumulation among various macrophyte species. Decrease in the amount of arsenic in the media was due to bioaccumulation of this metalloid by the macrophytes, as reflected by the presence of this metal in the plant tissues (*A. pinnata* accumulated 0.88 mg, *L. minor* 0.82 mg, *H. verticillata* 0.77.5 mg, and 0.85 mg per kg dry wt. of the biomass in the combination of all the three macrophytes) (**Table 4**). The arsenic concentration was below detectable limit in all the three untreated (control) plants. The sum of the arsenic detected in the media after phytoremediation and arsenic absorbed by the plant tissues was quite close to the 1.0 mg L^{-1}, which was the initial concentration prepared for the test solution (**Figures 2** and **3**).

Bharti and Banerjee [13] observed certain degree of difference in sum of the amounts of metals left behind in the phytoremediated coal mine effluent and metal accumulated in the plant tissues after phytoremediation. This was due to their sedimentation, adsorption to the clay particles and organic matters, co-precipitation with secondary minerals, cation-anion exchange, and complexation [7, 18]. In this case, such difference was not noticed because unlike coalmine effluent, the nature and concentration of contaminants in the arsenic solution

http://dx.doi.org/10.5772/intechopen.72238

Exposure period	Biomolecules	*L. minor*	*H. verticillata*	*A. pinnata*	*ALH*
Control (wild)	Chlorophyll	2.767 ± 0.203[a]	2.533 ± 0.116[a]	3.267 ± 0.392[a]	2.946 ± 0.008[a]
	Mg^{2+}	0.438 ± 0.0014[b]	0.556 ± 0.001[b]	0.519 ± 0.001[b]	0.607 ± 0.004[b]
	Proteins	75.033 ± 1.519[c]	94.833 ± 1.476[c]	82.867 ± 0.284[c]	84.833 ± 1.161[c]
1st day	Chlorophyll	2.633 ± 0.088[a]	2.173 ± 0.039[a]	2.75 ± 0.252[a]	2.836 ± 0.024[a]
	Mg^{2+}	0.434 ± 0.001[b]	0.514 ± 0.001[b]	0.516 ± 0.001[b]	0.584 ± 0.004[b]
	Proteins	70.306 ± 0.454[c]	90.153 ± 0.608[c]	80.42 ± 1.242[c]	80.886 ± 0.315[c]
2nd day	Chlorophyll	2.093 ± 0.064[a]	1.926 ± 0.039[d]	2.65 ± 0.078[a]	2.473 ± 0.041[a]
	Mg^{2+}	0.416 ± 0.001[b]	0.511 ± 0.001[b]	0.507 ± 0.001[b]	0.558 ± 0.007[b]
	Proteins	66.383 ± 1.732[d]	82.013 ± 0.325[e]	72.78 ± 0.015[d]	72.667 ± 0.576[d]
3rd day	Chlorophyll	1.997 ± 0.097[e]	1.786 ± 0.032[d]	1.84 ± 0.041[e]	1.867 ± 0.029[e]
	Mg^{2+}	0.407 ± 0.001[b]	0.492 ± 0.001[b]	0.504 ± 0.001[b]	0.542 ± 0.004[b]
	Proteins	62.356 ± 0.305[f]	76.82 ± 2.061[f]	66.956 ± 2.14[f]	66.903 ± 0.214[f]
4th day	Chlorophyll	1.873 ± 0.0218[e]	1.416 ± 0.088[d]	1.736 ± 0.044[e]	1.773 ± 0.012[e]
	Mg^{2+}	0.402 ± 0.001[b]	0.485 ± 0.001[b]	0.502 ± 0.001[b]	0.5073 ± 0.004[b]
	Proteins	56.946 ± 2.153[g]	70.053 ± 0.913[g]	62.703 ± 2.123[f]	64.623 ± 0.446[f]
5th day	Chlorophyll	1.713 ± 0.027[e]	1.166 ± 0.044[d]	1.573 ± 0.062[e]	1.156 ± 0.024[e]
	Mg^{2+}	0.396 ± 0.001[b]	0.480 ± 0.001[b]	0.496 ± 0.001[b]	0.477 ± 0.001[b]
	Proteins	54.573 ± 1.097[g]	64.403 ± 1.188[h]	57.906 ± 1.166[f]	58.453 ± 1.622[g]
6th day	Chlorophyll	1.503 ± 0.933[e]	0.926 ± 0.039[d]	1.383 ± 0.084[e]	1.306 ± 0.022[e]
	Mg^{2+}	0.378 ± 0.002[b]	0.473 ± 0.001[b]	0.492 ± 0.001[b]	0.455 ± 0.004[b]
	Proteins	48.383 ± 0.661[h]	56.687 ± 1.567[i]	44.38 ± 1.603[g]	50.686 ± 2.391[h]
7th day	Chlorophyll	1.283 ± 0.116[e]	0.796 ± 0.029[d]	1.133 ± 0.060[e]	1.836 ± 0.059[e]
	Mg^{2+}	0.368 ± 0.002[b]	0.465 ± 0.002[b]	0.489 ± 0.001[b]	0.438 ± 0.004[b]
	Proteins	42.233 ± 0.913[i]	48.226 ± 0.597[j]	34.86 ± 0.311[h]	42.576 ± 0.864[i]
10th day	Chlorophyll	0.673 ± 0.092[j]	0.503 ± 0.029[k]	0.693 ± 0.088[i]	0.653 ± 0.057[j]
	Mg^{2+}	0.309 ± 0.023[k]	0.409 ± 0.015[b]	0.471 ± 0.001[b]	0.382 ± 0.004[k]
	Proteins	35.593 ± 1.161[l]	39.156 ± 2.158[l]	28.343 ± 1.472[j]	35.806 ± 1.675[l]

[a–l]Values followed by different consequent alphabetic (a–l) superscript lowercase letters within the same column are significantly different at $p < 0.05$ level according to Dunnett's *t*-test, and the same letter within the column shows insignificant difference between the mean when compared to preceding value.

Table 2. Variation in different biomolecule (chlorophylls, magnesium ions, and proteins) contents (μg g^{-1} fresh wt.) of all the three macrophytes after different periods of exposure to 1.0 ppm of sodium arsenite solution.

were simple. A survey of **Figure 2** clearly shows that uptakes of arsenic in all the macrophytes are time dependent up to 10 days of treatment. **Figure 3** illustrated the relationship between arsenic uptake by the plants and its depletion from the contaminated medium. All the three macrophytes are useful for decontamination of arsenic. This study suggests that *A. pinnata* is

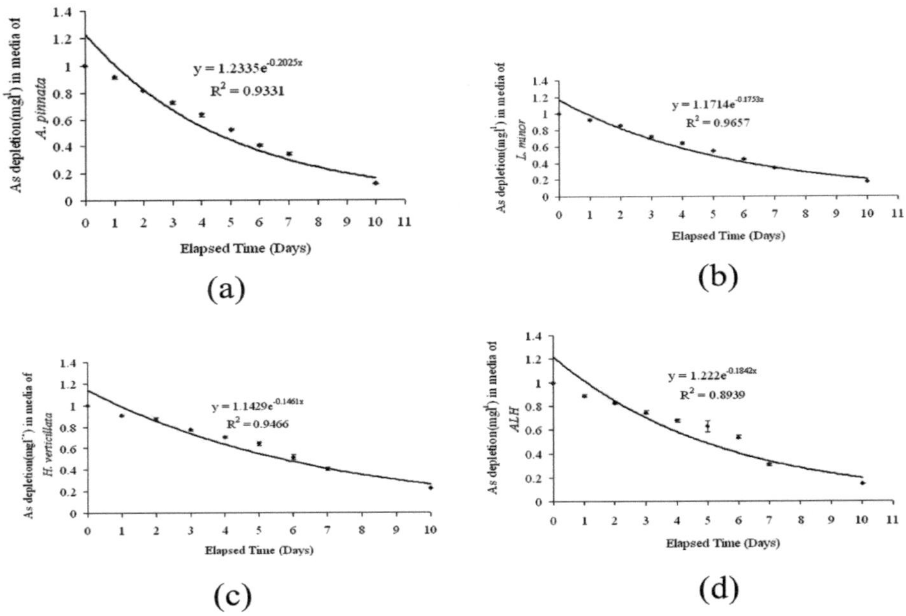

Figure 2. (a-d) Arsenic depletion curve shown by all three plants (*Azolla pinnata, Lemna minor* and *Hydrilla verticillata*) separately as well as in their combination (1:1:1) in 10 L of arsenic media (As=1000 μg L^{-1}) for 10 days.

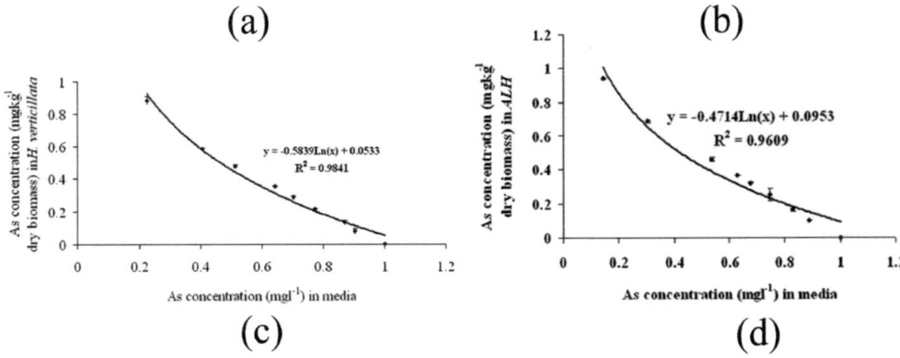

Figure 3. (a–d) Regression curves between arsenic concentration in plant tissues and media of macrophytes (media = grown in arsenic media) at different exposure day.

Macrophytes	Fresh wt. (g) taken at initial day of experiment	Changes of biomass (g) at the end of 7 days of experiment	Changes of biomass (g) at the end of 10 days of experiment
A. pinnata	10.00 ± 0.00[a]	11.55 ± 0.551[a] (+15.5%)	8.95 ± 0.550[b] (−10.5%)
L. minor	10.00 ± 0.00[a]	11.00 ± 0.550[a] (+10%)	9.50 ± 0.635[b] (−5%)
H. verticillata	10.00 ± 0.00[a]	11.25 ± 0.635[a] (+12.5%)	9.65 ± 0.650[b] (−3.5%)

[a, b]Values refer to the mean followed by standard error of mean. Means followed by the same letter in a column were not significantly different at $p < 0.05$.

[()]Denotes percentage change of biomass after 7 and 10 days.

Table 3. The changes of macrophytes fresh biomass during the experiment.

Period of exposure	Macrophytes			
	L. minor	*H. verticillata*	*A. pinnata*	*ALH*
Control (wild)	φ	φ	φ	φ
1st day	0.071 ± 0.004[a]	0.077 ± 0.004[a]	0.084 ± 0.005[a]	0.105 ± 0.001[a]
2nd day	0.134 ± 0.003[a,b]	0.133 ± 0.008[b]	0.176 ± 0.002[a]	0.169 ± 0.009[a]
3rd day	0.267 ± 0.012[b]	0.217 ± 0.002[c]	0.274 ± 0.002[b]	0.253 ± 0.020[b]
4th day	0.346 ± 0.008[b,c]	0.290 ± 0.003[d]	0.353 ± 0.022[b]	0.323 ± 0.001[c]
5th day	0.443 ± 0.012[c,d]	0.353 ± 0.002[d]	0.466 ± 0.001[b,c]	0.367 ± 0.001[d]
6th day	0.553 ± 0.008 [e]	0.476 ± 0.005[e]	0.593 ± 0.001[c,d]	0.462 ± 0.007[e]
7th day	0.657 ± 0.024[f]	0.586 ± 0.001[f]	0.654 ± 0.002 [d,e]	0.687 ± 0.003[f]
10th day	0.823 ± 0.020[g]	0.775 ± 0.005[g]	0.880 ± 0.003[g]	0.853 ± 0.006[g]

[a–g]Values followed by different consequent alphabetic (a–g) superscript lower case letters within the same column are significantly different at $p < 0.05$ level to Dunnett's *t*-test, and the same letter in a column was not significantly different between the mean when compared to just preceding value at $p < 0.05$.

ALH: Mixture of three macrophytes (*L. minor, H. verticillata,* and *A. pinnata*) in equal (1:1:1) ratio.

φBelow detection limit.

Table 4. Arsenic accumulation in different macrophytes at different periods of exposure.

the most efficient plant and can be used singly for this purpose. Phytoremediation beyond 10 days by the same plant will not have additional benefits.

Acknowledgements

The senior author gratefully thanks the University Grant Commission, Government of India, New Delhi, India for providing a Senior Research Fellowship. The authors also wish to thank Prof. A. K. Rai, Head, Department of Botany for providing atomic absorption spectrophotometer (AAS) facility.

Author details

Randhir Kumar* and Tarun Kumar Banerjee

*Address all correspondence to: randhir18bhu@gmail.com

Department of Zoology, Eco-physiology Unit, Banaras Hindu University, Varanasi, India

References

[1] Chowdhury TR, Basu GK, Mandal BK, Biswas BK, Samanta G, Chowdhury UK, Chanda CR, Lodh D, Roy SL, Saha KC, Roy S, Kabir S, Quamruzzaman Q, Chakraborti D. Arsenic poisoning in the Ganges delta. Nature. 1999;**401**:545-546

[2] Rahman MM, Chowdhury UK, Mukherjee SC, Mondal BK, Paul K Lodh D, Biswas BK, Chanda CR, Basu GK, Saha KC, Roy S, Das R, Palit SK, Quamruzzaman Q, Chakraborti D. Chronic arsenic toxicity in Bangladesh and West Bengal, India – A review and commentary. Journal of Toxicology. Clinical Toxicology. 2001;**39**:683-700

[3] Chakraborti D, Hussain A, Allauddin M. Arsenic: Environmental and health aspects with special reference to ground water in South Asia. Journal of Environmental Science and Health, Part A: Toxic Hazard Substance Environmental Engineering. 2003;**38**:11-15

[4] Matsui M, Nishigori C, Toyokuni S, Takada J, Akaboshi M, Ishikawa M. The role of oxidative DNA damage in human arsenic carcinogenesis: Detection of 8-hydroxy-2¢-deoxyguanosine in arsenic-related Bowen's disease. The Journal of Investigative Dermatology. 1999;**113**:26-31

[5] Dikshit AK, Pallamreddy K, Reddy LVP, Saha JC. Arsenic in ground water and its sorption by kimberlite tailings. Journal of Environmental Science and Health, Part A Environmental Science. 2000;**35**:65-85

[6] Lin T-H, Huang Y-L, Wang M-Y. Arsenic species in drinking water, hair, fingernails, and urine of patients with blackfoot disease. Journal of Toxicology and Environmental Health, Part A: Current Issues. 1998;**53**:85-93

[7] Mishra VK, Upadhyay AR, Pathak V, Tripathi BD. Phytoremediation of mercury and arsenic from tropical opencast coalmine effluent through naturally occurring aquatic macrophytes. Water, Air, and Soil Pollution. 2008;**192**:303-314

[8] Rahman MJ, Hasegawa H. Aquatic arsenic: Phytoremediation using floating macrophytes. Chemosphere. 2011;**83**:633-646

[9] Kumar R, Banerjee TK. Arsenic bioaccumulation in the nutritionallyimportant catfish *Clarias batrachus* exposed to the trivalent arsenic salt,sodium arsenite. Bulletin of Environmental Contamination and Toxicology. 2012;**89**:445-449

[10] Robinson BH, Mills TM, Petit D, Fung LE, Green SR, Clothier BE. Natural and induced cadmium-accumulation in poplar and willow: Implications for phytoremediation. Plant and Soil. 2000;**227**:301-306

[11] Arnon DE. Copper enzyme in isolated chloroplast, polyphenol oxidase in *Beta vulgaris*. Journal of Plant Physiology. 1949;**24**:1-15

[12] Lowry OH, Rosebrough NJ, Farr AL, Randall RJ. Protein measurement with the Folin phenol reagent. The Journal of Biological Chemistry. 1951;**193**:265-275

[13] Bharti S, Banerjee TK. Phytoremediation of the coalmine effluent. Ecotoxicology and Environmental Safety. 2012;**81**:36-42

[14] Satyakala G, Jamil K. Chromium induced biochemical changes in Eichhornia crassipes (Mart) solms and Pistia stratiotes. Bulletin of Environmental Contamination and Toxicology. 1992;**48**:921-928

[15] Delgado M, Bigeriego M, Guardiola E. Uptake of Zn, Cr and Cd by water hyacinths. Water Research. 1993;**27**:269-272

[16] Sharma P, Dubey RS. Lead toxicity in plants. Brazilian Journal of Plant Physiology. 2005;**17**:35-52

[17] Arvind P, Prasad MNV. Cadmium-zinc interactions in a hydroponic system using *Ceratophyllum demersum* L.: Adaptive ecophysiology, biochemistry and molecular toxicology. Brazilian Journal of Plant Physiology. 2005;**17**:3-20

[18] Khellaf N, Zerdaoui NM. Growth response of the duckweed *Lemna minor* to heavy metals pollution. Iran. Journal of Environmental Health Science and Engineering. 2009;**6**:161-166

Chapter 11

Phytoremediation and Physiological Effects of Mixed Heavy Metals on Poplar Hybrids

Romika Chandra and Kang Hoduck

Additional information is available at the end of the chapter

http://dx.doi.org/10.5772/intechopen.76348

Abstract

The effects of mixed heavy metals differ not only in different plants but also on the hybrids exposed. In this chapter, we focus on phytoremediation and the physiological effects of mixed heavy metals on four poplar hybrids. According to the results obtained from greenhouse pot experiments with mixed heavy metals, the photosynthetic and transpiration rates were affected by increased heavy metal concentrations. The concentration of heavy metals copper, chromium, cadmium, and zinc in the plant roots, stem and leaves varied with the concentration of mixed heavy metal as well as individual heavy metals. Based on the phytoextraction potential; hybrid 1 (Eco 28) was deduced as the best candidate for phytoremediation in mixed heavy metal contamination treatment. The results obtained are valuable in understanding how specific hybrids respond to mixed heavy metal stress especially when using them as bioindicators for phytoremediation experiments in multi-metal contaminated sites. Selection of new plants along with field trials over extended periods will increase the possibility of further enhancing and establishing phytoremediation technology in the future.

Keywords: phytoremediation, mixed heavy metals, poplar hybrids, physiological effects, phytoextraction potential

1. Introduction

1.1. Phytoremediation

Phytoremediation has gained considerable interest and support in the last decade. This environment-friendly green technology has gained its popularity over the years in terms of its success with other conventional techniques. The specific definitions of phytoremediation are various however the basic definition involves growing of plants in a contaminated matrix with the intentions of removing, transforming or stabilizing environmental contaminants (**Figure 1**) [1–4]. The generic term phytoremediation originates from the Greek prefix φυτο "phyto" – plant, attached to the Latin root "remedium" – to correct or remove, restoring balance, or remediating [5, 6].

Various physical, chemical, and biological techniques are available to remediate metal contaminated soils. Classical environmental cleanup methods are known as ex situ methods which are typically expensive and destructive include excavation, thermal treatment, chemical soil washing, soil incineration, volatilization, vitrification, chemical extraction, solidification, and landfills [5, 7] which either detoxifies or destroys the contaminant chemically or physically. As a result, the contaminant undergoes stabilization, solidification, immobilization, incineration or destruction [8]. These methods are not only labor intensive and expensive but also produce a residue rich in heavy metals which require further treatment. Moreover, these physiochemical technologies used for soil remediation render and create irreversible changes to soil properties altering the land usage as a medium for plant growth, as they remove all biological activities along with disturbing the native soil microflora [2, 5].

Hence, phytoremediation has been recognized as a cost-effective and eco-friendly method of remediating heavy metals from contaminated environments. From the five different processes involved in phytoremediation, phytoextraction has been identified as the superior

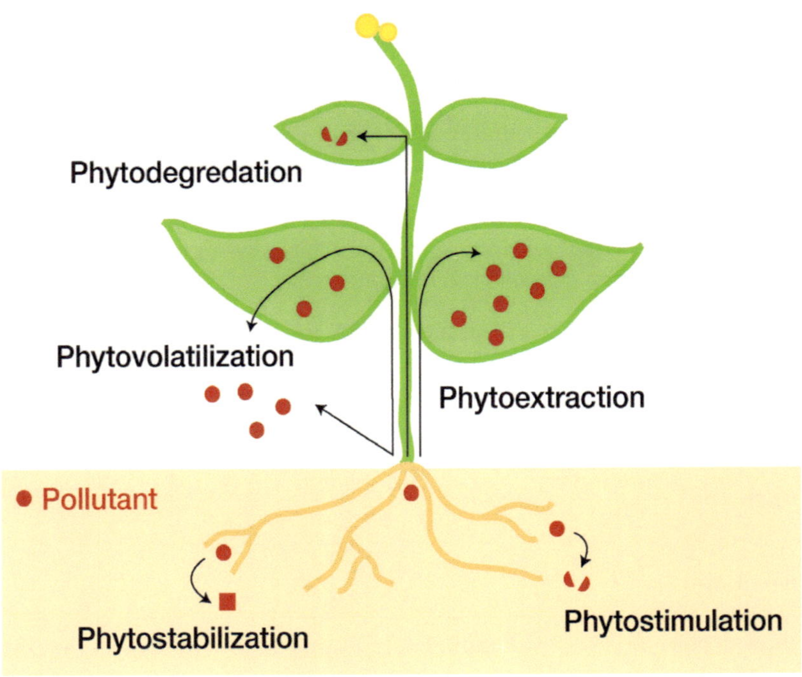

Figure 1. Possible pollutant fates during phytoremediation: the pollutant (represented by red circles) can be stabilized (phytostabilization) or degraded (phytostimulation) in the rhizosphere, sequestered or degraded (phytoextraction, phytodegradation) inside the plant tissue, or volatilized (phytovolatilization) in the air [4].

type where the plants extracts heavy metals from the contaminated soil [9]. Phytoextraction (also known as phytoaccumulation, phytoabsorption or phytosequestration) is the uptake of contaminants from the soil by plant roots and their translocation and accumulation in the above-ground biomass [8, 10, 11]. In addition, rapid growth with high biomass production, extensive root system, high survival and adaptation to low-quality soil substrates and high tolerance to excessive concentrations of heavy metals are properties exhibited by plants suitable for phytoremediation. In general, species and hybrids of poplar, jatropha, and willow are being exploited for the dual purpose of phytoremediation as well as energy production [4, 12–14]. Generally, plants have the potential to absorb metals from the substrate; however, few are capable of extracting, accumulating, and tolerating high concentrations of heavy metals in their system. The discovery of hyperaccumulator plants which are capable of absorbing heavy metals 50–500 times than normal plants has greatly contributed to the revolutionary advancement of phytoextraction technology [15].

1.2. Heavy metal pollution

Sources of pollution in the environment are widely due to global industrialization. Natural and anthropogenic sources are means through which heavy metals enter the environment. Significant natural sources include weathering of minerals, erosion and volcanic activity whereas anthropogenic sources include mining, smelting, electroplating, use of pesticides and fertilizers along with biosolids in agriculture, sludge dumping, industrial discharge, and atmospheric deposition [16–20]. Phytoremediation technology is applicable to a broad range of contaminants, including metals, radionuclides [21–23], and organic compounds such as chlorinated solvents, polycyclic hydrocarbons, pesticides, explosives, and surfactants [3].

From a chemical point of view, heavy metals are defined as elements with metallic properties and an atomic number of >20 and specific gravity of >5. The most common heavy metal contaminants are Cd, Cr, Cu, Hg, Pb, and Zn [2]. In this chapter the term "heavy metals (HMs)" will be referred to those potentially phytotoxic elements that are a major environmental concern due to their persistence in the environment and their impact on humans via the food chain. Heavy metals have adverse effects on human health and therefore heavy metal contamination deserves special attention [5]. HMs such as As, Cd, Hg, Pb, and Se are non-essential since they do not perform any known plant physiological function [24]. However, Co, Cu, Fe, Mn, Mo, Ni, and Zn are essential elements which are required for normal plant growth and metabolism [24]. These essential HM's at supra-optimal concentrations can lead to poisoning [25]. All essential metals are toxic at high concentrations since they cause oxidative stress due to the free radical formation and disrupt the function of pigments and enzymes by replacing essential metals [8, 26].

2. Poplar hybrids

Populus is a genus of 25–35 species of deciduous flowering plants in the family of Salicaceae, native mostly to the Northern Hemisphere. English names commonly applied include poplar, aspen, and cottonwood. This genus has a large genetic diversity and can grow from 15 to 50 m (49–164 ft.) tall, with trunks of up to 2.5 m (8 ft. 2 inch) in diameter. The genus *Populus* is divided into 6 sections on the basis of leaf and flower. Four different poplar clones were selected based on previous research as well as their response in terms of biomass in the field [27–29]. The responses of these four chosen clones were better with seasonal changes as well. These clones included:

1. Eco 28 – *Populus euramericana guinier*
2. DN 034 – *Populus deltoides* × *P. nigra*
3. TN 074 – *Populus trichocarpa* × *P. nigra*
4. TD 225 – *Populus trichocarpa* × *P. deltoids*

Early phytoremediation studies used hyperaccumulator species [30, 31] which are plants that are able to accumulate unusually high levels of metals in their tissue. Studies conducted on willows and poplars (Salicaceae family) showed that the efficiency of metal extraction is markedly lower compared to hyperaccumulators, even if on a large scale basis the removal of metals from soil could be higher [32]. According to Zacchini et al. [28], Salicaceae plants thrive in a wide range of soil and climatic conditions [33] and also express a good metal tolerance hence making them good candidates for phytoremediation work.

3. Poplar hybrids for phytoremediation

Four poplar hybrids were selected based on their genetic diversity, growth, and wellbeing in the field. Cuttings of approximately 15 cm from the hybrids were planted in 2 L pots filled with sandy loam soil. Two months after sprouting and root stabilization, the plantlets were treated with mixed heavy metals of concentrations ranging from 0 mg L^{-1} as control followed by 5, 50, 100, 200 and 500 mg L^{-1}.

The mixed heavy metals utilized included Chromium III chloride (Cr_3Cl), Copper (II) chloride ($CuCl_2$), Cadmium chloride ($CdCl_2$) and Zinc chloride ($ZnCl_2$). Individual HMs was separately prepared and equal amounts were mixed for each concentration. Each plantlet was treated with 20 ml of mixed HM only once during the 3 month treatment period while the plants were watered regularly. Photosynthesis and transpiration rates were measured using LCi-SD (ADC Bioscientific Ltd., Hoddesdon, UK) portable photosynthesis system before and during the treatment period along with the photosynthetic pigments.

After the 3 month treatment period, the plants were harvested for further heavy metal analysis. Plant leaves, stem, and roots were separated, thoroughly washed with running tap water and finally with distilled water after which morphological characteristics were noted and the plants were prepared for further biochemical and heavy metal analysis. Each concentration constituting of four replicates were analyzed for various parameters and an average was used for statistical analysis. Various plant parts including leaves, stems, and roots were dried at 50°C for 48 hours. These dried plant parts were ground using the 8000 mixer mill (SPEX. SamplePrep).

The physical and chemical properties including the levels of heavy metals were determined using the aqua regia method. Soil heavy metal analysis involved, a collection of the soil samples which were air-dried and sieved and digested using the aqua regia method. Soil samples 1.5 g were digested and the samples were analyzed for various HMs by ICP – AES analysis (iCAP 7000 Series ICP spectrometer, Thermo Fisher Scientific, Waltham, MA, USA).

The biological accumulation coefficient (BAC) was calculated as the ratio of heavy metal in the shoot to that in soil given in Eq. 1 [34]. Biotranslocation coefficient was determined as a ratio of heavy metals in plant shoot to that in plant root given in Eq. 2 [35, 36]. Eq. (3) shows the bioconcentration factor (BCF), calculated as metal concentration ratio in plant roots to soil; Eq. (4) shows the concentration index (CI), calculated as the concentration of heavy metals in the treated plant to control [35, 36].

These factors were used to determine the phytoextraction potential of the studied plants. Shoot in the equations refers to stem + leaves and HM for heavy metal.

$$BAC = HM_{shoot}/HM_{soil} \tag{1}$$

$$BTC = HM_{shoot}/HM_{root} \tag{2}$$

$$BCF = HM_{root}/HM_{soil} \tag{3}$$

$$CI = HM_{treated\ plant}/HM_{normal\ plant} \tag{4}$$

Data are expressed as mean ± standard error of mean (SEM and statistically analyzed by one-way analysis of variance (ANOVA) followed by Duncan's Multiple Range Tests using SPSS Version 21 (IBM Corp., USA). Statistical significance was accepted at $P < 0.05$.

3.1. Effects of heavy metals on photosynthesis and transpiration

Exposure of poplar hybrids to mixed HMs under greenhouse conditions showed an increase in the rate of photosynthesis with increased HM concentrations (**Table 1**).

The highest rate of photosynthesis was observed in Hybrid 1, 14.54 μmol m^{-2} s^{-1} at 500 mg L^{-1} mixed HMs. This was observed as the highest photosynthetic rate among all four hybrids. Fluctuations in hybrid 2 photosynthetic rates were observed across all concentrations with the lowest of 7.75 μmol m^{-2} s^{-1} at 5 mg L^{-1} and highest of 12.60 μmol m^{-2} s^{-1} at 200 mg L^{-1} HM concentration. The lowest photosynthetic rate of 2.61 μmol m^{-2} s^{-1} at 200 mg L^{-1}, increased 5 times to 10.08 at 500 mg L^{-1} was observed in hybrid 3. Hybrid 4, on the other hand, had a significantly higher photosynthesis rate of 8.20 μmol m^{-2} s^{-1} at 5 mg L^{-1} which decreased to 6.84 and 7.23 at 50 and 100 mg L^{-1} respectively. A clear pattern can be observed in terms of photosynthesis rate for all 4 poplar hybrids, the photosynthesis rate increases with increasing mixed HM concentrations with the majority being significant around 200–500 mg L^{-1} mixed heavy metal concentrations. This increase in the photosynthetic rate at high HM concentrations could be due to the

Heavy metal concentrations (mg L^{-1})	Photosynthesis rate – A (µmol m^{-2} s^{-1})			
	Hybrid 1 (Eco 28)	Hybrid 2 (DN 034)	Hybrid 3 (TN 074)	Hybrid 4 (TD 225)
0	10.21 ± 0.20^{bc}	10.12 ± 0.26^{b}	9.05 ± 0.34^{ab}	9.05 ± 0.17^{a}
5	8.82 ± 1.24^{c}	7.75 ± 0.26^{c}	9.72 ± 0.27^{a}	8.20 ± 0.54^{ab}
50	11.32 ± 0.18^{b}	11.43 ± 0.52^{ab}	7.11 ± 0.42^{c}	6.84 ± 0.63^{b}
100	8.56 ± 0.23^{c}	10.21 ± 0.93^{b}	8.12 ± 0.46^{bc}	7.23 ± 0.50^{b}
200	8.71 ± 0.21^{c}	12.60 ± 0.52^{a}	2.61 ± 0.50^{d}	8.88 ± 0.53^{a}
500	14.54 ± 0.70^{a}	11.73 ± 0.11^{a}	10.08 ± 0.20^{a}	7.70 ± 0.56^{ab}

Mean values ± SEM, (n = 10).

Means within columns followed by the same letters are not significantly different at $P < 0.05$ according to Duncan's multiple range test.

Values significantly different from control (0).

Table 1. Photosynthesis rates of four poplar hybrids treated with mixed heavy metals (Cd, Cr, Cu, and Zn) under greenhouse conditions.

ability of the hybrids to tolerate high concentration of HMs. This can be based on previous studies on higher plants and trees [37] and effects of chromium stress on photosynthesis [38] are due to carbon dioxide fixation, electron transport, photophosphorylation and enzyme activities [38] whereas cadmium stress leads to Fe(II) deficiency which seriously affected photosynthesis [39]. Hence, if the photosynthesis rates did not decrease this would mean that either the plant is capable of surviving in an environment with high HMs or due to competition with other heavy metals and other environmental factors limited heavy metal are absorption and translocation.

Variations in transpiration rates were observed across all hybrids, the highest value coincided with 50 mg L^{-1} for hybrid 1 at 11.23 mmol m^{-2} s^{-1} followed by 500 mg L^{-1} at 8.89 mmol m^{-2} s^{-1} (**Table 2**).

The highest transpiration rate at 9.99 mmol m^{-2} s^{-1} at 200 mg L^{-1} followed by 9.61 at 50 mg L-1 was observed for hybrid 2. Significantly different transpiration rates for control were observed at low mixed heavy metal concentrations of 5 and 100 mg L^{-1}. The transpiration rates of hybrid 3 were most significant at all concentration with the highest 10.87 mmol m^{-2} s^{-1} at 50 mg L^{-1} followed by 10.14 mmol m^{-2} s^{-1} at 100 mg L^{-1} and the lowest of 3.42 mmol m^{-2} s^{-1} at 200 mg L^{-1}. Hybrid 4 had the lowest transpiration rates at 50 mg L^{-1} (3.55 mmol m^{-2} s^{-1}) and the highest at 200 mg L^{-1} (8.14 mmol m^{-2} s^{-1}). No specific relationships were observed for transpiration rates across the different mixed HM concentrations in hybrid 4 however, the rates observed across all 4 hybrids of poplar significant. However, the differences varied for each hybrid; generally, a decrease in transpiration rate was observed with increase in mixed HM concentrations. According to Carlson et al., gas exchange measurements are often used to detect the most sensitive site of action [40]. In case of the studied hybrids even though the photosynthetic rate at high HM concentrations increased, the decrease in transpiration rate clearly suggests that the plants were stressed in certain ways. A more detailed study would help deduce the specific areas affected by the HMs.

Heavy metal concentrations (mg L^{-1})	Transcpiration rate – E (mmol m^{-2} s^{-1})			
	Hybrid 1 (Eco 28)	Hybrid 2 (DN 034)	Hybrid 3 (TN 074)	Hybrid 4 (TD 225)
0	7.45 ± 0.24^{c}	10.17 ± 0.41^{a}	8.29 ± 0.39^{b}	5.99 ± 0.16^{c}
5	5.96 ± 0.70^{d}	8.64 ± 0.52^{bc}	9.96 ± 0.1^{b}	6.38 ± 0.15^{bc}
50	11.23 ± 0.05^{a}	9.61 ± 0.26^{ab}	10.87 ± 0.44^{a}	3.55 ± 0.53^{d}
100	7.84 ± 0.28^{c}	8.77 ± 0.78^{abc}	10.14 ± 0.72^{a}	6.24 ± 0.61^{c}
200	5.30 ± 0.11^{d}	9.99 ± 0.38^{ab}	3.42 ± 0.53^{c}	8.14 ± 0.34^{ab}
500	8.89 ± 0.13^{b}	7.64 ± 0.1^{c}	9.53 ± 0.28^{ab}	7.52 ± 0.47^{a}

Mean values ± SEM, (n = 10).

Means within columns followed by the same letters are not significantly different at P < 0.05 according to Duncan's multiple range tests.

Values significantly different from control (0).

Table 2. Transpiration rates of four poplar hybrids treated with mixed heavy metals (Cd, Cr, Cu, and Zn) under greenhouse conditions.

Heavy metal stress alters many physiological and metabolic processes in plants. Based on this, the data presented in this chapter demonstrates that mixed HM exposure leads to a significant decrease in photosynthetic pigments in poplar hybrids. Chlorophyll content often measured to assess the impact of environmental stress, since changes in pigment content are linked to visual symptoms of plant illness and photosynthetic productivity [41]. In the present study, the photosynthetic rates decreased for all poplar hybrids across all HM concentrations except hybrid 1 (Eco 28). Decline in photosynthetic rates has been exhibited in other plants, due to a reduction in photosynthetic pigments by the HMs. In various plants, HMs such as Hg, Cu, Cr, Cd, and Zn have been found to decrease chlorophyll contents [42]. This decline in photosynthetic pigments is most probably due to the inhibition of the reductive steps in the biosynthetic pathways due to the high redox potential of many HMs. In addition, protochlorophyllide reductase the key enzyme, involved in the reduction of protochlorophyll to chlorophyll is well known to be inhibited HMs [43]. Various authors have reported a similar decrease in chlorophyll content under heavy metal stress in cyanobacteria, unicellular chlorophytes (*Chlorella*), gymnosperms such as *Picea abies* and angiosperms, such as *Zea mays*, *Quercus palustrus* and *Acer rubrum*, sunflower as well as almond [44–46]. There are few reports that show an enhancement of pigments after exposure to heavy metals [47].

Various studies have also been conducted on the effects of single heavy metals on different plant species. The effects of Cd and Pb on *Brassica juncea L.* exhibited a decline in growth, chlorophyll content and carotenoids, however, Cd was found to be more detrimental than Pb [48]. According to the study on physiological effects of Cd and Cu on peas (*Pisum sativum*), photosynthetic pigments and photosynthesis rates declined at all concentrations of Cd and only at high Cu concentrations [49].

The effects of mixed heavy metals, which compete with each other in the soil-water medium could be one of the reasons for the contradictory segments of the data. Root uptake and levels of accumulation in leaves vary depending on the hybrid, hence an overall expected decline in hybrids 2, 3 and 4. Whereas hybrid 1 had the highest photosynthetic rate, decreases for 5, 50 and 100 mg L^{-1} were observed. However, a slight increase which was lower than 0 mg L^{-1} was observed at 500 mg L^{-1} HM concentration. Hybrid 1 The significant increase in photosynthetic and transpiration rates in hybrid 1 is also supported by the increase in photosynthetic pigments. The overall BAC of HMs in hybrid 1 would be a contributing factor in understanding how hybrid 1 responds to high concentrations of mixed HMs. This highlights for a better understanding on the form of heavy metal ions in the soil solution and their interaction with the plant roots and eventually their absorption into the system and translocation to above ground parts.

Essential heavy metals (Cu and Zn) are constituents of many enzymes and proteins and are required for normal plant growth and development. However, greater concentrations of any HMs either essential or non-essential can lead to toxic symptoms and growth inhibition in most plants. Overall, a decrease in plant photosynthetic efficiency can be partly responsible for the decrease in plant growth and biomass production.

3.2. Phytoextraction potential of poplar hybrids

The biological accumulation coefficient (BAC), biological translocation coefficient (BTC), bio-concentration factor (BCF) and metal accumulation or concentration index (CI) are given in **Tables 3–6** respectively.

3.2.1. Biological accumulation coefficient (BAC)

Individual HMs showed variations in BAC in the studied hybrids. Copper and chromium BAC values were below 1.0 across all treatments in hybrid 1. However, Cd BAC values were at a high of 24.01 at 5 mg L^{-1} followed by 15.81 at 100 mg L^{-1}, decreasing by almost half to 7.50 and 8.13 for 200 and 500 mg L^{-1} respectively. BAC values for Zn were in the range of 2–3 with the highest significant value noted at 50 mg L^{-1}. The BAC values for hybrids 2, 3 and 4, Cu and Cr were all less than 1, whereas Cd and Zn were higher especially at 50 mg L^{-1} decreasing gradually to 500 mg L^{-1}. BAC values for hybrid 3 were higher for Cd and Zn.

3.2.2. Biological translocation coefficient (BTC)

Copper BTC were less than 1.0 for all hybrids across all HM concentrations. Chromium values fluctuated between 0.06 and 1.94. No significant differences were observed at higher heavy metal treatment concentrations. For Eco 28, 50 and 10 mg L^{-1} heavy metal treatment concentrations had BTC values >1.0, whereas for hybrid 2 and 3 the BTC values were greater than 1 at 200 mg L^{-1}. Cadmium BTC values greater than one ranging up to 3.0 for lower heavy metal concentrations only. The highest values were observed at 5 mg L^{-1} for hybrids 3 and 4. Zinc, on the other hand, had the highest BTC for all hybrids.

3.2.3. Bio-coefficient factor (BCF)

BCF values for Cu and Cr for all 4 hybrids were less than 1.0 across all treatments. However, hybrid 1 Zn BCF was slightly above 1.0, with the highest value of 1.21 at 100 mg L^{-1}. For hybrid 2 at 5 mg L^{-1} and hybrid 4 at 5, 50, and 200 mg L^{-1} the BCF value was greater than 1.0.

Hybrid	HM conc. (mg L^{-1})	BAC			
		Cu	Cd	Cr	Zn
Eco 28	0	0.10 ± 0.10^{b}	0.23 ± 0.23^{c}	0.83 ± 0.27^{a}	2.38 ± 0.35^{ab}
	5	0.12 ± 0.10^{ab}	nd	0.78 ± 0.19^{a}	2.73 ± 0.16^{ab}
	50	0.17 ± 0.25^{a}	24.01 ± 3.00^{a}	0.18 ± 0.44^{a}	3.16 ± 0.11^{b}
	100	0.13 ± 0.01^{ab}	15.81 ± 5.28^{ab}	0.14 ± 0.44^{a}	2.90 ± 0.15^{ab}
	200	0.12 ± 0.01^{ab}	7.50 ± 1.22^{bc}	0.12 ± 0.36^{a}	2.78 ± 0.26^{ab}
	500	0.11 ± 0.62^{b}	8.13 ± 3.73^{bc}	0.98 ± 0.43^{a}	2.12 ± 0.57^{b}
DN 034	0	0.15 ± 0.10^{a}	nd	0.22 ± 0.12^{a}	1.08 ± 0.19^{a}
	5	0.12 ± 0.01^{ab}	nd	0.09 ± 0.11^{a}	3.51 ± 0.22^{a}
	50	0.88 ± 0.19^{b}	27.72 ± 10.23^{a}	0.85 ± 0.27^{a}	3.10 ± 0.95^{a}
	100	0.96 ± 0.20^{b}	8.59 ± 2.08^{b}	0.10 ± 0.28^{a}	2.40 ± 0.71^{a}
	200	0.12 ± 0.01^{ab}	14.55 ± 6.12^{ab}	0.14 ± 0.38^{a}	3.11 ± 0.11^{a}
	500	0.12 ± 0.01^{ab}	6.52 ± 1.18^{b}	0.09 ± 0.20^{a}	3.18 ± 0.24^{a}
TN 074	0	0.15 ± 0.01^{a}	nd	0.58 ± 0.01^{a}	4.20 ± 0.35^{a}
	5	0.14 ± 0.13^{a}	nd	0.11 ± 0.06^{a}	4.81 ± 0.28^{a}
	50	0.13 ± 0.16^{a}	58.26 ± 26.00^{a}	0.13 ± 0.56^{a}	3.81 ± 0.13^{a}
	100	0.15 ± 0.17^{a}	17.30 ± 2.60^{b}	0.53 ± 0.01^{a}	4.34 ± 0.25^{a}
	200	0.14 ± 0.64^{a}	6.95 ± 1.16^{b}	0.23 ± 0.15^{a}	3.46 ± 0.87^{a}
	500	0.16 ± 0.02^{a}	11.08 ± 4.41^{b}	0.68 ± 0.21^{a}	4.89 ± 0.45^{a}
TD 225	0	0.09 ± 0.02^{ab}	nd	0.13 ± 0.04^{a}	2.25 ± 0.60^{a}
	5	0.12 ± 0.02^{ab}	nd	0.10 ± 0.02^{a}	2.42 ± 0.50^{a}
	50	0.15 ± 0.18^{a}	15.10 ± 2.49 ^{a}	0.19 ± 0.09^{a}	2.83 ± 0.27^{a}
	100	0.98 ± 0.15^{ab}	10.06 ± 1.46^{ab}	0.06 ± 0.01^{a}	1.93 ± 0.42^{a}
	200	0.07 ± 0.02^{b}	8.90 ± 4.28^{ab}	0.12 ± 0.05^{a}	1.48 ± 0.54^{a}
	500	0.93 ± 0.02^{ab}	6.81 ± 2.96^{bc}	0.15 ± 0.04^{a}	1.71 ± 0.49^{a}

Values are Mean ± SE, (n = 4). nd – not detected.
For each metal in each treatment values followed by the same letters are not significantly different at $P \leq 0.05$ according to Duncan's multiple range test.

Table 3. Biological accumulation coefficient (BAC) of Cu, Cd, Cr and Zn for the four poplar hybrids under greenhouse conditions.

3.2.4. Concentration index (CI)

The CI values were highest for the heavy metal Cd (**Table 6**) and increased with increasing heavy metal treatment concentrations across all studied hybrids. However, the highest CI was observed in hybrid 1 at 200 mg L^{-1} and 500 mg L^{-1} followed by hybrids 4, 3, and 2. Copper and chromium CI values were similar, with no significant differences between the two hybrids, whereas CI values for Zn were less than 1.0.

Hybrid	HM conc. (mg L^{-1})	BTC			
		Cu	Cd	Cr	Zn
Eco 28	0	0.35 ± 0.58^{ab}	3.40 ± 0.65^{a}	0.32 ± 0.72^{b}	2.44 ± 0.54^{a}
	5	0.28 ± 0.38^{b}	1.96 ± 0.44^{b}	0.79 ± 0.24^{b}	2.54 ± 0.45^{a}
	50	0.49 ± 0.10^{a}	1.25 ± 0.06^{bc}	1.94 ± 0.73^{a}	2.81 ± 0.31^{a}
	100	0.31 ± 0.56^{b}	1.30 ± 0.74^{bc}	1.19 ± 0.49^{ab}	2.42 ± 0.15^{a}
	200	0.26 ± 0.16^{b}	1.01 ± 0.26^{bc}	0.64 ± 1.44^{b}	2.76 ± 0.45^{a}
	500	0.90 ± 0.40 ^{c}	0.43 ± 0.19^{c}	0.28 ± 1.67^{b}	1.83 ± 1.01^{a}
DN 034	0	0.60 ± 0.32^{ab}	1.75 ± 0.60^{ab}	3.65 ± 1.79^{a}	5.25 ± 0.18^{a}
	5	0.49 ± 0.40^{bc}	2.61 ± 0.42^{a}	0.83 ± 0.27^{b}	5.20 ± 0.75^{a}
	50	0.35 ± 0.10^{c}	1.53 ± 0.70^{ab}	1.07 ± 0.56^{b}	4.28 ± 1.73^{a}
	100	0.33 ± 0.69^{c}	1.08 ± 0.26^{b}	0.78 ± 0.33^{b}	2.70 ± 0.70^{a}
	200	0.35 ± 0.50^{c}	1.08 ± 0.25^{b}	1.12 ± 0.33^{b}	3.93 ± 0.83^{a}
	500	0.30 ± 0.31^{c}	0.74 ± 0.70^{b}	0.69 ± 0.16^{b}	3.82 ± 0.22^{a}
TN 074	0	0.70 ± 0.85^{a}	4.82 ± 0.60^{a}	0.47 ± 0.18^{a}	5.97 ± 1.02^{c}
	5	0.57 ± 0.92^{ab}	2.86 ± 0.56^{b}	1.51 ± 2.64^{a}	4.47 ± 0.17^{c}
	50	0.33 ± 0.23^{bc}	1.43 ± 0.65^{c}	0.44 ± 0.30^{a}	12.80 ± 10.46^{c}
	100	0.50 ± 0.96^{ab}	1.18 ± 0.30^{c}	0.50 ± 0.46^{a}	115.97 ± 10.19^{a}
	200	0.35 ± 0.15^{bc}	0.60 ± 0.32^{c}	1.17 ± 0.83^{a}	47.53 ± 9.36^{b}
	500	0.13 ± 0.16^{c}	1.80 ± 0.32^{c}	0.06 ± 0.04^{a}	11.75 ± 10.62^{c}
TD 225	0	0.31 ± 0.0 4^{a}	2.53 ± 0.78^{a}	1.73 ± 0.85^{a}	3.23 ± 1.32^{a}
	5	0.29 ± 0.52^{a}	2.60 ± 0.93^{a}	1.12 ± 0.33^{a}	2.26 ± 0.43^{a}
	50	0.36 ± 0.58^{a}	2.55 ± 0.79^{a}	1.32 ± 0.77^{a}	2.83 ± 0.38^{a}
	100	0.23 ± 0.03^{ab}	0.88 ± 0.24^{ab}	0.55 ± 0.18^{a}	2.87 ± 0.85^{a}
	200	0.11 ± 0.24^{bc}	0.38 ± 0.07^{b}	0.57 ± 0.27^{a}	1.23 ± 0.38^{a}
	500	0.07 ± 0.03^{c}	0.47 ± 0.20^{b}	0.59 ± 0.27^{a}	0.99 ± 0.37^{a}

Values are Mean ± SE, (n = 4). nd – not detected.
For each metal in each treatment values followed by the same letters are not significantly different at $P < 0.05$ according to Duncan's multiple range test.

Table 4. Biological translocation coefficient (BTC) of Cu, Cd, Cr and Zn for the four poplar hybrids under greenhouse conditions.

The absorption, accumulation, and translocation of Cu, Cr, Cd, and Zn on the four studied poplar hybrids depend on the plant and soil environment which determines the availability of HMs. BTC and BAC values greater than 1 in addition to metal concentrations suggested by Baker and Brooks [15] would qualify a plant as a hyperaccumulator [36]. In the case of the studied hybrids, all four show translocation and accumulation potential for Cd and Zn

Hybrid	HM conc. ($mg\ L^{-1}$)	BCF			
		Cu	Cd	Cr	Zn
Eco 28	0	0.32 ± 0.66^{a}	0.58 ± 0.56^{b}	0.25 ± 0.58^{a}	1.18 ± 0.37^{a}
	5	0.45 ± 0.67^{a}	nd	0.10 ± 0.01^{b}	1.16 ± 0.57^{a}
	50	0.30 ± 0.10^{a}	13.44 ± 4.80^{a}	0.08 ± 0.28^{b}	1.12 ± 0.90^{a}
	100	0.46 ± 0.78^{a}	12.45 ± 1.90^{a}	0.13 ± 0.13^{b}	1.21 ± 0.89^{a}
	200	0.47 ± 0.15^{a}	8.27 ± 1.43^{a}	0.18 ± 0.15^{ab}	1.06 ± 0.98^{a}
	500	0.58 ± 0.19^{a}	6.99 ± 3.00^{ab}	0.17 ± 0.06^{ab}	0.62 ± 0.21^{a}
DN 034	0	0.24 ± 0.22^{c}	nd	0.58 ± 0.01^{b}	0.78 ± 0.48^{a}
	5	0.25 ± 0.01^{c}	nd	0.15 ± 0.49^{ab}	0.71 ± 0.84^{a}
	50	0.28 ± 0.12^{bc}	34.18 ± 15.58^{a}	0.12 ± 0.31^{ab}	0.86 ± 0.11^{a}
	100	0.28 ± 0.12^{bc}	8.06 ± 0.81^{b}	0.20 ± 0.70^{a}	0.83 ± 0.15^{a}
	200	0.35 ± 0.32^{ab}	12.18 ± 3.74^{b}	0.15 ± 0.33^{ab}	0.93 ± 0.23^{a}
	500	0.42 ± 0.24^{a}	8.62 ± 1.03^{b}	0.14 ± 0.01^{ab}	0.83 ± 0.06^{a}
TN 074	0	0.22 ± 0.05^{a}	nd	0.33 ± 0.42^{a}	0.75 ± 0.11^{ab}
	5	0.25 ± 0.06^{a}	nd	0.47 ± 0.25^{a}	1.08 ± 0.80^{a}
	50	0.23 ± 0.16^{a}	12.68 ± 8.20^{ab}	0.20 ± 0.98^{a}	0.57 ± 0.31^{c}
	100	0.32 ± 0.53^{a}	14.69 ± 2.08^{a}	0.11 ± 0.15^{a}	0.04 ± 0.04^{b}
	200	0.38 ± 0.74^{a}	8.68 ± 2.26^{ab}	0.20 ± 0.05^{a}	0.06 ± 0.01^{a}
	500	0.32 ± 0.38^{a}	11.08 ± 8.44^{ab}	0.17 ± 0.10^{a}	0.38 ± 0.34^{bc}
TD 225	0	0.28 ± 0.04^{c}	nd	0.10 ± 0.02^{b}	0.81 ± 0.14^{a}
	5	0.40 ± 0.01^{bc}	nd	0.11 ± 0.01^{b}	1.05 ± 0.16^{a}
	50	0.42 ± 0.02^{bc}	8.48 ± 3.05^{bc}	0.19 ± 0.03^{ab}	1.03 ± 0.12^{a}
	100	0.43 ± 0.01^{bc}	14.72 ± 5.49^{ab}	0.11 ± 0.02^{b}	0.75 ± 0.09^{a}
	200	0.62 ± 0.15^{ab}	19.74 ± 6.48^{a}	0.22 ± 0.02^{a}	1.16 ± 0.07^{a}
	500	0.55 ± 0.18^{bc}	4.68 ± 1.71^{bc}	0.18 ± 0.06^{ab}	0.72 ± 0.25^{a}

Values are Mean ± SE, (n = 4). nd – not detected.

For each metal in each treatment values followed by the same letters are not significantly different at $P < 0.05$ according to Duncan's multiple range test.

Table 5. Bio-coefficient factor (BCF) of Cu, Cd, Cr and Zn for the four poplar hybrids under greenhouse conditions.

whereas Cr translocation values are only significant for hybrid 1. Copper, on the other hand, has BTC and BAC values lower than 1 which is also supported by the heavy metal concentrations in the tissue (dry weight). The highest BCF (ratio of metal concentrations in the roots to that in soil) was for Cd for all hybrids and only for Zn in hybrid 1, which indicates the ability of the plant to accumulate targeted HMs from the soil medium. Phytoextraction efficiency is

Hybrid	HM Conc. (mg L^{-1})	CI			
		Cu	Cd	Cr	Zn
Eco 28	0				
	5	1.04 ± 0.13^{b}	2.88 ± 0.49^{d}	0.55 ± 0.48^{a}	0.89 ± 0.06^{a}
	50	0.78 ± 0.18^{b}	8.92 ± 1.66^{c}	0.82 ± 1.63^{a}	0.93 ± 0.07^{a}
	100	1.02 ± 0.14^{b}	17.29 ± 2.11^{b}	0.81 ± 0.10^{a}	0.97 ± 0.06^{a}
	200	1.20 ± 0.32^{b}	31.13 ± 1.62^{a}	0.96 ± 0.15^{a}	0.91 ± 0.05^{a}
	500	1.79 ± 0.73^{a}	33.65 ± 3.28^{a}	1.05 ± 0.31^{a}	0.78 ± 0.06^{a}
DN 034	0				
	5	1.02 ± 0.03^{b}	1.33 ± 0.03^{d}	1.78 ± 0.35^{a}	0.85 ± 0.04^{a}
	50	1.02 ± 0.10^{b}	5.00 ± 0.66^{cd}	1.30 ± 0.25^{a}	0.77 ± 0.16^{a}
	100	1.11 ± 0.04^{b}	6.89 ± 1.39^{c}	1.95 ± 0.54^{a}	0.82 ± 0.12^{a}
	200	1.25 ± 0.70^{b}	16.49 ± 2.33^{b}	1.97 ± 0.35^{a}	0.86 ± 0.04^{a}
	500	1.50 ± 1.48^{a}	29.62 ± 2.21^{a}	1.72 ± 0.17^{a}	0.83 ± 0.05^{a}
TN 074	0				
	5	1.14 ± 0.10^{b}	1.42 ± 0.24^{c}	3.95 ± 1.23^{a}	1.23 ± 0.18^{a}
	50	1.21 ± 0.85^{b}	4.60 ± 0.54^{c}	2.12 ± 0.66^{a}	0.86 ± 0.72^{b}
	100	1.30 ± 0.71^{b}	12.53 ± 0.89^{bc}	0.94 ± 0.11^{a}	0.87 ± 0.44^{b}
	200	1.55 ± 0.18^{b}	20.49 ± 3.25^{ab}	3.07 ± 1.09^{a}	0.94 ± 0.50^{b}
	500	2.39 ± 0.30^{a}	33.17 ± 8.13^{a}	2.20 ± 0.27^{a}	1.23 ± 0.11^{a}
TD 225	0				
	5	1.39 ± 0.03^{b}	2.94 ± 0.53^{d}	0.89 ± 0.15^{c}	1.52 ± 0.02^{a}
	50	1.49 ± 0.03^{b}	8.30 ± 1.01^{c}	1.85 ± 0.43^{a}	1.43 ± 0.12^{ab}
	100	1.47 ± 0.06^{b}	12.58 ± 1.56^{c}	0.67 ± 0.40^{c}	0.96 ± 0.13^{b}
	200	1.65 ± 0.18^{b}	25.98 ± 1.99^{b}	1.63 ± 0.33^{ab}	0.96 ± 0.21^{b}
	500	2.40 ± 0.19^{a}	35.69 ± 2.87^{a}	1.96 ± 0.20^{a}	1.03 ± 0.09^{b}

Values are Mean ± SE, (n = 4). nd – not detected.
For each metal in each treatment values followed by the same letters are not significantly different at $P < 0.05$ according to Duncan's multiple range test.

Table 6. Coefficient index (CI) of Cu, Cd, Cr and Zn for the four poplar hybrids under greenhouse conditions.

related to both plant metal concentration and dry matter yield hence, the ideal plant to remedy a contaminated site should be high yielding with the ability to tolerate and accumulate target contaminants [50].

Selection of poplar hybrids for phytoremediation was based on their general growth and performance. According to the data, the response of the hybrids to the different HMs and concentrations varied. The bioavailability of metals in trees and subsequent metal accumulation in its tissues can vary hugely depending on the metal contamination source and site conditions [51]. Based on the mean HM contents in the hybrids, the concentration of Zn is higher in leaves. The highest Zn content of 425.49 mg kg^{-1} dry weight for hybrid 3 followed by 309.52 mg kg^{-1} for hybrid 2 was observed (results not shown). Overall, Cu and Cr were neither well accumulated nor translocated in all hybrids based on the values of BAC, BTC, and BCF. On the other hand, Zn was the only HM that was accumulated in high concentrations especially in the leaves of all hybrids across all concentrations. The highest BAC values were for hybrid 3 at 500 mg L^{-1} whereas BTC values for hybrid 3 Zn were also the highest compared to all other clones, stating clearly the efficient accumulation as well as translocation ability of the hybrid. The BCF values for Cd and Zn for hybrid one were >1 for all concentrations except 500 mg L^{-1} Zn. Based on previous research [51–56] and the results presented in this chapter we can conclude that trees differ in their ability to absorb heavy metals from the soil. The translocation ability from the root to the shoots also vary under different conditions.

4. Conclusion and future prospects

In conclusion based on the physiochemical parameters analyzed, the individual heavy metal contents in each hybrid along with the phytoextraction potential indices, it can be deduced that hybrid 1 (Eco 28) can be selected as a suitable candidate for phytoremediation work with a focus on Cd and Zn phytoextraction capabilities. Hybrid 3 (TN 074) also shows potential as a phytoremediator, however, the studied physiochemical parameters were severely affected by exposure to high concentrations of mixed HMs. This also indicates that subsequent studies are needed to determine the potential of using hybrid 3 as a candidate for phytoremediation of mixed heavy metal contaminated sites. Phytoextraction has been advocated as an effective, eco-friendly and cost-effective technology for the remediation of soils contaminated with heavy metals. The success of phytoextraction depends on several factors which include the concentration of heavy metals in the soil, bioavailability of heavy metals for uptake, and the capability of the plant to absorb and accumulate metals in their tissues. The existing flora diversity needs to be exploited to screen out new and effective hyperaccumulators. In addition to these extensive field-based research for extended durations are also required to better understand heavy metal uptake and accumulation.

Acknowledgements

The authors would like to thank the entire team of Professor Kang Ho-Ducks laboratory for their continuous support and encouragement during the fieldwork and writing. We would also like to extend our appreciation and acknowledgment to Professor Goontawk Lee, Head of Soil Quality Analysis Center and Dr. Ho Sang Kang at the National Instrument Center for Environmental Management (NICEM), Seoul National University for their assistance and support in the soil and plant heavy metal analysis. This work was supported by the grant from the "Forest Science and Technology Development" (Project No. S211215L030110) funded by the Korea Forest Service, Republic of Korea.

Conflict of interest

No potential conflict of interest was reported by the authors.

Acronyms and abbreviations

A	photosynthesis rate
BAC	bioaccumulation coefficient
BCF	bio-coefficient factor
BTC	biotranslocation coefficient
Chl *a*	chlorophyll a
Chl *b*	chlorophyll b
CI	concentration index
E	transpiration rate

Author details

Romika Chandra[1]* and Kang Hoduck[2]

*Address all correspondence to: romika.chandra@gmail.com

1 Kunsan National University, Gunsan-si, Republic of Korea

2 Dongguk University, Goyang-si, Republic of Korea

References

[1] Belz KE. Phytoremediation: Overview [Internet]. 1997. Available from: http://www.webapps.cee.vt.edu/ewr/environmental/teach/gwprimer/phyto/phyto.html [Accessed: November 1, 2017]

[2] Tangahu BV, Abdullah SRS, Basri H, Idris M, Anuar N, Mukhlisin M. A review on heavy metals (As, Pb, and Hg) uptake by plants through phytoremediation. International Journal of Chemical Engineering. 2011;**2011**:1-31. DOI: 10.1155/2011/939161

[3] Paz-Alberto AM, Sigua GC. Phytoremediation: A green technology to remove environmental pollutants. American Journal of Climate Change. 2013;**2**:71-86

[4] Pilon-Smits E. Phytoremediation. Annual Review of Plant Biology. 2005;**56**:15-39

[5] Ali H, Khan E, Sajad MA. Phytoremediation of heavy metals—Concepts and applications. Chem. 2013;**91**:869-881

[6] Pantola RC, Alam A. Potential of Brassicaceae Burnett (Mustard family; angiosperms) in phytoremediation of heavy metals. International Journal of Scientific Research in Environmental Sciences. 2014;**2**:120-138

[7] Liu X, Li X, Chermaine Ong SM, Chu Z. Progress of phytoremediation: Focus on new plants and molecular mechanisms. Journal of Plant Biology and Soil Health. 2013;**1**:1-5

[8] Ghosh M, Singh SP. A review on phytoremediation of heavy metals and utilization of its by-products. Applied Ecology and Environmental Research. 2005;**3**:1-18

[9] Weerakoon S, Somaratne S. Phytoextractive potential among mustard (*Brassica juncea*) genotypes in Sri Lanka. Ceylon Journal of Science (Biological Sciences). 2010;**38**:85-93

[10] Kotrba P, Najmanova J, Macek T, Ruml T, Mackova M. Genetically modified plants in phytoremediation of heavy metal and metalloid soil and sediment pollution. Biotechnology Advances. 2009;**27**(6):799-810

[11] Garbisu C, Alkorta I. Basic concepts on heavy metal soil bioremediation. European Journal of Mineral Processing and Environmental Protection. 2003;**3**(1):58-66

[12] Abhilash PC, Powell JR, Singh HB, Singh BK. Plant–microbe interactions: Novel applications for exploitation in multipurpose remediation technologies. Trends in Biotechnology. 2012;**30**(8):416-420

[13] Prasad MN. Phytoremediation of metal-polluted ecosystems: Hype for commercialization. Russian Journal of Plant Physiology. 2003;**50**(5):686-701

[14] Chaudhry TM, Hayes WJ, Khan AG, Khoo CS. Phytoremediation—Focusing on accumulator plants that remediate metal-contaminated soils. Australasian Journal of Ecotoxicology. 1998;**4**(1):37-51

[15] Baker AJ, Brooks RR. Terrestrial higher plants which hyperaccumulate metallic elements. A review of their distribution, ecology and phytochemistry. Biorecovery, 1, 81-126. Bulletin of Environmental Contamination and Toxicology. 1989;**64**:489-496

[16] Modaihsh S, Al-Swailem M, Mahjoub M. Heavy metal contents of commercial inorganic fertilizer used in the Kingdom of Saudi Arabia. Agricultural and Marine Sciences. 2004;**9**:21-25

[17] Chehregani A, Malayeri BE. Removal of heavy metals by native accumulator plants. International Journal of Agricultural Biology. 2007;**9**:462-465

[18] Fulekar M, Singh A, Bhaduri AM. Genetic engineering strategies for enhancing phytoremediation of heavy metals. African Journal of Biotechnology. 2009;**8**:529-535

[19] Sabiha-Javied MT, Tufai M, Irfan N. Heavy metal pollution from phosphate rock used for the production of fertilizer in Pakistan. Microchemical Journal. 2009;**91**:94-99

[20] Wuana RA, Okieimen FE. Heavy metals in contaminated soils: A review of sources, chemistry, risks and best available strategies for remediation. ISRN Ecology. 2011;**2011**:1-20

[21] Saleh HM. Water hyacinth for phytoremediation of radioactive waste simulate contaminated with cesium and cobalt radionuclides. Nuclear Engineering and Design. 2012; **242**:425-432

[22] Saleh HM. Biological remediation of hazardous pollutants using water hyacinth—A review. Journal of Biotechnology Research. 2016;**2**(11):80-91

[23] Saleh HM, Bayoumi TA, Mahmoud HH, Aglan RF. Uptake of cesium and cobalt radionuclides from simulated radioactive wastewater by Ludwigia stolonifera aquatic plant. Nuclear Engineering and Design. 2017;**315**:194-199

[24] Jung MC. Heavy metal concentrations in soils and factors affecting metal uptake by plants in the vicinity of a Korean Cu-W mine. Sensors. 2008;**8**(4):2413-2423

[25] Rascio N, Navari-Izzo F. Heavy metal hyperaccumulating plants: How and why they do it? And what makes them so interesting? Plant Science. 2011;**180**:169-181

[26] Sarma H. Metal hyperaccumulation in plants: A review focusing on phytoremediation technology. Journal of Environmental Science and Technology. 2011;**4**:118-138

[27] Robinson BH, Green SR, Chancerel B, Mills TM, Clothier BE. Poplar for the phytomanagement of boron contaminated sites. Environmental Pollution. 2007;**150**:225-233

[28] Zacchini M, Pietrini F, Scarascia Mugnozza G, Iori V, Pietrosanti L, Massacci A. Metal tolerance, accumulation and translocation in poplar and willow clones treated with cadmium in hydroponics. Water Air and Soil Pollution. 2009;**197**:23-34

[29] Zacchini M, Iori V, Mugnozza GS, Pietrini F, Massacci A. Cadmium accumulation and tolerance in *Populus nigra* and *Salix alba*. Biologia Plantarum. 2011;**55**(2):383-386

[30] Baker AJM, Brooks RR. Terrestrial higher plants which accumulate metallic elements: A review of their distribution ecology and phytochemistry. Biorecovery. 1989;**1**:81-126

[31] Baker AJM, Reeves RD, Hajar ASM. Heavy metal accumularion and tolerance in British populations of the metallophyte *Thlaspi caerulescens* J & C Press (*Brassicaeceae*). The New Phytologist. 1994;**127**:61-68

[32] Greger M, Landberg T. Use of willow in phytoextraction. International Journal of Phytoremediation. 1999;**1**:115-123

[33] McGee AB, Schmierbach MR, Bazzaz FA. Photosynthesis and growth of populations of *Populus deltoids* from contrasting habitats. The American Midland Naturalist. 1981;**105**:305-311

[34] Yoon J, Cao X, Zhou Q, Ma LQ. Accumulation of Pb, Cu, and Zn in naïve plants growing on contaminated Florida site. The Science of the Total Environment. 2006;**368**(2):456-464

[35] Cui S, Zhou Q, Chao L. Potential hyperaccumulation of Pb, Zn, Cu and Cd in endurant plants distributed in an old smeltery, Northeast China. Environmental Geology. 2007;**51**(6):1043-1048

[36] Li MS, Luo YP, Su ZY. Heavy metal concentrations in soils and plant accumulation in a restored manganese mine land in Guangxi, South China. Environmental Pollution. 2007;**147**(1):168-175

[37] Van Assche F, Clijsters H. Multiple effects of heavy metal toxicity on photosynthesis. Effects of Stress on Photosynthesis, Springer. 1983:371-382

[38] Clijsters H, Van Assche F. Inhibition of photosynthesis by heavy metals. Photosynthesis Research. 1985;**7**(1):31-40

[39] Alcántara E, Romera FJ, Cañete M, De la Guardia MD. Effects of heavy metals on both induction and function of root Fe(III) reductase in Fe-deficient cucumber (*Cucumis sativus* L.) plants. Journal of Experimental Botany. 1994;**45**(12):1893-1898

[40] Carlson RW, Bazzaz FA, Rolfe GL. The effect of heavy metals on plants: II. Net photosynthesis and transpiration of whole corn and sunflower plants treated with Pb, Cd, Ni, and Tl. Environmental Research. 1975;**10**(1):113-120

[41] Parekh D, Puranik RM, Srivastava HS. Inhibition of chlorophyll biosynthesis by cadmium in greening maize leaf segments. Biochemie und Physiologie der Pflanzen. 1990; **186**:239-242

[42] Aggarwal A, Sharma I, Tripati BN, Munjal AK, Baunthiyal M, Sharma V. Metal toxicity and photosynthesis. In: Itoh S, Mohanty S, Guruorasad KN, editors. Photosynthesis: Overviews on Recent Progress & Future Perspectives. 1st ed. I K International Publishing House Pvt. Ltd; 2012. pp. 229-236

[43] Filippis LF, Pallaghy CK. Heavy metals: Sources and biological effects. Advances in Limnology[ERGEB. LIMNOL./ADV. LIMNOL.]; 1994

[44] Siedlecks A, Krupa Z. Interactions between cadmium and iron and its effects on photosynthetic capacity of primary leaves of *Phaseolus vulgaris*. Plant Physiology. 1996;**34**:833-841

[45] Elloumi N, Ben F, Rhouma ABB, Mezghani I, Boukhris M. Cadmium induced growth inhibition and alternation of biochemical parameters in almod seedlings grown in solution culture. Acta Physiologiae Plantarum. 2007;**29**:57-62

[46] Zengin FK, Munzuroglu O. Toxic effects of cadmium (Cd^{++}) on metabolism of sunflower (*Helianthus annuus* L.) seedlings. Acta Agriculturae Scandinavica Section B Soil and Plant Science. 2006;**56**:224-229

[47] Devi Prasad PV, Devi Prasad PS. Effects of cadmium, lead and nickel on three freshwater green algae. Water, Air, and Soil Pollution. 1982;**17**:263-268

[48] John R, Ahmad P, Gadgil K, Sharma S. Heavy metal toxicity: Effect on plant growth, biochemical parameters and metal accumulation by *Brassica juncea* L. International Journal of Plant Production. 2012;**3**(3):65-76

[49] Hattab S, Dridi B, Chouba L, Kheder MB, Bousetta H. Photosynthesis and growth responses of pea *Pisum sativum* L. under heavy metals stress. Journal of Environmental Sciences. 2009;**21**(11):1552-1556

[50] Turan M, Esringu A. Phytoremediation based on canola (*Brassica napus* L.) and Indian mustard (*Brassica juncea* L.) planted on spiked soil by aliquot amount of Cd, Cu, Pb, and Zn. Plant, Soil and Environment. 2007;**53**(1):7

[51] Pulford ID, Watson C. Phytoremediation of heavy metal-contaminated land by trees—A review. Environment International. 2003;**29**:529-540

[52] Pulford ID, Dickinson NM. 90 Phytoremediation technologies using trees. Trace Elements in the Environment: Biogeochemistry, Biotechnology, and Bioremediation. 2005;**14**:383

[53] Gerhardt KE, Huang XD, Glick BR, Greenberg BM. Phytoremediation and rhizoremediation of organic soil contaminants: Potential and challenges. Plant Science. 2009; **176**(1):20-30

[54] Pulford ID, Watson C, McGregor SD. Uptake of chromium by trees: Prospects for phytoremediation. Environmental Geochemistry and Health. 2001;**23**(3):307-311

[55] Cluis C. Junk-greedy greens: Phytoremediation as a new option for soil decontamination. BioTechology Journal. 2004;**2**:61-67

[56] Raskin I, Ensley BD. Phytoremediation of Toxic Metals. New Jersey, USA: John Wiley and Sons; 2000

Chapter 12

Phytoremediation: Halophytes as Promising Heavy Metal Hyperaccumulators

Kamal Usman, Mohammad A. Al-Ghouti and
Mohammed H. Abu-Dieyeh

Additional information is available at the end of the chapter

http://dx.doi.org/10.5772/intechopen.73879

Abstract

The continued accumulation of trace and heavy metals in the environment presents a significant danger to biota health, including humans, which is undoubtedly undermining global environmental sustainability initiatives. Consequently, the need for efficient remediation technologies becomes imperative. Phytoremediation is one of the most viable options in this regard. Hundreds of plants in laboratory experiments demonstrate the potential to remediate varying concentrations of heavy metals; however, the remediation capacity of most of these plants proved unsatisfactory under field conditions. The identification and selection of plants with higher metal uptake capacity or hyperaccumulators are one of the limitations of this technology. Additionally, the mechanism of heavy metal uptake by plants remains to be sufficiently documented. The halophyte plants are famous for their adaptation to harsh environmental conditions, and hence could be the most suitable candidates for heavy metal hyperaccumulation. The state of Qatar in the Gulf region encompasses rich resources of halophytes that have the potential for future investment toward human and environmental health. This chapter, therefore, gives an overview of phytoremediation, with emphasis on halophytes as suitable heavy metal hyperaccumulators for improved remediation of heavy metal–contaminated areas.

Keywords: halophytes, phytoremediation, heavy metals, hyperaccumulation

1. Introduction

Heavy metals and other organic compounds constitutes the major environmental contaminants, and the trials of phytoremediation to free pollutants from waste water and contaminated soil dates back to hundreds of years ago in plants such as the *Thlaspi caerulescens* and *Viola calaminaria,* which were reported to remediate high concentration of heavy metals [1]. Anthropogenic activities arising from industrialization largely contribute to the proliferation of these contaminants, either by direct leakage or accidents during transport of solid and liquid wastes from storage and industrial facilities [2, 3]. Strategies to clean up environmental contaminants, both organic and inorganic are either by physical, chemical and or biological treatments [4, 5]. However, physical and chemical methods are recognized for a number of disadvantages or limitations such as high cost and labor intensiveness. Additionally, chemical processes create another pollution and are especially costly since they generate heaps of sludge [6]. In view of this context, new and better approaches to clean up of metal contamination were thought up and became imperative, hence the exploration of various bio-based techniques. The use of biological agents is considered cheap, safer and has limited or no negative impact to the environment [7]. Bio-based remediation methods include bio-augmentation, bioremediation, bioventing, composting and phytoremediation. However, phytoremediation proves the most viable and useful alternative and has gain an increasing attention in recent times [8, 9]. The adverse and negative effects associated with these elements make them targets for phytoremediation [10]. Phytoremediation offer several advantages. It is cheap, promotes biodiversity, reduces erosion, less destructive and decreased energy consumption leading to reduced carbon dioxide emission [11]. To date, about 400 plant species were suggested to be metal hyper-accumulators [12]. However, few studies reported the toxicity of several metals combined [13], and while hyper-accumulation of nickel (Ni), cadmium (Cd), manganese (Mn), zinc (Zn) and selenium (Se) have been well established, the same is yet be available or demonstrated beyond doubt in plant species for copper (Cu), chromium (Cr), lead (Pb), thallium (Th) and cobalt (Co) metals. For instance, Cu is an important element for growth and general plant physiology, owing to its role as a cofactor to various types of enzymes involved in the transfer of electrons during metabolic processes, such as the tricarboxylic acid (TCA) cycle [14, 15]. However, at high concentrations, it is toxic to plants signaled by stunted growth, and although there is some physiological insight to Cu stress in plants, the responses are still vague at the functional level [16]. The accumulation of heavy metals in plant tissues results in a wide range of negative effects on growth. Although it affects seed germination, growth of seedlings and photosynthetic processes, which generally leads to the inhibition of the plants important enzymatic activity [17, 18], however, plants responds differently [19]. In dealing with the heavy metal stress, the root tissue is the first to be exposed to the associated toxins, and its cell wall has a mechanism of exchange that fixes the heavy metal ions, thereby limiting the transmission of the toxins to other plant tissues [20, 21]. Several studies reported many plants, including desert species as good phytoremediation agents, however, few are metal hyperaccumulators and their selection for efficient phytoremediation is still a challenge. This is demonstrated by slow growth, above ground biomass, root system and harvest [22]. Accordingly, successful heavy metal phytoremediation requirement of hyperaccumulation capacity in candidate plants position halophytes as suitable phytoremediators. This is due to their extensive stress tolerance mechanism, which enables them thrive in saline soil and in other desert conditions.

2. Phytoremediation

In simple terms, phytoremediation refers to a process where plants are employed to reduce or free up organic and inorganic contaminants from the environment [13] with the aid of associated microbes. The process by which contaminants are remediated differs; these may be in the form of removal, transfer, degradation and immobilization from either soil or water [23]. It is a unique approach capitalizing on plants roots ability for the initial uptake of pollutants, and eventually accumulating them onto the shoot tissue by translocation across the stem. Compared to other conventional treatment techniques, phytoremediation is new, with a great potential to providing the much-needed green technology solution to our deteriorating environment. To date, hundreds of plant species were suggested as potential phytoremediation agents [24].

2.1. Phytoremediation techniques

During phytoremediation, plants growing on soil or water contaminated with trace or heavy metals could absorb or tolerate these elements differently, depending on the physiological means involved and the kinds of metals present [25]. According to Halder and Ghosh [26] phytoremediation techniques are categorized into five; phytoextraction, phytofiltration, phytovolatilization, phytostabilization and phytotransformation.

2.1.1. Phytoextraction

Phytoextraction is a technique of phytoremediation where plants take up metals by translocation, and accumulate them in a form that can be extracted on its tissue [27]. It is one of the most common types of phytoremediation and the names; phytoabsorption, phytoaccumulation and phytosequestration are often used interchangeably to refer to phytoextraction [28]. It is considered as the major phytoremediation technique among all others for the removal of metals from contaminated water, sediment and soil. The efficiency of this remediation process depends on a number of factors from soil properties, metal bioavailability and speciation to the type of plant species. However, high concentration of absorbed metals usually ends up in the shoot biomass of the plant in harvestable form [12]. A number of recent studies reported various plant species that demonstrate phytoextraction strategy from both water and soil media [29–32].

Plants able to exhibit phytoextraction strategy in metal sequestration may potentially be hyper accumulators, referring to plants that consistently accumulate certain threshold of metal concentration in their shoot tissue, which varies according to the metals [22]. Generally, all hyper accumulators should possess characteristics such as high growth rate, widely branched shoot, high bioaccumulation and translocation capacity, high above ground biomass, easily cultivated and harvested [22, 33]. However, Ali, Khan [28] demonstrated two methods or approaches for metal phytoextraction in different plants, one producing less above ground biomass but significantly accumulate metals in high concentration and vice versa in the other plant species, with final metal accumulation in agreement with those of hyper accumulators. Consequently, hyper accumulation is more important in phytoremediation than volume of biomass produced, and this suggest the use of hyper accumulators as more acceptable since it has advantages such as safe disposal, cheap process and easy handling [28].

2.1.2. *Phytofiltration*

Phytofiltration or rhizofiltration, as used interchangeably, refers to the absorption or adsorption of contaminants from surface wastewater by plant roots thereby preventing them from leaching to the underground water [34]. It is a type of phytoremediation technique that can be demonstrated *in situ* by directly growing plants in the polluted water body [24]. Although it is commonly applicable using aquatic plant species [35], there are suggestions that the process may be applied to terrestrial plants, which remediate metals to precipitate with the aid of microbes root bio filter [36]. Indeed, root exudates cause metal precipitation which alters the rhizosphere pH level [37]. Many terrestrial plants including grasses grown in a hydroponic culture were shown to effectively remove metals such via phytofiltration [38]. In the same study, Indian mustard was especially reported to accumulate higher fold of metal concentration far beyond the initial concentration, and the removal is by tissue specific adsorption mediated by root metal concentration.

Quite a number of studies have shown many species of aquatic macrophytes that demonstrate phytofiltration potential. While experimenting for phytoremediation under different water conditions polluted with heavy metals, Liao and Chang [39] found that *Eichhonia crassipes* absorb and accumulates metal contaminants, it has also exhibit high growth rate and increased biomass production and thus considered a good phytofiltration agent. This plant species absorb high concentrations of Pb, Ni, Zn and Cu which accumulates much higher in the root tissue than the shoot, suggesting the important role of fibrous and tap root system found in the plant, which is one of the key characteristics of potential phytofiltration agent. In a similar study, other aquatic plant species including *Salvinia herzogii*, *E. crassipes*, *Pistia stratiotes* and *Hydromistia stolonifera* were shown to absorb high concentration of Cd with *P. stratiotes* accumulating higher Cd concentration and exhibiting faster growth rate, a feature attributed to possible complimentary mechanism for the enhanced metal uptake [40]. Absorption of Cd in the root of all the plants relates to the added concentration. In another study by Thayaparan, Iqbal [41] also reported that *Azolla pinnata* have shown a great potential in the removal of high Pb concentration by phytofiltration from polluted water. As in phytoextraction, potential phytofiltration agents should tolerate high metal concentration, exhibit fast and high growth rate as well as above ground biomass, however, in contrast to phytoextraction, they are expected to show limited translocation capacity of absorbed metals from root to shoot tissues [24]. For efficient phytofiltration, this is an advantage over phytoextraction, since low translocation of contaminants means reduced contamination of other parts of the plant.

2.1.3. *Phytostabilization*

In this technique, pollutants are converted into a less toxic or bioavailable form by the continuous precipitation of the plant rhizosphere. This is achieved either by surface run off prevention, erosion or leaching [27]. It is applicable in the stabilization of metals in contaminated soil, sediment or water environments, which ensures they are not transferred to the food chain from the soil by translocating to other parts of food crops or to the underground water. This is possible by sorption via the root, precipitation and subsequent metal reduction around the plant rhizosphere, for instance the toxic Cr^{6+} is converted to Cr^{3+}, which is less toxic [42, 43]. Variation exists as to how prone a metal is to phytostabilization and is subject to its chemical character. This is evidenced in a comparative study to evaluate metal accumulation capacity of two aquatic macrophytes *Phragmites australis* and *Typha domingensis*, where both are found to stabilize As and Hg but inefficient in the phytostabilization of other metals [44].

Although phytostabilization offer some advantages over other phytoremediation techniques, it is however limited to temporary measure to deal with pollutants contamination owing to the fact that metals are only inactivated and their movement restricted, but still remains in the contaminated environmental compartment [45]. It is useful in emergencies, since it can rapidly immobilize pollutants from soil, water or sediment. Equally important, it ensures that contaminants are not translocated to other plant tissues by trapping most of it in the plant root [46]. Considering the strategies employed in phytostabilization, plants that can appropriately fall under this mechanism is their ability to tolerate and immobilize metals and other contaminants, low translocation capacity from root to plant aerial parts and of course extensive and fibrous tap root system [7]. Among several studies that reported plants species with these characteristics [47–49] demonstrating the phytostabilization of Zn, Pb, Cu and Cd by different plants in soil and sediment polluted environments.

2.1.4. *Phytotransformation/phytodegradation*

Phytotransformation or phytodegradation is another technique of phytoremediation where contaminants and other nutrients are chemically modified through plant metabolism and render associated contaminants inactive in both plant root and shoot tissues [6]. Plant metabolic enzymes act on the surrounding contaminants, thereby transforming them to a less toxic form, plants rhizosphere microbes also aid in the transformation process of the compounds [50]. Although this mechanism is mostly against organic contaminants, inorganic compounds such as metals were also suggested, in which case a strategy akin to phytostabilization is employed to convert toxic metals to less toxic form [51]. However, this technique seem less efficient and reliable compared to others in that it requires longer period of time, strict soil characteristic such as depth and underground water availability and often require soil amendments.

2.1.5. *Phytovolatilization*

In phytovolatilization, contaminants are converted in to a volatile form and released to the air via plants leaves stomata [27, 34]. However, this mechanism merely transfers contaminants from one environmental compartment to another, which may somehow return back to the original source (soil) by precipitation and hence could be less popular to other phytoremediation techniques especially phytoextraction and phytofiltration [34, 52]. It is commonly employed when treating groups of highly volatile metals like Hg and As. Phytovolatilization of As involves the conversion of elemental As to selenoaminoacids, such as selenomethione, which is modified by methylation to a volatile and less toxic form, dimethylselenide [53].

3. Metal hyperaccumulator plants

Several plants species are known to tolerate high concentration of toxic metals. Tolerant species are best described as excluders, where metal uptake and translocation to different tissue parts are limited. While others that are capable of accumulating higher concentrations with improved translocation from the root to shoot part of the plant, thereby significantly reducing its availability in the soil, and they do so with no visible sign of toxicity effects. To date, heavy metals have no standard definition by recognized bodies in the area. Various researchers use different characteristics and levels in their description such as atomic mass and number, density, chemical character as well as their toxicity; however, there appears no connection between such properties [54]. According to Wang and Chen [55], three categories of heavy metals arising from both natural and artificial sources are of interest, these includes valuable metals e.g. Ag, Au, Pd, Pt, harmful metals e.g., As, Cu, Co, Cd, Cr, Hg, Ni, Pb, and radionuclides such as Am, Th, Ra and so on. The non-biodegradability and stable nature of heavy metals suggests increased exposure to living species including humans [54], periodic reviews of toxic metals effects are documented by many research groups [56–58].

When determining hyperaccumulators of toxic metal, the most important factor is the concentration of the metal ion threshold. Therefore, plants can be regarded as hyperaccumulators, when capable of accumulating toxic metals concentration to about 50 to 100 times more than non-hyperaccumulator plants [13, 59]. For instance, the threshold for Zn and Mn hyperaccumulation in plant shoot is pegged at 1% of dry biomass, 0.01% for Cd and 0.1% respectively for Ti, Se, Sb, Pb, Ni, Cu, Cr, Co, and As [13, 60]. To date, few plant species are classified as hyperaccumulators, the majority of them (3/4) are tolerant to Ni and belongs to the Brassicaceae family native to Western Asia and Southern Europe, with up to 48 species implicated in Ni accumulation of around 3% dry shoot mass [60–62]. There is increasing interest in plant hyperaccumulators in recent times, owing to their potential use in metal contaminated soil and water detoxification [25, 63].

3.1. The role of metal chelators in hyperaccumulation

The phytoremediation of heavy metals involve many physiological, biochemical and molecular activities. In this process, especially phytoextraction involves the accumulation and translocation of heavy metals to plant tissues. Plant metal chelators or phytochelatins (PCs) and metallothioneins (MTs) are the most common transporter proteins for heavy metal phytoremediation. MTs are cysteine rich proteins that are famous for metal binding and greatly assist in the process of sequestration of metals in ionic form [64]. PCs are glutathione synthase products and they binds to heavy metals thereby forming central part of the phytodetoxification mechanism [65, 66]. The induction of phytochelatins is induced by the activity of an enzyme, phytochelatins synthase, which is triggered by the activity of metal ions present [34, 67]. In an experiment to demonstrate the role of synthases, mutants in model plant *Arabidopsis thaliana* were shown to be hypersensitive to Hg and Cd, which is attributed to their inability to produce PCs [68]. On the other hand, MTs are genetically encoded metal binding peptides and usually bear low molecular weight. A number of studies demonstrate MTs role in the protection of plants against the toxicity of heavy metals in soil, sediment and water [65, 69, 70]. The expression of MTs and PCs, alongside organic acid synthesis, together functions in heavy metal uptake by plants and also their translocation to other tissue parts [42]. The expression of these natural chelators could be enhanced to increase the efficiency of heavy metal accumulation and translocation. Currently, there are many ongoing studies aimed at characterizing and identifying biomolecules involved in the transport and detoxification of heavy metals. This will aid in understanding the whole detoxification process involved in plants [28, 71], and to achieve this, the importance of comparative proteomic studies cannot be over-emphasized.

3.1.1. *The shoot proteome*

Plant shoot is an important tissue in phytoremediation process; it is especially responsible for accumulating the highest metals concentration when the subject plant employs phytoextraction technique, which is subject to the type of metal elements and bioavailability. In recent times, there has been an increased interest in the proteomics study of plant hyperaccumulators with the aim of characterizing and identifying proteins acting in metal sequestration and detoxification [72]. These are possible with the advancement in modern mass spectrometry techniques such as two-dimensional liquid chromatography matrix-assisted laser desorption/ionization time-of-flight (2D-LC/TOFMS), time of flight/mass spectrometry (TOF/MS), two-dimensional polyacrylamide gel electrophoresis (2D PAGE), and liquid chromatography- tandem mass spectrometry (LC–MS/MS). For instance, the proteome of many plants species including *Thlaspi caerulescens, Pteris vittata, Helianthus annuus* and *Agrostis tenuis* were recently searched for heavy metals detoxifying proteins; several key functional proteins were found that protect plants against oxidative stress, as well as those responsive to biotic and abiotic stress condition among others [12, 73, 74].

In the proteomics study of plant metal hyperaccumulators, comparison could be made, even when these studies are from different plants and metals. In 2005, [75] found that prolonged exposure of *Alyssum lesbiacum* to Ni in an optimized experimental condition induced only three proteins, and one of these proteins, iron superoxide dismutase (Fe-SOD), was demonstrated to have antioxidant activity [76], while the other two proteins were identified as chloroplast phosphoglycerate kinase and a transketolase both having a role in the carbohydrates metabolism. In *Anemone halleri*, photosynthetic protein (chlorophyll a/b binding) and membrane protein (photosystem II) were constantly translated and upregulated when treated with Zn and Cd, which is linked to the improved metabolic energy demand in this metal hyperaccumulating plant [77].

At high metal concentrations, increased proteins induction are involved in the defense against antioxidants and energy metabolism has been consistently observed; examples includes Renal Epithelial Protein (APX), Superoxide dismutase (SOD), cytochrome P450 and Glutathione S-transferase (GST). These suggest that, for the uptake, translocation and accumulation of heavy metals concentration on the shoot tissue, plants require the functional photosynthetic process as well as the activity of proteins that scavenge oxygen radical species [13]. Metabolic energy active proteins were also suggested to have important roles in metal tolerance by plants. The proteomes of *T. caerulescens* with variable tolerance to Cd and Zn metals were compared and there was a higher accumulation of the extrinsic subunit of photosystem II protein, which led to its stabilization in the more metal tolerant variant as against the less tolerant accession. In addition to GST and cytochrome P450 earlier mentioned, other proteins such as aspartate aminotransferase and thioredoxin are commonly found, and linked to the sequestration of xenobiotics including metals. GSTs have particularly been demonstrated to be up regulated in many other living species including bacteria and fungi treated with metals like Zn, Cu and Cd [78]; hence GST were suggested to confer resistance to toxic genes in these cells.

3.1.2. The root proteome

In plants, the root tissue is the first to be exposed to all potential toxicants whether in the soil or surface water and hence serve as the gateway route through which they can subsequently be translocated to other tissue parts. Plants diversity as to hyperaccumulators and non-hyperaccumulators exist, this is due to the fact that, while some species bear the complete mechanism of enhanced metal uptake and eventual translocation, others have limited sequestration capacity in their root vacuoles [79]. In non-hyperaccumulators plant roots, Zn transporters are only detectable in the absence of Zn, whereas in hyperaccumulators, there is constitutive expression of these proteins such as ZT1 even in Zn deficient condition [80, 81]. In *T. caerulescens*, the iron transporter protein IRT1 was found to be involved in Zn and Cd uptake [82], similarly, root proteome study of this hyperaccumulator and *A. lesbiacum* were conducted by Tuomainen, Tervahauta [83] to evaluate peptides involved in Zn and Cd hyperaccumulation. In these studies, various classes of proteins were identified, their availability and or abundance varies relative to metal exposure and accessions. As in the case with similar studies on shoot proteome of hyperaccumulators, ROS scavenging proteins were more abundant in the more metal tolerant accessions compared to the less metal tolerant species. It was concluded that the changes in the enzyme, superoxide dismutase (SOD) availability upon which Zn depends in the different accessions may be linked to ROS increase.

An important organelle, cell wall, in the plant root is also affected by its exposure to heavy metal stress. The putative protein, glycosyl hydrolase family 18, involved in the formation of cell wall structure was shown to be regulated in accordance to treatment conditions and accession. These proteins, which are particularly known to be involved in cell wall expansion, differ in terms of abundance between the root proteome of two accessions, which in turn also affect the capacity of metals uptake; higher Ni and Cd accumulation was observed in the variant with more protein abundance [13]. Despite the recent advancement in proteomics technology, root protein transporters are yet to be differentially identified. Indeed, this is in agreement with transcriptomics studies, with analyzed data suggesting the constitutive expression of metal genes transporters in plant metal hyperaccumulators [13, 81, 84].

3.2. The halophytes of Qatar are promising heavy metal hyperaccumulators

Some studies demonstrated the potential of several Qatari plants as good phytoremediation candidates, many among which are heavy metal hyperaccumulators. Examples includes species belonging to the genus *Zygophyllum*, which are as either metal tolerant or accumulators when tested on both polluted soil and wastewater media [85–89]. Others include Typha domingensis and *Phragmites australis* [2]. According to Carvalho and Martin [90], *Typha*

S/No	Plant species	Metal (s)	Metal accumulation (mg/kg)	References
1	*Atriplex halimus* subsp. *schweinfurthii*	Cadmium	606.51	[97]
	A. halimus L.	Cadmium	830	[98]
		Zinc	44	
2	*Arthrocnemum macrostachyum*	Lead	620	[99]
3	*Crucianella maritima*	Zinc	390	[99]
4	*Dittrichia viscosa*	Lead	270	[99]
5	*Tamarix smyrnensis* Bunge	Lead	800	[100]
		Cadmium	800	
6	*Typha domingensis*	Selenium	30	[90]
		Lead	59.13	[101]
7	*T. lotifolia* L	Cadmium	210	[102]
8	*Paspalum conjugatum* L. *Prosopis laevigata*	Lead	150	[103]

Table 1. Examples of phytoremediation studies using species belong to Qatari flora and/or their relatives.

domingensis remediated heavy metals from industrial waste water and solution cultures; similarly, members of the halophytes plant family Brassicaceae were reported to be important phytoremediation agents [3, 59, 91] and the tree plant *Prosopis juliflora* exhibited phytoremediation of heavy metals potential [92, 93]. Additionally, such as other plants, such as *Phragmites australis* that were previously shown to clean petroleum-polluted soils may be good candidates for the phytoremediation of typical oil and gas produced wastewater [94], others with similar potentials includes Medicago such as *Medicago sativa* and *Glycine max* which also demonstrated strong petroleum polluted soil phytoremediation activity [95, 96]. Some examples of other species tested for phytoremediation studies and their metal uptake capacity are summarized in **Table 1** above.

4. Conclusion

The accumulation of trace and heavy metals in the environment present a great risk to biota health. These contaminants are implicated in a wide range of human diseases and various long-term negative environmental consequences, thereby endangering overall sustainable development initiatives worldwide. Many conventional treatment strategies are widely practiced for the remediation of these contaminants. However, traditional remediation processes have many disadvantages, from complicating environmental pollution to high operational cost among others. Phytoremediation is one of the most promising alternatives in this regard, and laboratory experiments have demonstrated the capacity of hundreds of plants species to remediate different heavy metal contaminants. However, there still exist limitations in the application of this emerging technology. This may be linked to exposure to other stress factors in field conditions, and especially in extreme environments, which could significantly affect physiological function and general growth. An example is above the ground biomass accumulation, a key requirement for plants that is critical to phytoremediation success. The identification and selection of plants with higher metal uptake capacity or hyperaccumulators, even in the presence of other stress condition is therefore the objective of many phytoremediation studies in recent times. Additionally, our limited understanding on the molecular mechanism of heavy metal remediation, such as the exact role of transporter proteins is compounding progress in this area. However, it is obvious that several stress response molecules are key to the tolerance and or accumulation of heavy metal contaminants by potential phytoremediators. The halophytes are famous for their adaptation to stress environmental conditions, and hence could be the most suitable candidates in the search for appropriate heavy metal hyperaccumulators and consequent elucidation of mechanism of uptake. Indeed, these are significant steps essential to improving the efficiency of phytoremediation for large scale, field and industrial applications.

Acknowledgements

This book chapter was financially supported for publication by Qatar National Library (a member of The Qatar Foundation). The statements made herein are solely the responsibility of the authors. The authors wish to acknowledge the funding support of Academic Research Office at Qatar University for the grant (QUST-CAS-SPR-2017-2033).

Conflict of interest

The authors wish to declare no conflict of interest.

Author details

Kamal Usman, Mohammad A. Al-Ghouti and Mohammed H. Abu-Dieyeh*

*Address all correspondence to: dandelion@qu.edu.qa

Department of Biological and Environmental Sciences, College of Arts and Sciences, Qatar University, Doha, Qatar

References

[1] Lasat MM. Phytoextraction of toxic metals. Journal of Environmental Quality. 2002; **31**(1):109-120

[2] Yasseen BT, Al-Thani RF. Ecophysiology of wild plants and conservation perspectives in the State of Qatar. In: Stoytcheva M, Zlatev R, editor. Agricultural Chemistry. Croatia: InTech; 2013. pp. 210. ISBN: 978-953-51-1026-2

[3] Lone MI et al. Phytoremediation of heavy metal polluted soils and water: Progresses and perspectives. Journal of Zhejiang University. Science. B. 2008;**9**(3):210-220

[4] Baldwin SA et al. The microbial community of a passive biochemical reactor treating arsenic, zinc, and sulfate-rich seepage. Frontiers in Bioengineering and Biotechnology. 2015;**3**:27

[5] Hasegawa H, Rahman IMM, Rahman MA, editors. Environmental Remediation Technologies for Metal-Contaminated Soils. Tokyo: Springer; 2016. 254 p. DOI: 10.1007/978-4-431-55759-3. https://doi.org/10.1007/978-4-431-55759-3

[6] Tangahu BV et al. A review on heavy metals (As, Pb, and Hg) uptake by plants through phytoremediation. International Journal of Chemical Engineering. 2011;**2011**:31. DOI: 10.1155/2011/939161

[7] Doble M, Kumar A. Biotreatment of Industrial Effluents. Oxford: Elsevier; 2005. 336 p

[8] Witters N. Phytoremediation: An alternative remediation technology and a sustainable marginal land management option [Thesis]. Hasselt: Hasselt University; 2011

[9] Ullah A et al. Phytoremediation of heavy metals assisted by plant growth promoting (PGP) bacteria: A review. Environmental and Experimental Botany. 2015;**117**:28-40

[10] Monica RC, Cremonini R. Nanoparticles and higher plants. Caryologia. 2009;**62**(2):161-165

[11] Ali-Zade V, Alirzayeva E, Shirvani T. Plant resistance to anthropogenic toxicants: Approaches to phytoremediation. In: Plant Adaptation and Phytoremediation. Netherlands: Springer; 2010. pp. 173-192

[12] Walliwalagedara C et al. Differential expression of proteins induced by lead in the dwarf sunflower *Helianthus annuus*. Phytochemistry. 2010;**71**(13):1460-1465

[13] Visioli G, Marmiroli N. The proteomics of heavy metal hyperaccumulation by plants. Journal of Proteomics. 2013;**79**:133-145

[14] Himelblau E, Amasino RM. Delivering copper within plant cells. Current Opinion in Plant Biology. 2000;**3**(3):205-210

[15] Hall J. Cellular mechanisms for heavy metal detoxification and tolerance. Journal of Experimental Botany. 2002;**53**(366):1-11

[16] Lin C-Y et al. Comparison of early transcriptome responses to copper and cadmium in rice roots. Plant Molecular Biology. 2013;**81**(4-5):507-522

[17] Singh RP et al. Response of higher plants to lead contaminated environment. Chemosphere. 1997;**34**(11):2467-2493

[18] Sharma P, Dubey RS. Lead toxicity in plants. Brazilian Journal of Plant Physiology. 2005;**17**(1):35-52

[19] Liu D et al. Uptake and accumulation of lead by roots, hypocotyls and shoots of Indian mustard [*Brassica juncea* (L.)]. Bioresource Technology. 2000;**71**(3):273-277

[20] Allan DL, Jarrell WM. Proton and copper adsorption to maize and soybean root cell walls. Plant Physiology. 1989;**89**(3):823-832

[21] Branquinho C, Brown DH, Catarino F. The cellular location of Cu in lichens and its effects on membrane integrity and chlorophyll fluorescence. Environmental and Experimental Botany. 1997;**38**(2):165-179

[22] van der Ent A, et al., Hyperaccumulators of metal and metalloid trace elements: Facts and fiction. Plant and Soil, 2013. **362**(1-2): p. 319-334

[23] Ahmadpour P et al. Phytoremediation of heavy metals: A green technology. African Journal of Biotechnology. 2012;**11**(76):14036-14043

[24] da Conceição Gomes MA et al. Metal phytoremediation: General strategies, genetically modified plants and applications in metal nanoparticle contamination. Ecotoxicology and Environmental Safety. 2016;**134**:133-147

[25] Pilon-Smits EA, Freeman JL. Environmental cleanup using plants: Biotechnological advances and ecological considerations. Frontiers in Ecology and the Environment. 2006;**4**(4):203-210

[26] Halder S, Ghosh S. Wetland macrophytes in purification of water. International Journal of Environmental Sciences. 2014;**5**(2):432-437

[27] Nwoko CO. Trends in phytoremediation of toxic elemental and organic pollutants. African Journal of Biotechnology. 2010;**9**(37):6010-6016

[28] Ali H, Khan E, Sajad MA. Phytoremediation of heavy metals—Concepts and applications. Chemosphere. 2013;**91**(7):869-881

[29] Li J et al. Cadmium tolerance and accumulation in cultivars of a high-biomass tropical tree () and its potential for phytoextraction. Journal of Environmental Quality. 2010; **39**(4):1262-1268

[30] Simmons R et al. Towards practical cadmium phytoextraction with *Noccaea caerulescens*. International Journal of Phytoremediation. 2015;**17**(2):191-199

[31] Kacálková L, Tlustoš P, Száková J. Phytoextraction of risk elements by willow and poplar trees. International Journal of Phytoremediation. 2015;**17**(5):414-421

[32] Shaheen SM, Rinklebe J. Phytoextraction of potentially toxic elements by Indian mustard, rapeseed, and sunflower from a contaminated riparian soil. Environmental Geochemistry and Health. 2015;**37**(6):953-967

[33] Sharma P, Pandey S. Status of phytoremediation in world scenario. International Journal of Environmental Bioremediation & Biodegradation. 2014;**2**(4):178-191

[34] Sarma H. Metal hyperaccumulation in plants: A review focusing on phytoremediation technology. Journal of Environmental Science and Technology. 2011;**4**(2):118-138

[35] Olguín EJ, Sánchez-Galván G. Heavy metal removal in phytofiltration and phycoremediation: The need to differentiate between bioadsorption and bioaccumulation. New Biotechnology. 2012;**30**(1):3-8

[36] Salt DE et al. Phytoremediation: A novel strategy for the removal of toxic metals from the environment using plants. Nature Biotechnology. 1995;**13**(5):468-474

[37] Rai PK. Heavy metal pollution in aquatic ecosystems and its phytoremediation using wetland plants: An ecosustainable approach. International Journal of Phytoremediation. 2008;**10**(2):133-160

[38] Khilji S. Rhizofiltration of heavy metals from the tannery sludge by the anchored hydrophyte, *Hydrocotyle umbellata* L. African Journal of Biotechnology. 2008;**7**(20)

[39] Liao S, Chang W-L. Heavy metal phytoremediation by water hyacinth at constructed wetlands in Taiwan. Photogrammetric Engineering and Remote Sensing. 2004;**54**:177-185

[40] Maine MA, Duarte MV, Suñé NL. Cadmium uptake by floating macrophytes. Water Research. 2001;**35**(11):2629-2634

[41] Thayaparan M et al. Rhizofiltration of Pb by Azolla Pinnata. International Journal of Environmental Sciences. 2013;**3**(6):1811

[42] Wu G et al. A critical review on the bio-removal of hazardous heavy metals from contaminated soils: Issues, progress, eco-environmental concerns and opportunities. Journal of Hazardous Materials. 2010;**174**(1):1-8

[43] Ghosh M, Singh S. A review on phytoremediation of heavy metals and utilization of it's by products. Asian Journal on Energy & Environment. 2005;**6**(4):18

[44] Bonanno G. Comparative performance of trace element bioaccumulation and biomonitoring in the plant species *Typha domingensis, Phragmites australis* and *Arundo donax*. Ecotoxicology and Environmental Safety. 2013;**97**:124-130

[45] Vangronsveld J et al. Phytoremediation of contaminated soils and groundwater: Lessons from the field. Environmental Science and Pollution Research. 2009;**16**(7):765-794

[46] Abreu CA et al. Organic matter and barium absorption by plant species grown in an area polluted with scrap metal residue. Applied and Environmental Soil Science. 2012;**2012**

[47] Andreazza R et al. Evaluation of two Brazilian indigenous plants for phytostabilization and phytoremediation of copper-contaminated soils. Brazilian Journal of Biology. 2015; **75**(4):868-877

[48] Najeeb U et al. Enhancing the lead phytostabilization in wetland plant *Juncus effusus* L. through somaclonal manipulation and EDTA enrichment. Arabian Journal of Chemistry. 2014

[49] Zhang X, Zhang X, Huang K. Phytostabilization of acidic soils with heavy metal contamination using three forage grasses in combination with organic and inorganic amendments. Soil and Sediment Contamination: An International Journal, 2016(just-accepted): p. 00-00

[50] Dobos L, Carmen P. The most important methods for depollution of hydrocarbons polluted soils. Bulletin of University of Agricultural Sciences and Veterinary medicine Cluj-Napoca Agriculture. 2009;**66**(1)

[51] Kobraee S et al. Influence of micronutrient fertilizer on soybean nutrient composition. Indian Journal of Science and Technology. 2011;**4**(7):763-769

[52] Nikolić M, Stevović S. Family Asteraceae as a sustainable planning tool in phytoremediation and its relevance in urban areas. Urban Forestry & Urban Greening. 2015;**14**(4):782-789

[53] Wang J et al. Remediation of mercury contaminated sites–a review. Journal of Hazardous Materials. 2012;**221**:1-18

[54] Sathya A et al. Cultivation of sweet sorghum on heavy metal contaminated soils by phytoremediation approach for production of bioethanol. In: Bioremediation and Bioeconomy. 2016. pp. 271-292

[55] Wang J, Chen C. Biosorption of heavy metals by *Saccharomyces cerevisiae*: A review. Biotechnology Advances. 2006;**24**(5):427-451

[56] Wongsasuluk P et al. Heavy metal contamination and human health risk assessment in drinking water from shallow groundwater wells in an agricultural area in Ubon Ratchathani province, Thailand. Environmental Geochemistry and Health. 2014;**36**(1): 169-182

[57] Duruibe J, Ogwuegbu M, Egwurugwu J. Heavy metal pollution and human biotoxic effects. International Journal of Physical Sciences. 2007;**2**(5):112-118

[58] Nagajyoti P, Lee K, Sreekanth T. Heavy metals, occurrence and toxicity for plants: A review. Environmental Chemistry Letters. 2010;**8**(3):199-216

[59] Baker A, Brooks R. Terrestrial higher plants which hyperaccumulate metallic elements. A review of their distribution, ecology and phytochemistry. Biorecovery. 1989;**1**(2):81-126

[60] Krämer U. Metal hyperaccumulation in plants. Annual Review of Plant Biology. 2010;**61**: 517-534

[61] Milner MJ, Kochian LV. Investigating heavy-metal hyperaccumulation using *Thlaspi caerulescens* as a model system. Annals of Botany. 2008;**102**(1):3-13

[62] Baker AJ et al. Metallophytes: The unique biological resource, its ecology and conservational status in Europe, central Africa and Latin America. Ecology of Industrial Pollution. 2010:7-40

[63] Chaney RL et al. Improved understanding of hyperaccumulation yields commercial phytoextraction and phytomining technologies. Journal of Environmental Quality. 2007;**36**(5):1429-1443

[64] Kärenlampi S et al. Genetic engineering in the improvement of plants for phytoremediation of metal polluted soils. Environmental Pollution. 2000;**107**(2):225-231

[65] Fulekar M, Singh A, Bhaduri AM. Genetic engineering strategies for enhancing phytoremediation of heavy metals. African Journal of Biotechnology. 2009;**8**(4)

[66] Yurekli F, Kucukbay Z. Synthesis of phytochelatins in *Helianthus annuus* is enhanced by cadmium nitrate. Acta Botanica Croatica. 2003;**62**(1):21-25

[67] Cobbett CS. Phytochelatin biosynthesis and function in heavy-metal detoxification. Current Opinion in Plant Biology. 2000;**3**(3):211-216

[68] Memon AR, Schröder P. Implications of metal accumulation mechanisms to phytoremediation. Environmental Science and Pollution Research. 2009;**16**(2):162-175

[69] Jabeen R, Ahmad A, Iqbal M. Phytoremediation of heavy metals: Physiological and molecular mechanisms. The Botanical Review. 2009;**75**(4):339-364

[70] Sheoran V, Sheoran A, Poonia P. Role of hyperaccumulators in phytoextraction of metals from contaminated mining sites: A review. Critical Reviews in Environmental Science and Technology. 2010;**41**(2):168-214

[71] Seth CS. A review on mechanisms of plant tolerance and role of transgenic plants in environmental clean-up. The Botanical Review. 2012;**78**(1):32-62

[72] Visioli G, Marmiroli N. Proteomics of plant hyperaccumulators. In: Metal Toxicity in Plants: Perception, Signaling and Remediation. Berlin, Heidelberg: Springer; 2012. pp. 165-186

[73] Bona E et al. Proteomic analysis of *Pteris vittata* fronds: Two arbuscular mycorrhizal fungi differentially modulate protein expression under arsenic contamination. Proteomics. 2010;**10**(21):3811-3834

[74] Visioli G et al. Comparison of protein variations in Thlaspi Caerulescens populations from metalliferous and non-metalliferous soils. International Journal of Phytoremediation. 2010; **12**(8):805-819

[75] Ingle RA, Smith JAC, Sweetlove LJ. Responses to nickel in the proteome of the hyperaccumulator plant alyssum lesbiacum. BioMetals. 2005;**18**(6):627-641

[76] Freeman JL et al. Increased glutathione biosynthesis plays a role in nickel tolerance in Thlaspi nickel hyperaccumulators. The Plant Cell. 2004;**16**(8):2176-2191

[77] Farinati S et al. Proteomic analysis of *Arabidopsis halleri* shoots in response to the heavy metals cadmium and zinc and rhizosphere microorganisms. Proteomics. 2009;**9**(21): 4837-4850

[78] Waschke A et al. Identification of heavy metal-induced genes encoding glutathione S-transferases in the arbuscular mycorrhizal fungus Glomus intraradices. Mycorrhiza. 2006;**17**(1):1-10

[79] Maestri E et al. Metal tolerance and hyperaccumulation: Costs and trade-offs between traits and environment. Environmental and Experimental Botany. 2010;**68**(1):1-13

[80] Pence NS et al. The molecular physiology of heavy metal transport in the Zn/Cd hyperaccumulator *Thlaspi caerulescens*. Proceedings of the National Academy of Sciences. 2000;**97**(9):4956-4960

[81] Assunção A et al. Elevated expression of metal transporter genes in three accessions of the metal hyperaccumulator *Thlaspi caerulescens*. Plant, Cell & Environment. 2001;**24**(2): 217-226

[82] Lombi E et al. Influence of iron status on cadmium and zinc uptake by different ecotypes of the hyperaccumulator *Thlaspi caerulescens*. Plant Physiology. 2002;**128**(4):1359-1367

[83] Tuomainen M et al. Proteomics of *Thlaspi caerulescens* accessions and an inter-accession cross segregating for zinc accumulation. Journal of Experimental Botany. 2010;**61**(4): 1075-1087

[84] Hammond JP et al. A comparison of the *Thlaspi caerulescens* and *Thlaspi arvense* shoot transcriptomes. New Phytologist. 2006;**170**(2):239-260

[85] Hu R et al. Physiological responses and tolerance mechanisms to Pb in two xerophils: *Salsola passerina* Bunge and *Chenopodium album* L. Journal of Hazardous Materials. 2012; **205**:131-138

[86] Rejeb KB et al. Evaluation of the Cd2+ phytoextraction potential in the xerohalophyte *Salsola kali* L. and the impact of EDTA on this process. Ecological Engineering. 2013;**60**: 309-315

[87] Osman HE, Badawy RK. Effect of pollution on the chemical content and secondary metabolites of *Zygophyllum coccineum* and *Tamarix nilotica*. Egyptian Pharmaceutical Journal. 2013;**12**(1):73

[88] De La Rosa G et al. Role of ethylenediaminetetraacetic acid on lead uptake and translocation by tumbleweed (*Salsola kali* L.). Environmental Toxicology and Chemistry. 2007; **26**(5):1033-1039

[89] Lefèvre I, Corréal E, Lutts S. Cadmium tolerance and accumulation in the noxious weed *Zygophyllum fabago*. Botany. 2005;**83**(12):1655-1662

[90] Carvalho KM, Martin DF. Removal of aqueous selenium by four aquatic plants. Journal of Aquatic Plant Management. 2001;**39**:33-36

[91] Ebbs S et al. Phytoextraction of cadmium and zinc from a contaminated soil. Journal of Environmental Quality. 1997;**26**(5):1424-1430

[92] Prasad M, Tewari JC. Prosopis juliflora (Sw) DC: Potential for bioremediation and bioeconomy. In: Bioremediation and Bioeconomy. 2016. pp. 49-76

[93] Shukla O et al. Growth responses and metal accumulation capabilities of woody plants during the phytoremediation of tannery sludge. Waste Management. 2011;**31**(1):115-123

[94] Nie M et al. Understanding plant-microbe interactions for phytoremediation of petroleum-polluted soil. PLoS One. 2011;**6**(3):e17961

[95] Frick C, Germida J, Farrell R. Assessment of phytoremediation as an in-situ technique for cleaning oil-contaminated sites. In technical seminar on chemical spills. 1999. Environment Canada; 1998

[96] Njoku K, Akinola M, Oboh B. Phytoremediation of crude oil contaminated soil: The effect of growth of Glycine max on the physico-chemistry and crude oil contents of soil. Nature and Science. 2009;**7**(10):79-87

[97] Nedjimi B, Daoud Y. Cadmium accumulation in *Atriplex halimus* subsp. schweinfurthii and its influence on growth, proline, root hydraulic conductivity and nutrient uptake. Flora-Morphology, Distribution, Functional Ecology of Plants. 2009;**204**(4):316-324

[98] Lutts S et al. Heavy metal accumulation by the halophyte species Mediterranean saltbush. Journal of Environmental Quality. 2004;**33**(4):1271-1279

[99] Conesa H et al. Influence of soil properties on trace element availability and plant accumulation in a Mediterranean salt marsh polluted by mining wastes: Implications for phytomanagement. Science of the Total Environment. 2011;**409**(20):4470-4479

[100] Manousaki E, Kalogerakis N. Phytoextraction of Pb and Cd by the Mediterranean saltbush (*Atriplex halimus* L.): Metal uptake in relation to salinity. Environmental Science and Pollution Research. 2009;**16**(7):844-854

[101] Hegazy A, Abdel-Ghani N, El-Chaghaby G. Phytoremediation of industrial wastewater potentiality by *Typha domingensis*. International Journal of Environmental Science and Technology. 2011;**8**(3):639-648

[102] Fediuc E, Erdei L. Physiological and biochemical aspects of cadmium toxicity and protective mechanisms induced in *Phragmites australis* and *Typha latifolia*. Journal of Plant Physiology. 2002;**159**(3):265-271

[103] Paz-Alberto AM et al. Phytoextraction of lead-contaminated soil using vetivergrass (*Vetiveria zizanioides* L.), cogongrass (*Imperata cylindrica* L.) and carabaograss (*Paspalum conjugatum* L.). Environmental Science and Pollution Research International. 2007;**14**(7): 498-504

Chapter 13

Assessment of Heavy Metals in Landfill Leachate: A Case Study of Thohoyandou Landfill, Limpopo Province, South Africa

Joshua N. Edokpayi, Olatunde S. Durowoju and John O. Odiyo

Additional information is available at the end of the chapter

http://dx.doi.org/10.5772/intechopen.74009

Abstract

Landfilling of solid wastes has gained increasing acceptance due to the ease of disposal. However, such activity has consequences if the landfill site is not designed according to specification or does not have a leachate liner and collection system. Leachate possesses potential risk to surface and groundwater aquifer within the area surrounding the landfill site. The aim of this chapter is to assess the physicochemical parameters and heavy metal levels in leachate generated from a periurban landfill site situated in Thohoyandou, Limpopo Province, South Africa. Physicochemical parameters were measured onsite using standard methods, while heavy metals were analyzed with flame atomic absorption spectrometer (FAAS) after nitric acid digestion. pH, conductivity and turbidity values ranged from 6.97 to 7.68, 426 to 2288 μS/cm and 12.78 to 295.5 NTU, respectively. Most levels of the determined heavy metals exceeded the effluent discharge guideline limit of South African Department of Water Affairs. This could potentially spike their levels in surface and groundwater. Adequate measures should be put in place to manage the leachate generated from landfill sites.

Keywords: groundwater, heavy metals, landfill, leachate, potential risk, surface water

1. Introduction

1.1. General background

Solid waste management is of global concern in both developing and developed countries. Despite much awareness aimed at reducing the waste generated due to anthropogenic activities, there has been an increase in solid waste generation throughout the world [1–3]. This could partly be due to increase in population, industrialization and urbanization. Different efforts geared toward effective management of solid waste due to the perceived adverse health and environmental impacts have been reported [4–6].

Landfilling remains one of the most commonly used methods for solid waste management in most parts of the world. Its efficiency and safety coupled with cost make it the preferred method [7–9]. Several advances in landfill technology have been reported to enhance its suitability for solid waste management. Leachate is often generated from landfill processes due to the increasing presence of soil moisture and other favorable environmental factors [1, 5].

In most developing countries, the facility for leachate collection and treatment is often not part of the design of landfill sites. One of the adverse effects caused by solid waste disposal onto landfills is the contamination of surface and groundwater by leachate. The extent of such contamination depends on the quality of leachate generated from the landfill [4, 7]. The composition of a landfill waste varies from place to place and depends on a number of factors. Some of them include the location of the site, the socioeconomic status of the people the landfill is serving, the technology used in the landfill and the age of the waste generated including prevailing environmental factors such as weather conditions of the landfill site, all of which influences the composition of leachate [1, 4, 7].

Blight and Fourie [10] reported the variation of the solid waste component disposed in landfill sites (**Table 1**). This ranges from simple organic and inorganic materials to complex ones.

Traditionally, in most developing countries, the segregation of hazardous waste from non-hazardous waste before disposal into a landfill site is uncommon. Edokpayi et al. [11] reported the disposal of several household hazardous waste into sanitary landfill sites. When both kinds of wastes are thus disposed of, the leachate quality will be poorer than if the hazardous waste was pretreated before landfilling or treated using other available methods.

Components	**% Composition by undried mass**							
	USA	**UK**	**Cape Town**	**Johannesburg**	**Gaborone**	**Lima**	**Delhi**	**Abu Dhabi**
Ash, soil, building rubble, other	8–14	18	39*	1**	3	16	41	16
Paper and cardboard	34–44	29	19	23	12	14	9	6
Plastic	8–10	7	7*	13	5	7	1+	12
Metals	11–13	8	2+	10	6	4	1+	8
Glass	4–9	10	1+	8	6	3	1+	9
Organic, putrescible	12–22	25	32	45	68	56	47	49

*Building rubble landfilled, in which sand cover is included.
**Building rubble not landfilled, in which cover soil is not included.
+Scavenged before landfilling.

Table 1. Percentage of solid waste composition from different parts of the world [10].

1.2. Principles of leachate formation

Leachate is formed due to the interaction between the waste in the landfill and water from soil moisture, precipitation, snow and other liquid waste disposed of in the landfill site. Leachate can also be formed as a result of chemical and biochemical processes within the landfill. There is nonuniform and intermittent percolation of moisture through the solid waste in a landfill as a result of leachate generation [5, 12, 13]. This consequently leads to the removal of soluble fractions from the waste into the leachate. A wide variety of contaminants such as heavy metals, some other persistent organic pollutants like flame retardants, polycyclic aromatic hydrocarbons (PAHs), phenols, polychlorinated biphenyls (PCBs), chlorinated alkyl phosphate, polyfluoroalkyl and other fluorinated chemicals, pharmaceuticals and personal care products, polybrominated diphenyl ethers (PBDEs), polychlorinated dibenzofurans (PCDFs), polychlorinated dibenzodioxins (PCDDs) and polychlorinated diphenyl ethers (PCDEs) are also collected into the leachate as it drains through the pile of waste in the landfill [14–20]. This makes leachate a wastewater of high strength because of the high levels of various classes of contaminants it contains.

Leachate control should, therefore, be included in the design of solid waste management systems. The volume and flow rate of the leachates are dependent upon the percolation of water through the waste layers [5]. The transfer of contaminants into the leachate and biodegradation processes are affected by the flow patterns and velocity of the leachate. The general pattern of the generation of leachate in landfills was reported by Farquhar [21] as presented in **Figure 1**.

Percolation from the unsaturated zone into the saturated zone is thus possible, and this gives a pathway for groundwater contamination. Surface runoff through the landfill also provides a possible route to surface water contamination majorly during precipitation events. The generation of leachate from landfills if not properly managed can lead to several adverse environmental and health impacts.

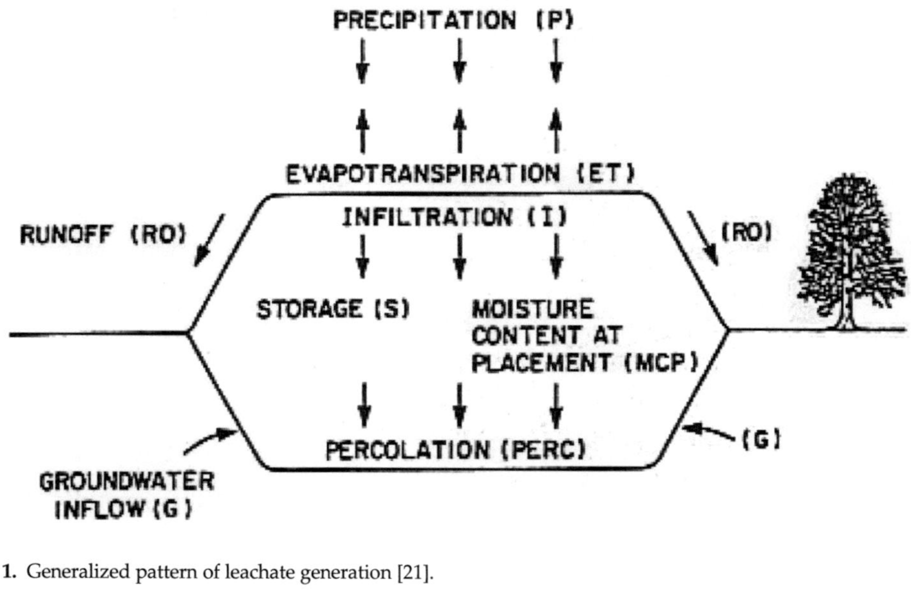

Figure 1. Generalized pattern of leachate generation [21].

1.3. Management and treatment of leachates

Leachate management is possible if the design of the landfill permits adequate lining, collection and storage systems. Landfill liners should be a part of the initial design of the landfill site. These liners are crucial as they act as a barrier preventing the leachate generated from the landfill to drain into the environment or groundwater aquifer [13, 22]. Various materials used as leachate liners include natural clays, geomembranes (constructed from various plastic materials such as polyvinyl chloride (PVC) and high-density polyethene (HDPE)), geotextiles, geonet and geosynthetic clay liners [13, 22–25]. The lining system can be made from a single material such as clay or a combination of clay and other materials. Basically, there are three types of lining systems which are single, composite or double lining systems. Various factors should be considered when selecting a lining system for a landfill as some lining materials can be easily degraded by chemicals in the leachate.

In developing countries, the lining of landfills and systems for collection and treatment of leachate are often found in some major cities, while same facilities in other cities, periurban or rural areas do not have a provision for leachate management. Collection of leachate in landfills is important to prevent it from accumulating up to the quantity that it can spill over, causing pollution in waterways and water courses [5, 12]. Poverty, lack of required expertise, corruption and misappropriation of funds could be some of the reasons that have limited adequate leachate management in the developing world.

Different methods have been reported for the management of leachate in the different parts of the world. The simpler method is to incorporate the leachate as an influent in a wastewater treatment facility [26]. This method is cheap since there will be no need for the construction of a new facility for the treatment. Conveyance pipe from the landfill to the wastewater treatment plant is sufficient, provided that the distance between both facilities is in close proximity. Depending on the leachate quality, pretreatment may be required before combining it with the influent of the municipal wastewater treatment works. However, in some cases, when a large quantity of leachate is generated on a daily basis, or the distance between both facilities is far apart, then onsite treatment of leachate will be the preferred option or the reticulation of the leachate back into the landfill. Several physical (evaporation, floatation, air stripping, filtration and membrane processes), chemical (coagulation and precipitation, adsorption, ion exchange, chemical oxidation) and biological methods (the use of ponds in various aerobic and anaerobic modes) have been reported for leachate treatment [26–30]. Cost, expertise, the age of the landfill, estimated volume of leachate generated, climatic condition and efficiency of the method often determine the treatment method of choice. Treatment of leachate using biological methods is more common in South Africa [26].

1.4. Study area

The landfill site with geographical coordinates 23°0′12.3″ S and 30°28′0.16″ E is located in Thulamela Municipality of Limpopo Province, South Africa (**Figure 2a**).

http://dx.doi.org/10.5772/intechopen.74009

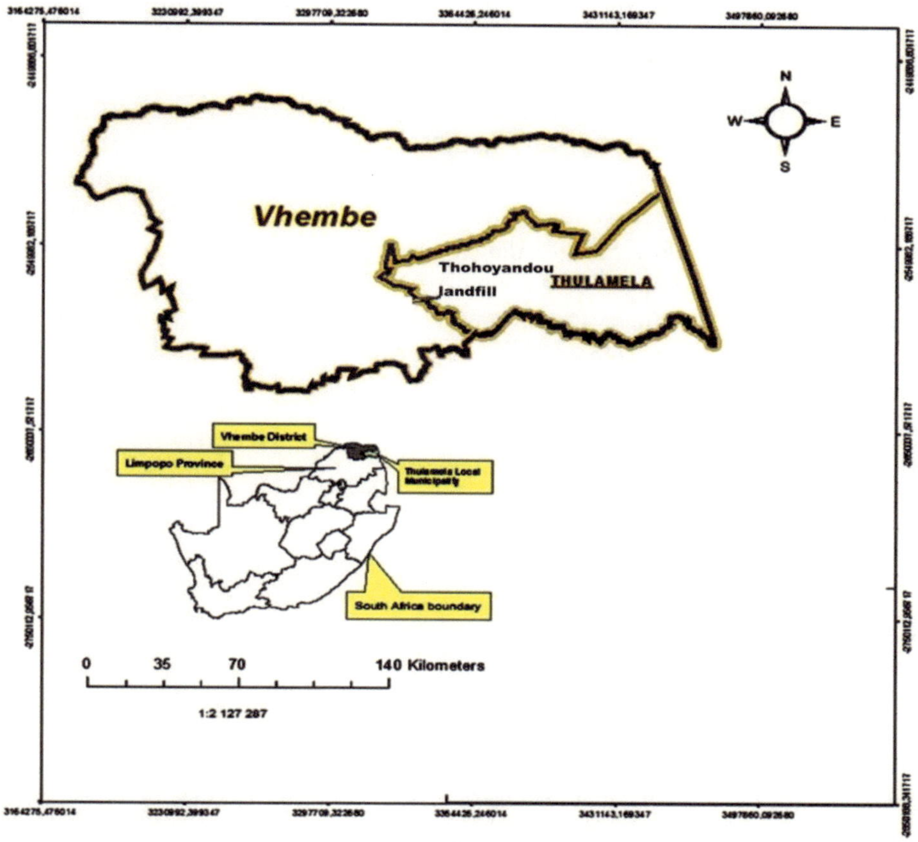

a

b

Figure 2. (a) Map of the study area. (b) Some pictures from Thohoyandou landfill site.

The site is located 180 km from the provincial city of Polokwane. In 2014, the facility received 210,000 tons of waste. Thohoyandou is a semiarid region with variations in temperature between 12 and 22°C in the dry season (winter) and between 20 and 40°C in the wet season (summer) [31]. Precipitation also varies between 340 and 2000 mm with an average rainfall of 800 mm. The region is classified as a humid subtropical dry forest biozone [32]. The landfill site is very close to the municipal wastewater treatment plant. Common surface water sources in the region include the Dzindi, Luvhuvhu and Mvudi Rivers. Thohoyandou landfill currently receives all kinds of sanitary wastes from the periurban center, which is Thohoyandou and other villages in the suburb. At the time of this study, there was no leachate collection and treatment system on the site. Residential areas are located around the landfill. Some other major land use around the landfill includes subsistence agriculture, schools and wastewater treatment plant. Some pictures taken from the landfill are presented in **Figure 2b**.

2. Materials and methods

2.1. Sampling

Six leachate samples were collected during the wet season of 2014 from various places within the premises of the landfill site. The samples were collected in a precleaned 1-L polyethylene sample container. The containers were washed with livid detergent and soaked with 1% nitric acid. Subsequently, the sample containers were rinsed with deionized water. Physicochemical parameters such as pH, conductivity and turbidity were measured in situ, while samples for the determination of heavy metals were preserved using concentrated nitric acid. The samples were transported on ice to Hydrology and Water Resources Department of the University of Venda. Four major metals (Calcium (Ca), Potassium (K), Magnesium (Mg) and Sodium (Na)) and 10 heavy metals (Aluminum (Al), Cadmium (Cd), Chromium (Cr), Cobalt (Co), Copper (Cu), Iron (Fe), Lead (Pb), Manganese (Mn), Nickel (Ni) and Zinc (Zn)) were selected and analyzed due to their availability in landfill leachate. The samples were preserved at −4°C until further analysis.

2.2. Instrumentation

pH and conductivity multimeter supplied by Fischer Thermoscientific was used for field measurements of pH and conductivity, respectively, while a turbidimeter supplied by HANA was employed to determine the turbidity of the samples. The measuring devices were calibrated following the protocol approved by the manufacturer. Flame atomic absorption spectrometer (FAAS) supplied by Perkin Elmer was used for the analysis of major and heavy metal content in the leachate samples. Calibration standards were prepared from AAS grade reagent for all the metals of interest.

http://dx.doi.org/10.5772/intechopen.74009

3. Results and discussion

The pHs of the leachate from all sites of sampling (**Table 2**) were found within the acceptable range of wastewater discharge onto the water resource by the South African Department of Water Affairs [33]. An average leachate pH of 7.39 was determined in this study which is suitable for methanogenic bacteria [34]. Various average pH values below and above this value have been reported for leachate in different parts of the developing world (**Table 3**). pH usually affects the speciation of metals in solutions with most metals easily speciated to their toxic forms in acidic pH below the value of 4, while some precipitate out of the solution at alkaline pH values.

Samples	pH	Conductivity (μS/m)	Turbidity (NTU)
L1	7.60	486	22.56
L2	7.47	485	28.83
L3	7.68	2288	295.5
L4	7.42	604	25.43
L5	7.20	721	33.15
L6	6.97	426	12.78

Table 2. Physicochemical parameters of leachate samples.

Parameters	Thohoyandou landfill (South Africa)	Alexandria landfill (Egypt) [34]	Semur landfill (Indian) [35]	El-Jadida landfill (Morocco) [36]
pH	7.39	8.9	6.9	8.8
Conductivity (μS/cm)	835	30,725	NR	26,000
Turbidity (NTU)	69.71	NR	NR	NR
Na (mg/L)	33.88	364	462	NR
K(mg/L)	22.94	5400	1241	NR
Ca (mg/L)	46.27	8.4	NR	190
Mg (mg/L)	16.58	302.5	NR	235

NR indicates values that were not reported.

Table 3. Comparison of the average physicochemical leachate characteristics in some developing countries.

Generally, the presence of inorganic charged particles such as carbonates, chlorides, sulfates and bicarbonate ions in water is often estimated in terms of the conductivity of the water. Conductivity oftentimes is also used as a surrogate for total dissolved solids [37]. **Table 2** shows the levels of conductivity determined from all the sampling sites. This level complied with the DWA wastewater discharge limit [33] and was much lower than the levels reported in Alexandria landfill leachate in Egypt (**Table 3**), Columbia (22,000 μS/cm) and Morocco (26,000 μS/cm) [34, 36, 38]. However, Rahim et al. [39] reported a lower conductivity level of 31.68 μS/cm for landfill leachate in Malaysia.

The turbidity of the leachate samples obtained was very high. Liquids with high turbidity often indicate the likelihood of silt, clay, fine organic matter and microscopic organisms that are suspended in the water. Site three recorded a very high turbidity level, much higher than the levels from other sites in the landfill.

Figure 3 shows the levels of major metals (Na, K, Mg and Ca) in the leachate of Thohoyandou landfill. The levels found are not higher than what can be disposed of into water bodies. The only concern is the continuous migration of salt-rich leachate into surface and groundwater sources which could increase their levels in the river and groundwater aquifers. Various sequences of major metals' occurrence in leachate varied greatly from one location to the other. The sequence for major metals determined in this study follows the trend: Ca > Na > K > Mg (**Table 3**), while at Alexandria landfill in Egypt, the sequence is K > Na > Mg > Ca [34]. Level of Mg is higher than Ca in the leachate of all the landfills reported except Thohoyandou landfill. Although rock type in groundwater aquifers contributes significantly to its salinity level, continuous percolation of metal-rich leachate may also contribute to increased groundwater salinity.

The levels of heavy metals determined in this study are presented in **Figures 4** and **5**. Variable levels of heavy metals were determined in the landfill site. On average, Fe, Mn and Al are present in higher concentrations than the other metals. A possible reason for the high level of these metals could be due to their relative abundance on Earth. Previous studies have shown that Fe and Al are present in higher concentrations in river water, soils and sediments around Thohoyandou when compared to other metals [32, 37, 40]. The relatively low levels of the

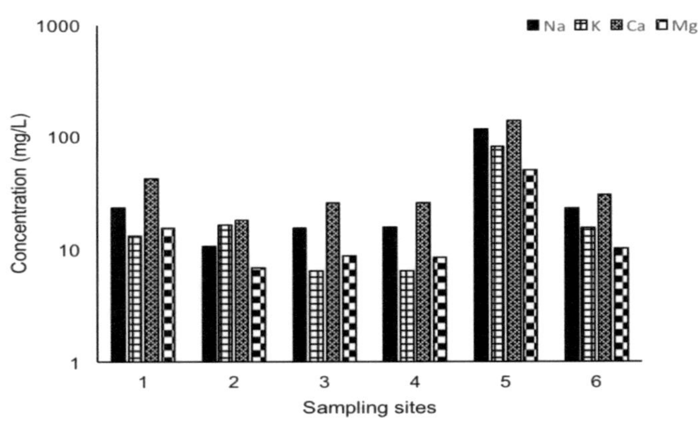

Figure 3. Concentrations of major metals in leachate samples.

http://dx.doi.org/10.5772/intechopen.74009

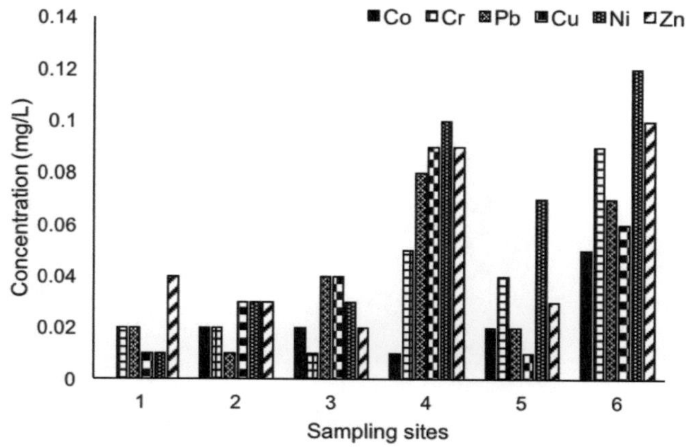

Figure 4. Levels of Co, Cr, Pb, Cu, Ni and Zn in different sampling sites of Thohoyandou landfill.

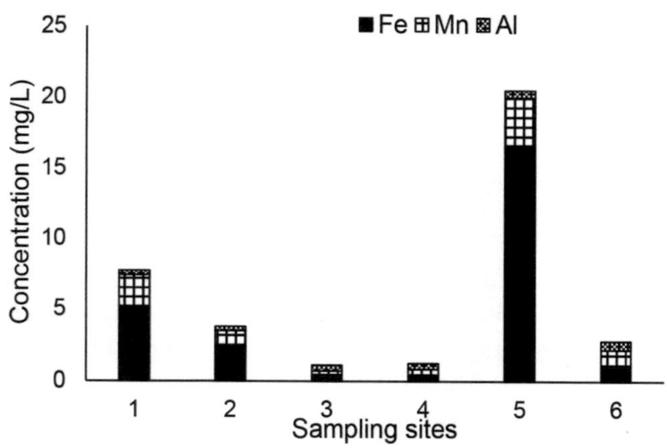

Figure 5. Levels of Fe, Mn and Al in different sampling sites of Thohoyandou landfill.

more hazardous heavy metals could suggest that the waste generated from this region is not largely from a highly industrialized area. It is, however, noteworthy to mention that most of the heavy metals were present in concentrations higher than the normal wastewater discharge guideline values except for Zn and Cr. Ni, Al and Co does not have guideline values for their discharge. **Table 4** shows the comparison of average heavy metal concentration in leachate from landfills in some developing countries. Fe recorded the highest level from all the countries except in Jebel Chakir, Landfill of Tunisia, where Fe was not part of the metals determined [34, 36, 37, 41, 42]. Mn and Zn were also part of the metals that generally existed in high concentration.

Heavy metals (mg/L)	Thohoyandou landfill (South Africa)	Alexandria landfill (Egypt) [34]	Semur landfill (Indian) [35]	El-Jadida landfill (Morocco) [36]	Alger landfill (Algeria) [41]	Jebel Chakir landfill (Tunisia) [42]	DWA wastewater discharge limit [33]
Al	0.42	NR	NR	NR	NR	NR	NR
Cd	0.02	0.10	0.02	0.03	NR	0.04	0.005
Cr	0.02	0.06	0.28	0.16	0.50	1.17	0.05
Co	0.02	NR	NR	0.21	NR	NR	NR
Cu	0.07	0.07	0.71	0.16	0.45	0.80	0.01
Fe	4.26	6.31	58.91	24	12.30	NR	0.3
Pb	0.03	0.02	1.20	NR	NR	0.35	0.01
Mn	1.50	0.84	NR	1.25	0.40	0.05	0.1
Ni	0.03	0.10	0.31	0.13	0.25	0.23	NR
Zn	0.08	0.75	1.29	0.75	0.70	0.75	0.1

NR indicates values that were not reported.

Table 4. Comparison of the average heavy metal characteristics in some developing countries.

4. Conclusion

This chapter has provided the much-needed baseline data of leachate for Thohoyandou landfill. The levels of Fe and Mn determined in the leachate of this study were very high as they exceeded the guideline value for wastewater discharge. Al concentration was also high. The most interesting result indicated that leachate from Thohoyandou landfill is of comparable quality to leachate from other developing countries.

Acknowledgements

The University of Venda is appreciated for providing the facilities to conduct this study.

Author details

Joshua N. Edokpayi*, Olatunde S. Durowoju and John O. Odiyo

*Address all correspondence to: joshua.edokpayi@univen.ac.za

Department of Hydrology and Water Resources, University of Venda, South Africa

References

[1] Al-Yaqout AF, Hamoda MF. Evaluation of landfill leachate in arid climate-a case study. Environment International. 2003;**29**:593-600. DOI: 10.1016/S0160-4120(03)00018-7

[2] Scott J, Beydoun D, Amal R, Low G, Cattle J. Landfill management, leachate generation, and leach testing of solid wastes in Australia and overseas. Critical Reviews in Environmental Science and Technology. 2005;**35**:239-332. DOI: 10.1080/10643380590917969

[3] Naveen BP, Mahapatra DM, Sitharam TG, Sivapullaiah PV, Ramachandra TV. Physico-chemical and biological characterization of urban municipal landfill leachate. Environmental Pollution. 2017;**220**:1-12. DOI: 10.1016/j.envpol.2016.09.002

[4] El-Fadel M, Findikakis AN, Leckie JO. Environmental Impacts of solid waste landfilling. Journal of Environmental Management. 1997;**50**:1-25. DOI: 10.1006/jema.1995.0131

[5] Zeiss C, Uguccioni M. Mechanisms and patterns of leachate flow in municipal solid waste landfills. Journal of Environmental Systems. 1994;**23**:247-270

[6] Kassim SM. The importance of recycling in solid waste management. Macromolecular Symposia. 2012;**320**:43-50. DOI: 10.1002/masy.201251005

[7] Capelo J, de Castro MAH. Measuring transient water flow in unsaturated municipal solid waste – A new experimental approach. Waste Management. 2007;**27**:811-819. DOI: 10.1016/j.wasman.2006.04.012

[8] El-Gohary FA, Kamel G. Characterization and biological treatment of pre-treated landfill leachate. Ecological Engineering. 2016;**94**:268-274. DOI: 10.1016/j.ecoleng.2016.05.074

[9] Krčmar D, Tenodi S, Grba N, Kerkez D, Watson M, Rončević S, Dalmacija B. Preremedial assessment of the municipal landfill pollution impact on soil and shallow groundwater in Subotica, Serbia. Science of the Total Environment. 2018;**615**:1341-1354. DOI: 10.1016/j.scitotenv.2017.09.283

[10] Blight GE, Fourie AB. Experimental landfill caps for semi-arid and arid climates. Waste Management & Research–Journal of International Solid Wastes and Public Cleansing Association, ISWA. 2005;**23**:113-125. DOI: 10.1177/0734242X05052458

[11] Edokpayi JN, Odiyo JO, Durowoju OS, Adetoro A. In: Daniel Mmereki, editor. Household Hazardous Waste Management in Sub-Saharan Africa. Rijeka, Croatia: InTech; 2017. DOI: 10.5772/66292. Available from: https://www.intechopen.com/books/household-hazardous-waste-management/household-hazardous-waste-management-in-sub-saharan-africa

[12] El-Fadel M, Bou-Zeid E, Chahine W, Alayli B. Temporal variation of leachate quality from pre-sorted and baled municipal solid waste with high organic and moisture content. Waste Management. 2002;**22**:269-282

[13] Hughes KL, Christy AD, Heimlich JE. Landfill Types and Liner Systems. United States of America: Ohio State University; 2013

[14] Adeniyi A, Dayomi M, Siebe P, Okedeyi O. An assessment of the levels of phthalate esters and metals in the Muledane open dump, Thohoyandou, Limpopo Province, South Africa. Chemistry Central Journal. 2008;**2**:9. DOI: 10.1186/1752-153X-2-9

[15] Chimuka L, Nefale F, Masevhe A. Determination of phenols in water samples using a supported liquid membrane extraction probe and liquid chromatography with photodiode array detection. South African Journal of Chemistry. 2007;**60**:102-108

[16] Huset CA, Barlaz MA, Barofsky DF, Field JA. Quantitative determination of fluorochemicals in municipal landfill leachates. Chemosphere. 2011;**82**:1380-1386. DOI: 10.1016/j.chemosphere.2010.11.072

[17] Busch J, Ahrens L, Sturm R, Ebinghaus R. Polyfluoroalkyl compounds in landfill leachates. Environmental Pollution. 2010;**158**:1467-1471. DOI: 10.1016/j.envpol.2009.12.031

[18] Eggen T, Moeder M, Arukwe A. Municipal landfill leachates: A significant source for new and emerging pollutants. Science of the Total Environment. 2010;**408**:5147-5157. DOI: 10.1016/j.scitotenv.2010.07.049

[19] Clarke BO, Anumol T, Barlaz M, Snyder SA. Investigating landfill leachate as a source of trace organic pollutants. Chemosphere. 2015;**127**:269-275. DOI: 10.1016/j.chemosphere.2015.02.030

[20] Dudzinska MR, Czerwinski J. Persistent Organic Pollutants (POPs) in leachates from municipal landfills. International Journal of Environmental Engineering. 2011;**3**:253

[21] Farquhar GJ. Leachate: Production and characterization. Canadian Journal of Civil Engineering. 1989;**16**:317-325. DOI: 10.1139/l89-057

[22] Sarabian T, Rayhani MT. Hydration of geosynthetic clay liners from clay subsoil under simulated field conditions. Waste Management. 2013;**33**:67-73. DOI: 10.1016/j.wasman.2012.08.010

[23] Benson CH, Kucukkirca IE, Scalia J. Properties of geosynthetics exhumed from a final cover at a solid waste landfill. Geotextiles and Geomembranes. 2010;**28**:536-546. DOI: 10.1016/j.geotexmem.2010.03.001

[24] Divya PV, Viswanadham BVS, Gourc JP. Influence of geomembrane on the deformation behaviour of clay-based landfill covers. Geotextiles and Geomembranes. 2012;**34**:158-171. DOI: 10.1016/j.geotexmem.2012.06.002

[25] Salim IA, Miller CJ, Howard JL. Sorption isotherm-sequential extraction analysis of heavy metal retention in landfill liners. Soil Science Society of America Journal. 1996;**60**:107-114. DOI: 10.2136/sssaj1996.03615995006000010018x

[26] DWA Guidelines for Lechate Control. Pretoria, South Africa: Department of Water Affairs

[27] Castrillón L, Fernández-Nava Y, Ulmanu M, Anger I, Marañón E. Physico-chemical and biological treatment of MSW landfill leachate. Waste Management. 2010;**30**:228-235. DOI: 10.1016/j.wasman.2009.09.013

[28] Peng Y. Perspectives on technology for landfill leachate treatment. Arabian Journal of Chemistry. 2017;**10**:S2567-S2574. DOI: 10.1016/j.arabjc.2013.09.031

[29] Foo KY, Hameed BH. An overview of landfill leachate treatment via activated carbon adsorption process. Journal of Hazardous Materials. 2009;**171**:54-60. DOI: 10.1016/j.jhazmat.2009.06.038

[30] Raghab SM, Abd El Meguid AM, Hegazi HA. Treatment of leachate from municipal solid waste landfill. HBRC Journal. 2013;**9**:187-192. DOI: 10.1016/j.hbrcj.2013.05.007

[31] Edokpayi JN, Odiyo JA, Msagati TAM, Potgieter N. Preliminary monitoring of faecal indicator organisms of surface water: A case study of Mvudi River, South Africa. Journal of Science and Technology (Ghana). 2016;**36**:33-37. DOI: 10.4314/just.v36i1.6

[32] Edokpayi JN, Odiyo JO, Popoola OE, Msagati TAM. Assessment of trace metals contamination of surface water and sediment: A case study of Mvudi River, South Africa. Sustainability. 2016;**8**:135. DOI: 10.3390/su8020135

[33] DWA Wastewater limit values applicable to discharge of wastewater into a river resource. National Water Act, Government Gazzette No 20528; 1999; p. Pretoria, South Africa

[34] Abd El-Salam MM, Abu-Zuid G. Impact of landfill leachate on the groundwater quality: A case study in Egypt. Journal of Advanced Research. 2015;**6**:579-586. DOI: 10.1016/j.jare.2014.02.003

[35] Nagarajan R, Thirumalaisamy S, Lakshumanan E. Impact of leachate on groundwater pollution due to non-engineered municipal solid waste landfill sites of erode city, Tamil Nadu, India. Journal of Environmental Health Science & Engineering. 2012;**9**:35. DOI: 10.1186/1735-2746-9-35

[36] Chofqi A, Younsi A, Lhadi EK, Mania J, Mudry J, Veron A. Environmental impact of an urban landfill on a coastal aquifer (El Jadida, Morocco). Journal of African Earth Sciences. 2004;**39**:509-516. DOI: 10.1016/j.jafrearsci.2004.07.013

[37] Edokpayi JN, Odiyo JO, Olasoji SO. Assessment of heavy metal contamination of Dzindi River, in Limpopo Province, South Africa. International Journal of Natural Sciences Research. 2014;**2**:185-194

[38] Olivero-Verbel J, Padilla-Bottet C, De la Rosa O. Relationships between physicochemical parameters and the toxicity of leachates from a municipal solid waste landfill. Ecotoxicology and Environmental Safety. 2008;**70**:294-299. DOI: 10.1016/j.ecoenv.2007.05.016

[39] Rahim BEA, Yusoff I, Samsudin AR, Yaacob WZW, Rafek AGM. Deterioration of groundwater quality in the vicinity of an active open-tipping site in West Malaysia. Hydrogeology Journal. 2010;**18**:997-1006. DOI: 10.1007/s10040-009-0567-3

[40] Edokpayi JN, Odiyo JO, Msagati TAM, Popoola EO. Removal efficiency of faecal indicator organisms, nutrients and heavy metals from a peri-urban wastewater treatment plant in Thohoyandou, Limpopo Province, South Africa. International Journal of Environmental Research and Public Health. 2015;**12**:7300-7320. DOI: 10.3390/ijerph120707300

[41] Belkacemi R, Kerbachi M. Characterization and evolution of the landfill leachate of Oued Smar, Alger. Techniques Sciences Methods. 1996;**11**:615-618

[42] Abdelwaheb A, Moncef Z, Hamed BD. Landfill leachate generation and its impact on water at an urban landfill(Jebel Chakir, Tunisia). Hydrology: Current Research. 2012;**3**:1-6. DOI: 10.4172/2157-7587.1000128

Section 5

Adsorption of Heavy Metals

Chapter 14

The Thermodynamics of Heavy Metal Sorption onto Lignocellulosic Biomass

Carlos Escudero-Oñate and Isabel Villaescusa

Additional information is available at the end of the chapter

http://dx.doi.org/10.5772/intechopen.74260

Abstract

The sorption equilibrium and thermodynamics of Cu(II), Ni(II), Pb(II), and Cd(II) onto grape stalks (GS), a lignocellulosic waste from wine production industries, have been investigated. Different equilibrium models have been assessed to describe the experimental sorption equilibrium profile in the range of 5–60°C. Maximum sorption capacities have been calculated by means of Langmuir equilibrium model and mean free sorption energies through the Dubinin-Radushkevich (D-R) model. Mean free energies suggest that metal sorption takes place mainly through an ion exchange mechanism, except for Pb(II), where an additional contribution connected to a stronger bond might take place. The calculation of thermodynamic parameters, ΔG^0, ΔH^0 and ΔS^0, puts into evidence that the sorption of all the metals onto GS is a spontaneous and exothermic process that occurs with an increase of randomness at the solid/liquid interface.

Keywords: sorption, divalent metals, lignocellulosic sorbent, isotherm, thermodynamic

1. Introduction

Metals can enter the environment through a large variety of processes such as weathering of soils and rocks, volcanoes, and from a variety of anthropogenic activities [1, 2]. From the anthropogenic sources, modern industry is, to a large degree, a major responsible of environmental pollution. They are frequently released into the soil and water as from various polluting sources, such as foundries, tanneries, textile, microelectronic, fertilizer and pesticide industries, mining activity and other industrial activities [3]. These inorganic species occur naturally as ions, compounds and complexes, and they can lead to health problems and degradation of natural environments due to their toxicity and persistent character. Developing sustainable and environmentally friendly technologies to remove toxic metal from industrial effluents is a relevant topic nowadays.

Several remediation techniques to remove metal ions from aqueous solutions are available, which range from traditional physico-chemical methods to emerging bioremediation methods [3–10]. Methods that have been successfully deployed in industrial environments include the use of physico-chemical processes such as chemical precipitation, ion exchange, oxidation/reduction, reverse osmosis and electrochemical treatment [2, 11–13]. These methods however exhibit a set of drawbacks such as high acquisition and operation costs, low performance at relatively low—but still relevant—concentration of metals and being sources of secondary pollution [14–16]. To overcome the aforementioned issues, bioremediation-based methods have appeared as potential candidates in the treatment of heavy metal effluents. Bioremediation methods include bioaccumulation, biosorption and phytoremediation. Biosorption in some cases has demonstrated an outstanding potential, comparable to the performance obtained in ion exchange-based methods.

In regular sorption studies, the performance of a sorbent is evaluated by studying the kinetics of the process and assessing the amount sorbed versus the sorbate concentration in solution at equilibrium to get the isotherm curve. Obtaining the characteristics sorption isotherms themselves do not provide automatically any information about the reaction involved in the sorption phenomenon [17]. The study of the effect of temperature on the sorption process and the evaluation of the thermodynamic properties such as Gibb's free energy, enthalpy and entropy of the process provide valuable information about the strength of the interactions between sorbate and sorbent and the energy associated with the sorption process. The standard free energy of the reaction (ΔG^0, $J{\cdot}mol^{-1}$) is the difference between the initial state (free solute compound)and the final equilibrated state (sorbed compound), and the parameter is related to the spontaneity of the sorption process. Negative values of ΔG^0 indicate that the process is spontaneous. The magnitude of the enthalpy of the process (ΔH^0, $kJ{\cdot}mol^{-1}$) gives an idea about the type of sorption interactions (physical or chemical). Whilst in physisorption-based processes enthalpy range is comprised between 2.1 and 20.9 $kJ{\cdot}mol^{-1}$, higher enthalpy values are characteristic from chemisorption (20–800 $kJ{\cdot}mol^{-1}$) [18]. The value of the change of enthalpy $\Delta H^0 < 0$ or $\Delta H^0 > 0$ also suggests the character exothermic or endothermic of the process, respectively.

The change of entropy (ΔS^0) reflects essentially the variation on the disorder of a system (on macroscopic level) along a process. A positive value of this parameter indicates increased randomness at the solid/solution interface that may also include some changes in the sorbent and sorbate structure. Moreover, $\Delta S^0 > 0$ implies an increase in the degree of freedom of the adsorbed species. The negative value of change of entropy ($\Delta S^0 < 0$) suggests that the adsorption process involves an associative mechanism. Also a negative value of ΔS^0 implies a decreased disorder at the sorbent/solution interface during the sorption process causing the sorbate species to escape from the solid phase to the solution phase.

In this chapter, the sorption equilibrium of Cu(II), Ni(II), Pb(II), and Cd(II) onto a lignocellulosic material, grape stalks (GS), has been investigated. The studies were performed at different temperatures and allowed gathering relevant thermodynamic parameters to better describe the interactions established between the divalent metals and the sorbent.

2. Effect of temperature on Cu(II), Ni(II), Pb(II) and Cd(II) equilibrium

The residues of grape stalk (GS) obtained from a wine production industry were washed with distilled water, cut in small pieces, dried and ground to obtain a sorbent with a particle size range 0.25–0.50 mm. The sorption equilibrium of Cu(II), Ni(II), Pb(II) and Cd(II) in single metal solutions onto GS was explored at different temperatures within the range of 5–60°C. The characteristic sorption isotherms were obtained contacting 0.1 g of GS powder with 15 mL of different Cu(II), Ni(II), Pb(II) and Cd(II) solutions within the initial concentration range of 5–1000 mg·L^{-1}. Stoppered glass tubes were employed, and the initial pH was adjusted to 5.2. For an accurate temperature control, the samples were placed in an incubator (ICP-500, Memmert). After equilibration of the sorbent biomaterials with the metal solutions for 24 hours, the samples were filtered and acidified adding 100 μL of ultra-high-quality HNO_3, and metal concentration in solution was determined by flame atomic absorption spectrophotometry (Varian SpectrAA 220FS).

From the experimental sorption equilibrium results, thermodynamic parameters related to the affinity and energy of the sorbent-sorbate interaction were obtained and discussed.

2.1. Sorption isotherms

Equilibrium relationships between adsorbent and adsorbate are described by adsorption isotherms and reflect the relationship between the quantity adsorbed and that remaining in solution at a given temperature [19]. Sorption isotherms provide essential information for optimization of the adsorption mechanism pathways since they are expression of the surface properties and capacities of the sorbents. They become therefore relevant tools in the design of sorption systems, since they help understanding how sorbates interrelate with the sorbent materials [20].

Cu(II), Ni(II), Pb(II) and Cd(II) sorption isotherms onto GS for the different temperatures explored were obtained by plotting the amount of metal adsorbed per GS sorbent mass unit (q_e; mol·g^{-1}) as a function of the remaining metal concentration in solution (C_e; mol·L^{-1}). The amount of sorbed metal was computed according to the equation presented below:

$$q_e = \frac{(C_0 - C_f)}{m} * V \quad (1)$$

where C_0 and C_f are the initial and final metal concentration in solution, respectively (mol·L^{-1}), m(g) is the sorbent mass (g) and V(L) represents the volume of the solution. The characteristic isotherms are presented in **Figure 1**. This figure clearly demonstrates that the amount of metal sorbed increases as it does the remaining metal concentration in solution until a maximum value is achieved. In the studied range, the temperature seemed not to have a clear effect on the maximum sorption capacity of GS for the divalent metals, except for Cu(II). In this case, the increase of temperature involved an increase on the maximum sorption capacity at equilibrium from 0.22 mmol·g^{-1} (at 5°C) to approximately 0.28 (at 60°C). The experimental Cu(II), Ni(II), Pb(II) and Cd(II) sorption equilibrium results onto GS were submitted to Langmuir, Freundlich and D-R models. The different models and the results obtained are presented and discussed in the next section.

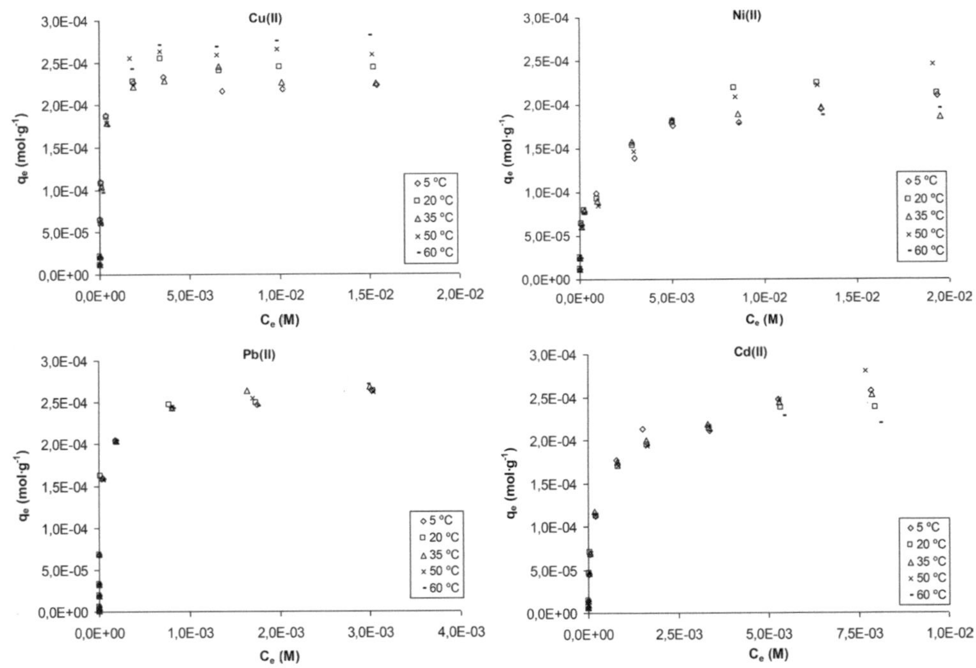

Figure 1. Sorption isotherms of Cu(II), Ni(II), Pb(II) and Cd(II) onto GS. T: 5–60°C.

2.2. Modeling and calculation of sorption equilibrium parameters

The equilibrium adsorption isotherms are one of the most important data that help understanding the sorption mechanism/s and provide fundamental insight for optimization and scale-up of sorption-based processes. Among the different isotherm models available, three of the most representative and largely employed have been chosen for this study: Langmuir, Freundlich and D-R isotherms.

The Langmuir model involves homogenous distribution of sorption sites and is based on the next set of assumptions: (i) the maximum sorption capacity corresponds to a saturated monolayer of solute in the surfaces, (ii) all the active sites are equivalent and the sorption energy remains constant, and (iii) there is no migration of adsorbed species in the plane of the surfaces. From this model, the maximum uptake, q_{max} (mol·g^{-1}), and the Langmuir constant, K_L (L·mol^{-1}), can be obtained. While q_{max} reflects the maximum uptake of the sorbent, the parameter K_L is a constant related to the energy of adsorption that quantitatively reflects the affinity between the sorbent and the sorbate. Fitting the experimental dataset to the Langmuir model, the effect of the temperature and of the nature of the metal on the different sorption equilibriums can be ascertained. By means of the Langmuir constant, it can also be discussed whether an adsorption system is favorable or unfavorable. The essential feature of the Langmuir isotherm can be expressed by means of the parameter R_L, a dimensionless constant referred to as separation factor or equilibrium parameter. R_L is calculated using the following equation [15, 21–27]:

$$R_L = \frac{1}{1 + K_L \cdot C_0} \tag{2}$$

being K_L the Langmuir constant (L·mol^{-1}) and C_0 the initial metal concentration (mol·**L**$^{-1}$). The R_L parameter is considered as a reliable indicator of the adsorption being the next four the possible situations [15, 25] (**Table 1**).

The Freundlich model is based in an empirical equation largely employed in the description of sorption processes in heterogeneous systems. On its linear form, the Freundlich equation takes the form:

$$\log q_e = \log K_F + \frac{1}{n}\log C_e \tag{3}$$

where K_F and 1/n are empirical constants indicate of the relative sorption capacity and sorption intensity, respectively.

The experimental dataset was also submitted to the D-R isotherm model. The selection of this model was supported by the fact that it is considered as more general than Langmuir and it does not rely necessarily in the formation of a homogenous monolayer surface or a constant adsorption potential. D-R model has been used by several authors to distinguish between physical and chemical adsorption onto different biomaterials [28, 29]. This model can be linearized and described by the next equation:

$$\ln q_e = \ln q_m - \beta\varepsilon^2 \tag{4}$$

being β is a constant related to the mean free energy of adsorption per mole of the adsorbate (mol^2·J^{-2}), q_m is the theoretical saturation capacity of the monolayer and ε is the Polanyi potential. The expression of this last parameter is RTln(1 + (1/C_e)), being R (8.314 J·mol^{-1}·K^{-1}) the gas constant and T (K) the absolute temperature. Hence, by plotting ln(q_e) against ε^2, it is possible to generate the value of q_m (mol·g^{-1}) from the intercept and the value of β from the slope.

The constant β provides information about the mean free energy, E (J·mol^{-1}). The parameter E is defined as the free energy change required to transfer 1 mol of sorbate from the solution to the solid surface and can be calculated using the relationship [30, 31]:

$$E = \frac{1}{\sqrt{2\beta}} \tag{5}$$

The fitting of the experimental sorption datasets to the linearized expressions of Langmuir, Freundlich and D-R adsorption isotherms for the different temperatures explored is presented

R_L value	Type of isotherm
>1	Unfavorable
1	Linear
$0 < R_L < 1$	Favorable
0	Irreversible

Table 1. The isotherm characteristics according to the R_L value.

in **Figures 2–4**, respectively. The separation factor (R_L) calculated for the different initial metal concentrations has been also plotted and is presented in **Figure 5**.

From the linear plots of the different isotherm models, the characteristic sorption parameters of the divalent metals onto GS were calculated and are presented in **Table 2**. The separation factor (R_L) has been also plotted for the different initial metal concentrations and temperatures (**Figure 5**). As it may be observed, all the R_L values are found in the range $0 < R_L < 1$, indicating that the sorption of all the metals onto GS is a favorable process regardless on the initial concentration.

According to the R^2 values presented in **Table 2**, the best fitting to the experimental dataset is provided by the Langmuir model. In general, the calculated values of maximum capacity and affinity M(II)-GS obtained through this model indicate that there is not a dramatic effect of the temperature on the sorption process. Cu(II), however, seems exhibiting a slight increase on maximum sorption capacity when the temperature is increased.

When Q_{max} is compared at a standard temperature of 20°C, it can be observed that a very similar capacity (about $2.5{\cdot}10^{-4}$ $mol{\cdot}g^{-1}$) is achieved regardless of the metal. The Q_{max} values obtained at 20°C through the Langmuir model for the four metals is in agreement with previously reported data [32, 33]. The effect of temperature on the strength of the interaction sorbent-sorbate will be explored later by calculating specifically the thermodynamic parameters of the adsorption process.

The modeling of the experimental dataset according to the D-R equation was also able to provide a good fitting of the experimental trends observed. This model allowed computing the mean

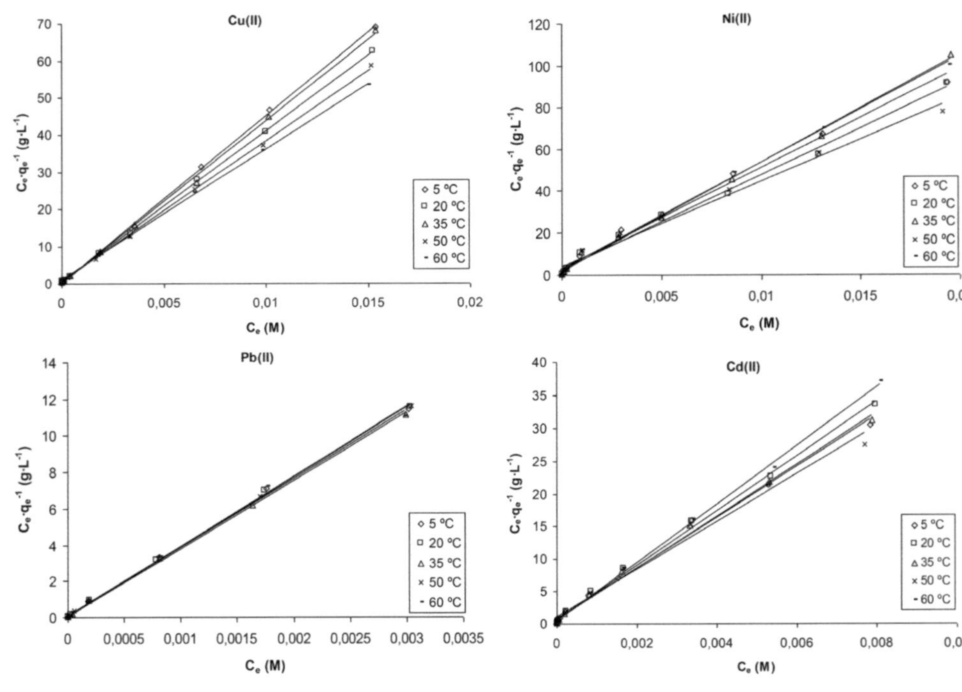

Figure 2. Langmuir model fitting of Cu(II), Ni(II), Pb(II) and Cd(II) sorption results in the temperature range of 5–60°C.

http://dx.doi.org/10.5772/intechopen.74260

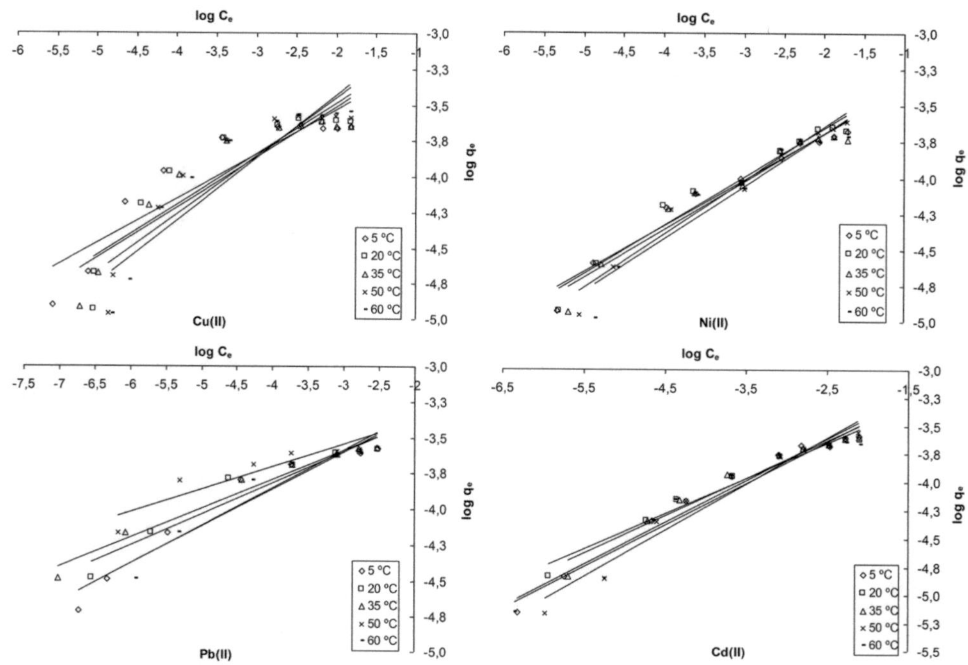

Figure 3. Freundlich model fitting of Cu(II), Ni(II), Pb(II) and Cd(II) sorption results in the temperature range of 5–60°C.

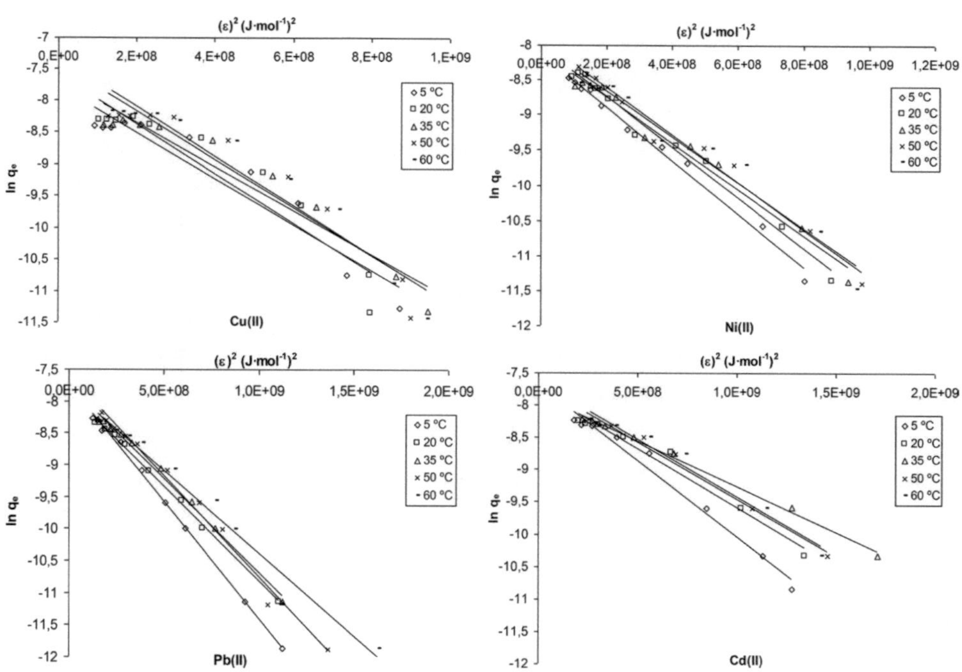

Figure 4. Dubinin-Radushkevich model fitting of Cu(II), Ni(II), Pb(II) and Cd(II) sorption results in the temperature range of 5–60°C.

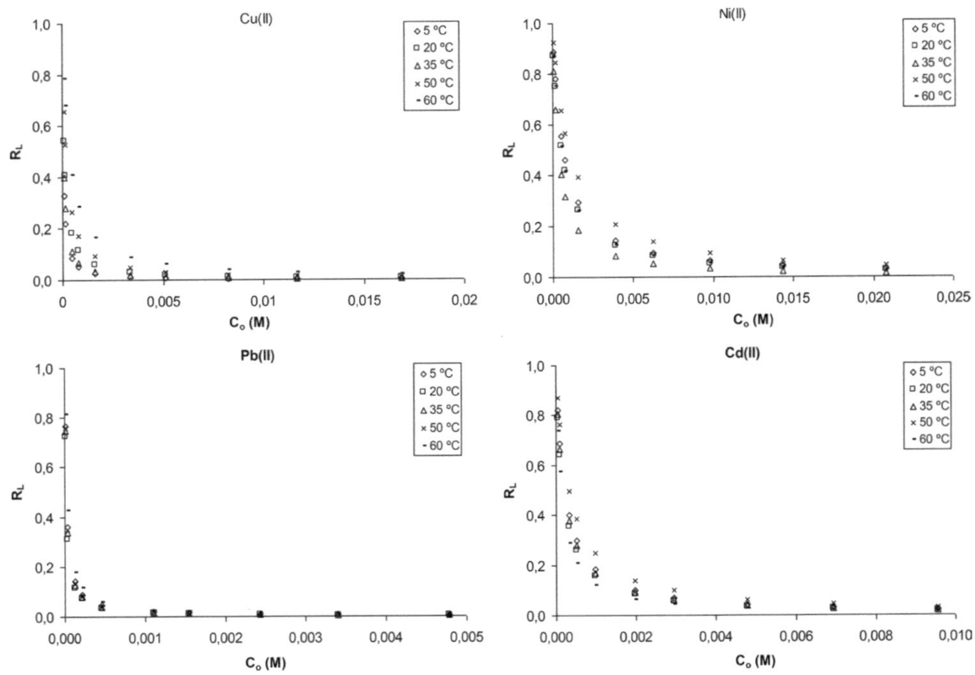

Figure 5. Variation of adsorption intensity (R_L) for Cu(II), Ni(II), Pb(II) and Cd(II) with the initial metal concentration in the temperature range of 5–60°C.

free energy of adsorption E (kJ·mol^{-1}) according to Eq. (5). This parameter provides useful information that allows classifying the adsorption mechanism as chemical ion exchange or physical adsorption. If the magnitude of E is between 8 and 16 kJ·mol^{-1}, the adsorption process follows a chemical ion exchange [15, 30]. On the other hand, values of E < 8 kJ·mol^{-1} indicate that the adsorption process is of a physical nature [34]. Values higher than 16 kJ·mol^{-1} would be indicative of more energetic interactions than the corresponding to an ion exchange process. As it can be seen in **Table 2**, the values of adsorption mean free energies are in the range 8 < E (kJ·mol^{-1}) < 16 for Cu(II), Ni(II) and Cd(II) sorption at all the temperatures and for Pb(II) at the lowest one, 5°C. These results point out that Cu(II), Ni(II) and Cd(II) sorption onto GS at all the studied temperatures and Pb(II) sorption at 5°C proceeds mainly via ion exchange. For Pb(II) at the temperature of 20°C and higher, the values in the range 16.64 < E < 18.81 indicate that there is an extra contribution to sorption by ion exchange and stronger Pb(II)-GS bonds are being formed.

With the dataset generated in the equilibrium experiments performed at different temperatures, the characteristic thermodynamic parameters of Cu(II), Ni(II), Pb(II) and Cd(II) sorption onto GS were calculated.

2.3. Thermodynamic parameters of adsorption

The temperature dependence of the sorption process is associated with several thermodynamic parameters that allow concluding whether the process is spontaneous or not. The Gibbs free energy change, ΔG^0, is an indicative of the spontaneity of a chemical reaction, and

http://dx.doi.org/10.5772/intechopen.74260

		Langmuir			Freundlich			Dubinin-Radushkevich			
Metal	**T (°C)**	**$Q_{max}\cdot 10^4$ ($mol\cdot g^{-1}$)**	**$K_L\cdot 10^{-4}$ ($L\cdot mol^{-1}$)**	**R^2**	**1/n**	**K_F**	**R^2**	**$Q_{max}\cdot 10^4$ ($mol\cdot g^{-1}$)**	**$\beta\cdot 10^9$ ($mol^2\cdot kJ^{-2}$)**	**E ($kJ\cdot mol^{-1}$)**	**R^2**
Cu(II)	**5**	2.22	2.34	0.999	0.31	2.04	0.836	4.20	3.61	11.76	0.913
	20	2.46	0.97	0.999	0.36	2.26	0.824	5.04	3.89	11.34	0.895
	35	2.29	1.72	0.998	0.31	2.02	0.858	4.67	3.44	12.06	0.931
	50	2.69	0.60	0.999	0.41	2.55	0.840	6.07	3.78	11.50	0.919
	60	2.85	0.32	0.999	0.44	2.75	0.859	6.64	3.88	11.35	0.926
Ni(II)	**5**	2.09	0.15	0.992	0.32	2.09	0.976	2.93	3.79	11.49	0.913
	20	2.22	0.18	0.993	0.34	2.17	0.958	3.24	3.58	11.82	0.895
	35	1.93	0.28	0.997	0.33	2.16	0.941	3.06	3.28	12.35	0.931
	50	2.45	0.10	0.986	0.37	2.36	0.958	3.55	3.37	12.18	0.919
	60	1.97	0.18	0.997	0.37	2.32	0.914	3.31	3.23	12.44	0.926
Pb(II)	**5**	2.61	4.71	0.999	0.37	2.35	0.972	4.67	2.36	14.55	0.913
	20	2.60	5.86	0.999	0.31	2.06	0.972	4.22	1.81	16.64	0.895
	35	2.68	5.23	0.999	0.33	2.15	0.957	3.92	1.41	18.81	0.931
	50	2.61	4.95	0.994	0.41	2.55	0.957	4.59	1.76	16.84	0.919
	60	2.65	3.55	0.997	0.37	2.34	0.940	4.81	1.77	16.80	0.926
Cd(II)	**5**	2.57	0.46	0.993	0.26	1.83	0.938	4.37	3.66	11.69	0.913
	20	2.40	0.56	0.997	0.22	1.65	0.926	3.84	2.95	13.03	0.895
	35	2.52	0.51	0.995	0.20	1.59	0.966	4.33	2.93	13.06	0.931
	50	2.73	0.31	0.985	0.16	1.44	0.850	5.02	3.15	12.60	0.919
	60	2.24	0.76	0.998	0.26	1.81	0.921	4.12	2.58	13.92	0.926

Table 2. Adsorption isotherm constants for the adsorption of cu(II), Ni(II), Pb(II) and cd(II) onto GS as a function of temperature.

therefore, it is an important criterion when it comes to spontaneity assessment of a sorption process. Both energy and entropy factors must be considered in order to determine the Gibbs free energy of the process. Reactions occur spontaneously at a given temperature if ΔG^0 has a negative value, and this parameter can be determined from the following equation:

$$\Delta G^0 = -RT \ln K_L \tag{6}$$

being R the ideal gas constant (8.314 $J \cdot mol^{-1} \cdot K^{-1}$) and T the absolute temperature (K). Besides, the standard Gibbs free energy can be defined in terms of enthalpy (ΔH^0) and entropy (ΔS^0) using the equation:

$$\Delta G^0 = \Delta H^0 - T \Delta S^0 \tag{7}$$

Including the Langmuir equilibrium constant (K_L) in the Van't Hoff equation, the enthalpy and the entropy of the process can be calculated:

$$\ln K_L = \frac{\Delta S^0}{R} - \frac{\Delta H^0}{RT} \tag{8}$$

From the equation presented above, the values of ΔH^0 and ΔS^0 can be determined using the slope and the intercept of the plot of ln K_L versus 1/T. ΔG^0, ΔH^0 and ΔS^0 were calculated for the different temperatures, and the results are presented in **Table 3**.

The negative values of ΔG^0 observed for all the M(II)-GS systems indicate that the sorption process is feasible and spontaneous. The negative ΔH^0 values obtained for the sorption of all the metals indicate that the sorption process is also exothermic. It is worth noting that, from the four metals, the sorption of Cu(II) is the process that involves a higher exchange of energy, releasing about 17.5 kJ per mol of sorbed metal. The sorption of other three metals involved a much lower energy release, varying from about 6 $kJ \cdot mol^{-1}$ in the case of Cd(II) to just 1.30 $kJ \cdot mol^{-1}$ in the case of Ni(II).

On the other hand, the positive values of ΔS^0 indicated that the randomness at the solid/liquid interface increases during the adsorption of these divalent metal ions onto GS [22]. A probable explanation for the observed increase of the disorder can be based on the fact that the adsorbed water molecules (which are displaced by the adsorbate species when metals are transferred from the liquid to the solid phase) gain more translational energy than the energy lost by the adsorbate ions [35].

To display the effect of the temperature on the spontaneity of the sorption process, the values of ΔG^0 obtained for the different metals have been plotted as a function of temperature in **Figure 6**.

As it may be observed in **Figure 6**, ΔG exhibits a general decreasing trend when the temperature of the system increased. The thermal effect in the spontaneity of the sorption process can be also assessed through the numerical values of the slopes of the plot for the different metals: Cu(II), −19.26; Ni(II), −56.01; Pb(II), −76.60; and Cd(II), −76.57. The negative values

Metal	T (°C)	ΔG^0 ($kJ\cdot mol^{-1}$)	ΔH^0 ($kJ\cdot mol^{-1}$)	ΔS^0 ($J\cdot mol^{-1}\cdot K^{-1}$)
Cu(II)	5	−23.27		
	20	−22.38		
	35	−24.98	−17.48	19.99
	50	−23.36		
	60	−22.31		
Ni(II)	5	−16.95		
	20	−18.26		
	35	−20.34	−1.30	57.45
	50	−18.54		
	60	−20.79		
Pb(II)	5	−24.88		
	20	−26.76		
	35	−27.83	−3.45	78.32
	50	−29.04		
	60	−29.02		
Cd(II)	5	−19.52		
	20	−21.05		
	35	−21.87	−5.99	57.34
	50	−21.63		
	60	−24.76		

Table 3. Thermodynamic parameters of Cu(II), Ni(II), Pb(II) and Cd(II) sorption onto GS at different temperatures.

obtained indicate that the sorption process is spontaneous, being Pb(II) and Cd(II) the most temperature-sensitive metals.

The values of variation of enthalpy, entropy and Gibbs free energy presented in **Table 3** allow establishing a different set of comparisons between the different metals. So, in basis to the energy released when the metal is adsorbed, the next ranking can be drafted out:

ΔH^0: Cu > Cd > Pb > Ni

In basis to the increase of randomness that metal sorption provokes in the system:

ΔS^0: Pb > Ni ≈ Cd > Cu

And lastly, in basis to a more general criterion of spontaneity of the sorption process for temperatures within 5 to 50°C:

ΔG^0 (5–50°C): Pb > Cu > Cd > Ni.

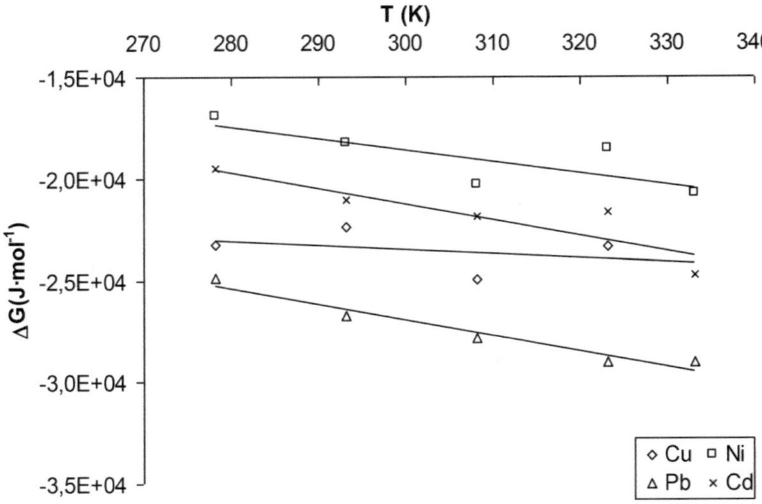

Figure 6. Variation of Gibbs free energy with temperature for in all the M(II)-GS systems.

Sorbent	Metal	ΔG^0 ($kJ \cdot mol^{-1}$)	ΔH^0 ($kJ \cdot mol^{-1}$)	ΔS^0 ($J \cdot mol^{-1}$)	Reference
Grape stalks	Cu(II)	−22.38	−17.48	19.99	This work
	Ni(II)	−18.26	−1.30	57.45	
	Pb(II)	−27.83	−3.45	78.32	
	Cd(II)	−21.05	−5.99	57.34	
Cellulosic waste orange peel	Cu(II)	−12.48	−19.55	−24.12	[24]
Corn silk (*Zea mays* L)	Cu(II)	−17.10	10.75	95.08	[36]
	Zn(II)	−16.68	7.83	83.68	
Pseudomonas putida	Pb(II)	−21.20	−18.69	8.4	[37]
	Cu(II)	−16.50	23.12	128	
Hazelnut shells	Ni(II)	−15.05	32.49	158.54	[35]
	Pb(II)	−20.94	21.41	142.11	
	Cd(II)	−20.80	12.21	110.73	
Almond shells	Ni(II)	−12.36	47.29	199.00	
	Pb(II)	−21.57	50.55	233.32	
	Cd(II)	−17.45	17.76	118.28	
Capsicum annuum	Cu(II)	−15.93	−7.63	28.35	[22]
Modified spent Chrysanthemum	Cu(II)	−25.19	−11.42	47.0	[25]
Rapeseed biomass	Pb(II)	−35.33	10.05	155.0	[38]

Sorbent	Metal	ΔG^0 ($kJ \cdot mol^{-1}$)	ΔH^0 ($kJ \cdot mol^{-1}$)	ΔS^0 ($J \cdot mol^{-1}$)	Reference
Bacillus pumilus sp. *AS1*	Pb(II)	−0.56	7.53	0.027	[39]
Loquat (*Eriobotrya japonica*) leaves	Cd(II)	−8.21	29.73	125.44	[40]
Sargassum filipendula	Cd(II)	−3.82	0.26	0.87	[41]
Penicillium simplicissimum	Cd(II)	−18.27	20.03	130.9	[42]
	Zn(II)	−17.08	25.42	145.5	
	Pb(II)	−20.04	39.13	202.5	
Sporopollenin	Cu(II)	−7.54	17.55	85.58	[43]
	Pb(II)	−13.78	31.97	150.98	
	Cd(II)	−8.85	13.99	76.64	

Table 4. Comparison of the sorption thermodynamic parameters obtained for GS with these observed for other biomass.

It has to be remarked however that for the highest temperature, 60°C, an inversion on the spontaneity of the sorption process takes place between Cd(II) and Cu(II), getting therefore the ranking the next form:

ΔG^0 (60°C): Pb > Cd > Cu > Ni.

The thermodynamic results gathered in our study have been compared to those reported by other authors. A summary of the most relevant results found in a bibliographic survey are presented in **Table 4**.

As it can be observed, all the authors reported negative values of ΔG^0 and most of them also positive values of ΔS^0. These results clearly indicate that sorption is a spontaneous process that mostly takes place with an increase of the randomness of the system. On the other hand, the ΔH^0 values reported were either positive or negative. Sorption processes showing negative enthalpy values would be the sorption of Cu(II), Ni(II), Pb(II) and Cd(II) onto GS (this work), Pb(II) sorption onto *Pseudomonas putida* and Cu(II) sorption onto both, *Capsicum annuum* and modified spent chrysanthemum. On the other hand, positive enthalpy values were reported for Cu(II) sorption onto *Pseudomonas putida,* Ni(II), Pb(II) and Cd(II) sorption onto both, hazelnut and almond shells, or Cu(II), Pb(II) and Cd(II) sorption onto sporopollenin. Thus, these results indicate that metal sorption might take place through release or absorption of energy to or from the system.

3. Conclusions

The sorption of Cu(II), Ni(II), Pb(II) and Cd(II) onto grape stalk follows a Langmuirian sorption trend in the whole range of temperature explored. Freundlich and Dubinin-Radushkevich models were also capable of providing a reasonably satisfactory description of the sorption equilibrium. The mean free energy E calculated by means of the Dubinin-Radushkevich model demonstrated that sorption of Cu(II), Ni(II) and Cd(II) onto grape stalk proceeds mainly via ion exchange. In the case of Pb(II), an extra contribution to the ion exchange at temperatures higher than 5°C was observed. This extra contribution would be based on the establishment of stronger Pb(II)-GS interactions. The enthalpy and entropy variation in the sorption process demonstrates that Cu(II), Ni(II), Pb(II) and Cd(II) sorption onto grape stalks is a spontaneous exothermic process that involves an increase of the randomness of the system.

The thermodynamic parameters of metal sorption onto GS allowed establishing different rankings: based on the energy released in the sorption process, ΔH^0, Cu > Cd > Pb > Ni; on the increase of randomness provoked by the sorption process, ΔS^0, Pb > Ni ≈ Cd > Cu; and, finally, generalizing in base to the spontaneity of the overall process, ΔG^0 (5–50°C), Pb > Cu > Cd > Ni.

Conflict of interest

The authors certify that they have no conflict of interest.

Author details

Carlos Escudero-Oñate[1]* and Isabel Villaescusa[2]

*Address all correspondence to: carlos.escudero@niva.no

1 Norwegian Institute for Water Research (NIVA), Gaustadalléen, Oslo, Norway

2 Chemical Engineering Department, Escola Politècnica Superior, Universitat de Girona, c/Mª Aurèlia Capmany, Girona, Spain

References

[1] Deniz F, Karabulut A. Biosorption of heavy metal ions by chemically modified biomass of coastal seaweed community: Studies on phycoremediation system modeling and design. Ecological Engineering. 2017;**106**:101-108

[2] Vijayaraghavan K, Balasubramanian R. Is biosorption suitable for decontamination of metal-bearing wastewaters? A critical review on the state-of-the-art of biosorption processes and future directions. Journal of Environmental Management. 2015;**160**:283-296

[3] Vendruscolo F, da Rocha Ferreira GL, Antoniosi Filho NR. Biosorption of hexavalent chromium by microorganisms, International Biodeterioration & Biodegradation. 2017; **119**:87-95

[4] Lesmana SO, Febriana N, Soetaredjo FE, Sunarso J, Ismadji S. Studies on potential applications of biomass for the separation of heavy metals from water and wastewater. Biochemical Engineering Journal. 2009;**44**:19-41

[5] Senthilkumaar S, Bharathi S, Nithyanandhi D, Subburam V. Biosorption of toxic heavy metals from aqueous solutions. Bioresource Technology. 2000;**75**:163-165

[6] Ahmad A, Bhat AH, Buang A. Biosorption of transition metals by freely suspended and Ca-alginate immobilised with Chlorella Vulgaris: Kinetic and equilibrium modeling. Journal of Cleaner Production. 2018;**171**:1361-1375

[7] Wu M, Liang J, Tang J, Li G, Shan S, Guo Z, Deng L. Decontamination of multiple heavy metals-containing effluents through microbial biotechnology. Journal of Hazardous Materials. 2017;**337**:189-197

[8] Renu MA, Singh K, Upadhyaya S, Dohare RK. Removal of heavy metals from wastewater using modified agricultural adsorbents. Materials Today: Proceedings. 2017; **4**:10534-10538

[9] Castro L, Blázquez ML, González F, Muñoz JA, Ballester A. Biosorption of Zn(II) from industrial effluents using sugar beet pulp and F. Vesiculosus: From laboratory tests to a pilot approach. Science of the Total Environment. 2017;**598**:856-866

[10] Demey H, Vincent T, Guibal E. A novel algal-based sorbent for heavy metal removal. Chemical Engineering Journal. 2018;**332**:582-595

[11] Sheng PX, Ting Y-P, Chen JP. Biosorption of heavy metal ions (Pb, cu, and cd) from aqueous solutions by the marine alga Sargassum sp. in single- and multiple-metal systems. Industrial & Engineering Chemistry Research. 2007;**46**:2438-2444

[12] Jiang L, Zhou W, Liu D, Liu T, Wang Z. Biosorption isotherm study of Cd2+, Pb2+ and Zn2+ biosorption onto marine bacterium Pseudoalteromonas sp. SCSE709-6 in multiple systems. Journal of Molecular Liquids. 2017;**247**:230-237

[13] Carolin CF, Kumar PS, Saravanan A, Joshiba GJ, Naushad M. Efficient techniques for the removal of toxic heavy metals from aquatic environment: A review. Journal of Environmental Chemical Engineering. 2017;**5**:2782-2799

[14] Ronda A, Martín-Lara MÁ, Blázquez G, Bachs NM, Calero M. Copper biosorption in the presence of lead onto olive stone and pine bark in batch and continuous systems. Environmental Progress & Sustainable Energy. 2014;**33**:192-204

[15] Wei W, Wang Q, Li A, Yang J, Ma F, Pi S, Wu D. Biosorption of Pb (II) from aqueous solution by extracellular polymeric substances extracted from Klebsiella sp. J1: Adsorption behavior and mechanism assessment. Scientific Reports. 2016;**6**

[16] Gupta VK, Rastogi A. Equilibrium and kinetic modelling of cadmium(II) biosorption by nonliving algal biomass Oedogonium sp. from aqueous phase. Journal of Hazardous Materials. 2008;**153**:759-766

[17] Limousin G, Gaudet JP, Charlet L, Szenknect S, Barthès V, Krimissa M. Sorption isotherms: A review on physical bases, modeling and measurement. Applied Geochemistry. 2007;**22**:249-275

[18] Chowdhury PSS. Insight into adsorption thermodynamics. In: Thermodynamics. Rijeka, Croatia: INTECH; 2011, pp. 349-364

[19] El-Khaiary MI. Least-squares regression of adsorption equilibrium data: Comparing the options. Journal of Hazardous Materials. 2008;**158**:73-87

[20] Anastopoulos I, Karamesouti M, Mitropoulos AC, Kyzas GZ. A review for coffee adsorbents. Journal of Molecular Liquids. 2017;**229**:555-565

[21] Karunanithi R, Ok YS, Dharmarajan R, Ahmad M, Seshadri B, Bolan N, Naidu R. Sorption, kinetics and thermodynamics of phosphate sorption onto soybean stover derived biochar. Environmental Technology & Innovation. 2017;**8**:113-125

[22] Özcan A, Özcan AS, Tunali S, Akar T, Kiran I. Determination of the equilibrium, kinetic and thermodynamic parameters of adsorption of copper(II) ions onto seeds of Capsicum Annuum. Journal of Hazardous Materials. 2005;**124**:200-208

[23] Saha GC, Hoque MIU, Miah MAM, Holze R, Chowdhury DA, Khandaker S, Chowdhury S. Biosorptive removal of lead from aqueous solutions onto Taro (Colocasiaesculenta(L.) Schott) as a low cost bioadsorbent: Characterization, equilibria, kinetics and biosorption-mechanism studies. Journal of Environmental Chemical Engineering. 2017;**5**:2151-2162

[24] Guiza S. Biosorption of heavy metal from aqueous solution using cellulosic waste orange peel. Ecological Engineering. 2017;**99**:134-140

[25] Yi Y, Lv J, Zhong N, Wu G. Biosorption of Cu2+ by a novel modified spent chrysanthemum: Kinetics, isotherm and thermodynamics. Journal of Environmental Chemical Engineering. 2017;**5**:4151-4156

[26] Wang N, Jin R-N, Omer AM, Ouyang X-k. Adsorption of Pb(II) from fish sauce using carboxylated cellulose nanocrystal: Isotherm, kinetics, and thermodynamic studies. International Journal of Biological Macromolecules. 2017;**102**:232-240

[27] Khodabandehloo A, Rahbar-Kelishami A, Shayesteh H. Methylene blue removal using Salix Babylonica (weeping willow) leaves powder as a low-cost biosorbent in batch mode: Kinetic, equilibrium, and thermodynamic studies. Journal of Molecular Liquids. 2017;**244**:540-548

[28] Pleşa Chicinaş R, Bedelean H, Stefan R, Măicăneanu A. Ability of a montmorillonitic clay to interact with cationic and anionic dyes in aqueous solutions. Journal of Molecular Structure. 2018;**1154**:187-195

[29] Singh H, Chauhan G, Jain AK, Sharma SK. Adsorptive potential of agricultural wastes for removal of dyes from aqueous solutions. Journal of Environmental Chemical Engineering. 2017;**5**:122-135

[30] Shaker MA. Thermodynamics and kinetics of bivalent cadmium biosorption onto nanoparticles of chitosan-based biopolymers. Journal of the Taiwan Institute of Chemical Engineers. 2015;**47**:79-90

[31] Rangabhashiyam S, Sujata L, Balasubramanian P. Biosorption characteristics of methylene blue and malachite green from simulated wastewater onto Carica papaya wood biosorbent. Surfaces and Interfaces. 2017. https://doi.org/10.1016/j.surfin.2017.09.011

[32] Martínez M, Miralles N, Hidalgo S, Fiol N, Villaescusa I, Poch J. Removal of lead(II) and cadmium(II) from aqueous solutions using grape stalk waste. Journal of Hazardous Materials. 2006;**133**:203-211

[33] Villaescusa I, Fiol N, Martínez Ma, Miralles N, Poch J, Serarols J. Removal of copper and nickel ions from aqueous solutions by grape stalks wastes. Water Research. 2004; **38**:992-1002

[34] Onyango MS, Kojima Y, Aoyi O, Bernardo EC, Matsuda H. Adsorption equilibrium modeling and solution chemistry dependence of fluoride removal from water by trivalent-cation-exchanged zeolite F-9. Journal of Colloid and Interface Science. 2004;**279**:341-350

[35] Bulut Y, Tez Z. Adsorption studies on ground shells of hazelnut and almond. Journal of Hazardous Materials. 2007;**149**:35-41

[36] Petrović M, Šoštarić T, Stojanović M, Petrović J, Mihajlović M, Ćosović A, Stanković S. Mechanism of adsorption of Cu2+ and Zn2+ on the corn silk (Zea Mays L.). Ecological Engineering. 2017;**99**:83-90

[37] Uslu G, Tanyol M. Equilibrium and thermodynamic parameters of single and binary mixture biosorption of lead (II) and copper (II) ions onto pseudomonas putida: Effect of temperature. Journal of Hazardous Materials. 2006;**135**:87-93

[38] Morosanu I, Teodosiu C, Paduraru C, Ibanescu D, Tofan L. Biosorption of lead ions from aqueous effluents by rapeseed biomass. New Biotechnology. 2017;**39**:110-124

[39] Sayyadi S, Ahmady-Asbchin S, Kamali K, Tavakoli N. Thermodynamic, equilibrium and kinetic studies on biosorption of Pb+2 from aqueous solution by Bacillus Pumilus sp. AS1 isolated from soil at abandoned lead mine. Journal of the Taiwan Institute of Chemical Engineers. 2017;**80**:701-708

[40] Awwad AM, Salem NM. Kinetics and thermodynamics of cd(II) biosorption onto loquat (Eriobotrya Japonica) leaves. Journal of Saudi Chemical Society. 2014;**18**:486-493

[41] Verma A, Kumar S, Kumar S. Statistical modeling, equilibrium and kinetic studies of cadmium ions biosorption from aqueous solution using S. Filipendula. Journal of Environmental Chemical Engineering. 2017;**5**:2290-2304

[42] Fan T, Liu Y, Feng B, Zeng G, Yang C, Zhou M, Zhou H, Tan Z, Wang X. Biosorption of cadmium(II), zinc(II) and lead(II) by Penicillium simplicissimum: Isotherms, kinetics and thermodynamics. Journal of Hazardous Materials. 2008;**160**:655-661

[43] Ünlü N, Ersoz M. Adsorption characteristics of heavy metal ions onto a low cost biopolymeric sorbent from aqueous solutions. Journal of Hazardous Materials. 2006;**136**:272-280

Chapter 15

Removal of Heavy Metals Using Adsorption Processes Subject to an External Magnetic Field

Ma. del Rosario Moreno Virgen,
Omar Francisco González Vázquez,
Virginia Hernández Montoya and
Rigoberto Tovar Gómez

Additional information is available at the end of the chapter

http://dx.doi.org/10.5772/intechopen.74050

Abstract

Adsorption is a broadly used process for the removal of heavy metals and the world trend is directed to the application of new technologies to intensify existing processes. The properties of the magnetic field (intensity and arrangement) and the intrinsic magnetic properties of the adsorbent and the adsorbate are decisive for satisfactory results. The intensity of the magnetic field is important, because this implies that the greater number of spins present will align with the magnetic field according to the magnetic nature present, allowing the mobility of the adsorbate and generating heterogeneity on the surface of the adsorbent. Similarly, the arrangement of the magnetic field will determine the direction of the magnetic field lines. The application of a magnetic field as an alternative for the intensification of the adsorption process based on the consideration that the magnetic field is safe, environmentally friendly and economic.

Keywords: heavy metals, adsorption, magnetic field

1. Introduction

The importance of removing dissolved heavy metals from water is a primary concern for society because heavy metals represent a risk to both public and environmental health. Heavy metals are toxic and carcinogenic, and they can easily enter the food chain [1, 2].

According to the Environmental Protection Agency (EPA), heavy metals are considered priority pollutants and must be eliminated or reduced from any water body that may or may not come into contact with the environment [3, 4]. There are a variety of conventional techniques capable of removing heavy metals from water, which have already been tested successfully, such as precipitation, ion-exchange, reverse osmosis, membrane separation, adsorption, and so on [5, 6]. Specifically, adsorption is one of the most promising and frequently used methods due to easy operation, high efficiency and economic benefits [7, 8]. The static magnetization coupled to an adsorption process has attracted special attention because it is an easy-to-implement, low-cost and environmentally friendly method. Recent studies have presented an improvement in the adsorption process when a magnetic field is applied [9], which is attributed to different factors and possible modifications that can change properties of the solution and material under magnetic field effect [10]. However, there are several parameters and operating conditions that are determinant within the adsorption process subjected to a static magnetic field such as the physicochemical and magnetic nature of the adsorbate and the adsorbent and the configuration and intensity of the magnetic field. The process by which the magnetic field affects the adsorbate-adsorbent system is complex, and parameters such as zeta potential, magnetic susceptibility, ionic radius and hydration of the ion have allowed to elucidate the possible mechanism.

2. Mineral adsorbents

Table 1 shows mineral adsorbents used in the removal of heavy metals from water such as Cd(II), Zn(II), Pb(II), Cu(II), Cr(VI) and Ni(II). Particularly, the Fe_3O_4/PANI/MnO_2 core-shell hybrid was the most efficient in the removal of Cd(II), showing an adsorption capacity of 154 mg/g [11]; followed by hydrotalcite-like compounds (ZnAl-diethylenetriaminepentaacetic acid and ZnAl-meso-2,3-dimercaptosuccinic acid) with values of 112.4 and 49.5 mg/g, respectively [12]. On the other hand, the metakaolin-based geopolymer showed the higher adsorption capacity of Zn(II) (74.5 mg/g) [13], in comparison with natural bentonit and bentonite (52.9 and 39.5 mg/g, respectively) and other adsorbents [11, 13, 14]. Also, for the removal of Cr(VI), the processed diatomite/MCM-41 composite was the most efficient with adsorption capacity of 70.9 mg/g [15]; followed by diatomite-supported magnetite nanoparticles with a value of 69.2 mg/g [16]. As well, different adsorbents were used to study the removal of Cu(II) and the better was the hydrotalcite-like compound ZnAl-meso-2,3-dimercaptosuccinic acid with an adsorption capacity of 69.9 mg/g [12]; this material was also used to remove Cd(II) with significant results. The Zn(II), Cr(VI), Cu(II) and Ni(II) are the heavy metals more studied with inorganic or mineral adsorbents. Respect to removal of Ni(II), natural bentonite obtained the higher adsorption capacity (50.0 mg/g) [13]. Finally, other heavy metal studied was the Pb(II) with the best value of 404.0 mg/g again using the hydrotalcite-like compound ZnAl-meso-2,3-dimercaptosuccinic acid [12].

Adsorbent	Removed species	Adsorption capacity (mg/g)	Ref
Fe_3O_4/PANI/MnO_2 core-shell hybrids	Cd(II)	154	[11]
	Zn(II)	[a]88.0 (17.6 mg/g)	
	Pb(II)	[a]99.1 (19.8 mg/g)	
	Cu(II)	[a]99.1 (19.8 mg/g)	
Processed diatomite/MCM-41 composite	Cr(VI)	70.9	[15]
Diatomite-supported magnetite nanoparticles	Cr(VI)	69.2	[16]
Tannin-immobilized activated clay	Cr(VI)	24.1	[17]
Modified natural zeolites	Cr(VI)	3.5–8.8	[18]
Modified rectorite	Cr(VI)	21	[19]
geopolymer/alginate hybrid spheres	Cu(II)	60.8	[20]
Aminopropyltriethoxysilane/sepiolite	Cu(II)	8.9	[21]
Multiwalled carbon nanotubes	Cu(II)	50.3	[22]
Zeolite	Cu(II)	9.0	[23]
Ca-alginate	Cu(II)	42.7	[24]
Bentonite/chitosan	Cu(II)	9.8	[25]
Inorganic-organic clays	Cr(VI)	16.6–17.9	[26]
Hydrotalcite-like compound ZnAl-NO_3			[12]
ZnAl-dtpa (diethylenetriaminepentaacetic acid)	Cu(II)	48.3	
	Pb(II)	145.0	
	Cd(II)	49.5	
ZnAl-dmsa (meso-2,3-dimercaptosuccinic acid)	Cu(II)	69.9	
	Pb(II)	404.0	
	Cd(II)	112.4	
Na-bentonite	Zn(II)	23.1	[13]
Polyphosphate-modified kaolinite clay	Zn(II)	27.8	[13]
Natural zeolite	Zn(II)	13.4	[13]
Kaolinite	Zn(II)	7.2	[13]
Metakaolinite	Zn(II)	12.4	[13]
Vermiculite	Zn(II)	31.7	[13]
Bentonite	Zn(II)	39.5	[13]
Zeolite	Zn(II)	18.7	[13]
Natural bentonite	Zn(II)	52.9	[13]

Adsorbent	Removed species	Adsorption capacity (mg/g)	Ref
Metakaolin-based geopolymer	Zn(II)	74.5	[13]
Vermiculite	Ni(II)	25.3	[13]
Natural bentonite	Ni(II)	50.0	[13]
Zeolite	Ni(II)	1.98	[13]
Na-bentonite	Ni(II)	24.2	[13]
Ca-bentonite	Ni(II)	6.3	[13]
Natural clinoptilolite	Ni(II)	8.7	[13]
Montmorillonite	Ni(II)	21.1	[13]
Metakaolin-based geopolymer	Ni(II)	42.6	[13]
Mixed silica-alumina oxide	Cu(II)	0.1	[14]
	Ni(II)	0.2	
	Zn(II)	4.2	

[a]Removal percentage.

Table 1. Mineral adsorbents used in the adsorption of heavy metals from water.

3. Mineral carbons

The mineral carbons used to study adsorption of heavy metals from water are peat, lignite, anthracite, graphite and principally the bituminous carbon, which is the most used. The bituminous carbon has been physically activated with steam and modified by impregnation with calcium, phosphoric acid, sulfur and nitric acid, and subsequently, a thermal treatment has been applied [27–32].

Table 2 concentrates the mineral carbons used in the adsorption of heavy metals from water reported in the literature. There are few reports about the use of this kind of adsorbents and

Mineral carbon	Modification method used	Modifying agent	Removed species	Adsorption capacity (mg/g)	Ref
Bituminous carbon	Impregnation-thermal treatment	Calcium-H_3PO_4	Zn(II)	3.4	[27]
			Cd(II)	8.0	
			Ni(II)	5.0	
Bituminous carbon	—	—	Cd(II)	5.1	[28]
			Cu(II)	13.3	
			Cr(VI)	10.5	
Bituminous coal (AC-109)	Impregnation-thermal treatment	Sulfur	Hg^0	1.6	[29]
			Hg(II)	0.7	

Mineral carbon	Modification method used	Modifying agent	Removed species	Adsorption capacity (mg/g)	Ref
Bituminous coal (AC-107)	Impregnation-thermal treatment	Sulfur	Hg^0	1.9	[29]
			Hg(II)	0.7	
Bituminous coal (AC-C2)	Impregnation-thermal treatment	HNO_3	Hg^0	1.8	[29]
			Hg(II)	0.7	
Bituminous charcoal	Physical activation	Steam	Cu(II)	[a]0.1	[30]
			Cr(VI)	[a]0.3	
			Zn(II)	[a]0.3	
Lignite charcoal	Physical activation	Steam	Cu(II)	[a]0.1	[30]
			Cr(VI)	[a]0.3	
			Zn(II)	[a]0.3	
Bituminous coal	Physical activation	Water vapor	Cd(II)	1.7	[31]
			Zn(II)	2.0	
Bituminous carbon	Impregnation-thermal treatment	Calcium-H_3PO_4	Zn(II)	1.6	[32]
			Cd(II)	2.7	
			Ni(II)	2.3	

[a]Industrial effluent.

Table 2. Mineral carbons used in the adsorption of heavy metals from water.

the bituminous carbon is the most studied. Specifically, the better adsorbent to remove Zn(II), Cd(II) and Ni(II) is the bituminous carbon modified with calcium and H_3PO_4 (3.4, 8.0 and 5.0 mg/g, respectively) [27]. On the other hand, the unmodified bituminous carbon shows the higher adsorption capacity of Cu(II) and Cr(VI) with values of 13.3 and 10.5 mg/g, respectively [28]. Additionally, it is important to mention that bituminous charcoal and lignite charcoal were used in adsorption studies of a metallurgic effluent containing Cu(II), Cr(VI) and Zn(II), and the adsorption capacities were low [30]. Finally, Hg^0 and Hg(II) were studied employing bituminous coal modified with sulfur and HNO_3 and adsorption capacities of 1.9 and 0.7 mg/g, respectively, were obtained [29].

4. Magnetic adsorbents

The term "***magnetic adsorbent***" is commonly used to describe the materials derived from iron (particles of γ-Fe_2O_3, Fe_3O_4 and $CoFe_2O_3$) or composites prepared from different types of matrix modified with iron compounds, which are used in the adsorption of water contaminants [33]. Particularly, for the adsorption of heavy metals from water, the principal magnetic adsorbents used are the composites, which are prepared with different methodologies using iron compounds and a specific matrix such as activated carbon, biochar, carbon nanotubes, α-cellulose, chitosan, cotton, resins, silica, starch, polymers, different type of biomass, and so on (see **Table 3**) [34–55]. The methods used in the preparation of magnetic composites can be

Matrix	Modified method	Iron compounds	Removed species	Adsorption capacity (mg/g)	Ref
Activated carbon	Co-precipitation	$FeSO_4{\cdot}7H_2O$, $FeCl_3{\cdot}6H_2O$	Cu(II)	[a]99.8	[34]
Activated carbon	Co-precipitation	$Fe(NO_3)_3{\cdot}9H_2O$	As(III)	38.8	[35]
			As(V)	51.3	
Activated carbon	Co-precipitation	$FeCl_2$	As(V)	1.95	[36]
Activated carbon	Reduction reaction	FeO_4S	As(III)	18.19	[37]
			As(V)	12.02	
Carbon nanotubes	Co-precipitation	$FeSO_4{\cdot}4H_2O$, $FeCl_3{\cdot}6H_2O$	Ni(II)	51.62	[38]
Biochar from *Astragalus mongholicus*	Co-precipitation	$FeSO_4{\cdot}4H_2O$, $FeCl_3{\cdot}6H_2O$	Cr(VI)	44.74	[39]
α-Cellulose	Extrusion dropping technology	Fe_3O_4	Cu(II)	47.573	[40]
			Pb(II)	37.994	
			Zn(II)	20.800	
Chitosan	Co-precipitation	$FeCl_3{\cdot}6H_2O$, $FeSO_4$	Cu(II)	188	[41]
			Cd(II)	159	
			Zn(II)	72	
			Cr(VI)	46	
Chitosan	Hydrothermal	$FeCl_3{\cdot}6H_2O$	Pb(II)	97.97	[42]
			Cu(II)	83.65	
Chitosan	Crosslinking method	Fe_3O_4	Cu(II)	143.27	[43]
Cotton	Impregnation-thermal treatment	$Fe(NO_3)_3{\cdot}9H_2O$	Cr(VI)	3.74	[44]
Cyclosorus interruptus	Co-precipitation	$FeSO_4{\cdot}7H_2O$, $FeCl_3{\cdot}6H_2O$	Pb(II)	90.1	[33]
Dowex 50 WX4 resin	Co-precipitation	$FeCl_3{\cdot}6H_2O$, $FeCl_2{\cdot}4H_2O$	Cr(VI)	400	[45]
			Ni(II)	384	
			Cu(II)	416	
			Cd(II)	398	
			Pb(II)	380	
Graphene oxide	Co-precipitation	$FeSO_4{\cdot}7H_2O$, $FeCl_3{\cdot}6H_2O$	Hg(II)	400	[46]
Macroalgal biomass (Kelp)	Impregnation-thermal treatment	$FeCl_3{\cdot}6H_2O$	Cd(II)	34.89	[47]
			Cu(II)	46.74	
			Zn(II)	95.49	

Matrix	Modified method	Iron compounds	Removed species	Adsorption capacity (mg/g)	Ref
Mesoporous silica	Thermal decomposition	$FeCl_3{\cdot}6H_2O$	Fe(III)	[a]79.9	[48]
Palygorskite	Co-precipitation	$FeCl_3{\cdot}6H_2O$, $FeCl_2{\cdot}4H_2O$	Pb(II)	26.7	[49]
Pine bark	Impregnation-thermal treatment	$Fe(NO_3)_3{\cdot}9H_2O$	Pb(II)	25.294	[50]
			Cd(II)	14.960	
Poly p-phenylenediamine	Co-precipitation	$FeCl_3{\cdot}6H_2O$, $FeSO_4{\cdot}7H_2O$	As(V)	35.14	[51]
Poly(vinylidene fluoride)	Impregnation-thermal treatment	$Fe(NO_3)_3{\cdot}9H_2O$	Cr(VI)	1423.4	[52]
Silica	Co-precipitation	$FeCl_2{\cdot}4H_2O$, $FeCl_3{\cdot}6H_2O$	Pb(II)	270	[53]
			Hg(II)	303	
			Pb(II)	256.4	
Starch	Crosslinking method	Fe_3O_4	Pb(II)	[a]83.1	[54]
			Cu(II)	[a]66.5	
Waste orange peel	Co-precipitation	$FeSO_4{\cdot}7H_2O$, $FeCl_3{\cdot}6H_2O$	As(III)	11.12	[55]

[a]Removal percentage.

Table 3. Magnetic composites used in the adsorption of heavy metals and metalloids from water.

grouped in the following categories according to the data reported in the literature [56]: (1) chemical co-precipitation, (2) thermal decomposition, (3) hydrothermal, (4) polyol process, (5) sol-gel and (6) chemical reduction. In this context, it is relevant to mention that the chemical co-precipitation is the most used method for the preparation of magnetic composites and this can be carried out in different routes, which include the following: (1) mixture of iron solutions ($FeSO_4{\cdot}7H_2O$ and $FeCl_3{\cdot}6H_2O$ or $FeCl_3{\cdot}6H_2O$ and $FeCl_2{\cdot}4H_2O$ or $Fe(NO_3)_3{\cdot}9H_2O$) with the matrix, and then, the addition of ammonium or sodium hydroxide or $NH_3{\cdot}H_2O$ to precipitate the iron nanoparticles on the matrix surface [33, 34, 36, 38, 39, 41, 45, 49, 55]. (2) The preparation of nanoparticles of iron (Fe_3O_4) by precipitation using $FeCl_3{\cdot}6H_2O$ and $FeCl_2{\cdot}4H_2O$ with additives such as $CaCl_2{\cdot}2H_2O$, $Na_3PO_4{\cdot}12H_2O$, EDTA [57, 58]. (3) First, the precipitation of nanoparticles of iron (Fe_3O_4), second, the mixture of Fe_3O_4 with materials such as *Cyclosorus interruptus*, silica and p-phenylenediamine-thiourea-formaldehyde polymer; and finally, the addition of 3-aminopropyltriethoxysilane or tetraethyl orthosilicate or potassium persulfate, depending on the final application of composite [33, 51, 53].

The second most popular method used in the preparation of magnetic composites includes the impregnation of the matrix with iron compounds such as $Fe(NO_3)_3{\cdot}9H_2O$ and $FeCl_3{\cdot}6H_2O$ and a thermal treatment of impregned materials using drying on a regular oven overnight [44], pyrolysis in conventional systems [47, 50, 52, 59] or pyrolysis in microwave system [60].

Table 3 summarizes the different types of magnetic composites used in the removal of some heavy metals and metalloids from water such as Pb(II), Cu(II), Ni(II), Cd(II), Zn(II), Cr(VI), Fe(III), Hg(II), As(II) and As(V). In general, the composites prepared with the resin Dowex 50 WX4 were the most efficient in the removal of Ni(II), Cu(II), Cd(II) and Pb(II), showing an adsorption capacity of 384, 416, 398 and 380 mg/g, respectively [45]. These values are higher than the data registered for other composites obtained from carbon nanotubes [34], α-cellulose [40], chitosan [41–43], silica [53], starch [54] and different types of biomass such as waste orange peel, pine bark and *Cyclosorus interruptus* (see **Table 3**) [33, 50, 55]. Additionally, it is relevant to mention that for the removal of Cr(VI), the composite obtained with poly(vinylidene fluoride) shows the higher adsorption capacity (1423.4 mg/g) in comparison with the samples prepared from chitosan (46 mg/g), cotton (3.74 mg/g) and Dowex 50 WX4 resin (400 mg/g) [41, 44, 45]. In this context, the adsorption of As(III) and As(V) has also been studied employing magnetic composites prepared with activated carbon principally [35–37]. For example, the composite prepared with bituminous coal carbon is showing a high adsorption capacity of As(III) and As(V) (38.8 and 51.3 mg/g, respectively), in comparison with the materials prepared with an inorganic matrix as the poly p-phenylenediamine (34.15 mg of As(V)/g) [35, 51]. Finally, it is important to mention that one of the most toxic heavy metals, Hg(II), was studied employing magnetic composites obtained with silica and with graphene oxide, highlighting the latter material with an adsorption capacity of 400 mg/g [46].

In summary, the adsorption of heavy metals has been studied using magnetic composites prepared principally by co-precipitation method, however, a homogeneous comparison is not possible because all the adsorption conditions and physicochemical properties of magnetic adsorbents are different (porosity, chemical functionally, etc.). But, the investigation in this area is increasing in past years and the data reported in this chapter will serve as a basis for future publications.

5. Adsorbate properties and adsorption conditions

The adsorption of heavy metals on different adsorbents occurs via ion-exchange, complexation, coordination, chelation, microprecipitation, electrostatic interaction and/or combination of these mechanisms [61]. Consequently, these adsorption mechanisms are defined by the physicochemical characteristics of adsorbents (texture and chemical functionality), the adsorption conditions (pH, temperature, mass of adsorbent, treated volume, type of adsorption system, etc.) and the properties of adsorbates. Particularly, the principal properties of heavy metals considered in the adsorption mechanism are as follows: (1) ionic radius, (2) electronegativity, (3) atomic weight, (4) hydration radius, (5) oxidation number and (6) electrostatic potential.

According to the data reported in the literature, the ionic radius and the electronegativity are the two principal properties used in the interpretation of heavy metal adsorption mechanisms. For example, **Figure 1** shows the adsorption results of Pb(II), Cd(II), Cu(II), Ni(II) and Zn(II) as function of ionic radius. The adsorbents used were the red marine alga *Kappaphycus*

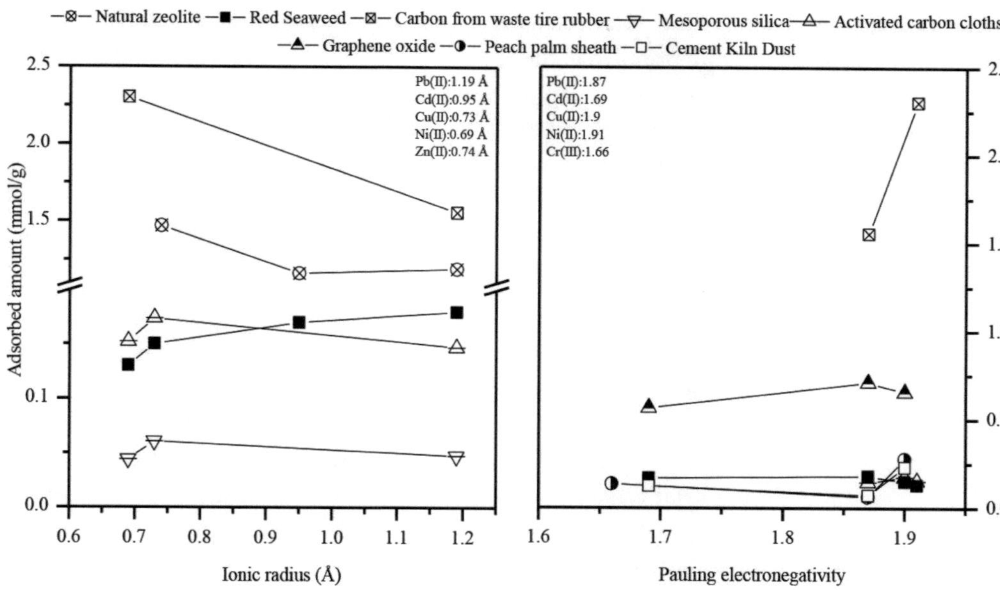

Figure 1. Adsorption results of heavy metals from water of different adsorbents in function of ionic radius and Pauling electronegativity.

alvarezii, carbon from waste tire rubber, mesoporous silica, natural zeolite and activated carbon cloths [61–65]. In general, a clear behavior is not defined and only in the adsorption of Pb(II), Cd(II), Cu(II) and Ni(II) on red marine alga, a clear correlation was obtained between the adsorbed amount of metal ions and the ionic radius. In this case, the adsorbed amount was higher for the heavy metal with greater ionic radius (Pb:1.19 Å), showing a maximum adsorption capacity of 0.51 mmol/g by the mechanism of electrostatic interaction [61]. The opposite behavior was observed for the adsorption of Pb(II), Cd(II), Ni(II) and Zn(II), using natural zeolite and carbon prepared from waste tire rubber as an adsorbent, where the adsorbed amount decreased as the ionic radius increased [62, 68]. Also, the hydration radius has been used to understate the adsorption of heavy metals and the results also do not follow a specific trend [66].

On the other hand, the Pauling electronegativity of heavy metals has also been correlated with the adsorption of heavy metals. **Figure 1** shows the adsorption results of Pb(II), Cd(II), Cu(II), Ni(II) and Cr(III) employing different adsorbents in function of Pauling electronegativity [61, 62, 65, 67–69]. Particularly, it is interesting to note that a defined behavior was observed in the adsorption of Pb(II) and Ni(II) using the carbon from waste tire rubber as an adsorbent [62]. Specifically, this adsorbent has the higher adsorption capacity of Ni(II) and Pb(II), and the adsorption was higher for the more electronegative metal. This behavior can be associated with the chemical functionality of carbon because this material was activated with hydrogen peroxide and consequently, the presence of O─H group of phenols and C═O group of esters favored the adsorption of heavy metals [62]. Additionally, for the adsorption of Pb(II), Cd(II), Cu(II) and Cr(III) on cement kiln dust and peach palm sheath, the electronegativity does not keep a relation with the absorbed quantity of metals [68, 69]. However, it is important to mention that the Cr has a different oxidation number (III) in comparison with the other heavy metals (II). In this context, there are few reports that have studied the effect of oxidation number of heavy metals in the adsorption process and particularly, the works related with the adsorption by complexation mechanism have considered this property. For example, in the removal of Ni^{2+}, Fe^{2+} and Fe^{3+} (with external hydroxyl functional groups) was demonstrated that high-valence species complex more quickly than the low-valence due to the ability to accept more of the ligand's electron donation, where the adsorption followed the trend $Ni^{2+} > Fe^{3+} > Fe^{2+}$ [70].

Finally, it is relevant to mention that there are many factors which affect heavy metal adsorption such as initial concentration, temperature, adsorbent dose, pH, contact time and stirring speed. However, the pH plays a vital role in deciding the maximum adsorption capacity of the adsorbent such as activated carbons, carbon nanotubes, graphene, biosorbents, low-cost adsorbents (vegetal and industrial wastes), silica, chitosan, zeolites, alumina, clay, algal biomass, red mud, magnetic composites, and so on. The pH affects the surface charge of the adsorbent, the degree of ionization and speciation of the surface functional groups and metal ions [71–74]. In general, the adsorption of metallic species increases with increasing pH in certain range because at low pH, there a competition between the metallic species and the H^+ ions of aqueous solution.

6. Magnetic ordering and zeta potential

The behavior of the materials when they are exposed to a magnetic field is different, depending on their physicochemical and magnetic nature, this behavior can be observed in **Table 4**. [75].

On the other hand, superparamagnetism is observed in very small particles of transition metals and their compounds, particularly their oxides. It can be used to characterize fine dispersions of metal, alloy and their oxides and has applications in several areas. Parameters such as the magnetization per cubic centimeter or magnetization per gram are better parameters than the susceptibility for describing superparamagnetism and magnetically ordered ferromagnetic and ferromagnetic materials. The applied magnetic field is expressed in oersteds (Oe). Furthermore, the magnetic moment in ferromagnetic and ferromagnetic materials refers to the saturation moment and not the effective moment. Ferromagnetic metals, such as Fe, Co, Ni and insulators such as γ-Fe_2O_3 and Fe_3O_4, show the well-known hysteresis curve; in their unmagnetized state, the unpaired electrons associated with each atom have a net magnetic moment or magnetization, which is the vector sum of all unpaired electrons in that domain. In this sense, it is important to explain that when ferromagnetic and ferromagnetic materials are heated above a critical Curie temperature, they change over to paramagnetic behavior; thus the hysteresis disappears. In contrast, ideal superparamagnetic systems, when sufficiently cooled below a critical blocking temperature will experience a very slow relaxation time. Their net magnetic moment will align parallel to the applied field and appear to behave as if they had an apparent ferromagnetic behavior. This aspect will result in hysteresis of

Magnetic behavior	Description
Diamagnetic	A magnetic field acting on any atom induces a magnetic dipole for the entire atom by influencing the magnetic moment caused by the orbiting electrons. These dipoles can oppose the magnetic field, causing the magnetization to be less than zero. A diamagnetic behavior gives a negative susceptibility. Materials such as copper, silver, gold and alumina are diamagnetic at room temperature
Paramagnetic	A net magnetic moment due to electronic spin is associated with each atom when materials have unpaired electrons. When a magnetic field is applied, the dipoles line up with the field, causing a positive magnetization. Because the dipoles do not interact, extremely large magnetic fields are required to align all of the dipoles; so, the effect is lost as soon as the magnetic field is removed. This behavior can be observed in metals such as aluminum and titanium
Ferromagnetic	This type of behavior is promoted by the unfilled energy levels in the *3d* level of iron, nickel and cobalt. In this kind of materials, the permanent unpaired dipoles easily line up with the imposed magnetic field due to the exchange interaction. Large magnetizations are obtained even for small magnetic fields
Antiferromagnetic	In this behavior, the magnetic moments produced in neighboring dipoles line up in opposition to one another in the magnetic field, even though the strength of each dipole is very high. Examples of these materials are manganese, chromium, MnO and NiO
Ferrimagnetic	Different ions have different magnetic moments, like in ceramic materials. In a magnetic field, the dipoles of ion *A* may line up with the field, while dipoles of ion *B* oppose the field. But because the strengths of the dipoles are not equal, a net magnetization results. Ceramics called ferrites perfectly represent this behavior

Table 4. Classification and magnetic behavior of materials.

"apparent" ferromagnetic behavior. Conversely, above the critical blocking temperature, the hysteresis will disappear and the clusters will show a unique curve with no hysteresis [76].

It is possible to classify the materials from the magnetic point of view as a function of the number of Bohr magnetons, whose parameter can be calculated from the Langevin function, first derived for the noninteracting paramagnetic spins and to the noninteracting:

$$\sigma/\sigma_s = coth(\mu_c H/kT) - (kT/\mu_c H) \tag{1}$$

where σ is the magnetization per gram; σ_s is the magnetization of saturation; μ_c is the magnetic moment of the cluster; H is the applied magnetic field; k is the Boltzmann constant; T is the temperature (K).

A parameter used to corroborate the effect of the magnetic field on an adsorbate-adsorbent system is the zeta potential. Some studies have reported this parameter, such as Zhang et al. in 2004 where the zeta potential was measured in magnetized and non-magnetized Ca-rectorite suspensions observing that in the magnetized solutions the zeta potential was larger than in the nonmagnetized ones when the Cu concentration was zero. In contrast, the magnetic treatment reduced the zeta potential in Ca-rectorite suspensions containing Cu [77]. This same behavior was observed in a study made with Na-rectorite dispersions magnetized and not magnetized in the presence and absence of Zn [78].

In the literature, there is a little information about the mechanism of magnetic effects on adsorbate-adsorbent systems. However, one of the clearest suggests that magnetic exposure alters the arrangement of water molecules, ions and hydrated ions adsorbed on the surface of the particle in such a way that the effective adsorbed layer becomes thicker, then the slipping plane shifts outward from the particle surface so that the magnitude of zeta potential will be reduced, and the apparent size of the particles becomes larger so that their diffusivity will be reduced [79].

7. Importance of the physicochemical and magnetic properties of the adsorbates in the adsorption process assisted by an external magnetic field

The physicochemical properties of adsorbates play a very important role in the adsorption phenomenon in general. Properties such as ionic radius, electronegativity, valence, charge and hydration number are considered fundamental for the selection of the adsorbent material and the understanding of the adsorption mechanism, since these parameters help to determine forces and mobilities [80]. In particular, adsorbates such as heavy metals are present in a positive ionic state, this implies that there is an instability in charges; however, being dissolved in water, they remain stable in the solution because they balance their charge with the electronic cloud of oxygen present in the water molecule, which is highly negative, thus the ion is surrounded by water molecules quasi-stabilized; this phenomenon is called ion hydration. The hydration of the ion is fundamental in the adsorption process, since it establishes the adsorption force that each of the ions will have on the surface of an adsorbent material, in addition, it determines the advantage over other ions to occupy active sites, which explains the Gouy-Chapman theory.

The Gouy-Chapman theory states that ions of equal charge are adsorbed with equal force on the surface of the adsorbent; however, the hydration radius determines the adsorption strength, being those with lower hydrated radius more strongly retained; this implies an increase in the size of the ion, reducing its mobility [81]. Based on this, the hydration of each ion will be different, since it will be surrounded by a certain number of water molecules depending on the size of the ion and its charge, and this number of water molecules surrounding the ion is called the number of water molecules of hydration.

To know the hydration number of an ion, it is necessary to study the mobility of the ion within the solution, this mobility is mainly affected by two forces: the electric force Fe (Eq. (2)) and the viscous force Fv (Eq. (3)):

$$Fe = z_i e_0 x \tag{2}$$

where $z_i e_0$ is the electric charge; x is the electric field.

$$Fv = 6\pi r \eta v \tag{3}$$

where r is the ionic radius; η is the viscosity; v is the ion speed.

The phenomenon of hydration is explained when the ion reaches a constant speed, which implies that the electric and viscous forces are equalized (Eq. (4)).

$$r = \frac{z_i e_0 x}{6\pi\eta v} \tag{4}$$

Bearing in mind that the electric mobility u is the result of the velocity ratio on the electric field (Eq. (5)), we find the hydrated radius r_h (Eq. (6)):

$$u = \frac{v}{x} \tag{5}$$

$$r_h = \frac{z_i e_0}{6\pi\eta u} \tag{6}$$

Finally, the hydration number n_h is found by differentiating the hydrated radius r_h with the crystallographic radius of the r_{Cris} ion divided by the molecular radius of the water r_{H_2O} (Eq. (7)):

$$n_h = \frac{r_h^3 - r_{Cris}^3}{r_{H_2O}^3} \tag{7}$$

Ionic hydration is mainly due to thermodynamic phenomena [82]; however, studies have been reported where changes in the hydration of the ions attributable to exposure to an external magnetic field were observed, including variation in ion displacement and complex formation [83]. Alterations in ionic hydration due to exposure to a magnetic field can have a greater or lesser effect depending on the thickness of the hydration layer of each ion and the thermodynamic conditions of hydration, suggesting that magnetic field treatment causes changes in the structure of the water that hydrates the ions [84]. It is theorized that the magnetic field favors the hydration of ionic species due to the creation of normal Lorentz forces that help to penetrate the boundary layer of natural hydration generating a molecular arrangement that allows greater hydration [85]. In addition, the transport of the adsorbates to the active sites is promoted due to the impulse generated by the Lorentz forces and the electrostatic forces of Van der Waals, achieving an advantage over viscous forces [86]. On the other hand, it has been shown that divalent cations, such as heavy metals, are more susceptible to the presence of a magnetic field, since they seem to promote their polarization, as well as showing a "magnetic memory" effect that it can break when the temperature increases [87, 88], and at the same time, it has been observed that the magnetic memory effect can disappear nonlinearly in time depending on the configuration of the applied magnetic field and the exposure time to the magnetic field [89]. The alteration in the polarity of the hydrated ions due to the exposure to a magnetic field also generates changes in the zeta potential (ζ) of the solutions, influencing the pH values and encouraging the adsorption of metallic species [77–79, 90]. **Figure 2** explains the improvement in ionic hydration under the influence of a magnetic field which is due to a better accommodation of the water molecules surrounding the ion, not only in the first hydration layer, but also in the second layer although with a less force than in the first.

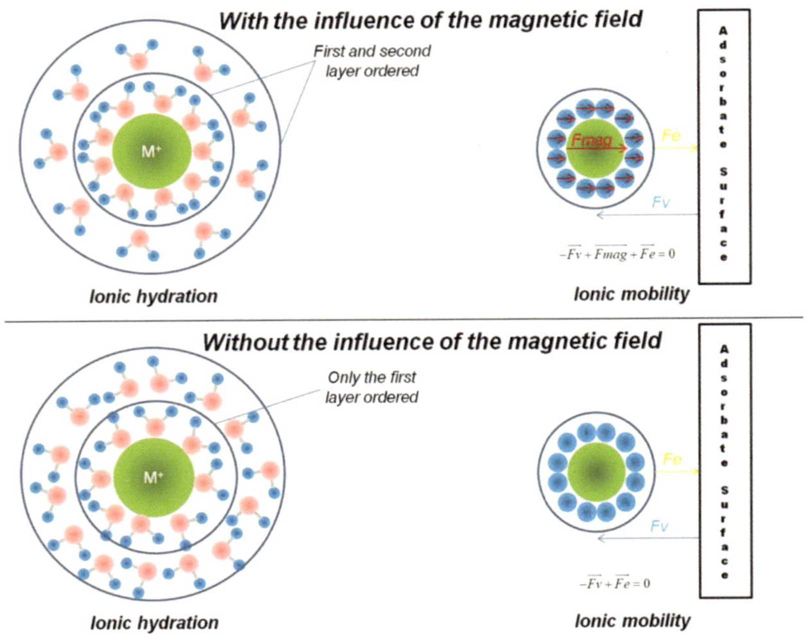

Figure 2. Mechanism of adsorption of heavy metals with and without the presence of a magnetic field.

The improvement in mobility of the hydrated ion under the influence of the magnetic field is due to the fact that the magnetic susceptibility of both the ion and the water molecules generates a component of magnetic force (*Fmag*) that added to the electrostatic force (*Fe*), which manage to overcome the viscous force (*Fv*), making the ion more likely to collide with the surface of the adsorbent.

The interactions that occur between the ions and the magnetic field can be explained by understanding the magnetic susceptibility; this is an intrinsic property that has all the matter. Magnetic susceptibility (χ_{Mag}) is a magnitude that represents the sensitivity to the magnetization of a material influenced by a magnetic field, determined by the sum of a diamagnetic component or diamagnetic susceptibility (χ_{Dia}) and a paramagnetic component or paramagnetic susceptibility (χ_{Para}) [91] (Eq. (8)), understanding that the diamagnetic matter opposes the flow of the magnetic field and the paramagnetic matter is oriented in the same direction of the magnetic field.

$$\chi_{Mag} = \chi_{Para} + \chi_{Dia} \tag{8}$$

Substances that are magnetized in the same direction of the magnetic or paramagnetic field have an odd electronic orbital and their magnetic susceptibility is positive, while those that are magnetized against the magnetic field or diamagnetic have their electronic orbital even and have negative magnetic susceptibility [92, 93], and each substance has a different value of magnetic susceptibility, inclusive, the same substance with different oxidation state can vary in its magnitude of magnetic susceptibility. In the case of ionic species such as heavy

http://dx.doi.org/10.5772/intechopen.74050

metals, they have distinct values of magnetic susceptibility between them, coupled with this, the water molecules that surround a hydrated ion also have a magnetic susceptibility, so as a whole, generates a new magnitude of magnetic susceptibility, being the sum of the susceptibility of the solitary ion and that of each of the water molecules that surround it in the main layer. Based on this, the net susceptibility of the hydrated ion will depend on the magnetic susceptibility value of the ion χ_{Mag} and the multiplication of the magnetic susceptibility of water χ_{H_2O} by the hydration number n_h (Eq. (9)).

$$\chi_h = \chi_{mag\ ion} + [(n_h)(\chi_{H_2O})] \tag{9}$$

As an example, the susceptibilities of three heavy metals are shown in **Table 5**: Pb, Cd and Zn, where the Pb has a magnetic susceptibility lower than that of the Cd and Zn; contrary to the values of ionic radio and electronegativity; however, we found that the hydration index n_h of Pb is lower than that of Cd and Zn, which is consistent with the Gouy-Chapman theory. On the other hand, the value of the magnetic susceptibility of water, which, like that of heavy metals is negative, which implies its diamagnetism and with which it will be possible to determine the hydrated magnetic susceptibility, resulting to be now larger than the hydrated magnetic susceptibilities of both Cd and Zn, compared to that of Pb.

As can be seen in the specific case of Pb, Cd and Zn are diamagnetic species despite being heavy metals; however, many transition metal complexes exhibit both diamagnetic and paramagnetic complexes. For example, the compounds of the complex ion $[Co(NH_3)6]^{3+}$ do not have unpaired electrons, but the compounds of $[CoF_6]^{3-}$ ion have four per metal ion. Both complexes contain Co(III) with $3d^6$ electronic configuration [94], although both Zn and Cd are elements of electronic configuration *d*, they exhibit diamagnetic behavior because they are at the limit of groups IIB and tend to behave similar to group IIIA.

On the other hand, a simple way to know the magnetic ordering of ionic species in aqueous solution may be the coloration of the salt from which they start, since in heavy metal compounds, orbitals often are divided into two sets of separate orbitals by the division energies, which corresponds to wavelengths of light in the visible region. The absorption of visible light causes electronic transitions between orbitals in these sets, which imply a correlation between the paramagnetism and the salt coloration, while the diamagnetism of some metallic nitrates is colorless [94].

	χ_{mag} (1 × 10^{-6} emu/mol)	r_{ion} (pm)	n_h (min–max) [2]	χ_h (1 × 10^{-6} emu/mol)
Pb	−23	1.2	4–7.5	−120.275
Cd	−19.7	0.97	10–12.5	−181.825
Zn	-9.5	0.74	10–12.5	−171.625
H_2O	−12.97	—	—	—

Table 5. Physicochemical and magnetic parameters of different ions and hydrated ions.

Overall, a strong interdependence of the physicochemical and magnetic properties of ionic adsorbates can be observed and how they interact with exposure to a magnetic field.

8. Generation and configuration of a magnetic field applied to an adsorption process

Knowing the characteristics of the magnetic field that will interact with both the adsorbate and the adsorbent is overriding, three design parameters can be mentioned in the application of a magnetic field for assistance in the adsorption process: (1) type of source, (2) arrangement or trajectory of field lines and (3) magnetic field strength.

The source of the magnetic field can be classified by its generation in three different ways: alternating current (AC) electromagnets, direct current (DC) electromagnets and permanent magnets.

In general, the electromagnets consist of a solenoid made from cable made of an electrically conductive material, these solenoids generate a magnetic field in their cavity to the passage of electric current. The design of the solenoid goes from linear, annular and cylindrical forms according to the application, where the magnetic field strength that could generate depends on the number of turns, intensity and current potential [95].

The AC electromagnets are distinguished by passing electric current in which the magnitude and direction of the current vary in cyclical periods. This type of electromagnets have been used to generate magnetic fields applied in multiple ways ranging from heating by induction micellar particles to remove dyes from water [96], the acceleration in the germination of seeds under the application of a magnetic pulse [97], to be used as an incentive for the adsorption of volatile organic compounds [98]. They present some advantages in the generation of temperatures by induction and work with magnetic pulses of high frequency; however, it has some other disadvantages such as the creation of a heterogeneous and discontinuous magnetic field.

DC electromagnets are usually the most used for assistance in processes to eliminate water pollutants, and they are completely similar to AC electromagnets, with the difference that the field they generate is homogeneous and continuous due to the nature of the electric current, although they can also generate pulsations [99–101]. Studies have been carried out on the application of an electromagnet in adsorption processes, where it has been found that the field generated by them decreases the equilibrium time [101], increases the efficiency of dye removal by orienting the molecules and creating micromagnets from magnetic adsorbent materials [102]. It was also found that there is an effect of the pulsed magnetic field on the morphology of the surface of the adsorbents and the intensification of the adsorption activity [103]. However, the heating of the solenoid generates a disadvantage because the temperature control in adsorption studies is fundamental, so it would be necessary to use a cooling system for temperature control [9], raising the equipment costs.

Within the electromagnets, there is a new modality that is also used for water purification processes; it is the superconducting magnets that generate the high-gradient magnetic field. The superconducting magnets are capable of generating a magnetic field from 2 to 10 T [95, 104].

Finally, permanent magnets, which are manufactured from ceramic and nonceramic ferromagnetic materials that have iron, nickel, cobalt, rare earth and even silica as the main elements [105–108].

The magnetic force that permanent magnets could present depends on two factors: the coercive field that presents the material and the field strength with which it is loaded in its manufacture where it is possible to find a wide variety of intensities that can be higher than 2 T per unit of length. In addition to this, they have the advantage of offering a reduction in the operational cost related to the consumption of electric power for the generation of magnetic field [95]. In addition, both the polarity of the magnet and the versatility of its shape can be an advantage in the design of devices for assisting the adsorption process. On the other hand, its fragility, oxidation in environments with humidity and limitation in size are rejected for industrial application. Like electromagnets, permanent magnets have been used for their application in adsorption processes, showing changes in kinetics [108]; modification in the magnetic ordering of materials [109], modification in the magnetic ordering in addition to the removal of metals from groundwater without modifying components of Ca and Mg materials [110].

Like the source, the configuration of the magnetic field is significant in the design of an adsorption system assisted by magnetic fields, since the influence, gradient of intensity and direction will determine the type of effect in some of the adsorption phases. The configuration of the magnetic field will influence in various ways ranging from orienting, hydrating and redirecting the adsorbate, to magnetize, activate and prevent the degeneration of the adsorbent [111–115].

In the case of adsorption in continuous systems, the application of the magnetic field could be based on the orientation of the magnetic field, which could be parallel or perpendicular to the directions of the fluid flow [95]. Particularly in the use of permanent magnets, they have been used in the treatment of electrolytes where they were arranged around the treatment circuit, so that the applied field was orthogonal to the direction of fluid flow [116].

For adsorption studies in batch systems, there are multiple configurations. Experiments have been carried out in which a pair of permanent magnets is arranged in parallel, where the space between them is the adsorption system showing favorable results [117]. This could be explained thanks to the sum of magnetic intensities, which is a phenomenon that occurs when it has two or more magnets in parallel and their polarities are attracted, and this is only effective at short distances and this distance is a function of the intensities particular of each permanent magnet involved. On the other hand, the variation of adsorption results was investigated by applying different magnetic dispositions taking as an output variable the removal of the adsorbate, where three different configurations in which permanent magnets were disposed and evaluated, particularly the behavior of the fluidized adsorbent particles was evaluated. In the contaminated solution, obtaining different degrees of removal of a dye, this study explains that the effect of the Lorentz forces modify the trajectories of the adsorbent particles inside the container, opposing the turbulent forces created by the mechanical agitation [112]. On the other hand, it has been found that there is a relationship between the magnetic configuration and the gradient of magnetic intensities, in addition to the arrangement of magnetic fields that can influence the solution, the adsorbent material and both [113].

The evaluation of different magnetic systems can be designed from computer simulations which can quantify and visually inspect the trajectories of the field lines, the gradient of magnetic intensities and the spatial distribution profiles of the magnetic flux density [84, 118–120], in such a way that it is possible to be certain that the adsorption system will be affected by the magnetic field.

Considering all the information presented in this chapter, it is possible to affirm that the magnetic field is a viable tool within the adsorption processes for the removal of heavy metals in an efficient and environmentally friendly way; however, there are different parameters that are essential to take into account to achieve the expected results. The mechanism under which the magnetic field acts on the adsorbate-adsorbent system is quite complex and depends on several variables within the system.

Author details

Ma. del Rosario Moreno Virgen*, Omar Francisco González Vázquez,
Virginia Hernández Montoya and Rigoberto Tovar Gómez

*Address all correspondence to: moreno_virgen@yahoo.com.mx

Tecnológico Nacional de México, Instituto Tecnológico de Aguascalientes,
Aguascalientes, Mexico

References

[1] Alijani H, Beyki MH, Shariatinia Z, Bayat M, Shemirani F. A new approach for one step synthesis of magnetic carbon nanotubes/diatomite earth composite by chemical vapor deposition method: Application for removal of lead ions. Chemical Engineering Journal. 2014;**253**:456-463. DOI: 10.1016/j.cej.2014.05.021

[2] Chen G, Qiao C, Wang Y, Yao J. Synthesis of magnetic gelatin and its adsorption property for Cr(VI). Industrial and Engineering Chemistry Research. 2014;**53**:15576-15581. DOI: 10.1021/ie502709u

[3] Agency for Environmental Protection (EPA) Federal Law on Control of Water Pollution "Clean Water Act". United States of America. 1977

[4] Hernández-Montoya V, Pérez-Cruz MA, Mendoza-Castillo DI, Moreno-Virgen MR, Bonilla-Petriciolet A. Competitive adsorption of dyes and heavy metals on zeolitic structures. Journal of Environmental Management. 2013;**116**:213-221. DOI: 10.1016/j.jenvman.2012.12.010

[5] Tang WW, Zeng GM, Gong JL, Liang J, Xu P, Zhang C, Huang BB. Impact of humic/fulvic acid on the removal of heavy metals from aqueous solutions using nanomaterials: A review. Science of the Total Environment. 2014;**468**:1014-1027. DOI: 10.1016/j.scitotenv.2013.09.044

[6] Lesmana SO, Febriana N, Soetaredjo FE, Sunarso J, Ismadji S. Studies on potential applications of biomass for the separation of heavy metals from water and wastewater. Biochemical Engineering Journal. 2009;**44**:19-41. DOI: 10.1016/j.bej.2008.12.009

[7] Zeng G, Liu Y, Tang L, Yang G, Pang Y, Zhang Y, Zhou Y, Li Z, Li M, Lai M, He X, He Y. Enhancement of Cd(II) adsorption by polyacrylic acid modified magnetic mesoporous carbon. Chemical Engineering Journal. 2015;**259**:153-160. DOI: 10.1016/j.cej.2014.07.115

[8] Masoumi A, Ghaemy M, Bakht AN. Removal of metal ions from water using poly(MMA-co-MA)/modified-Fe_3O_4 magnetic nanocomposite: Isotherm and kinetic study. Industrial and Engineering Chemistry Research. 2014;**53**:8188-8197. DOI: 10.1021/ie5000906

[9] Ferreira J, Oliveira L, Ragozoni M, Paulo da Silva J, Ramalho T. Adsorption of aromatic compounds under magnetic field influence. Water, Air, and Soil Pollution. 2012;**223**:3545-3551. DOI: 10.1007/s11270-012-1132-6

[10] Tireli A, Firmino M, Ferreira L, Guimaraes I, Guerreiro M, Paulo J. Influence of magnetic field on the adsorption of organic by clays modified with iron. Applied Clay Science. 2014;**97**:1-7. DOI: 10.1016/j.clay.2014.05.014

[11] Zhang J, Han J, Wang M, Guo R. Fe_3O_4/PANI/MnO_2 core–shell hybrids as advanced adsorbents for heavy metal ions. Journal of Materials Chemistry A. 2017;**5**:4058-4066. DOI: 10.1039/c6ta10499a

[12] Pavlovic I, Pérez MR, Barriga C, Ulibarri MA. Adsorption of Cu^{2+}, Cd^{2+} and Pb^{2+} ions by layered double hydroxides intercalated with the chelating agents diethylenetriamine-pentaacetate and meso-2,3-dimercaptosuccinate. Applied Clay Science. 2009;**43**:125-129. DOI: 10.1016/j.clay.2008.07.020

[13] Kara I, Yilmazer D, Akar ST. Metakaolin based geopolymer as an effective adsorbent for adsorption of zinc(II) and nickel(II) ions from aqueous solutions. Applied Clay Science. 2017;**139**:54-63. DOI: 10.1016/j.clay.2017.01.008

[14] Wawrzkiewicz M, Wisniewska M, Wołowicz A, Gun'ko VM, Zarko VI. Mixed silica-alumina oxide as sorbent for dyes and metal ions removal from aqueous solutions and wastewaters. Microporous and Mesoporous Materials. 2017;**250**:128-147. DOI: 10.1016/j.micromeso.2017.05.016

[15] Selim AQ, Mohamed EA, Mobarak M, Zayed AM, Seliem MK, Komarneni S. Cr(VI) uptake by a composite of processed diatomite with MCM-41: Isotherm, kinetic and thermodynamic studies. Microporous and Mesoporous Materials. 2018;**260**:84-92. DOI: 10.1016/j.micromeso.2017.10.041

[16] Yuan P, Liu D, Fan M, Yang D, Zhu R, Ge F, Zhu J, He H. Removal of hexavalent chromium [Cr(VI)] from aqueous solutions by the diatomite-supported/unsupported magnetite nanoparticles. Journal of Hazardous Materials. 2010;**173**:614-621. DOI: 10.1016/j.jhazmat.2009.08.129

[17] Li W, Tang Y, Zeng Y, Tong Z, Liang D, Cui W. Adsorption behavior of Cr(VI) ions on tannin-immobilized activated clay. Chemical Engineering Journal. 2012;**193-194**:88-95. DOI: 10.1016/j.cej.2012.03.084

[18] Zeng Y, Woo H, Lee G, Park J. Removal of chromate from water using surfactant modified Pohang clinoptilolite and Haruna chabazite. Desalination. 2010;**257**:102-109. DOI: 10.1016/j.desal.2010.02.039

[19] Hong H, Jiang W, Zhang X, Tie L, Li Z. Adsorption of Cr(VI) on STAC-modified rectorite. Applied Clay Science. 2008;**42**:292-299. DOI: 10.1016/j.clay.2008.01.015

[20] Ge Y, Cui X, Liao C, Li Z. Facile fabrication of green geopolymer/alginate hybrid spheres for efficient removal of Cu(II) in water: Batch and column studies. Chemical Engineering Journal. 2017;**311**:126-134. DOI: 10.1016/j.cej.2016.11.079

[21] Catauro M, Bollino F, Papale F, Lamanna G. Investigation of the sample preparation and curing treatment effects on mechanical properties and bioactivity of silica rich metakaolin geopolymer. Materials Science and Engineering C: Materials for Biological Applications. 2014;**36**:20-24. DOI: 10.1016/j.msec.2013.11.026

[22] Tofighy MA, Mohammadi T. Adsorption of divalent heavy metal ions from water using carbon nanotube sheets. Journal of Hazardous Materials. 2011;**185**:140-147. DOI: 10.1016/j.jhazmat.2010.09.008

[23] Erdem E, Karapinar N, Donat R. The removal of heavy metal cations by natural zeolites. Journal of Colloid and Interface Science. 2004;**280**:309-314. DOI: 10.1016/j.jcis.2004.08.028

[24] Algothmi WM, Bandaru NM, Yu Y, Shapter JG, Ellis AV. Alginate-graphene oxide hybrid gel beads: An efficient copper adsorbent material. Journal of Colloid and Interface Science. 2013;**397**:32-38. DOI: 10.1016/j.jcis.2013.01.051

[25] Grisdanurak N, Akewaranugulsiri S, Futalan CM, Tsai WC, Kan CC, Hsu CW, Wan MW. The study of copper adsorption from aqueous solution using crosslinked chitosan immobilized on bentonite. Journal of Applied Polymer Science. 2012;**125**:E132-E142. DOI: 10.1002/app.35541

[26] Rathnayake SI, Martens WN, Xi Y, Frost RL, Ayoko GA. Remediation of Cr (VI) by inorganic-organic clay. Journal of Colloid and Interface Science. 2017;**490**:163-173. DOI: 10.1016/j.jcis.2016.11.070

[27] Tovar-Gómez R, Rivera-Ramírez DA, Hernández-Montoya V, Bonilla-Petriciolet A, Durán-Valle CJ, Montes-Morán MA. Synergic adsorption in the simultaneous removal of acid blue 25 and heavy metals from water using a $Ca(PO_3)_2$-modified carbon. Journal of Hazardous Materials. 2012;**199-200**:290-300. DOI: 10.1016/j.jhazmat.2011.11.015

[28] Schier de Lima L, Machado Araujo MD, Pércio Quináia S, Migliorine DW, Garcia JR. Adsorption modeling of Cr, Cd and Cu on activated carbon of different origins by using fractional factorial design. Chemical Engineering Journal. 2011;**166**:881-889. DOI: 10.1016/j.cej.2010.11.062

[29] Hsi HC, Rood MJ, Massoud RA, Chang YM. Effects of sulfur, nitric acid, and thermal treatments on the properties and mercury adsorption of activated carbons from bituminous coals. Aerosol and Air Quality Research. 2013;**13**:730-738. DOI: 10.4209/aaqr.2012.07.0177

[30] Schier de Lima L, Pércio Quináia S, Melquiades FL, de Biasi GEV, Garcia JR. Characterization of activated carbons from different sources and the simultaneous adsorption of Cu, Cr, and Zn from metallurgic effluent. Separation and Purification Technology. 2014;**122**:421-430. DOI: 10.1016/j.seppur.2013.11.034

[31] Aguayo-Villarreal IA, Hernández-Montoya V, Bonilla-Petriciolet A, Tovar-Gómez R, Ramírez-López EM, Montes-Morán MA. Role of acid blue 25 dye as active site for the adsorption of Cd^{2+} and Zn^{2+} using activated carbons. Dyes and Pigments. 2013;**96**:459-466. DOI: 10.1016/j.dyepig.2012.08.027

[32] Tovar-Gómez R, Moreno-Virgen MR, Moreno-Pérez J, Bonilla-Petriciolet A, Hernández-Montoya V, Durán-Valle CJ. Analysis of synergistic and antagonistic adsorptionof heavy metals and acid blue 25 on activated carbon from ternary systems. Chemical Engineering Research and Design. 2015;**93**:755-772. DOI: 10.1016/j.cherd.2014.07.012

[33] Zhou J, Liu Y, Zhou X, Ren J, Zhong C. Magnetic multi-porous bio-adsorbent modified with amino siloxane for fast removal of Pb(II) from aqueous solution. Applied Surface Science. 2018;**427**:976-985. DOI: 10.1016/j.apsusc.2017.08.110

[34] Ebrahimi Zarandi MJ, Reza Sohrabi M, Khosravi M, Mansouriieh N, Davallo M, Khosravan A. Optimizing Cu(II) removal from aqueous solution by magnetic nanoparticles immobilized on activated carbon using Taguchi method. Water Science and Technology. 2016;**74**:38-47. DOI: 10.2166/wst.2016.152

[35] Chen W, Parette R, Zou J, Cannon FS, Dempsey BA. Arsenic removal by iron-modified activated carbon. Water Research. 2007;**41**:1851-1858. DOI: 10.1016/j.watres.2007.01.052

[36] Chang Q, Wei L, W-C Y. Preparation of iron-impregnated granular activated carbon for arsenic removal from drinking water. Journal of Hazardous Materials. 2010;**184**:515-522. DOI: 10.1016/j.jhazmat.2010.08.066

[37] Zhu H, Jia Y, Wu X, Wang H. Removal of arsenic from water by supported nano zero-valent iron on activated carbon. Journal of Hazardous Materials. 2009;**172**:1591-1596. DOI: 10.1016/j.jhazmat.2009.08.031

[38] He J, Shang H, Zhang X, Sun X. Synthesis and application of ion imprinting polymer coated magnetic multi-walled carbon nanotubes for selective adsorption of nickel ion. Applied Surface Science. 2018;**428**:110-117. DOI: 10.1016/j.apsusc.2017.09.123

[39] Shang J, Pi J, Zong M, Wang Y, Li W, Liao Q. Chromium removal using magnetic biochar derived from herb residue. Journal of the Taiwan Institute of Chemical Engineers. 2016;**68**:289-294. DOI: 10.1016/j.jtice.2016.09.012

[40] Luo X, Lei X, Cai N, Xie X, Xue Y, Yu F. Removal of heavy metal ions from water by magnetic cellulose-based beads with embedded chemically modified magnetite nanoparticles and activated carbon. Sustainable Chemistry & Engineering. 2016;**4**:3960-3969. DOI: 10.1021/acssuschemeng.6b00790

[41] Horst MF, Álvarez M, Lassalle VL. Removal of heavy metals from wastewater using magnetic nanocomposites: Analysis of the experimental conditions. Separation Science and Technology. 2016;**51**:550-563. DOI: 10.1080/01496395.2015.1086801

[42] Liu T, Han X, Wang Y, Yan L, Du B, Wei Q, Wei D. Magnetic chitosan/anaerobic granular sludge composite: Synthesis, characterization and application in heavy metal ions removal. Journal of Colloid and Interface Science. 2017;**508**:405-414. DOI: 10.1016/j.jcis.2017.08.067

[43] Yi X, He J, Guo Y, Han Z, Yang M, Jin J, Gu J, Ou M, Xu X. Encapsulating Fe_3O_4 into calcium alginate coated chitosan hydrochloride hydrogel beads for removal of Cu (II) and U (VI) from aqueous solutions. Ecotoxicology and Environmental Safety. 2018;**147**:699-707. DOI: 10.1016/j.ecoenv.2017.09.036

[44] Zhu J, Gu H, Guo J, Chen M, Wei H, Luo Z, Colorado HA, Yerra N, Ding D, Ho TC, Haldolaarachchige N, Hopper J, Young DP, Guo Z, Wei S. Mesoporous magnetic carbon nanocomposite fabrics for highly efficient Cr(VI) removal. Journal of Materials Chemistry A. 2014;**2**:2256-2265. DOI: 10.1039/C3TA13957C

[45] Lasheen MR, El-Sherif IY, El-Wakeel ST, Sabry DY, El-Shahat MF. Heavy metals removal from aqueous solution using magnetite Dowex 50WX4 resin nanocomposite. Journal of Materials and Environmental Science. 2017;**8**:503-511

[46] Zhou C, Zhu H, Wang Q, Wang J, Cheng J, Guo Y, Zhou X, Bai R. Adsorption of mercury(II) with an Fe_3O_4 magnetic polypyrrole–graphene oxide nanocomposite. RSC Advances. 2017;**7**:18466-18479. DOI: 10.1039/c7ra01147d

[47] Son E-B, Poo K-M, Chang J-S, Chae K-J. Heavy metal removal from aqueous solutions using engineered magnetic biochars derived from waste marine macro-algal biomass. Science of the Total Environment. 2018;**615**:161-168. DOI: 10.1016/j.scitotenv.2017.09.171

[48] Meng C, Zhikun W, Qiang LV, Li C, Shuangqing S, Songqing H. Preparation of amino-functionalized Fe_3O_4@$mSiO_2$ core-shell magnetic nanoparticles and their application for aqueous Fe^{3+} removal. Journal of Hazardous Materials. 2018;**341**:198-206. DOI: 10.1016/j.jhazmat.2017.07.062

[49] Rusmin R, Sarkar B, Tsuzuki T, Kawashima N, Naidu R. Removal of lead from aqueous solution using superparamagnetic palygorskite nanocomposite: Material characterization and regeneration studies. Chemosphere. 2017;**186**:1006-1015. DOI: 10.1016/j.chemosphere.2017.08.036

[50] Kumar Reddy DH, Lee SM. Magnetic biochar composite: Facile synthesis, characterization and application for heavy metal removal. Colloid Surface A. 2014;**20**:96-103. DOI: 10.1016/j.colsurfa.2014.03.105

[51] Elwakeel KZ, Al-Bogami AS. Influence of Mo(VI) immobilization and temperature on As(V)sorption onto magnetic separable polyp-phenylenediamine-thiourea-formaldehyde polymer. Journal of Hazardous Materials. 2018;**342**:335-346. DOI: 10.1016/j.jhazmat.2017.08.007

[52] Cao Y, Huang J, Peng X, Cao D, Galaska A, Qiu S, Liu J, Khan AM, Young DP, Ryu JE, Feng H, Yerra N, Guo Z. Poly(vinylidene fluoride) derived fluorine-doped magnetic carbon nanoadsorbents for enhanced chromium removal. Carbon. 2017;**115**:503-514. DOI: 10.1016/j.carbon.2017.01.033

[53] Vojoudi H, Badiei A, Bahar S, Ziarani GM, Faridbod F, Ganjali MR. A new nano-sorbent for fast and efficient removal of heavy metals from aqueous solutions based on modification of magnetic mesoporous silica nanospheres. Journal of Magnetism and Magnetic Materials. 2017;**441**:193-203. DOI: 10.1016/j.jmmm.2017.05.065

[54] Wang S, Zhang C, Chang Q. Synthesis of magnetic crosslinked starch-graftpoly(acrylamide)-co-sodium xanthate and its application in removing heavy metal ions. Journal of Experimental Nanoscience. 2017;**12**:270-284. DOI: 10.1080/17458080.2017.1321793

[55] Shehzad K, Xie C, He J, Cai X, Xu W, Liu J. Facile synthesis of novel calcined magnetic orange peel composites for efficient removal of arsenite through simultaneous oxidation and adsorption. Journal of Colloid and Interface Science. 2018;**511**:155-164. DOI: 10.1016/j.jcis.2017.09.110

[56] Noor NM, Othman R, Mubarak NM, Abdullah EC. Agricultural biomass-derived magnetic adsorbents: Preparation and application for heavy metals removal. Journal of the Taiwan Institute of Chemical Engineers. 2017;**78**:168-177. DOI: 10.1016/j.jtice.2017.05.023

[57] Keochaiyom B, Wan J, Zeng G, Huang D, Xue W, Hu L, Huang C, Zhang C, Cheng M. Synthesis and application of magnetic chlorapatite nanoparticles for zinc (II), cadmium (II) and lead (II) removal from water solutions. Journal of Colloid and Interface Science. 2017;**505**:824-835. DOI: 10.1016/j.jcis.2017.06.056

[58] Ghasemi E, Heydari A, Sillanpää M. Superparamagnetic Fe_3O_4@EDTA nanoparticles as an efficient adsorbent for simultaneous removal of Ag(I), Hg(II), Mn(II), Zn(II), Pb(II) and Cd(II) from water and soil environmental samples. Microchemical Journal. 2017;**131**:51-56. DOI: 10.1016/j.microc.2016.11.011

[59] Rahman Khattak MMU, Zahoor M, Muhammad B, Khan FA, Ullah R, AbdEI-Salam NM. Removal of heavy metals from drinking water by magnetic carbon nanostructures prepared from biomass. Journal of Nanomaterials. 2017;**1**:1-10. DOI: 10.1155/2017/5670371

[60] Yap MW, Mubarak NM, Sahu JN, Abdullah EC. Microwave induced synthesis of magnetic biochar from agricultural biomass for removal of lead and cadmium from wastewater. Journal of Industrial and Engineering Chemistry. 2017;**45**:287-295. DOI: 10.1016/j.jiec.2016.09.036

[61] Praveen RS, Vijayaraghavan K. Optimization of Cu(II), Ni(II), Cd(II) and Pb(II) biosorption by red marine alga Kappaphycus alvarezii. Desalination and Water Treatment. 2014;**55**:1-9. DOI: 10.1080/19443994.2014.927334

[62] Gupta VK, Ganjali MR, Nayak A, Bhushan B, Agarwal S. Enhanced heavy metals removal and recovery by mesoporous adsorbent prepared from waste rubber tire. Chemical Engineering Journal. 2012;**197**:330-342. DOI: 10.1016/j.cej.2012.04.104

[63] Lin L-C, Thirumavalavan M, Wang Y-T, Lee J-F. Surface area and pore size tailoring of mesoporous silica materials by different hydrothermal treatments and adsorption of heavy metal ions. Colloid Surface A. 2010;**369**:223-231. DOI: 10.1016/j.colsurfa.2010.08.032

[64] Castaldi P, Santona L, Enzo S, Melis P. Sorption processes and XRD analysis of a natural zeolite exchanged with Pb^{2+}, Cd^{2+} and Zn^{2+} cations. Journal of Hazardous Materials. 2008;**156**:428-434. DOI: 10.1016/j.jhazmat.2007.12.040

[65] Faur-Brasquet C, Reddad Z, Kadirvelu K, Le Cloirec P. Modeling the adsorption of metal ions (Cu^{2+}, Ni^{2+}, Pb^{+2}) onto ACCs using surface complexation models. Applied Surface Science. 2002;**196**:356-365. DOI: 10.1016/S0169-4332(02)00073-9

[66] Chen SB, Ma YB, Chen L, Xian K. Adsorption of aqueous Cd^{2+}, Pb^{2+}, Cu^{2+} ions by nano-hydroxiapatite: Single-and multi-metal competitive adsorption study. Geochemical Journal. 2010;**44**:233-239. DOI: 10.2343/geochemj.1.0065

[67] Yang K, Chen B, Zhu X, Xing B. Aggregation, adsorption and morphological transformation of graphene oxide in aqueous solutions containing different metal cations. Environmental Science & Technology. 2016;**50**:11066-11075. DOI: 10.1021/acs.est.6b04235

[68] Massocatto CL, de Andrade M, Honorato AC, Caetano J, Tarley CRT, Gonçalves-Júnior A, Colauto NB, Linde Colauto GA, Cardoso Dragunski D. Biosorption of Pb^{2+}, Cr^{3+}, and Cu^{2+} by peach palm sheath modified colonized by *Agaricus Blazei*. Desalination and Water Treatment. 2016;**57**:19927-19938. DOI: 10.1080/19443994.2015.1107503

[69] El Zayat M, Elagroudy S, El Haggar S. Removal of some heavy metals in selected wastewater using cement kiln dust. In: World Environmental and Water Resources Congress. 2015. pp. 2459-2469

[70] Castillo VA, Barakat MA, Ramadan MH, Woodcock HL, Kuhn JN. Metal ion remediation by polyamidoamine dendrimers: A comparison of metal ion, oxidation state, and titania immobilization. International Journal of Environmental Science and Technology. 2014;**11**:1497-1502. DOI: 10.1007/s13762-013-0346-5

[71] Fu F, Wang Q. Removal of heavy metal ions from wastewaters: A review. Journal of Environmental Management. 2011;**92**:407-418. DOI: 10.1016/j.jenvman.2010.11.011

[72] Bisht R, Agarwal M, Singh K. Heavy metal removal from wastewater using various adsorbents: A review. Journal of Water Reuse and Desalination. 2017;**07**(4):387-419. DOI: 10.2166/wrd.2016.104

[73] Deliyanni EA, Kyzas GZ, Triantafyllidis KS, Matis KA. Activated carbons for the removal of heavy metal ions: A systematic review of recent literature focused on lead and arsenic ions. Open Chemistry. 2015;**13**:699-708. DOI: 10.1515/chem-2015-0087

[74] Aeisyah Abas SN, Shah Ismail MH, Kamal ML, Izhar S. Adsorption process of heavy metals by low-cost adsorbent: A review. World Applied Sciences Journal. 2013;**28**:1518-1530. DOI: idosi.wasj.2013.28.11.1874

[75] Askeland DR. Ciencia e ingeniería de los materiales. 3rd ed. Mexico: International Thompson; 1998

[76] Drago RS. Magnetism. Physical Methods for Chemists. 2nd ed. United States: Saunders College Publishing; 1992

[77] Zhang G, Yang X, Liu Y, Jia Y, Yu G, Ouyang S. Copper(II) adsorption on Ca-rectorite, and effect of static magnetic field on the adsorption. Journal of Colloid and Interface Science. 2004;**278**:265-269. DOI: 10.1016/j.jcis.2004.05.046

[78] Zhang G, Liu Y, Xie Y, Yang X, Hu B, Ouyang S, Liu H, Wang H. Zinc adsorption on Na-rectorite and effect of static magnetic field on the adsorption. Applied Clay Science. 2005;**29**:15-21. DOI: 10.1016/j.clay.2004.09.001

[79] Higashitani K, Iseri H, Okuhara K, Kage A, Hatade S. Magnetic effects on zeta potential and diffusivity of nonmagnetic colloidal particles. Journal of Colloid and Interface Science. 1995;**172**:383-388. DOI: 10.1006/jcis.1995.1268

[80] Elwakeel KZ. Removal of Cr(VI) from alkaline aqueous solutions using chemically modified magnetic chitosan resins. Desalination. 2010;**250**:105-112. DOI: 10.1016/j.desal.2009.02.063

[81] Bockris JO, Reddy AKN. Modern Electrochemistry 1. 2nd ed. Boston: Kluwer Academic Publishers; 2002

[82] Higashitani K, Oshitani J. Magnetic effects on thickness of adsorbed layer in aqueous solutions evaluated directly by atomic force microscope. Journal of Colloid and Interface Science. 1998;**204**:363-378. DOI: 10.1006/jcis.1998.5590

[83] Gehr R, Zhai ZA, Finch JA, Rao SR. Reduction of soluble mineral concentrations in $CaSO_4$ saturated water using a magnetic field. Water Research. 1995;**29**:933-940. DOI: 10.1016/0043-1354(94)00214-R

[84] Holysz L, Szczes A, Chibowski E. Effects of a static magnetic field on water and electrolyte solutions. Journal of Colloid and Interface Science. 2007;**316**:996-1002. DOI: 10.1016/j.jcis.2007.08.026

[85] Lipus LC, Krope J, Crepinsek L. Dispersion destabilization in magnetic water treatment. Journal of Colloid and Interface Science. 2001;**236**:60-66. DOI: 10.1006/jcis.2000.7392

[86] Jiang X, Qiao J, Lo IMC, Wang L, Guan X, Lu Z, Zhou G, Xu C. Enhanced paramagnetic Cu^{2+} ions removal by coupling a weak magnetic field with zero valent iron. Journal of Hazardous Materials. 2015;**283**:880-887. DOI: 10.1016/j.jhazmat.2014.10.044

[87] Silva IB, Queiroz Neto JC, Petri DFS. The effect of magnetic field on ion hydration and sulfate scale formation. Colloids and Surfaces A: Physicochemical and Engineering Aspects. 2015;**465**:175-183. DOI: 10.1016/j.colsurfa.2014.10.054

[88] Oshitani J, Yamada D, Miyahara M, Higashitani K. Magnetic effect on ion-exchange kinetics. Journal of Colloid and Interface Science. 1999;**210**:1-7. DOI: 10.1006/jcis.1998.5952

[89] Holysz L, Chibowski M, Chibowski E. Time-dependent changes of zeta potential and other parameters of in situ calcium carbonate due to magnetic field treatment. Colloids and Surfaces A: Physicochemical and Engineering Aspects. 2002;**208**:231-240. DOI: 10.1016/S0927-7757(02)00149-8

[90] Higashitani K, Oshitani J. Measurements of magnetic effects on electrolyte solutions by atomic force microscope. Institution of Chemical Engineers. 1997;**75**:115-119. DOI: 10.1205/095758297528887

[91] Bain A, Berry JF. Diamagnetic corrections and Pascal's constants. Journal of Chemical Education. 2008;**85**:532

[92] Lide DR. Magnetic susceptibility of the elements and inorganic compounds. Handbook of Chemistry and Physics. 87th ed. Boca Raton: Taylor and Francis; 2005. pp. 130-135

[93] Hollens WRA, Spencer JF. The magnetic susceptibility of cadmium compounds. Journal of the Chemical Society. 1935;**0**:495

[94] Whitten KW, Davis RE, Stanley GG, Peck ML. Chemistry. 9th ed. Belmont, CA. USA: Mary Finch; 2010

[95] Ambashta RD, Sillanpää M. Water purification using magnetic assistance: A review. Journal of Hazardous Materials. 2010;**180**:38-49. DOI: 10.1016/j.jhazmat.2010.04.105

[96] Zhang Q, Lu T, Bai DM, Lin DQ, Yao SJ. Self-immobilization of a magnetic biosorbent and magnetic induction heated dye adsorption processes. Chemical Engineering Journal. 2016;**284**:972-978. DOI: 10.1016/j.cej.2015.09.047

[97] Moon JD, Chung HS. Acceleration of germination of tomato seed by applying AC electric and magnetic fields. Journal of Electrostatics. 2000;**48**:103-114. DOI: 10.1016/S0304-3886(99)00054-6

[98] Samonin VV, Podvyaznikov ML, Chentsov MS, Spiridonova EA, Kiseleva VL. Effect of AC magnetic field on adsorption of benzene and ethanol vapors by activated carbons. Russian Journal of Applied Chemistry. 2012;**85**:1176-1181. DOI: 10.1134/S107042721208006X

[99] Yavuz H, Çelebi SS. Effects of magnetic field on activity of activated sludge in wastewater treatment. Enzyme and Microbial Technology. 2000;**26**:22-27. DOI: 10.1016/S0141-0229(99)00121-0

[100] Novickij V, Grainys A, Švediene J, Markovskaja S, Paškevičius A, Novickij J. Microsecond pulsed magnetic field improves efficacy of antifungal agents on pathogenic microorganisms. Bioelectromagnetics. 2014;**35**:347-353. DOI: 10.1002/bem.21848

[101] Duan L, Guo S, Yang J. Study on the effect of a magnetic field on Pb (II) removal using modified chitosan. Advances in Chemical Engineering and Science. 2012;**2**:101-107. DOI: 10.4236/aces.2012.21011

[102] Tireli AA, Marcos FCF, Oliveira LF, Guimarães IR, Guerreiro MC, Silva JP. Influence of magnetic field on the adsorption of organic compound by clays modified with iron. Applied Clay Science. 2014;**97-98**:1-7. DOI: 10.1016/j.clay.2014.05.014

[103] Bel'chinskaya LI, Khodosova NA, Bityutskaya LA. Adsorption of formaldehyde at mineral nanoporous sorbents exposed to a pulse magnetic field. Protection of Metals and Physical Chemistry of Surfaces. 2009;**45**:203-206. DOI: 10.1134/S2070205109020130

[104] Li XL, Yao KL, Liu HR, Liu ZL. The investigation of capture behaviors of different shape magnetic sources in the high-gradient magnetic field. Journal of Magnetism and Magnetic Materials. 2007;**311**:481-488. DOI: 10.1016/j.jmmm.2006.07.040

[105] Szymura S, Sojka L. The effect of silicon on the structure and properties of Fe-Cr-Co permanent magnet alloys. Journal of Magnetism and Magnetic Materials. 1986;**53**:379-389. DOI: 10.1016/0304-8853(86)90184-8

[106] Herbst JF, Croat JJ. Neodymium-iron-boron permanent magnets. Journal of Magnetism and Magnetic Materials. 1991;**100**:57-78. DOI: 10.1016/0304-8853(91)90812-O

[107] Sagawa M, Hirosawa S, Yamamoto H, Fujimura S, Matsuura Y. Nd–Fe–B permanent magnet materials. Japanese Journal of Applied Physics. 1987;**26**:785-800. DOI: 10.1143/JJAP.26.785

[108] Jia YY, Zhang GK, Hu B, Dong W. Adsorption capacity of kaolinite for copper (II) under magnetic field. Journal of Wuhan University of Technology-Materials Science Edition. 2004;**19**:52-54. DOI: 10.1007/BF03000168

[109] Patkowski J, Myśliwiec D, Chibowski S. Adsorption of polyethyleneimine (PEI) on hematite. Influence of magnetic field on adsorption of PEI on hematite. Materials Chemistry and Physics. 2014;**144**:451-461. DOI: 10.1016/j.matchemphys.2014.01.019

[110] Cotton GB, Navratil JD, Bradley H. Magnetic adsorption method for the treatment of metal contaminated aqueous waste. Waste Management Symposia. 1999;**2**:1-11

[111] Brown P, Bushmelev A, Butts CP, Eloi J, Grillo I, Baker PJ, Schmidt AM, Eastoe J. Properties of new magnetic surfactants. Langmuir. 2013;**29**:3246-3251. DOI: 10.1021/la400113r

[112] Hao X, Liu H, Zhang G, Zou H, Zhang Y, Zhou M, Gu Y. Magnetic field assisted adsorption of methylene blue onto organo-bentonite. Applied Clay Science. 2012;**55**:177-180. DOI: 10.1016/j.clay.2011.11.019

[113] González Vázquez OF, Moreno Virgen MR, Hernández Montoya V, Tovar Gómez R, Alcántara Flores JL, Pérez Cruz MA, Montes Morán MA. Adsorption of heavy metals in the presence of a magnetic field on adsorbents with different magnetic properties. Industrial and Engineering Chemistry Research. 2016;**55**:9323-9331. DOI: 10.1021/acs.iecr.6b01990

[114] Zubov VE, Kudakov AD, Levshin NL, Fedulova TS. The influence of adsorption of water molecules on magnetic susceptibility of amorphous ferromagnets. Sensors and Actuators A. 2001;**91**:214-217. DOI: 10.1016/S0924-4247(01)00494-0

[115] Ozeki S, Miyamoto J, Ono S, Wakai C, Watanabe T. Water-solid interactions under steady magnetic fields: Magnetic-field-induced adsorption and desorption of water. The Journal of Physical Chemistry. 1996;**100**:4205-4212. DOI: 10.1021/jp9528774

[116] Tombácz E, Ma C, Busch KW, Busch MA. Effect of a weak magnetic field on hematite sol in stationary and flowing systems. Colloid & Polymer Science. 1991;**269**:278-289. DOI: 10.1007/BF00665502

[117] Mohammed RR, Ketabachi MR, McKay G. Combined magnetic field and adsorption process for treatment of biologically treated palm oil mill effluent (POME). Chemical Engineering Journal. 2014;**243**:31-42. DOI: 10.1016/j.cej.2013.12.084

[118] Petkovska L, Ieee SM, Cvetkovski G. FEM Based Simulation of a Permanent Magnet Synchronous Motor Performance Characteristics. In: IEEE 5th International Power Electronics and Motion Control Conference. 2006. DOI: 10.1109/IPEMC.2006.4777986

[119] Zakaria Z, Saiful M, Mansor B, Jahidin AH, Shuhanaz M, Azlan Z, Rahim RA. Simulation of magnetic flux leakage (MFL) analysis using FEMM software. In: IEEE Symposium on Industrial Electronics and Applications. 2010. pp. 481-486. DOI: 10.1109/ISIEA.2010.5679417

[120] Vergallo C, Ahmadi M, Mobasheri H, Dini L. Impact of inhomogeneous static magnetic field (31.7–232.0T) exposure on human neuroblastoma SH-SY5Y cells during cisplatin administration. PLoS One. 2014;**9**:1-19. DOI: 10.1371/journal.pone.0113530

Chapter 16

Biosorbents in the Metallic Ions Determination

Absolon C. da Silva Júnior, Alessa G. Siqueira,
Carolina A. de Sousa e Silva,
Jordana de Assis N. Oliveira, Nívia M.M. Coelho and
Vanessa Nunes Alves

Additional information is available at the end of the chapter

http://dx.doi.org/10.5772/intechopen.76081

Abstract

This chapter provides an overview and discusses analytical strategies for metallic ions determination using solid phase extraction. Solid phase extraction (SPE) is a much-used technique for extraction and/or concentration of complex samples, so that the analytes present in low concentration were detected mainly using chromatographic methods. However, in recent years, this technique has been widely used in the development of methodologies for metallic ions determination in the deferential samples. This technique shows simplicity and rapidity comparing with other conventional techniques, liquid–liquid extraction, cloud point extraction and others. Solid phase extraction procedures become even more interesting when commercial adsorbents are exchanged for others with higher adsorptive capacity, selectivity, flexibility, economy and low environmental impacts. For this purpose, some inorganic, organic and several natural adsorbents are used. New approaches to obtain adsorbent materials from natural sources such as fungi, bacteria, industrial residues and composting materials have received attention. These materials have been used in the development of analytical methods with varied proposals, such as preconcentration or speciation of metal ions.

Keywords: adsorbents, solid phase extraction, metallic ions

1. Introduction

Metallic ions are found in the most diverse types of samples in concentrations ranging from mg L^{-1} to those considered as trace. So, daily involuntary we contact various metallic ions, whether in air, water, industrial products (cosmetics, makeups, remedies, deodorant, tinctures used in tattoos, etc.) and even in foods (fertilizer, packaging, pesticides, etc.).

These ions may be necessary for maintaining vital functions of organisms, since their absence can cause many diseases, as they can also be extremely toxic if ingested even at low concentrations. Mercury, cadmium, arsenic, chromium and lead, are among the metallic ions that most cause health problems [1].

Considering the toxicity or necessity of these species by the organisms, the development of analytical methodologies capable of determining and enabling their monitoring in different types of samples becomes primordial.

The atomic absorption spectrometry with flame atomization (FAAS) was utilized for many years as one of the most commonly techniques used for determining inorganic ions in the most diverse samples. However, it presents some difficulties related primarily to the detection limit and selectivity. These difficulties can be circumvented using extraction and preconcentration techniques, such as liquid–liquid extraction, co-precipitação, solid phase extraction (SPE) [2].

Solid-phase extraction (SPE) was introduced in 1976 to meet the disadvantages presented by liquid–liquid extraction and today is the most popular method of sample preparation [3]. This technique consists of the mass transfer of a liquid phase (sample) for a solid phase, called the adsorbent, and the bulk of the adsorbers used in these procedures are of synthetic origin. The main objectives of SPE are the removal of interference from the matrix, the concentration and the isolation of the analytes [3]. The state of the art of the SPE indicates that there is a tendency to develop miniaturized methods, based on flow-injection analysis systems (FIA), Sequential injection Analysis (SIA) and Lab-on-A-Valve (Lov) [4].

The adsorbents used in solid phase extraction, as ion exchangers were up to mid-1934, natural compounds called *lignitos*. After this period, several researchers were dedicated to the development of new ionic exchangers, such as polymerization resins. In the work presented by Reis and collaborators in 1996, approximately 90% of the research's involving extraction or preconcentration of metallic ions were performed using polymeric resins such as adsorbers [5].

Today, the number of scientific articles using commercial absorbents in solid phase extraction systems continues around 90% [6]. In this sense, in the face of the ever-increasing search for methodologies that meet the principles of green chemistry, the study of alternative adsorbents in the SPE procedures, has been highlighted.

So, called alternative adsorbent materials have been applied successfully on metal adsorptive process. In its majority are considered materials of easy acquisition and in some cases, taken as by-products of industrial processes, such as sugarcane bagasse, onion skin, peanut shells, corn cobs and rice husks [7]. Therefore, the choice of an alternative adsorbing material is not only based on its adsorption capability, but also at its low cost, its ease of obtaining and abundance in nature. In this sense, this chapter makes a revision of the main absorbers used in the solid phase extraction techniques presented in recent years.

2. Adsorbents used in solid phase extraction systems

2.1. Lignocellulosic adsorbents

The lignocellulosic materials are copious renewable sources, and usually underutilized. The agribusiness, for example, generates numerous sources of biomass that are not adequately and/or properly harnessed, turning them into industrial waste [8]. Thus, the development of methods that use these wastes has been highlighted in recent years, with the lignocellulosic one of the most widely used adsorbent materials, because of its high adsorptive capacity for metallic ions and organic compounds.

The lignocellulosic biomass is the plant dry matter [9], and it is composed mainly of a mixture of lignin and holocellulose (total carbohydrate content in a plant). The typical chemical composition of the lignocellulosic materials is 48% carbon, 6% hydrogen and 45% oxygen. Proteins, organic acids and other not structuring materials are minority components [10]. In this way, it can be considered as lignocellulosic materials; rice straw, wheat straw, cotton stalk, rice hull, sugarcane bagasse, corncob, sawdust and others [9].

The ability of these materials to interact with metallic ions and organic compounds is not fully understood, however, it can be said that the adsorption capacity is related to the presence of functional groups such as hydroxyl, and amines. These functional groups are present in the pulp, hemicellulose and lignin structures. It is important to note that this process is not based on a single mechanism, the sequestration of the metal ion can occur through complex mechanisms, including ion exchange and complexation. It being probable that at least some of these mechanisms act simultaneously in varying degrees depending on the biomass and the metal ion [11].

Davis et al. [12] indicated that ion-exchange is an important concept in biosorption, because it explains many of the observations made during heavy metal uptake experiments. In this context, the term ion-exchange does not explicitly identify the mechanism of heavy metal binding to biomass, and electrostatic or London–van der Waals forces should be considered as the precise mechanism of chemical binding, i.e., ionic and covalent bonds.

Therefore, the lignocellulosic materials (due its composition, availability and low cost) are becoming widely explored in the development of new analytical methodologies for metallic ions determination in trace concentrations in various types of samples. It can be found several papers in the literature which discuss the use of lignocellulosic materials as natural adsorbents. **Table 1** shows the use of lignocellulosic adsorbents in the development of analytical methods.

The *Moringa oleifera* is one of the most widely used lignocellulosic adsorbents in the development of analytical methodologies. Native to northern India, it currently grows in many regions including Africa, Southeast Asia, the Pacific and Caribbean Islands and South America [18, 19]. Research focused on the use of *M. oleifera* seeds and fruits in water purification and

Material	Treatment	Sample	Analyte	L.O.D (μg L-1)	Preconcentration factor	References
Moringa oleifera seeds	—	Beer	Pb	7.5	93	[13]
Rice husks	—	Water and Red Wine	Cd and Pb	1.14 and 14.1	72.4 and 46	[14]
Coffee	—	Water	Cu	0.22	—	[15]
Chinese herb Pang Da Hai	—	Water	Pb and Cd	0.0096 and 0.0032		[16]
Peel of *Citrus reticulata* (mandarin orange)	—	Alcoholic beverages	Ni	3.1	12	[17]

Table 1. Lignocellulosic adsorbents used in development of preconcentration and metal ion determination.

the treatment of turbid water is the best-known application [11]. It is considered a good natural adsorbent in the adsorption of trace metals; it can be used in the development of analytical methods.

The seeds were used in the determination of silver in water [20], inorganic arsenic in natural waters [21], cadmium and zinc in alcohol fuel [22, 23] and lead in beer samples [13]. Its husks were used in the selective extraction of chromium in water [24], in the copper determination in gasoline samples [25], in zinc determination in cachaça, and in the determination of lead in chicken feed [27].

For the determination of Ag(I) in aqueous solutions, Araújo and collaborators [20] describes the use of solid phase extraction on preconcentration system by flow injection, and detection by flame atomic absorption spectrometry (FAAS). In this system, 35 mg of *Moringa oleifera* seeds was used to build a mini-column. With the aim of finding the best conditions for the determination of silver, the preconcentration system of analysis by flow injection was optimized. Once the system was optimized, the analytical characteristics were estimated on preconcentration factor equals 35, detection limit of 0.22 μg L^{-1}, quantification limit of 0.73 μg L^{-1}, linear range between 5 and 20 μg L^{-1}, and precision of 3.8% (5 μg L-1, n = 7). Thus, showing an excellent repeatability. The method was employed in mineral and tap water samples, and the precision was assessed by the analysis of a certified reference material for water (APS-1071 NIST) and by recuperation tests which presented values between 94 and 101%. Consequently, the authors inferred that the method had satisfactory detection limits and precision.

In work developed by Alves and collaborators [22], it was made a methodology for on-line preconcentration and Cd(II) determination in alcohol fuel samples using *Moringa oleifera* seeds as bioadsorbent with detection by FAAS. The flow and chemical variables were optimized and the best values were found when using 10 mL of an enriched solution with 50 μg g^{-1} of Cd(II) on pH 9.1, 0.5 mol L^{-1} of HNO_3 as eluent, 200 mg of adsorbent to fill the mini column, sample flow rate of 6.0 mL min-1 and buffering solution concentration of 0.05 mol L^{-1}.

With the optimized parameters, the analytical characteristics were obtained with a detection limit of 5.50 μg L^{-1} and precision below 2.3% (35.0 μg L^{-1}, n = 9), linear range between 5 and 150 μg L^{-1}. The developed method was successfully applied on alcohol fuel samples and the precision was evaluated through the recuperation test, which had a variation of 97.5–100%.

Carmo and collaborators [25] describe a similar methodology [22, 24, 26] for solid phase extraction using *Moringa oleifera* husks as natural adsorbent for direct Cu(II) determination in gasoline samples, with detection by FAAS. Pure gasoline is almost insoluble in water, whilst the ethanol is completely miscible in gasoline and in water in any concentration. This is the reason why the authors used ethanol to dilute the gasoline samples in the ratio of 50% (v/v); which eases the copper ion sorption process in the sample. In this context, the parameters that influenced the adsorption/desorption of Cu(II) by the *Moringa oleifera* husks in gasoline samples were evaluated using 15.0 mL of 10 μg L^{-1} Cu(II) solution. The best conditions were found on pH 4.0, 0.5 mol L^{-1} of HNO_3 as eluent, sample flow rate of 6.0 mL min^{-1}, and 96.0 mg of adsorbent mass. After finding the best values, the analytical characteristics of the method were analyzed. The results obtained were a preconcentration factor equals 14, detection limit of 0.75 μg L^{-1}, linear range between 0.75 and 500 μg L^{-1}, and precision of 1.86% (50.0 μg L^{-1}, n = 9).

Oliveira and collaborators [27] reported a similar system to that used by Araújo and collaborators [20] for the lead determination in chicken feed samples. Metallic ions such as, Cr, Cu, Fe, Zn, Ag, Ca, Pb and Mo can be found in chicken feed [28]. Once ingested, these ions can remain in the chicken meat and be transferred indirectly to the people. Several analytical techniques can be used for determining metallic ions, however the complexity of the matrix can bring some difficulties. Thus, in this chapter, the authors applied a system of extraction/preconcentration for lead determination in chicken feed using *Moringa oleifera* husks as bioadsorbent where the detection was performed using FAAS. The off-line preconcentration system was optimized, and the best conditions were obtained on pH 9, 30 mg of adsorbent mass, 100 μL of HNO_3 0.5 mol L^{-1} as eluent, elution and preconcentration flow rate of 2.0 mL min^{-1}. The analytical performance for developed method was evaluated through the application of the optimized system. As a result, it was found a detection limit of 1.66 μL^{-1}, linear range between 5.0 and 1000 μg L^{-1}, and precision of 1,63%. It was found a preconcentration factor of 550 with a consume rate of 0.02 mL.

In addition to the determination of metallic ions at a trace level, the bioadsorbents have been successfully employed in the development of methods used for speciation analysis. Speciation analysis is particularly tough, due the high complexity of the matrix and the low concentration of these elements. Some matrices require a high dilution factor to minimize interference, which leads to an increased limit of detection (LOD) of the technique [29]. Alves and collaborators [21] developed a selective extraction method for inorganic arsenic in natural waters. *Moringa oleifera* seeds were used as a biosorbent for the solid phase extraction and detected by graphite furnace atomic absorption spectroscopy (GFAAS). The As(III) and As(V) species behavior was evaluated in a broad pH band. Its results showed that the As(III) had a better adsorption on pH 7, whereas the As(V) had lower adsorption values (being mainly free in solution), which allowed it to be determined directly by GFAAS. With the aim of having an effective adsorption of As(III), the method was optimized. It was then registered that 1.0 g of adsorbent was sufficient to keep it in an agitation time of 1 hour. The limit detection for As(V) was 6.3 μg L^{-1} and its precision was below 1.23%. The recuperation results (92%) were done by the use of four different water samples, which were mixed with the determined concentration of As(III) and As(V). The precision was proven by employing the Drinking Water Standard Solution (APS 1071). As a result, the authors able to conclude that the determination method is an efficient and attractive option for the determination of inorganic arsenic in water samples.

In the study developed by Alves and Coelho [24], analytical methodology applied for selective extraction of inorganic chromium in water, using a system of analysis by flow injection like the used by the previous group [23] was evaluated. For the selective extraction and the chromium species preconcentration, the *Moringa oleifera* husks were used as biosorbent and the detection was made through the FAAS technique. The *Moringa oleifera* husks showed itself efficient in the separation of the Cr(III) and Cr(VI) species. The trivalent specie is adsorbed in the pH range between 7 and 9, whereas the Cr(VI) is adsorbed in the pH band between 1 and 2. The methodology was then optimized, it evaluated the necessary parameters for the attainment of an effective Cr(III) extraction; the optimal values were found on a 7 pH, 58.0 mg of adsorbent mass, 0.01 mol L^{-1} of HCI as eluent, and sample flow rate of 4.5 mL min^{-1}. The same conditions were applied for the determination of Cr(VI), however the pH band of the solution was between 1 and 2. For this method, the detection limits found for Cr(III) and Cr(VI) were 1.92 and 2.45 mg L^{-1}, and precision of 1.63 and 0.08%, respectively [24]. The use of lignocellulosic adsorbent presented several advantages, among them was the reduced cost analysis, since the material presents stability to 100 preconcentration/eluation cycles.

In similar work, Sousa Neto [30] used *Luffa cylindrica* fibers for inorganic chromium separation. The material is positively charged surface in solutions with pH below 6.0 while above this value, the surface is negatively charged. From this evidence, a study was performed to evaluate the effect of pH on the chromium inorganic species adsorption and found that at pH 4.0 it is possible to separate the chromium species (III and VI) (**Figure 1**). The analytical performance was evaluated by the parameters, detection limit (LOD), quantification limit (LOQ), accuracy (D.P.R) and the values were 19.2; 63.99 μg L^{-1} and 0.19%, respectively, with the linear range of the method equal to 63.90–2000 μg L^{-1}. Finally, the accuracy of the method was evaluated from analysis of water certificate refer`ence material (APS-1071) and sediment (APS-1066). Recovery tests were performed on samples of water, energy and isotonic values have been found in the range from 87.60.

Ribeiro et al. [17] used the Mandarin peel (*Citrus reticulata*) as natural adsorbent to develop the analytical methodology using solid phase extraction for the direct determination of nickel in alcoholic beverages (whiskes and cachaça). In this work, it was used the online flow injection analysis system like that one previously mentioned in other works with detection by FAAS [17]. It was used 10 mL of a Ni(II) 30.0 μg L^{-1} solution, and the best conditions were found on 50 mg of adsorbent mass, pH 6, sample flow rate of 4.3 mL min^{-1}, and HCI 1.0 mol L^{-1} with flow rate of 2.0 mL min^{-1} as eluent. Consequently, the analytical characteristics of the methodology were linear range between 10 and 75 mg L^{-1}, detection limit of 3.1 μg L^{-1}, preconcentration factor equals 12, and standard deviation of 0.9%.

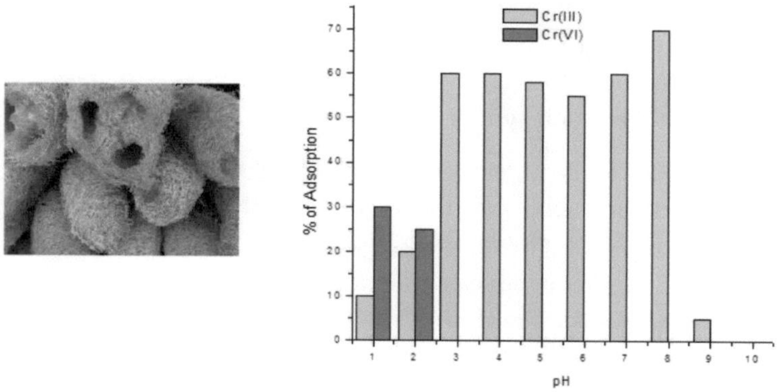

Figure 1. Inorganic chromium adsorption on *Luffa cylindrica* fibers.

Liu and collaborators [16] assessed the Chinese herb Pang Da Hai (PDH) wastes, *Sterculia lychnophora* hance seeds was used as natural adsorbent in the development of an analytical method for the determination of lead and cadmium in solid waste samples, with detection by FAAS. The parameters that influenced in the preconcentration method development were studied, and the best condition for both analyzed ions was 0.1 mol L^{-1} of NHO_3 as eluent and pH 6.5. The adsorption capacities of PDH for Pb(II) and Cd(II) were 27.1 and 17.5 mg g^{-1}, respectively. The standard deviation related to the developed method was 10% for both metals; it was used a solution in the concentration of 100 μg L^{-1} and 20 mg of adsorbent. The method was validated by the analysis with the certified standard (GBW 08301), and the values found in the analyzed samples comply with the values stipulated by the reference material; they have a relative error between 2.5 and 11%. The detection limits for both Pb and Cd were 0,096 and 0.032 μg L^{-1}, respectively. The method was applied in real samples of tap and river water, and the recuperation results were satisfactory for all analyzed samples.

2.2. Mineral adsorbents and humic acid

The minerals are inorganic compounds formed naturally and in general have an established chemical structure, responsible by the physical properties. Although they are produced by nature, these minerals can be synthesized in the laboratory by ensuring a higher purity and uniformity of crystalline structure These natural resources are of great value for industrial production, and are used as raw material in different areas such as in the production of household appliances, electrical wires, jewelry, construction materials, in addition to serving as a source of energy.

Due to its chemical–physical characteristics, minerals are either natural or synthetic, have properties of ionic exchangers, which allow their use successfully in the adsorption of metallic ions and organic compounds. Among the minerals most commonly used adsorption processes are the montmorillonites, the vermiculitas, zeolites, carbon nanotubes and same oxides, such as graphene oxides, iron oxides, and titanium oxides [31–33]. Typically, the application of these minerals is in adsorption processes for environmental remediation, particularly for treatment and recovery of waters where heavy metals and other pollutants as much as organic inorganic compounds can be removed.

In the development of analytical methodologies for the determination and pre-concentration of metallic ions, few studies have been developed. However, in recent times these materials have gained a lot of attention for be a low-cost alternative and easy access since they may be found in nature in abundance and synthesized simply.

Synthetic zeolites, for example, have been used as adsorbent material on copper ions determination and preconcentration through the formation of a chelate of copper in minicolunas of synthetic zeolite lynde type (Na-LTA) and faujasite Type X (Na-Fau). To evaluate the performance of these materials, a flow injection system was used for water samples from different sources. The preconcentration factors ranged from 35 to 125 for the LTA and 30–65 to Na-Fau. The accuracy was evaluated by recovery experiments and analysis of certified reference materials. The limits of detection, the standard deviation and relative linear calibration charts were 0.1 ng mL^{-1}, 2.6% and 2.5–30 ng mL^{-1} to Na-LTA and 0.4 ng mL^{-1}, 2.8% and 2.5–40 ng mL^{-1} for In-Fau respectively. The results indicate that the zeolites are promising materials for application in copper extraction preconcentration in trace levels with detection by flame atomic absorption spectrometry (FAAS) [34].

The adsorption capacity of natural zeolites was evaluated through the development of an analytical methodology for cobalt ions determination in natural waters through the formation of a neutral chelate with pyrrolidine acid ammonium diocarbamato (APDC). The neutral chelate formed was retained in a column packed with natural zeolites and then eluted with iso-methyl-butyl ketone (MIBK) and detected by flame atomic absorption spectrometry (FAAS). The detection limit and the relative standard deviation (%RSD) was 1.9–2.3 μg L^{-1} and 2.3–4.5%, respectively. The sampling frequency of 27 h^{-1} and preconcentration factor of 130–160 have been achieved, to a 6 mL sample volume, indicating a high analyte retention on natural zeolites [33].

The use of these adsorbents increases more and more, because efficient synthesis procedures, purification and characterization ensures the availability of pure materials and well characterized. Soylak and Unsal [35], have developed a separation and preconcentration procedure based on solid phase extraction for Fe(III) and Cr(III) ions, using single-walled carbon nanotubes (SWCNTs). The ions were recovered quantitatively analyzed at pH 8.0 using 30 mg of carbon nanotube. The detection limits were 2.12 and 4.08 μg L^{-1}, for iron and chromium, respectively. The method was validated by analyzing of water and food samples certificates.

For chromium speciation Tuzen and Soylak [36] used carbon nanotubes (MWNTs) multilayer in a solid phase extraction procedure and the formation of a chelate by using of ammonium pyrrolidine dithiocarbamate (APDC). The procedure had some validated analytical parameters including pH, solvent type and effects of array. The detection limit was 0.90 g L^{-1} for Cr(VI). The adsorption capacity of the multilayer carbon nanotube was 9.50 mg g^{-1} Cr(VI). The presented method was applied to the speciation of chromium in natural water showing a 95% recovery factor indicating that the proposed methodology using multilayer carbon nanotubes is more efficient in terms of selectivity, detection limit and preconcentration factor when compared to other methods of preconcentration of chrome described in the literature.

Iron oxides are found in nature in the hematite form (αFe_2O_3), goethite (αFeOOH) and magnetite (Fe_3O_4). In recent times, the processes involving adsorption using these oxides as adsorbent materials have attracted great interest, mainly due to the high adsorptive capacity and magnetic property of these materials. These solids can be used in preconcentration methods, magnetic separation and molecular identification of biomolecules. Separation techniques using magnetic solids also offer some advantages over centrifugation, filtration and solid phase extraction once these steps are simplified [37]. Parham and Rahbar [38], used magnetic iron oxide nanoparticles (MIONs) as adsorbent to develop a methodology for extracting and determination of fluoride ion in water samples. The method is based on the discoloration of the Fe(III)-SCN formed with fluoride ions extracted and sequentially, monitoring by spectrophotometry at λ_{max} = 458 nm. The detection and quantification limit were 0.042 and 0.015 μg mL^{-1}, respectively. In this method the preconcentration factor achieved was 50 and the method was applied to the determination of fluoride in water samples and concentration values obtained were compared with the default values that are presenting SPADNS according to the method of the comparison. The no use of toxic compounds, the short procedure (about 8 min per sample) and a low LD are the main advantages of the proposed method.

Metal oxides such as iron oxide cited previously, zirconia oxide and titanium oxide obtained in the nanoscale have become important for various lines of research due to its special properties such as high chemical activity and mainly the good capacity to adsorb a wide variety of metallic ions. In addition, the particle size is another factor that contributes to the use of those materials in adsorption processes, because fine particles and nanoscale enable high adsorption speeds when compared the classic adsorbents [39].

Most of the nano adsorbents are used with their surfaces modified by complexing agents or microorganisms to increase the capacity to adsorb analytes. However, this modification limits the repeated use of adsorbent in adsorption processes. There are synthesized materials in the form of mixed oxides that enable their use without requiring any modification by another reagent. The nano compound of boron trioxide and titanium dioxide (B_2O_3/TiO_2) is an example of these mixed oxides. Kalfa and contributors [40] developed a methodology for cadmium separation and preconcentration in tap water and tea samples using this nano compound. Under the best experimental conditions, preconcentration factor and the detection limit were determined being 50 and 1.44 μg L^{-1}, respectively. The precision of the method was confirmed by analysis of certified reference materials (tea leaves GBW-07605). The results showed good agreement with the certified values (< 10%) relative error. Cadmium recovery in optimum conditions was 96 ± 3% at a 95% confidence level. The use of solid phase extraction enables the development of a simple method, selective, precise, fast and economical for preconcentration and determination of cadmium.

Some hybrid materials based on silica (SiO_2) have been synthesized by mixed metal oxides form as the titanium dioxide (TiO_2) and used to preconcentrate a large quantity of metallic ions the trace level. Lima and contributors [41] describe the analytical performance of a hybrid composite material of SiO_2, Al_2O_3 and TiO_2. The material was prepared by a sol–gel process to act as adsorbent in copper preconcentration system. The retention capacity was 1.4 mg g^{-1} in dynamic conditions. The detection and quantification limits were of 0.50 and 1.4 μg L^{-1}, respectively, and the calibration curve is linear in the range of 5.0–245.0 μg L^{-1} (r = −0999). The relative standard deviation is 3.20 (for n = 7 to a concentration of Cu(II) of 10 μg L^{-1}). The method was applied to the determination of copper ions in water, vegetables and alcohol fuel using flame atomic absorption spectrometry (FAAS).

Among the wide variety of materials that can be used as adsorbents, humic acid, although little reported in the literature has a good application in the area of analytical chemistry. Humic substances are natural materials and called major components from the decomposition of organic matter. This decomposition is subdivided into three parts: humic acid, fulvic acid and humina [42]. These substances have a high capacity of cationic exchange for different species due to the presence of carbonyl groups, phenolic hydroxyl, and alcoholics in their structure as well, metals to interact with the humic substances, are retained by active sites of adsorption.

Pereira & Arruda [43] used vermicompost, humic substances produced from the metabolism of organic residues in the soil by earthworms, and humic acid commercial for Cd(II) and Pb(II) preconcentration using flow systems coupled to the flame atomic absorption spectrometry (FAAS). The Cd(II) ion was preconcentrated for 4 minutes (flow rate of 4.0 mL min^{-1}) and eluted with 220 mL of nitric acid 3.0 mol L^{-1}. Under these conditions, the preconcentration factor obtained was 46 and 27 to vermicompost and humic acid, respectively. Except when used nitric acid 1.0 mol L^{-1} (for humic acid), all conditions of Pb(II) preconcentration were identical to those of Cd(II), and the preconcentration factors obtained were 62 and 83 to vermicompost and humic acid, respectively. This method was applied in mineral and tap water samples.

Bianchin and collaborators [44], evaluated the Cd(II) determination using vermicompost in a solid phase extraction methodology with a system of flow injection analysis and was detected by flame atomic absorption spectrometry. Optimal extraction conditions were obtained using the sample pH in the range of 7.3–8.3, buffered with tris (hydroxymethyl) aminomethane the 50 mmol L^{-1}, a sample flow rate of 4 mL min^{-1}, and 160 mg of adsorbent mass. With the optimized conditions, the preconcentration factor, the detection limit and the analytical frequency were estimated as 32; 1.7 μg L^{-1} and 20 samples per hour respectively. The analytic curve was linear to 5 at least 50 gL^{-1}, with correlation coefficient of 0.998 and a relative standard deviation of 2.4% (35 μgL^{-1}, n = 7). The method was applied in ethanol fuel samples and the accuracy was assessed by testing of recovery, with a 94% recovery varying from 100% indicating that the worm compost is an efficient and stable material on cadmium preconcentration in complex samples.

2.3. Fungi and bacteria

The process that employs biological materials as microbial biomass for the accumulation of organic and inorganic substances from diluted solutions is the biosorption. Initially research on biosorption described the use of biological materials for the metal removal [45]. However, today, algae, fungi, bacteria and yeast are considered an important part of the natural absorbers used in the development of analytical methodologies.

Depending on the metabolic activity, the bioadsorbents can be divided into two classes: the one where there is only a rapid interaction of the metallic species with the surface of the material and the one where the metallic species initially exert a rapid superficial interaction with the bioadsorbent alive. Subsequently, a slower and more complex process is established, initiated by the connection of the metallic species with the cellular membrane, followed by the transport through this and, finally, resulting in intracellular reactions such as methylation, reduction and oxidation [46].

Industrial processes for the production of antibiotics, enzymes and fermentation processes in general generate a large amount of biological material as by-products that are normally discarded and can be used in the adsorption process [47].

The microbial biomass has in their cell walls several chemical groups: polysaccharides, proteins or lipids that link by means of the functional groups (hydroxyl, sulfhydryl, carboxyl, amine and phosphate), and is responsible for attracting and retaining your metals biomass, therefore they possess mechanisms responsible for biosorption, and can be one or a combination of this, such as ion exchange, complexation, coordination, adsorption, electrostatic interaction, chelation and micro precipitation [48].

Several micro-organisms have been studied in determination of metal ions. Fungi are micro-organisms considered to be good bioadsorbents, since they are easily reproduced by producing a high amount of biomass, besides being microorganisms capable of adapting in extreme environments [48].

Mendil et al. [49] developed a preconcentration system using *Penicillium italicum* to determine Cd(II), Mn(II), Pb(II), Fe(II), Ni(II) and Co(II) in trace levels. The adsorption occurs in pH range 8.5–9.5. After optimizing the method, the detection limit was 0.41 and 1.60 $\mu g\ L^{-1}$, for cadmium and iron, respectively. The method was successfully applied for the determinations of analytes in natural water, cultivated mushroom, lichen (*Bryum capillare Hedw*), moss (*Homalothecium sericeum*) and refined table salt samples. The proposed method is safe, fast and handy.

The fungus Penicillium was used by Baytak and collaborators [50] to determine copper, zinc and lead in the dam water, waste water, spring water, parsley and carrots. The fungi were immobilized in pumice stone, and the ideal conditions for adsorption, such as solid phase mass, pH, solution flow rate and elution solution was studied. The analyte was determined by flame atomic absorption spectrometry (FAAS). The preconcentration factor obtained was 50 and the detection limits of detection were 1.8; 1.3 and 5.8 for copper, zinc and lead, respectively. Already the accuracy of the method was evaluated in samples of tea standard reference (GBW-07605).

To determine arsenic in drinking water samples, Shahlaei and collaborators [51] used dead fungal biomass (*Aspergillus niger*) in a solid phase extraction (SPE) system. We evaluated the effects of parameters such as time of biosorption, sample volume, pH, type and concentration of the eluent and effect of interfering ions. The recovery was evaluated by precision between 99 and 102%. The detection limit found for the method was of 1 mg mL^{-1}.

Alpodogan and collaborators [52] also used dead fungal biomass from *Aspergillus niger* loaded into TiO_2 nanoparticles for copper, nickel, cadmium, manganese, and chromium pre-concentration in milk samples, with determination by flame atomic absorption spectrometry (FAAS). The preconcentration factor was 600, detection limits for copper, nickel, cadmium, manganese, chromium and lead were 0.21; 0.34; 018; 0.08; 0.53; 0.61 g L^{-1}, respectively.

Among the fungi, yeasts are the most exploited scientifically, due to the fact that eukaryotic organisms are easier to manipulate, serving as an excellent model for studies. Ohta and collaborators [53] evaluated a preconcentration method for cooper determination in trace level in river water, using yeast *Saccharomyces cerevisiae*.

Another microorganism fairly used in the metallic ions adsorption is the bacteria. Bacteria have adsorptive capacity due to the presence of nitrogen (N) and oxygen (O) ligands in their cell walls, where mechanism of micro precipitation between metal and bacteria, in addition to presenting also an electrostatic attraction and exchange by ion groups in your cell wall [54]. Pagnanelli and collaborators [55] analyze the adsorption capacity of biomass of *Sphaerotilus natans* for lead, copper, zinc and cadmium in aqueous solutions. The biosorption was tested in different pHs, experimental results showed the positive effect of the increase of pH on the uptake of metals analyzed, showing that biomass had significant capacity to adsorb the ions studied.

2.4. Functionalized biosorbent

The functionalization of supporting materials with bio-organic species of sorption capacity for metallic ions has attracted extensive attention in the field of analytical chemistry. The purpose of this procedure is to improve the preconcentration of metallic ions commonly found in food, vegetables, drinking water and/or river samples at concentrations considered trace and ultra-trace.

This functionalization can be accomplished by structural modifications of biomass surfaces or by immobilization of bio-organic species on the surfaces of biosynthetic, inorganic, polymeric materials and biomass itself. In addition, the surfaces of these bio-organic compounds can be treated with acid and base solutions or organic and inorganic solvents which in many cases increase the adsorption, selectivity and recovery capacity of those metals ions. In this context, several bio-organic systems functionalized and/or immobilized on solid supports will be discussed that are used in the processes of adsorption of metal ions [56].

Yang and contributors [57] described in their article that various efforts have been made to generate selectivity and adsorbent capacity through chemical modification and genetic engineering, due to the low selectivity of biological cells towards certain metal species and a low capacity for anionic metal species in neutral pH. In this section, we will describe some methods that have been used with great success in metal ion analyzers after functionalization or immobilization of the biological cells and other biosorbents on a support material.

The main biological organisms used as functionalizing agent on supporting material in analytical chemistry for trace metal determination are shown in **Table 2**. They offer several advantages, such as ease in immobilization on the surface of supporting material, their biodegradability, natural abundance, low cost, and simple production.

Bathocuproine (BCP) or 2,9-dimethyl-4,7-diphenyl-1,10-phenanthroline is well-known reagent that respond selectively to Cu by forming a colored 2:1 complex with copper. It was successfully used to determine the Pb(II), Fe(III), Co(II), Cr(III) and Zn(II) in water and food samples after preconcentration [59]. Teodoro and collaborators [60] in their study, functionalized bamboo fiber with bathocuproine for determination of copper in foods using FAAS. Copper preconcentration was performed in an off-line system consisting of a glass column containing bamboo fiber functionalized with bathocuproine and peristaltic pump. This method made it possible to separate, preconcentrate and determine the copper ion in wine, cassava flour, corn flour, oregano and coriander with concentrations much lower than mg L^{-1}, that is, in the $\mu g\ L^{-1}$ for liquid samples and $\mu g\ g^{-1}$ for solid samples.

Biological organisms	Types
Bacteria	Bacillus sp., Streptococcus sp., Streptomyces sp., Escherichia sp., Enterobacter sp., Klebsiella sp.
Fungi	Filamentous fungi (*Aspergillus niger*, Alternaria sp., *Rhizopus oryzae*), unicellular yeasts (Saccharomyces sp., Yamadazyma sp.), Pleurotus sp., Helvella sp., Coriolus sp., Agaricus sp.,
Plant-derivatives	*Musa paradisiaca, Citrus reticulata, Glycine max, Zea mays, Sorghum bicolor, Agave sisalana perrine, Bambusa vulgaris*
Other biological organisms	Algae (Chlorella sp.), resting eggs, sea sponge, eggs shell membrane

Source: Adapted from Escudero and collaborators [58].

Table 2. Types of biological organisms used as functionalizing agent on surface of supporting material.

In the study developed by Baytak and collaborators [61] the fungus *Rhizopus oryzae* functionalized on the almond bark (natural cellulose) was used to preconcentrate and determine the Cu(II), Fe(III), Mn(II) and Zn(II) in water, fish and vegetable tissues samples. The glass column (150 mm × 8.0 mm i.d) containing the Rhizopus oryzae-loaded almond shells was used during the preconcentration of the analytes. Analyte detection was based on high resolution continuous source atomic absorption spectrometry. According to the author, the proposed method presented several advantages, such as precision, accuracy, preconcentration factor, reuse, environmental friendly material and did not require chelating or complexing agents during ion analyzes.

Boas and collaborators [62] used macadamia nuts waste to investigate the adsorption of copper ions. They proved that by chemical modifications of the residues of macadamia nuts was treated by NaOH solution and citric acid, and other residues treated by NaOH solution showed a better adsorption of Cu^{2+} ions compared to "in natura" residues. The adsorption process consisted of the addition of 0.5 g of the "in natura" and modified residues in a solution of Cu^{2+} ions at 100 mg L^{-1} at optimized pHs and after 24 hours the supernatants were analyzed with flame atomic absorption spectrophotometer (FAAS).

Studies have shown that sisal fiber, obtained from the plant Agave sisalana perrine of Mexican origin, can be used for preconcentration of metals. In a recent study by Dias and collaborators [63] first used the sisal fiber functionalized with 3-aminomethylalizarin-N,N-diacetic acid or Alizarine Fluorine Blue (complexing reagent) to preconcentrate and determine the copper ion in tobacco leaves. According to these results, the concentration of copper in tobacco leaves was very low.

According to Chen and collaborators [64], the eggshell contains on its surface carboxylic groups that reduce the adsorption of arsenic ions on the surface of the shell. They performed a chemical modification of the eggshell by methyl esterification to improve the bioadsorption efficiency of the arsenic ion and verified a high selectivity for these ions.

When using the key words sorghum and bioadsorption, in search engines of different journals, we can find several researches where sorghum is used as adsorbent material to preconcentrate, determine and remove heavy metals present in natura. A recent work has used the sorghum agricultural residue in natura and chemically modified with phosphoric acid to determine the cadmium, copper, manganese and lead ions in samples of black tea and river water using flame atomic absorption spectrometry (FAAS) as detection device [65]. In this work, the authors verified that the agricultural waste of chemically modified sorghum obtained an adsorption capacity 4.5 times greater when compared to the in natura waste. After this observation, they used the modified residue as solid phase for preconcentration and determination of the Cd, Cu, Mn and Pb ions present in tea and river water samples. According to their results, the chemically modified sorghum presented great potential in the separation and preconcentration of the metallic ions in real samples.

Passive sampler disks based on diffusion in thin films by concentration gradient were prepared with *Saccharomyces cerevisiae* yeast functionalized in agarose gel (polysaccharide extracted from red seaweed) to preconcentrate and determine Cd(II) with ICP-OES [66]. These disks were prepared with different proportions of gel and yeast that aimed at an optimized composition for the determination of the Cd(II) ion in river and sea water. In the study the yeast *Saccharomyces cerevisiae* was chosen due to previous works that found an excellent ability to separate and preconcentrate metal ions.

3. Conclusion

Considering the constant presence of metallic ions in people's daily lives, it is extremely necessary to develop methods capable of determining these, even at low concentrations. In recent years there has been increasing interest in developing methods that seek to comply with the terms of chemical "green", these methods involve the use of materials leading to a more clean, healthy and sustainable environment.

The development of methods that use alternative adsorbents has been highlighted in recent years, because of its high adsorptive capacity for metallic ions and organic compounds. In this sense, the study and development of new materials with low cost and the possibility of reuse has been an interesting area in analytical chemistry.

Acknowledgements

The authors are grateful for financial support from the government agencies Conselho Nacional de Desenvolvimento Científico e Tecnológico (CNPq), Fundação de Amparo à Pesquisa do Estado de Minas Gerais (FAPEMIG), Fundação de Amparo à Pesquisa do Estado de Goiás (FAPEG) and Coordenação de Aperfeiçoamento de Pessoal de Nível Superior (CAPES).

Conflict of interest

The authors declare there is no conflict of interest.

Author details

Absolon C. da Silva Júnior[1], Alessa G. Siqueira[1], Carolina A. de Sousa e Silva[1], Jordana de Assis N. Oliveira[1], Nívia M.M. Coelho[2] and Vanessa Nunes Alves[1*]

*Address all correspondence to: vanessanalves@gmail.com

1 Universidade Federal de Goiás – Regional Catalão, Catalão, Goias, Brazil

2 Universidade Federal de Uberlândia – Av, Uberlândia, Minas Gerais, Brazil

References

[1] Soares ARS. Desenvolvimento de Métodos para Determinação de Chumbo e Níquel em Produtos Cosméticos e Cabelo por GF AAS [dissertation]. Belo Horizonte: Universidade Federal de Minas Gerais; 2012

[2] Ferreira SLC, Lemos VA, Moreira BC, Costa ACS, Santelli RE. An online continuous flow system for copper enrichment and determination by flame atomic absorption spectrometry. Analytica Chimica Acta. 2000;**403**:259-264. DOI: 10.1016/S0003-2670(99)00632-7

[3] Jardim ICSF. Extração em Fase Sólida: Fundamentos Teóricos e Novas Estratégias para Preparação de Fases Sólidas. Scientia Chromatographica. 2010;**2**:13-25

[4] Pacheco PH, Gil RA, Cerutti SE, Smichowski P, Martinez L. Biosorption: A new rise for elemental solid phase extraction methods. Talanta. 2011;**85**:2290-2300. DOI: 10.1016/j.talanta.2011.08.043

[5] Reis BF, Miranda CES, Baccan N. Pré-concenttação empregando extração fase líquida-fase sólida em sistemas de análise em fluxo com detecção espectrométrica. Química Nova. 1996;**19**:623-635

[6] Science direct [Internet]. 2017. Available from: http://www.sciencedirect.com. [Accessed: Dec 1, 2017]

[7] Tarley CRT, Silviera G, Santos WNL, Matos GD, Silva EGP, Bezerra MA, Miró M, Ferreira SLC. Chemometric tools in electroanalytical chemistry: Methods for optimization based on factorial design and response surface methodology. Microchemical Journal. 2009;**92**:58-67. DOI: 10.1016/j.microc.2009.02.002

[8] Demirbas A, Demirbas MF. Importance of algae oil as a source of biodiesel. Energy Convers. Manage. 2011;**52**:163-170. DOI: 10.1016/j.enconman.2010.06.055

[9] Cai J, He Y, Yu X, Banks SW, Yang Y, Zhang X, Yu Y, Liu R, Bridgwater AV. Review of physicochemical properties and analytical characterization of lignocellulosic biomass. Renewable and Sustainable Energy Reviews. 2017;**76**:309-322. DOI: 10.1016/j.rser.2017.03.072

[10] Correia DS, Santos FHR, Correia MEF. Enzimas oxidativas microbianas envolvidas na biodegradação da lignocelulose: produção, características bioquímicas e importância biotecnológica. Embrapa Agrobiologia. 2011;**1**:284

[11] Araújo CST, Carvalho DC, Rezende HC, Almeida ILS, Coelho LM, Coelho, NMM, Marques TL, Alves, VN. Chapter 10: Bioremediation of waters contaminated with heavy metals using Moringa oleifera seeds as biosorbent. In: Applied Bioremediation – Active and Passive Approache. Intech; 2013. pp. 227-255

[12] Davis TA, Mucci A, Volesky B. A review of the biochemistry of heavy metal biosorption by brown algae. Water Research. 2003;**37**:4311-4330. DOI: 10.1016/S0043-1354(03)00293-8

[13] Alves VN, Borges SSO, Neto WB, Coelho NMM. Determination of low levels of lead in beer using solid-phase extraction and detection by flame atomic absorption spectrometry. Journal of Automated Methods and Management in Chemistry. 2011;**1**:1-6. DOI: 10.1155/2011/464102

[14] Tarley CRT, Ferreira SLC, Arruda MAZ. Use of modified rice husks as a natural solid adsorbent of trace metals: Characterization and development of an on-line preconcentration system for cadmium and lead determination by FAAS. Microchemical Journal. 2004;**77**:163-175. DOI: 10.1016/j.microc.2004.02.019

[15] Dávila-Guzmán NE, Cerino FJ, Rangel-Mendez JR, Flores PED. Copper biosorption by spent coffee ground: Equilibrium, kinetics, and mechanism. Clean Soil, Air, Water. 2013;**41**:557-564

[16] Liu Y, Chang X, Guo Y, Meng S. Biosorption and preconcentration of lead and cádmium on waste Chinese herb pang Da Hai. Journal of Hazardous Materials. 2006;**B135**:389-394. DOI: 10.1016/j.jhazmat.2005.11.078

[17] Ribeiro GC, Coelho LM, Coelho NMM. Determination of nickel in alcoholic beverages by FAAS after online preconcentration using mandarin peel (*Citrus reticulada*) as biosorbent. Journal of the Brazilian Chemical Society. 2013;**24**(6):1072-1078. DOI: 10.5935/0103-5053.20130138

[18] Bhatti HN, Mumtaz B, Hanif MA, Nadeem R. Removal of Zn(II) ions from aqueous solution using Moringa oleifera Lam. (horseradish tree) biomass. Process Biochemistry. 2007;**42**:547-553. DOI: 10.1016/j.procbio.2006.10.009

[19] Reddy DHK, Seshaiah K, Reddy AVR, Rao MM, Wang MC. Biosorption of Pb^{2+} from aqueous solutions by Moringa oleifera bark: Equilibrium and kinetic studies. Journal of Hazardous Materials. 2010;**174**:831-838. DOI: 10.1016/j.jhazmat.2009.09.128

[20] Araújo CST, Alves VN, Rezende HC, Coelho NMM. Development of a flow system for the determination of low concentrations of silver using Moringa oleífera seeds as biosorbent and flame atomic absorption spectrometry. Microchemical Journal. 2010;**96**:82-85. DOI: 10.1016/j.microc.2010.02.002

[21] Alves VN, Neri TS, Borges SSO, Carvalho DC, Coelho NMM. Determination of inorganic arsenic in natural waters after selective extraction using Moringa oleífera seeds. Ecological Engineering. 2017;**106**:431-435. DOI: 10.1016/j.ecoleng.2017.06.007

[22] Alves VN, Mosqueta R, Coelho NMM, Bianchin JN, Roux KCDP, Martendal E, Carasek E. Determination of cádmium in alcohol fuel using Moringa oleífera seeds as a biosorbent in an on-line system coupled to FAAS. Talanta. 2010;**80**:1133-1138. DOI: 10.1016/j.talanta.2009.08.040

[23] Alves VN, Mosquetta R, Carasek E, Coelho NMM. Determination of Zn(II) in alcohol fuel by flame atomic absorption spectrometry after on-line preconcentration using a solid phase extraction system. Journal of Analytical Chemistry. 2012;**67**(5):448-454. DOI: 10.1134/S1061934812050139

[24] Alves VN, Coelho NMM. Selective extraction and preconcentration of chromium using Moringa oleífera husks as biosorbent and flame atomic absorption spectrometry. Microchemical Journal. 2013;**109**:16-22. DOI: 10.1016/j.microc.2012.05.030

[25] Carmo SN, Damásio FQ, Alves VN, Marques TL, Coelho NMM. Direct determination of copper in gasoline by flame atomic absorption spectrometry after sorption and preconcentration os Moringa oleífera husks. Microchemical Journal. 2013;**110**:320-325. DOI: 10.1016/j.microc.2013.04.010

[26] Alves VN, Borges SSO, Coelho NMM. Direct zinc determination in Brazilian sugar cane spirit by solid phase extraction using Moringa oleifera husks in a flow system with detection by FAAS. International Journal of Analytical Chemistry. 2011;**1**:1-8

[27] Oliveira JAN, Siqueira LMC, Sousa Neto JA, Coelho NMM, Alves VN. Preconcentration system for determination of lead in chicken feed using Moringa oleífera husks as a biosorbent. Microchemical Journal. 2017;**133**:327-332. DOI: 10.1016/j.microc.2017.04.001

[28] Gattás G, Barbosa FF. Cobre na nutrição de aves e suínos. Revista Eletrônica Nutrine. 2004;**1**:133-177

[29] Coelho LM, Arruda MAZ. Preconcentration procedure using cloud point extraction in the presence of electrolyte for cadmium determination by flame atomic absorption spectrometry. Spectrochimica Acta, Part B: Atomic Spectroscopy. 2005;**60**:743-748. DOI: 10.1016/j.sab.2005.02.016

[30] Sousa Neto JA. Avaliação e desenvolvimento de um método para extração seletiva de cromo utilizando a luffa cylindrica como bioadsorvente [dissertation]. Catalão: Universidade Federal de Goiás; 2016

[31] Anjos VE, Rohwedder JR, Cadore S, Abate G, Grassi MT. Montmorillonite and vermiculite as solid phases for the preconcentration of trace elements in natural waters: Adsorption and desorption studies of As, Ba, Cu, Cd, Co, Cr, Mn, Ni, Pb, Sr, V, and Zn. Applied Clay Science. 2014;**289**:296-299. DOI: 10.1016/j.clay.2014.07.013

[32] Ortega EP, Ramos RL, Barron JM. Role of electrostatic interactions in the adsorption of cadmium(II) from aqueous solution onto vermiculite. Applied Clay Science. 2014;**10**(17-88):89. DOI: 10.1016/j.clay.2013.12.012

[33] Peña PY, Rondón W. Mini-columns packed with natural zeolites of Cuba for determination of cobalt using the ligand APDC in a flow injection system coupled to a FAAS. Rev. Téc. Ing. Univ. Zulia. 2014;**215**:224

[34] Peña PY, López W, Burguera JL, Burguera M, Gallignani M, Brunetto R, Carrero P, Rondon C, Imbert F. Synthetic zeolites as sorbent material for on-line preconcentration of copper traces and its determination using flame atomic absorption spectrometry. Analytica Chimica Acta. 2000;**249**:258-403. DOI: 10.1016/S0003-2670(99)00566-8

[35] Soylak M, Unsal YE. Chromium and iron determinations in food and herbal plant samples by atomic absorption spectrometry after solid phase extraction on single-walled carbon nanotubes (SWCNTs) disk. Food and Chemical Toxicology. 2010;**1511**:1515-1548. DOI: 10.1016/j.fct.2010.03.017

[36] Tuzen M, Soylak M. Multiwalled carbon nanotubes for speciation of chromium in environmental samples. Journal of Hazardous Materials. 2007;**219**:225-147. DOI: 10.1016/j.jhazmat.2006.12.069

[37] Arteaga KA, Rodriguez JA, Barrado E. Magnetic solids in analytical chemistry: A review. Analytica Chimica Acta. 2010;**157**:165-674. DOI: 10.1016/j.aca.2010.06.043

[38] Parham H, Rahbar N. Solid phase extraction–spectrophotometric determination of fluoride in water samples using magnetic iron oxide nanoparticles. Talanta. 2009;**664**:669-680. DOI: 10.1016/j.talanta.2009.07.045

[39] Türker AR. New sorbents for solid-phase extraction for metal enrichment. Clean. 2007;**548**:557-535. DOI: 10.1002/clen.200700130

[40] Kalfa OM, Yalçinkaya Ö, Türker AR. Synthesis of nano B_2O_3/TiO_2 composite material as a new solid phase extractor and its application to preconcentration and separation of cadmium. Journal of Hazardous Materials. 2009;**455**:461-166. DOI: 10.1016/j.jhazmat.2008.11.112

[41] Lima FG, Ohara MO, Clausen DN, Nascimento RD, Ribeiro ES, Segatelli MG, Bezerra MA, Tarley CRT. Flow injection on-line minicolumn preconcentration and determination of trace copper ions using an alumina/ titanium oxide grafted silica matrix and FAAS. Microchimica Acta. 2012;**61**:70-178. DOI: 10.1007/s00604-012-0807-4

[42] Landgraf MD, Alves MR, Silva SC, Rezende MOO. Caracterização de ácidos húmicos de vermicomposto de esterco bovino compostado durante 3 e 6 meses. Química Nova. 1999;**22**:484-486. DOI: 10.1590/S0100-40421999000400003

[43] Pereira MG, Arruda MAZ. Preconcentration of Cd(II) and Pb(II) using humic substances and flow systems coupled to flame atomic absorption spectrometry. Microchimica Acta. 2004;**215**:222-146. DOI: 10.1007/s00604-004-0231-5

[44] Bianchin JN, Martendal E, Mior R, Alves VN, Araújo CST, Coelho NMM, Carasek E. Development of a flow system for the determination of cadmium in fuel alcohol using vermicompost as biosorbent and flame atomic absorption spectrometry. Talanta. 2009;**333**:336-378. DOI: 10.1016/j.talanta.2008.11.012

[45] Aksu Z. Equilibrium and kinetic modelling of biosorption of Remazol Black B by Rhizopus arrhizus in a batch system: Effect of temperature. Process Biochemistry. 2000;**36**:431-439

[46] Madrid Y, Camara C. Biological substrates for metal preconcentration and speciation. Trends in Analytical Chemistry. 1997;**16**:36-44

[47] Crini G. Non-conventional low-cost adsorbents for dye removal: A review. Bioresource Technology. 2006;**97**:1061-1085. DOI: 10.1016/j.biortech.2005.05.001

[48] Veglio F, Beolchini F. Removal of metals by biosorption : A review. Hydrometallurgy. 1997;**44**:301-316. DOI: 10.1016/S0304-386X(96)00059-X

[49] Mendil D, Tuzen M, Soylak MA. Biosorption system for metal ions on Penicillium italicum – Loaded on Sepabeads SP 70 prior to flame atomic absorption spectrometric determinations. Journal of Hazardous Materials. 2008;**152**:1171-1178. DOI: 10.1016/j.jhazmat.2007.07.097

[50] Baytak S, Kenduzler E, Turker AR, Gok N. Penicillium digitatum immobilized on pumice stone as a new solid phase extractor for preconcentration and/or separation of trace metals in environmental samples. Journal of Hazardous Materials. 2008;**153**:975-983. DOI: 10.1016/j.jhazmat.2007.09.049

[51] Shahlaei M, Pourhossein A. Determination of arsenic in drinking water samples by electrothermal atomic absorption spectrometry after preconcentration using the biomass of *Aspergillus niger* loaded on activated charcoal. Journal of Chemistry. 2014;**1**:1-6

[52] Alpdogan G, Zor SD, Peksel A. Analysis of some trace metals in milk samples by FAAS after Preconcentration by *Aspergillus niger* loaded on TiO_2 nanoparticles. Current Analytical Chemistry. 2016;**12**:642-650

[53] Ohta K, Tanahasi H, Suzuki T, Kanec S. Preconcentration of trace copper with yeast for river water analysis. Talanta. 2001;**53**:715-720. DOI: 10.1016/S0039-9140(00)00502-6

[54] Buratto AP, Costa RD, Ferreira ES. Aplicação de biomassa fúngica de Pleurotus ostreatus em processo de biossorção de íons cobre(II). Eng. Sanit. Ambiente. 2012;**17**:413-420

[55] Pagnanelli F, Esposito A, Toro L, Vegliò F. Metal speciation and pH effect on Pb, Cu, Zn and Cd biosorption onto Sphaerotilus natans: Langmuir-type empirical model. Water Research. 2003;**37**:627-633. DOI: 10.1016/S0043-1354(02)00358-5

[56] Bag H, Turker AR, Lale M. Determination of Cu, Zn, Fe, Ni and Cd by flame atomic absorption spectrophotometry after preconcentration by Escherichia coli immobilized on sepiolite. Talanta. 2000;**51**:1035-1043. DOI: 10.1016/S0039-9140(00)00289-7

[57] Yang T, Chen M-L, Wang J-H. Genetic and chemical modification of cells for selective separation and analysis of heavy metals of biological or environmental significance. TrAC Trends in Analytical Chemistry. 2015;**66**:90-102. DOI: 10.1016/j.trac.2014.11.016

[58] Escudero LB, Avila Maniero M, Agostini E, Smichowski PN. Biological substrates: Green alternatives in trace elemental preconcentration and speciation analysis. Trac-Trends in Analytical Chemistry. 2016;**80**:531-546. DOI: 10.1016/j.trac.2016.04.002

[59] Mendil D, Karatas M, Tuzen M. Separation and preconcentration of Cu(II), Pb(II), Zn(II), Fe(III) and Cr(III) ions with coprecipitation method without carrier element and their determination in food and water samples. Food Chemistry. 2015;**177**:320-324. DOI: 10.1016/j.foodchem.2015.01.008

[60] Teodoro MTF, de S, Dias F, da Silva DG, Bezerra MA, Dantas AF, Teixeira LSG, et al. Determination of copper total and speciation in food samples by flame atomic absorption spectrometry in association with solid-phase extraction with bamboo (Bambusa vulgaris) fiber loaded with bathocuproine. Microchemical Journal. 2017;**132**:351-357. DOI: 10.1016/j.microc.2017.01.033

[61] Baytak S, Mert R, Turker AR. Determination of Cu(II), Fe(III), Mn(II) and Zn(II) in various samples after preconcentration with Rhizopus oryzae loaded natural cellulose (almond bark). International Journal of Environmental Analytical Chemistry. 2014;**94**:975-987. DOI: 10.1080/03067319.2014.900679

[62] Boas NV, Casarin J, Caetano J, Gonçalves Junior AC, Tarley CRT, Dragunski DC. Biossorção de cobre utilizando-se o mesocarpo e o endocarpo da macadâmia natural e quimicamente tratados. Revista Brasileira de Engenharia Agrícola e Ambiental. 2012;**16**:1359-1366. DOI: 10.1590/S1415-43662012001200014

[63] Dias Fde S, Bonsucesso JS, Oliveira LC, dos Santos WN. Preconcentration and determination of copper in tobacco leaves samples by using a minicolumn of sisal fiber (*Agave sisalana*) loaded with *Alizarin fluorine* blue by FAAS. Talanta. 2012;**89**:276-279. DOI: 10.1016/j.talanta.2011.12.027

[64] Chen ML, Gu CB, Yang T, Sun Y, Wang JH. A green sorbent of esterified egg-shell membrane for highly selective uptake of arsenate and speciation of inorganic arsenic. Talanta. 2013;**116**:688-694. DOI: 10.1016/j.talanta.2013.07.061

[65] Uçar G, Bakircioglu D, Kurtulus YB. Determination of metal ions in water and tea samples by flame-AAS after preconcentration using sorghum in nature form and chemically activated. Journal of Analytical Chemistry. 2014;**69**:420-425. DOI: 10.1134/S1061934814050098

[66] Menegario AA, Tonello PS, Durrant SF. Use of *Saccharomyces cerevisiae* immobilized in agarose gel as a binding agent for diffusive gradients in thin films. Analytica Chimica Acta. 2010;**683**:107-112. DOI: 10.1016/j.aca

Section 6

Toxicity and Health Risk of Heavy Metals

Chapter 17

Heavy Metals in Urban Dust

Fumiyuki Nakajima and Rupak Aryal

Additional information is available at the end of the chapter

http://dx.doi.org/10.5772/intechopen.74205

Abstract

Urban dust is contaminated by pollutants from various diffused sources and is often difficult to be controlled. It elevates the background heavy metal exposure level to humans and ecosystems in and around urban area and may give an apparent or hidden impact. In this chapter, the recent studies on heavy metals in the urban dust are summarized. First, the heavy metal pollution of the dust in the atmosphere and on the road surface is described. Then, the process of road runoff and subsequent infiltration to the soil is discussed from the viewpoint of heavy metal mobility. Finally, the ecotoxicity of road dust is shown, and the causal pollutants in the dust are discussed.

Keywords: urban road dust, road runoff, infiltration, ecotoxicity, atmospheric particulate matter

1. Introduction

For the last few years, Global Environmental Goals (GEGs) agreed by a number of countries associated to the United Nations have shown commitments to address and improve air pollution and air quality, biodiversity, chemical and waste, energy, environmental governance, forests, freshwater, land, and oceans and seas. In the environmental issues, increasing metal concentration is one of the core issues [1]. There are a number of heavy metals of concerns that include arsenic (As), copper (Cu), zinc (Zn), lead (Pb), cadmium (Cd), and mercury (Hg).

Heavy metals usually occur in soil at low concentrations generated by the natural process such as weathering of rocks and minerals, volcanic eruption, etc. [2]. For the last few decades, because of human interference in the natural geochemical cycle, heavy metal input to our environment is speeding up [3]. Heavy metals become toxic and are regarded as contaminants in the environments when their rates of generation via man-made cycles are more rapid relative to natural ones. The metal accumulation in both urban and rural environments above the background concentrations may affect ecosystems including plants and animals and has been causing a serious threat to human health. Some of the serious examples of metal toxicity include Minamata disease (methylmercury poisoning) [4] and itai-itai disease (cadmium poisoning) [5].

This chapter focuses on heavy metals in the urban dust. Urban dust contamination is caused by various diffused sources and is often difficult to be controlled. Although the degree of contamination is usually less than the direct industrial and mining pollution, it elevates the background heavy metal exposure level to humans and ecosystems in and around the urban area and may give an apparent or hidden impact. In this chapter, heavy metals in the urban dust, in the atmosphere, and on the road surface are summarized first. Then, the process of road runoff and subsequent infiltration to the soil is discussed from the viewpoint of heavy metal mobility. Finally, the recent research progress on the ecotoxicity of road dust is described.

2. Heavy metals in atmosphere and on road surface

2.1. Atmospheric particulate matter

The urban atmosphere is subjected to large inputs of anthropogenic contaminants arising from both stationary (power plants, industries, incinerators, and residential heating) and mobile sources (automobile- and transportation-related activities on road) [2] and natural surrounding sources (windblown dust, weathering of rocks and minerals, sea salt, volcanic eruptions, etc.). These pollutants are associated to particulate, liquid, and the vapor phases. They are subsequently transported to the Earth's surface through dry and wet deposition. The air pollutants can be inorganic, organic, and/or microbiological. Among the pollutants, inorganic heavy metals such as copper, zinc, cadmium lead, arsenic, and chromium are a major concern due to their toxicity and potentially carcinogenic characteristics [6].

Aryal et al. studied the heavy metals (Ni, Cu, Zn, Pb, and Cd) in the particles (<10 μm) in the urban atmosphere in industrial city of Ulsan, South Korea [7]. **Figure 1** shows the concentration

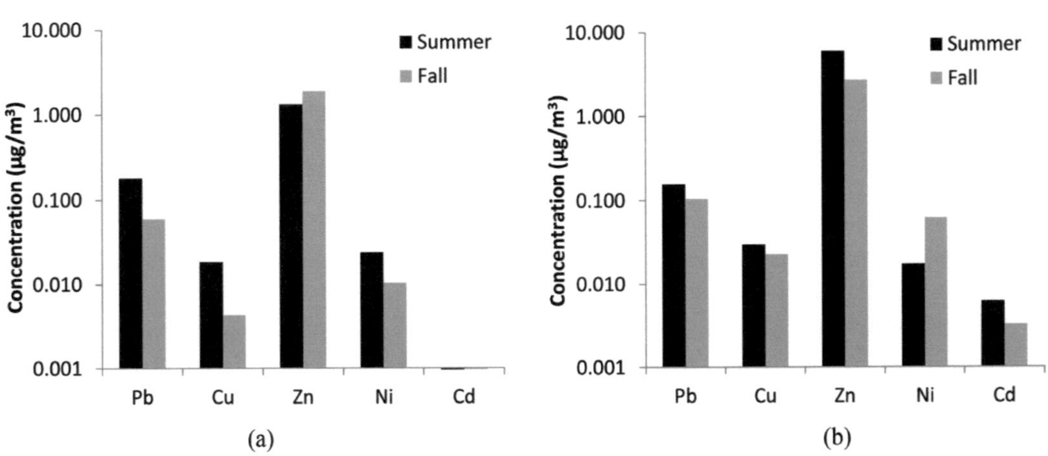

Figure 1. Metal concentrations (μg/m³) in the atmospheric particulate matter (<10 μm) in (a) petrochemical area and (b) nonferrous metal industry area [6].

of heavy metals in atmospheric particulate in summer and fall seasons at the two sampling sites. The result indicated the highest Zn content among the five elements and higher heavy metal concentrations in summer except for Ni in the nonferrous metal industry area (likely due to the input from the surrounding area). Their detailed analysis of the size-fractionated dust particles showed that the particles were in bimodal distribution with one peak corresponding to the particle size range of 1.1–4.7 μm and the other to the range of 4.7–10 μm. Among the metals measured, Pb was found size dependent, whereas Zn was homogeneously mixed in all sizes.

Many cities are frequently impacted by dust storms traveling from a long distance. Examples include yellow sand-dust storms in Asia and Sahara dust storms in Northern Europe, etc. Since the dust storm travels a long distance, they may pick up many hazardous pollutants available on the path. The composition of such dust particles is of a great concern. Aryal et al. investigated physical and chemical properties of particles in the dust storm that hit east coast cities of Australia including Sydney on 23 September 2009 and blanketed for a number of hours [8]. The dust storm was the worst since 1942 for Sydney dwellers. The dust particle analysis by a wide-angle X-ray scattering (WAXS) and electron diffraction X-ray showed that it contained mostly crustal elemental oxides such as SiO_2, Al_2O_3, and Fe_2O_3. **Figure 2** shows the elemental composition obtained from electron diffraction X-ray spectra and morphology of the dust particle using scanning image microscopy (**Figure 2**). The ratio of Al/Si in the Sydney dust storm was ~0.39 which indicated desert origin. They also analyzed the organic matter in the dust and concluded that the dust was unlikely to contain the biohazardous compounds to human health.

2.2. Road dust

Heavy metal deposit on the urban and rural road surface can cause significant and serious contamination. The vehicular fuel exhaust, abrasion of vehicle parts including tire and body, and industrial emissions are the sources of toxic metals [9, 10]. Road dust scatters to the atmosphere, causing air pollution and transported to wider areas [11]. Such resuspended road dust can enter the human body through direct ingestion of dust, inhalation of dust particles through the mouth and nose, and dermal absorption, threatening people's health [12]. Road dust is transported also to water environment by atmospheric deposition and by

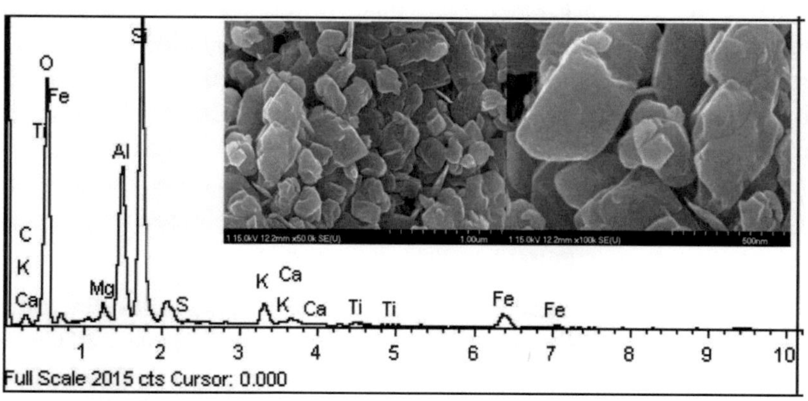

Figure 2. X-ray spectrum of the Sydney dust storm particles [7].

surface runoff of rainwater. The contaminated dust may deteriorate the ecological health in the receiving waterbodies. Therefore, to protect the human and ecological health, attention should be paid to the heavy metals in road dust in large cities.

There are a number of studies of road dust and heavy metals. A study on road dust and heavy metals in Winterthur, Switzerland, revealed that heavy metals are accumulated in various dust sizes [13]. Another study reported heavy metal concentrations in road dust in Ulsan, South Korea, in various particle sizes (<75, 75–180, 180–800, and 800–2000 μm) [1]. The heavy metal contents overall ranged from 0.4 to 21 μg/g for Cd, 25 to 204 μg/g for Cu, 6 to 58 μg/g for Ni, 43 to 343 μg/g for Pb, and 79 to 326 μg/g for Zn. The heavy metal concentration was found in the order of: industrial complex > industrial vicinity > heavy traffic > residential > background.

3. Heavy metals in road runoff and infiltration

3.1. Road runoff

Urban runoff, especially from the roadways (road runoff), often contains significant loads of dissolved and particulate metals along with other pollutants such as hydrocarbons and nutrients [14–19]. Unlike organic compounds, metal elements are not degraded in the environment and constitute an important class of constituents generated by traffic activities due to their non-degradability.

During wet weather period, the surface-deposited metals are washed off to the pipe network and finally to the receiving waterbodies including rivers, lakes, ponds, reservoirs, etc. The deposited pollutants are washed off mostly at the early stage of the runoff, called first flush [20]. Under heavy traffic conditions, the depositions of toxic pollutants are usually more concerned rather than those of other nonpoint pollutants such as nitrogen and phosphorus.

A monitoring on highway runoff in Winterthur, Switzerland, was conducted from September to December 2000 [13, 21]. The study covered nine runoff events, and samples were collected first 3 mm rainfall in order to capture the first flush. Furumai et al. reported that the runoff samples with higher SS concentrations had higher heavy metal concentrations [13]. The particle-bound Zn, Cu, and Pb concentrations were 155–524, 29–69, and 13–46 μg/L, respectively. The heavy metal concentrations were close to other studies [22–27]. Aryal et al. reported the fractionated samples in this monitoring campaign [21]. The collected serial total suspended solid (TSS) samples were fractionated into five size fractions (<20, 20–45, 45–106, 106–250, and >250 μm) to understand the runoff behavior of the particles. The wash-off behavior of fine and coarse fractions was quite different. With respect to the TSS, a coarse fraction (>106 μm) showed power growth relationship, whereas the finer fractions (<106 μm) showed clustering at lower range TSS and almost saturated at higher TSS region (**Figure 3**). This result indicated that particle sizes were very important in metal transport in runoff.

Aryal and Lee also studied the relationship between heavy metal content in particles and SS concentrations in urban runoff [28]. They found that all the runoff samples had low heavy metal content in particles at the beginning of the rainfall with higher SS concentration. As the time elapsed, the heavy metal content increased with lower SS concentration. The content of heavy

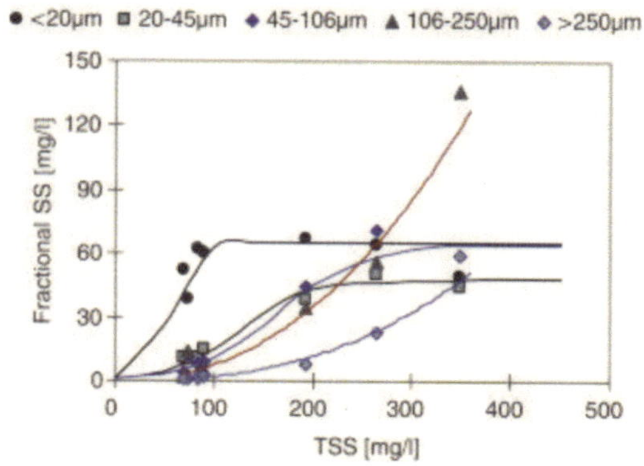

Figure 3. Relationship between total suspended solid (TSS) concentrations and size-fractionated SS concentrations in highway runoff [20].

metals reached the lowest level when SS concentration reached the highest value. The high contents of heavy metals at low SS concentration were believed to be due to higher contribution of fine particles to the total suspended solids (**Figure 4**).

3.2. Stormwater infiltration

The urban runoff is managed in the city via different means. Among the stormwater management system, stormwater infiltration facilities are one of them for flood control, designed to reduce the peak flow by infiltrating water to the subsurface [29]. However, there is a considerable concern of groundwater contamination by heavy metals since these infiltration system structures are usually not designed with any consideration for pollutant retention [30]. The environment in the facilities including pH, oxidation reduction potential (oxygen availability), minerals, and organic matter defines the mobility of deposited metals [31–35]. Within the sediment, different particle sizes have different heavy metals accumulating behavior [36].

Aryal et al. surveyed more than 200 infiltration inlets for road runoff in Tokyo [37]. They measured sediment depth, collected sediment samples, and measured heavy metals in total,

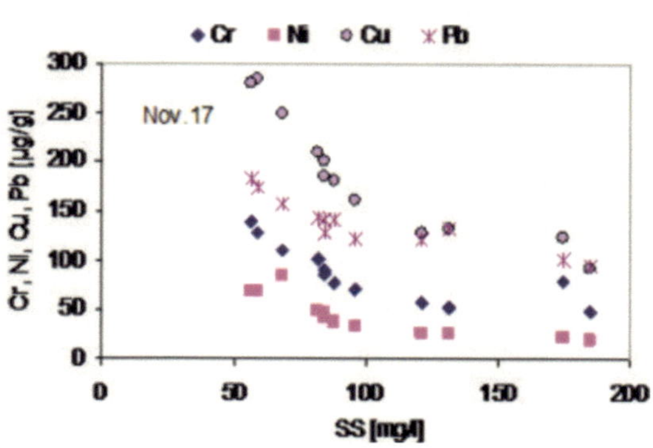

Figure 4. Relationship between SS concentrations in the runoff samples and heavy metal content in SS (adopted from [28]).

in different particle sizes as well as speciation of heavy metals. The sediment accumulation depth varied from almost few centimeters to more than 14 cm. The particle size analysis of the sediment showed a heterogeneous distribution of size at the different depth. The heavy metal contents in the inlet sediment ranged from Cr 7 to 143 μg/g, Mn 383 to 1009 μg/g, Co 9 to 31 μg/g, Ni 10 to 84 μg/g, Cu 51 to 316 μg/g, Zn 288 to 1974 μg/g, and Pb 24 to 230 μg/g. The heavy metal contents in the sediment were mostly close to the road dust but in some cases lower than the background soils. **Table 1** shows the heavy metal contents in the sediments at two inlets. The study in the inlet sediment showed lower in content at the bottom sediment than the top. This reflects that heavy metals may have leached to subsurface soil [37, 38].

Organic matter is one of the most important factors in heavy metal mobility in soil and sediment. Organic matter associated with the particulate matter may contain wide ranges of organics with potentially different sources. Humic substances are often the dominant organic fraction in soils and sediments [39]. They consist of heterogeneous mixtures of molecules having molecular weights ranging from a few hundred to several hundred thousand Daltons and are polyfunctional (each molecule may have a large number of different complexing sites, e.g., carboxylic, phenolic, phthalic, salicylic, and amine functional groups) [40–42]. Due to the large variation in structural properties, different organic matters can possess different strengths in terms of metal bonding. Also, the competitiveness among metals to bind to organic matter may vary [32, 43, 44]. It is reported that the general order of affinity of metal cations complexed by organic matter is Cu > Cd > Fe > Pb > Ni > Co > Mn > Zn [45]. Organic matter with a different functional group shows different affinities to the same heavy metals. This means that a single metal may show different affinities to different organic substances and may attach to the site where it is attracted more.

Aryal et al. studied road dust organic matter in different particle sizes and the organics' possible role in binding heavy metals [46]. The road dust was collected from industrial, business, and commercial districts and residential areas in the city of Ulsan, South Korea. Their study used fluorescent spectroscopy in combination with parallel factor analysis (PARAFAC). The study found that the finest fraction contained more microbial humic-like substances, whereas the coarse fraction was enriched with fulvic acid. The organic matter in two fractions (75–180 and 180–800 μm) showed dual characteristics. It was also observed that Zn had high affinity with aromatic protein, humic substances, and microbial by-products, whereas Cd showed a negative correlation with fulvic acids.

	Cr		Mn		Co		Ni		Cu		Zn		Pb	
Inlet	**T**	**B**	**T**	**B**	**T**	**B**	**T**	**B**	**T**	**B**	**T**	**B**	**T**	**B**
#4	31	94	842	765	31	25	80	62	259	214	1827	1350	271	220
#172	74	67	965	699	33	28	67	66	221	201	1071	791	96	85

Note: T, top (0–3 cm) and B, bottom (12–15 cm).

Table 1. Heavy metal content (μg/g) in the top and bottom sediment of two infiltration inlets of road runoff in Tokyo (#4 and #172) [36].

The fate of heavy metals in infiltration systems is highly dependent upon the sorptive and desorptive behavior of organic matter received from various sources such as road surfaces, parks, and agricultural areas. The organic matter found in the infiltration inlet sediment seemed changed with time due to prolonged deposition. Aryal et al. investigated the potential role of organic matter in binding and mobilizing heavy metals in the deposited sediment of infiltration facilities in Tokyo by using fluorescent spectroscopy combined with parallel factor analysis [47]. The surface sediment (few cm deep) showed a variation in organic matter and heavy metal content place by place. The sediment samples in lower layers had been relatively similar in different infiltration facilities. The result suggested that the sediment surface was used to be disturbed by scouring and mixing while receiving the runoff. The heavy metal content was found lower at the higher depth indicating possible release to the subsurface soil. The study also found that Cu and Zn had a positive linear relationship in the sediment samples which reflected that both of them had a common source. The PARAFAC analysis identified three major components with different spectral shapes and volumes in different excitation emission locations. The components reflected UV A humic-like substances (defined in [48]) and fresh organic matter [49], UV C-like organic matter having higher molecular weight humic substances [50, 51], and commercial humic acid [52]. While correlating organic components with heavy metals, they observed that UV A-type substances had a negative relationship with all metals, whereas commercial humic-type substances had a positive relationship with heavy metals.

4. Ecotoxicity of urban dust

As described above, urban road dust contains toxic heavy metals as well as other various toxic chemicals such as polycyclic aromatic hydrocarbons (PAHs), perfluorinated surfactants (PFSs), etc. [53–55]. Stormwater runoff brings the road dust to the receiving waterbody and poses the harmful effect to the aquatic lives. In this section, recent studies on ecotoxicity of urban road dust are reviewed, and the heavy metal contribution to the effect is discussed.

Watanabe et al. and Khanal et al. reported the toxicity of the dilution series of road dust in Tokyo to calculate LC20 and LC50 by direct contact test with a benthic crustacean (ostracod, *Heterocypris incongruens*) (ISO 14371) [56, 57]. Those results are summarized in **Table 2**. Watanabe et al. showed the apparent lethal toxicity in six out of seven road dust samples tested although the dose-response relationship was shown only for the St. 6 in **Table 2** [56]. Six out of 10 samples in [57] were toxic as shown in **Table 2**, and the rest of the four samples had <50% mortality (so that LC20 and LC50 were not calculated).

Niyommaneerat et al. recently developed a new method of sediment chronic toxicity evaluation using ostracod reproduction and applied the method to urban road dust [58]. The road dust sample in [58] caused delayed egg production in 6.25% (v/v) dilution although LC20 for 6-day mortality test was 14% (v/v). Acute chronic ratio (ACR) was calculated as 6.8, and the authors estimated that 179 times of dilution was necessary to make the most toxic road dust (St.8 in [57]; see **Table 2**) nontoxic in the reproduction endpoint. Such dilution is not always achieved in the cities as the road dust contaminated the sediment in 3.7% in a river receiving road runoff in Tokyo [59].

Road dust sample name in reference	Toxicity to benthic ostracod *H. incongruens*		Ref.
	LC20 (% (v/v))*	LC50 (% (v/v))*	
St. 6	20	30	[56]
St. 1	13	31	[57]
St. 2	9	41	
St. 3	10	13	
St. 4	49	67	
St. 7	13	24	
St. 8	1.6	3.8	

*The unit is a volumetric percentage of road dust in the tested sediment. The sediment was composed of road dust and nontoxic reference sediment (quartz sand [56] and commercially available reference soil [57]).

Table 2. Toxicity of urban road dust to benthic ostracod *Heterocypris incongruens*.

Urban road dust, originally in dry form, is likely to change its toxic nature by wetting process. Watanabe et al. showed the toxicity of water-soluble (easily desorbable) components and remaining solids (particle-bound components), separately [56]. As a result, the primary toxic compounds of road dust existed mainly in the water-soluble form rather than the particle-bound form. Interestingly, after 7 days of incubation in water, the solid part also showed significant lethal toxicity. The similar phenomena of toxicity change in solid phase were also confirmed in other road dust samples [60]. The causal substances of the toxicity observed after several days are still unknown.

Toxicity change was also observed in seawater [61]. Road dust toxicity was tested in different salinities with the estuarine amphipod *Grandidierella japonica*. For three road dust samples, the toxicity in 3.5% salinity was significantly higher than in 0.5 and 2.0% salinity. The toxicity change may be caused by the change of bioavailability and/or toxicity of chemicals in the road dust.

In most of the cases, it is generally difficult to identify the causal substances of toxic environmental samples. Watanabe et al. and Khanal et al. applied toxicity identification evaluation (TIE) approach to the road dust toxicity assessment by benthic ostracod *H. incongruens* [62, 63]. In both of the studies, the two different types of adsorbents were used to manipulate the toxic substances in road dust, namely, Ambersorb/XAD for hydrophobic organic matter and Chelex/SIR-300 for cationic metals. Watanabe et al. showed that the adsorbent for hydrophobic organic matter was effective to reduce the toxicity of all the three road dust tested and concluded that the heavy metals were not considered to be major toxicants in the dust sample [62]. In contrast, Khanal et al. indicated that the adsorbent effectiveness depended on the dust samples (**Figure 5**) and one of the dust samples (St. 7 in **Figure 5**) became less toxic only by the adsorbent for cationic metals [63].

One of the limitations in the TIE approach is that the adsorbent mainly removes the dissolved toxicant and may not reduce the dietary exposure through the ingestion of contaminated particles in the sediment tests. To assess the dietary exposure pathway, we should have dose-response relationship for the contaminated solids. However, such information is very limited

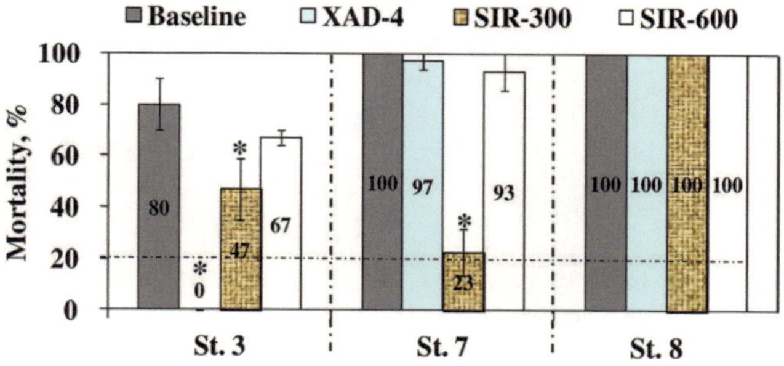

Figure 5. Toxicity reduction by adsorbent addition to urban road dust [63]. Baseline indicates toxicity of urban road dust without adsorbent addition. XAD-4, SIR-300, and SIR-600 are adsorbents for hydrophobic organic matter, cationic metals, and ammonia, respectively. The toxicity was evaluated in a 6-day ostracod test.

even for the sediment toxicity test species. Sevilla et al. showed such dose-response relationship for a benthic ostracod to heavy metal-contaminated food algae (**Figure 6**) [64, 65]. This will be useful to interpret the toxicity test result. For example, Watanabe et al. showed the overall toxicity of urban road dust but could not discuss the dietary pathway of heavy metal exposure [56]. The toxic urban road dust contained heavy metals in 1300–2980 μg Zn/g, 279–1200 μg Cu/g, and 0.64–2.0 μg Cd/g. Assuming that the heavy metal bioavailability of the dust and algae is similar (probably it is a safer side assessment), zinc and copper may cause the mortality of the ostracod species.

Biomarker is a promising tool to identify the heavy metal exposure from the road dust. But for relatively small-sized benthic organisms used in sediment toxicity tests, it is technically difficult to obtain the sufficient amount of tissues to quantify the conventional biomarker such as metallothionein. Recent development of molecular technique enables us to detect the biological response to toxic substances more comprehensively as gene expression. Hiki et al. reported a challenge in exploring biomarker genes of benthic amphipod *G. japonica* to zinc, copper, and nicotine [66]. They applied a technique of cDNA-amplified fragment length polymorphism (AFLP) to the non-model species whose genome sequence information was not available. The unique loci to each exposure treatment were identified, and they considered

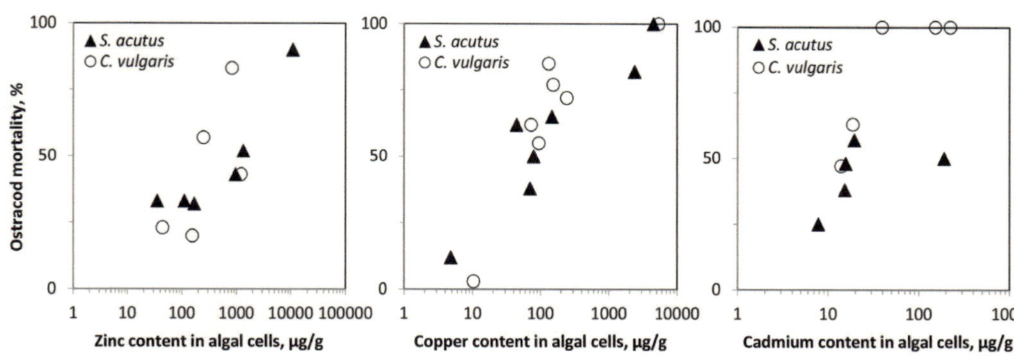

Figure 6. Dose-response relationship for benthic ostracod *Heterocypris incongruens* to heavy metal-contaminated food algae (*Scenedesmus acutus* and *Chlorella vulgaris*) (adopted from [64]).

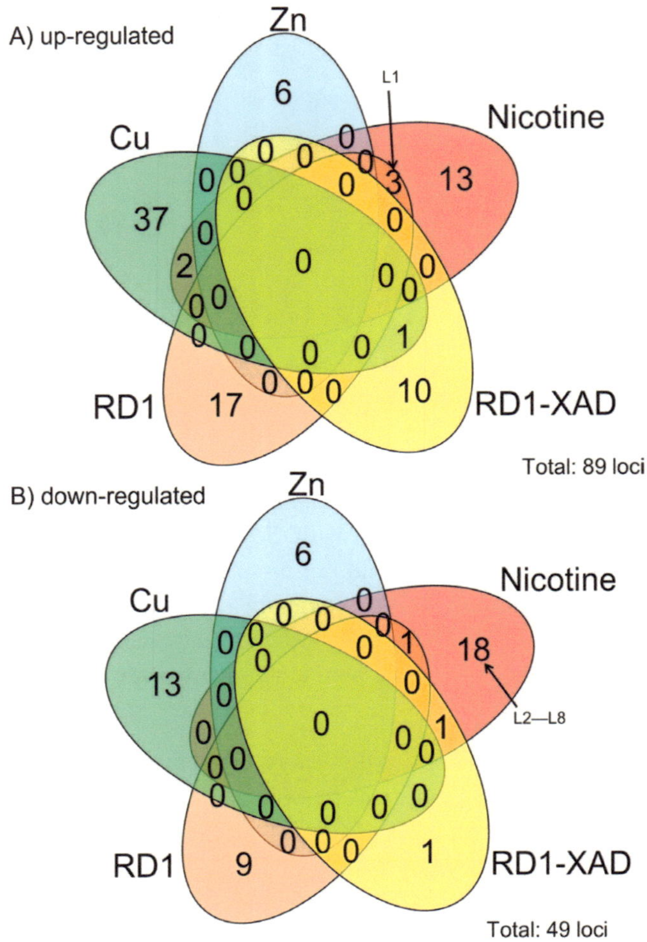

Figure 7. Venn diagram showing the number of upregulated loci compared with control (A) and downregulated ones (B) [66].

them as potentially useful biomarkers of exposure to each toxicant. They applied the tool to a road dust sample and showed that there were no common gene expressions in zinc/copper treatment and road dust exposure (**Figure 7**). It suggested that, although zinc and copper existed in the road dust, they seems little bioavailable to the species under the test condition.

5. Conclusions

This chapter summarizes the recent research progress on heavy metals in urban dust, especially in urban road dust from the viewpoint of its mobility and toxicity. Urban dust varies locally in composition, and consequently, the nature of its mobility and toxicity differs. Ecotoxicity of road dust has been shown in many of road dust samples, but further research is needed to clarify the causal toxicant and its persistency in the environment.

Author details

Fumiyuki Nakajima[1]* and Rupak Aryal[2]

*Address all correspondence to: nakajima@env.t.u-tokyo.ac.jp

1 Department of Urban Engineering, The University of Tokyo, Tokyo, Japan

2 Natural and Built Environments Research Centre, University of South Australia, Adelaide, Australia

References

[1] GEGs. Global Environmental Goals. United Nations Environment Program [Internet]. 2010. Available from: http://geodata.grid.unep.ch/gegslive/ [Accessed 2017-12-12]

[2] Bradl HB. Chapter 1. Sources and origins of heavy metals. In: Bradl HB, editor. Heavy Metals in the Environment: Origin, Interaction and Remediation. Interface Science and Technology. Vol. 6. The Netherlands: Elsevier; 2005. pp. 1-27. DOI: 10.1016/S1573-4285(05)80020-1

[3] Vidal O, Rostom F, François C, Giraud G. Global trends in metal consumption and supply: The raw material-energy nexus. Elements. 2002;**13**:319-324. DOI: 10.2138/gselements.13.5.319

[4] McAlpine D, Araki S. Minamata disease an unusual neurological disorder caused by contaminated fish. The Lancet. 1958;**272**(7047):629-631. DOI: 10.1016/S0140-6736(58)90348-9

[5] Kaji M. Itai-itai disease: Lessons for the way to environmental regeneration. In: Fujigaki Y, editor. Lessons from Fukushima: Japanese Case Studies on Science, Technology and Society. Cham: Springer International Publishing; 2015. pp. 141-165. DOI: 10.1007/978-3-319-15353-7_7

[6] Tchounwou PB, Yedjou CG, Patlolla AK, Sutton DJ. Heavy metal toxicity and the environment. In: Luch A, editor. Molecular, Clinical and Environmental Toxicology. Experientia Supplementum. Vol. 101. Basel: Springer; 2012. pp. 133-164. DOI: 10.1007/978-3-7643-8340-4_6

[7] Aryal R, Kim A, Lee BK, Kamruzzaman M, Beecham S. Characteristics of atmospheric particulate matter and metals in industrial sites in Korea. Environment and Pollution. 2013;**2**(4):10-21. DOI: 10.5539/ep.v2n4p10

[8] Aryal R, Kandel D, Acharya D, Chong MN, Beecham S. Unusual Sydney dust storm and its mineralogical and organic characteristics. Environmental Chemistry. 2012;**9**(6): 537-546. DOI: 10.1071/EN12131

[9] Li X, Poon CS, Liu PS. Heavy metal contamination of urban soils and street dusts in Hong Kong. Applied Geochemistry. 2001;**16**(11):1361-1368. DOI: 10.1016/S0883-2927(01)00045-2

[10] Johansson C, Norman M, Burman L. Road traffic emission factors for heavy metals. Atmospheric Environment. 2009;**43**(31):4681-4688. DOI: 10.1016/j.atmosenv.2008.10.024

[11] Apeagyei E, Bank MS, Spengler JD. Distribution of heavy metals in road dust along an urban-rural gradient in Massachusetts. Atmospheric Environment. 2011;**45**(13):2310-2330. DOI: 10.1016/j.atmosenv.2010.11.015

[12] Zheng N, Liu J, Wang Q, Liang Z. Health risk assessment of heavy metal exposure to street dust in the zinc smelting district, northeast of China. Science of the Total Environment. 2010;**408**(4):726-733. DOI: 10.1016/j.scitotenv.2009.10.075

[13] Furumai H, Aryal RK, Nakajima F. Profile analysis of polycyclic aromatic hydrocarbons and heavy metals in size fractionated highway dust and runoff samples. In: Proceedings of 9th International Conference on Urban Drainage; 8-13 September 2002; Portland. Reston: ASCE; 2012. pp. 1-13. DOI: 10.1061/40644(2002)205

[14] Launay MA, Dittmer U, Steinmetz H. Organic micropollutants discharged by combined sewer overflows—Characterisation of pollutant sources and stormwater-related processes. Water Research. 2016;**104**:82-92. DOI: 10.1016/j.watres.2016.07.068

[15] Krein A, Schorer M. Road runoff pollution by polycyclic aromatic hydrocarbons and its contribution to river sediments. Water Research. 2000;**34**(16):4110-4115. DOI: 10.1016/S0043-1354(00)00156-1

[16] Han Y, Lau SL, Kayhanian M, Stenstrom MK. Characteristics of highway stormwater runoff. Water Environment Research. 2006;**78**(12):2377-2388. DOI: 10.2175/106143006X95447

[17] Zhao H, Li X, Wang X. Heavy metal contents of road-deposited sediment along the urban–rural gradient around Beijing and its potential contribution to runoff pollution. Environmental Science & Technology. 2011;**45**(17):7120-7127. DOI: 10.1021/es2003233

[18] Birch H, Mikkelsen PS, Jensen JK, Holten Lützhøft HC. Micropollutants in stormwater runoff and combined sewer overflow in the Copenhagen area, Denmark. Water Science and Technology. 2011;**64**:485-493. DOI: 10.2166/wst.2011.687

[19] Wang S, He Q, Ai H, Wang Z, Zhang Q. Pollutant concentrations and pollution loads in stormwater runoff from different land uses in Chongqing. Journal of Environmental Sciences. 2013;**25**(3):502-510. DOI: 10.1016/S1001-0742(11)61032-2

[20] Herngren L, Goonetilleke A, Ayoko GA. Understanding heavy metal and suspended solids relationships in urban stormwater using simulated rainfall. Journal of Environmental Management. 2005;**76**(2):149-158. DOI: 10.1016/j.jenvman.2005.01.013

[21] Aryal RK, Furumai H, Nakajima F, Boller M. Dynamic behavior of fractional suspended solids and particle-bound polycyclic aromatic hydrocarbons in highway runoff. Water Research. 2005;**39**(20):5126-5134. DOI: 10.1016/j.watres.2005.09.045

[22] Legret M, Pagotto C. Evaluation of pollutant loadings in the runoff waters from a major rural highway. Science of the Total Environment. 1999;**235**:143-150. DOI: 10.1016/S0048-9697(99)00207-7

[23] Drapper D, Tomlinson R, Williams P. Pollutant concentrations in road runoff: Southeast Queensland case study. Journal of Environmental Engineering. 2000;**126**:313-320. DOI: 10.1061/(ASCE)0733-9372

[24] Gromaire MC, Garnaud S, Saad M, Chebbo G. Contribution of different sources to the pollution of wet weather flows in combined sewers. Water Research. 2001;**35**:521-533. DOI: 10.1016/S0043-1354(00)00261-X

[25] Zhao H, Li X, Wang X, Tian D. Grain size distribution of road-deposited sediment and its contribution to heavy metal pollution in urban runoff in Beijing, China. Journal of Hazardous Materials. 2010;**183**(1-3):203-210. DOI: 10.1016/j.jhazmat.2010.07.012

[26] Göbel P, Dierkes C, Coldewey WG. Storm water runoff concentration matrix for urban areas. Journal of Contaminant Hydrology. 2007;**91**:26-42. DOI: 10.1016/j.jconhyd.2006.08.008

[27] Helmreich B, Hilliges R, Schriewer A, Horn H. Runoff pollutants of a highly trafficked urban road—Correlation analysis and seasonal influences. Chemosphere. 2010;**80**:991-997. DOI: 10.1016/j.chemosphere.2010.05.037

[28] Aryal RK, Lee BK. Characteristics of suspended solids and micropollutants in first-flush highway runoff. Water, Air, and Soil Pollution: Focus. 2009;**9**(5-6):339-346. DOI: 10.1007/s11267-009-9243-9

[29] Furumai H, Jinadasa HKPK, Murakami M, Nakajima F, Aryal RK. Model description of storage and infiltration functions of infiltration facilities for urban runoff analysis by a distributed model. Water Science and Technology. 2005;**52**(5):53-60

[30] Mikkelsen PS, Häfliger M, Ochs M, Jacobsen P, Tjell JC, Boller M. Pollution of soil and groundwater from infiltration of highly contaminated stormwater—A case study. Water Science and Technology. 1997;**36**(8-9):325-330. DOI: 10.1016/S0273-1223(97)00578-7

[31] Tyler LD, McBride MB. Mobility and extractability of cadmium, copper, nickel, and zinc in organic and mineral soil columns. Soil Science. 1982;**134**(3):198-205. DOI: 10.1097/00010694-198209000-00009

[32] Elliott HA, Liberati MR, Huang CP. Competitive adsorption of heavy metals by soils. Journal of Environmental Quality. 1986;**15**(3):214-219. DOI: 10.2134/jeq1986.00472425001500030002x

[33] Lo KSL, Yang WF, Lin YC. Effects of organic matter on the specific adsorption of heavy metals by soil. Toxicological & Environmental Chemistry. 1992;**34**(2-4):139-153. DOI: 10.1080/02772249209357787

[34] Yong RN, Phadungchewit Y. pH influence on selectivity and retention of heavy metals in some clay soils. Canadian Geotechnical Journal. 1993;**30**(5):821-833. DOI: 10.1139/t93-073

[35] Mesquita ME, Carranca C, Menino MR. Influence of pH on copper-zinc competitive adsorption by a sandy soil. Environmental Technology. 2002;**23**(9):1043-1050. DOI: 10.1080/09593332308618352

[36] Marsalek J, Marsalek PM. Characteristics of sediments from a stormwater management pond. Water Science and Technology. 1997;**36**(8-9):117-122. DOI: 10.1016/S0273-1223(97)00610-0

[37] Aryal RK, Murakami M, Furumai H, Nakajima F, Jinadasa HKPK. Prolonged deposition of heavy metals in infiltration facilities and its possible threat to groundwater contamination. Water Science and Technology. 2006;**54**(6-7):205-212. DOI: 10.2166/wst.2006.584

[38] Hossain MA, Furumai H, Nakajima F, Aryal RK. Heavy metals speciation in soakaways sediment and evaluation of metal retention properties of surrounding soil. Water Science and Technology. 2007;**56**(11):81-89. DOI: 10.2166/wst.2007.746

[39] MacCarthy P, Malcolm RL, Clapp CE, Bloom PR. An introduction to soil humic substances. In: MacCarthy P, Clapp CE, Malcolm RL, Bloom PR, editors. Humic Substances in Soil and Crop Sciences: Selected Readings. Madison: SSSA; 1990. pp. 1-12. DOI: 10.2136/1990.humicsubstances.c1

[40] Kramer RW, Kuawinski EB, Zang X, Green-Church KB, Jones RB, Freitas MA, Hatche PG. Studies of the structure of humic substances by electrospray ionization coupled to a quadrupole-time of flight (QQ-TOF) mass spectrometer. In: Ghabbour EA, Davis G, editors. Humic Substances: Structures, Models and Function. Cambridge: Royal Society of Chemistry; 2001. pp. 95-107

[41] Chin YP, Aiken G, O'Loughlin E. Molecular weight, polydispersity, and spectroscopic properties of aquatic humic substances. Environmental Science & Technology. 1994;**28**(11):1853-1858. DOI: 10.1021/es00060a015

[42] Buffle J, Filella M. Physico-chemical heterogeneity of natural complexants: Clarification. Analytica Chimica Acta. 1995;**313**(1-2):144-150. DOI: 10.1016/0003-2670(95)00204-D

[43] Atanassova I. Competitive effect of copper, zinc, cadmium and nickel on ion adsorption and desorption by soil clays. Water, Air, and Soil Pollution. 1999;**113**(1-4):115-125. DOI: 10.1023/A:1005076821325

[44] Jalali M, Moradi F. Competitive sorption of Cd, Cu, Mn, Ni, Pb and Zn in polluted and unpolluted calcareous soils. Environmental Monitoring and Assessment. 2013;**185**(11):8831-8846. DOI: 10.1007/s10661-013-3216-1

[45] Adriano D, Vangronsveld J, Bolan N, Wenzel W. Heavy metals. In: Encyclopedia of Soils in the Environment. New York, USA: Springer; 2005. pp. 175-182. DOI: 10.1016/B0-12-348530-4/00196-X

[46] Aryal RK, Duong TTT, Lee BK, Hossain MA, Kandel D, Kamruzzaman M, Beecham S, Chong MN. Organic matter composition variability in road sediment and its role in binding heavy metals. Sustainable Environment Research. 2014;**24**(2):81-91

[47] Aryal R, Furumai H, Nakajima F, Beecham S, Kandasamy J. Characterisation of prolonged deposits of organic matter in infiltration system inlets and their binding with heavy metals: A PARAFAC approach. Water, Air, and Soil Pollution. 2015;**226**(6). DOI: 10.1007/s11270-015-2426-2

[48] Cory RM, McKnight DM. Fluorescence spectroscopy reveals ubiquitous presence of oxidized and reduced quinones in dissolved organic matter. Environmental Science & Technology. 2005;**39**(21):8142-8149. DOI: 10.1021/es0506962

[49] Klapper L, McKnight DM, Fulton JR, Blunt-Harris EL, Nevin KP, Lovley DR, Hatcher PG. Fulvic acid oxidation state detection using fluorescence spectroscopy. Environmental Science & Technology. 2002;**36**(14):3170-3175. DOI: 10.1021/es0109702

[50] Stedmon CA, Markager S, Bro R. Tracing dissolved organic matter in aquatic environments using a new approach to fluorescence spectroscopy. Marine Chemistry. 2003;**82**(3-4):239-254. DOI: 10.1016/S0304-4203(03)00072-0

[51] Stedmon CA, Markager S. Resolving the variability in dissolved organic matter fluorescence in a temperate estuary and its catchment using PARAFAC analysis. Limnology and Oceanography. 2005;**50**(2):686-697. DOI: 10.4319/lo.2005.50.2.0686

[52] Pingqing F, Fengchang W, Congqiang L. Fluorescence excitation-emission matrix characterization of a commercial humic acid. Chinese Journal of Geochemistry. 2004;**23**(4):309-318. DOI: 10.1007/BF02871302

[53] Takada H, Onda T, Harada M, Ogura N. Distribution and sources of polycyclic aromatic hydrocarbons (PAHs) in street dust from the Tokyo metropolitan area. Science of the Total Environment. 1991;**107**:45-69. DOI: 10.1016/0048-9697(91)90249-E

[54] Pengchai P, Furumai H, Nakajima F. Source apportionment of polycyclic aromatic hydrocarbons in road dust in Tokyo. Polycyclic Aromatic Compounds. 2004;**24**(4-5): 713-789. DOI: 10.1080/10406630490487828

[55] Murakami M, Takada H. Perfluorinated surfactants (PFSs) in size-fractionated street dust in Tokyo. Chemosphere. 2008;**73**:1172-1177. DOI: 10.1016/j.chemosphere.2008.07.063

[56] Watanabe H, Nakajima F, Kasuga I, Furumai H. Toxicity evaluation of road dust in the runoff process using a benthic ostracod *Heterocypris incongruens*. Science of the Total Environment. 2011;**409**:2366-2372. DOI: 10.1016/j.scitotenv.2011.03.001

[57] Khanal R, Furumai H, Nakajima F. Toxicity assessment of size-fractionated urban road dust using ostracod *Heterocypris incongruens* direct contact test. Journal of Hazardous Materials. 2014;**264**:53-64. DOI: 10.1016/j.jhazmat.2013.10.058

[58] Niyommaneerat W, Nakajima F, Tobino T, Yamamoto K. Development of a chronic sediment toxicity test using the benthic ostracod *Heterocypris incongruens* and their application to toxicity assessments of urban road dust. Ecotoxicology and Environmental Safety. 2017;**143**:266-274. DOI: 10.1016/j.ecoenv.2017.05.011

[59] Kumata H, Yamada J, Masuda K, Takada H, Sato Y, Sakurai T, Fujiwara K. Benzothiazolamines as tire-derived molecular markers: Sorptive behavior in street runoff and application to source apportioning. Environmental Science & Technology. 2002;**36**(4):702-708. DOI: 10.1021/es0155229

[60] Khanal R. Characterization of Toxicants in Size-Fractionated Urban Road Dust Using Ostracod *Heterocypris incongruens* Direct Contact Test and Toxicity Identification Evaluation Procedure [Thesis]. Tokyo: The University of Tokyo; 2013

[61] Hiki K, Nakajima F. Effect of salinity on the toxicity of road dust in an estuarine amphipod *Grandidierella japonica*. Water Science and Technology. 2015;**72**(6):1022-1028. DOI: 10.2166/wst.2015.304

[62] Watanabe H, Nakajima F, Kasuga I, Furumai H. Application of whole sediment toxicity identification evaluation procedures to road dust using a benthic ostracod *Heterocypris incongruens*. Ecotoxicology and Environmental Safety. 2013;**89**:245-251. DOI: 10.1016/j.ecoenv.2012.12.003

[63] Khanal R, Furumai H, Nakajima F. Characterization of toxicants in urban road dust by toxicity identification evaluation using ostracod *Heterocypris incongruens* direct contact test. Science of the Total Environment. 2015;**530-531**:96-102. DOI: 10.1016/j.scitotenv.2015.05.090

[64] Nakajima F, Sevilla JB, Kasuga I. Dose-response relationship of dietary heavy metals to mortality of ISO sediment toxicity test species *Heterocypris incongruens*. In: Proceedings of 50th Environmental Engineering Research Forum; 19-21 November 2013; Sapporo. Tokyo: Japan Society of Civil Engineers; 2013. pp. 160-162 (in Japanese)

[65] Sevilla JB, Nakajima F, Kasuga I. Comparison of aquatic and dietary exposure of heavy metals Cd, Cu and Zn to benthic ostracod *Heterocypris incongruens*. Environmental Toxicology and Chemistry. 2014;**33**(7):1624-1630. DOI: 10.1002/etc.2596

[66] Hiki K, Nakajima F, Tobino T. Application of cDNA-AFLP to biomarker exploration in a non-model species *Grandidierella japonica*. Ecotoxicology and Environmental Safety. 2017;**140**:206-213. DOI: 10.1016/j.ecoenv.2017.02.037

Chapter 18

Health Risk Assessment of Heavy Metals on Primary School Learners from Dust and Soil within School Premises in Lagos State, Nigeria

Olatunde S. Durowoju, Joshua N. Edokpayi, Oluseun E. Popoola and John O. Odiyo

Additional information is available at the end of the chapter

http://dx.doi.org/10.5772/intechopen.74741

Abstract

This chapter is aimed at evaluating learner's health risk based on the concentration of toxic metals (Pb, Cr, Cd and Mn) in soil/dust from playgrounds/classrooms in selected primary schools in Lagos State. Samples were divided into four groups based on the density of the locations. Concentration of toxic metals in samples were determined by Graphite Furnace Atomic Absorption Spectrophotometer (GFA-EX7) technique after microwave digestion. The result showed that some of the heavy metals in the soil were higher than permissible limits set by DPR, FEPA and WHO. The soil/dust were contaminated with Cr, Cd and Pb but Mn was within permissible limit. Due to exposure to playground soil and classroom dust, hazardous index (HI) for non-carcinogenic/carcinogenic risk in children was estimated. HI value indicated that the heavy metal pollution may pose no obvious non-cancer health risk to children learning in such schools. However, children via ingestion pathway are exposed to the greatest carcinogenic risk followed by the inhalation pathway. The cancer risk for learners was found to be 3.2×10^{-2} (1 in 31 individuals). Hence, there is need for local environmental authorities to be warned about the potential health risks caused by heavy metals in playground/classroom.

Keywords: classroom, health risk, hazard index, heavy metals, Lagos State, playgrounds

1. Introduction

Lagos State is the most crowded city in Nigeria, the second quickest developing city in Africa, and the seventh in the world [1]. The rapid development of industrialization and urbanization in recent decades has resulted in the high emissions of both metal and organic pollutants, inevitably rendering the environment especially defenseless to ecological degradation and pollution [2–4]. Heavy metals are nondegradable and accumulate in the environment with no known homeostasis mechanism for their removal [5]. High level of heavy metals may affect human health by hindering normal functioning of organs/systems such as liver, kidney, central nervous system, bones, among others or acting as cofactors in other diseases [5, 6]. A few metals are fundamental to life and assume key parts as wellsprings of vitamins, and minerals in the working of body organs. All living organisms require different measures of metals but at higher concentrations, it becomes toxic [7]. Upon absorption by the body, heavy metals keep on accumulating in vital organs like the brain, liver, bones, and kidneys, for years or decades causing serious health consequences [8].

Substantial metals like copper (Cu), cobalt (Co), and zinc (Zn) are basically required for typical body development and elements of living beings, while the high groupings of different metals like cadmium (Cd), chromium (Cr), manganese (Mn), and lead (Pb) are considered profoundly dangerous for human and aquatic life [9]. A measure of Cr is required for typical body capacities; while its high fixations may cause poisonous quality, including liver and kidney issues and genotoxic cancer-causing agent [10–12]. By and large, high concentrations of Mn and Cu in drinking water can cause mental illnesses, for example, Alzheimer and Manganism [12], which mostly affects the intellectual functions of 10-year-old children [13]. Pb is additionally an exceptionally dangerous and cancer-causing metal and may cause perpetual wellbeing dangers, including migraine, crabbiness, stomach torment, nerve harms, kidney harm, circulatory strain, lung tumor, stomach growth, and gliomas [14–16]. As the children are most susceptible to Pb toxicity, their introduction to large amounts of Pb cause serious wellbeing complexities, for example, behavioral unsettling influences, memory disintegration, and decreased capacity to comprehend as well as anemia for long-term exposure [16]. Pb likewise initiates renal tumors and disturbs the normal functioning of kidneys, joints, reproductive, and nervous systems [17].

Soils and dust have become a very good diagnostic tool of environmental conditions that influence human health [18]. Chemical composition of soil and dust has been conducted in many studies during the last 10 years. Special attention has been devoted to studies on urban parking playgrounds and on road deposited dust [19–25]. Soil particles directly or indirectly transform into house dust, and can be ingested by adults and children through various means [26]. Dermal contact, ingestion, and inhalation are the fundamental route of exposure to toxic metals in urban environment [26, 27]. Probability of exposure to unfavorable impacts of soil ingestion is higher children than adults [26]. Children fundamentally interact with soil and clean in classrooms and play areas. Children could ingest a lot of harmful metals from soil, dust, and air [27]. Because of their low resilience to poisons and hand-to-mouth pathways, the wellbeing hazard is high in this populace [28, 29]. Hence, the control of conceivably destructive substances in soil is of high significance and must be kept at low levels in the areas frequented by children [30].

School playgrounds and classrooms are places specifically designed to allow children to play, keep them happy, and develop their learning abilities [31]. They are designed to enhance the inter-relationships among kids, provide enjoyment and recreation, and develop physical fitness and flexibility. Present day, playing grounds are all the more frequently furnished with metal made sandbox, wilderness exercise center, carousel, slide, and different materials that can upgrade recreational exercises [31]. Safety, particularly on playgrounds, relates to prevention of injuries, but less attention on the heavy metals health risk. The health risk is associated with exposure to toxicants. Human Exposure to Soil Pollutants (HESP) model is often used to survey dangers of exposures to carcinogenic and non-carcinogenic sources and is only in view of information from soil contamination [32]. The risk assessment for heavy metals is estimated by comparing the daily oral intake of heavy metals to the normal standard daily oral intake. The target hazard quotient (HQ), which according to USEPA [32], is the proportion of the estimated exposure to a contaminant, the measurement at which no unfavorable impact occurs, is the current method for non-cancer risk assessment. If the proportion is estimated above 1, then potential well-being impacts are conceivable.

Although direct exposure forms the basis of human health risk assessments, it is not discussed as widely in the literature as the first pathway (indirect exposure) [33–35]. Studies have, in any case, demonstrated that open-air exercises where people come into contact with metal-contaminated soil could also represent an important exposure pathway for intake of heavy metals for humans, particularly children [33–35]. The aim of this chapter was to evaluate risk assessment to children's health based on the content of toxic metals in soil from playgrounds and classroom dust in selected primary schools in Lagos State. This study included the investigation of several toxic metals such as Pb, Mn, and Cr in surface soil and dust samples from playgrounds and classrooms of some selected primary schools in Lagos State. Therefore, the concentration of heavy metals and risk assessment were assessed in this chapter, to ascertain the potential health risks children are subjected to, in playgrounds and classrooms at selected schools in Lagos State.

1.1. Study area

Lagos State (**Figure 1**) represents the most urbanized state in Nigeria with about 300 industries on 12 industrial estates, representing more than 60% of all industries activities in the country. It is the smallest state in Nigeria, yet it has the second biggest populace of more than 9 million [1]. The rate of populace development is around 275,000 people for every annum with a populace thickness of 2594 people for each km^2. Lagos State primary schools list was acquired from the Lagos State Primary School Education Board (SPEB), Maryland. Two schools were selected in each of the 20 local government areas (namely Agege, Ajeromi-Ifelodun, Apapa, Alimosho, Amuwo-Odofin, Badagry, Eti-Osa, Epe, Ifako-Ijaye, Ikeja, Ojo, Oshodi-Isolo, Surulere, Mushin, Shomolu, Lagos Mainland, Lagos Island, Kosofe, Ikorodu, and Ibeju-Lekki), one in highly populated area and the other in the low-density area. One of the two primary schools selected was from the planned areas (low traffic density areas, government reserved areas or residential estates) and the other from the unplanned area (high traffic density, commercial, or industrial areas).

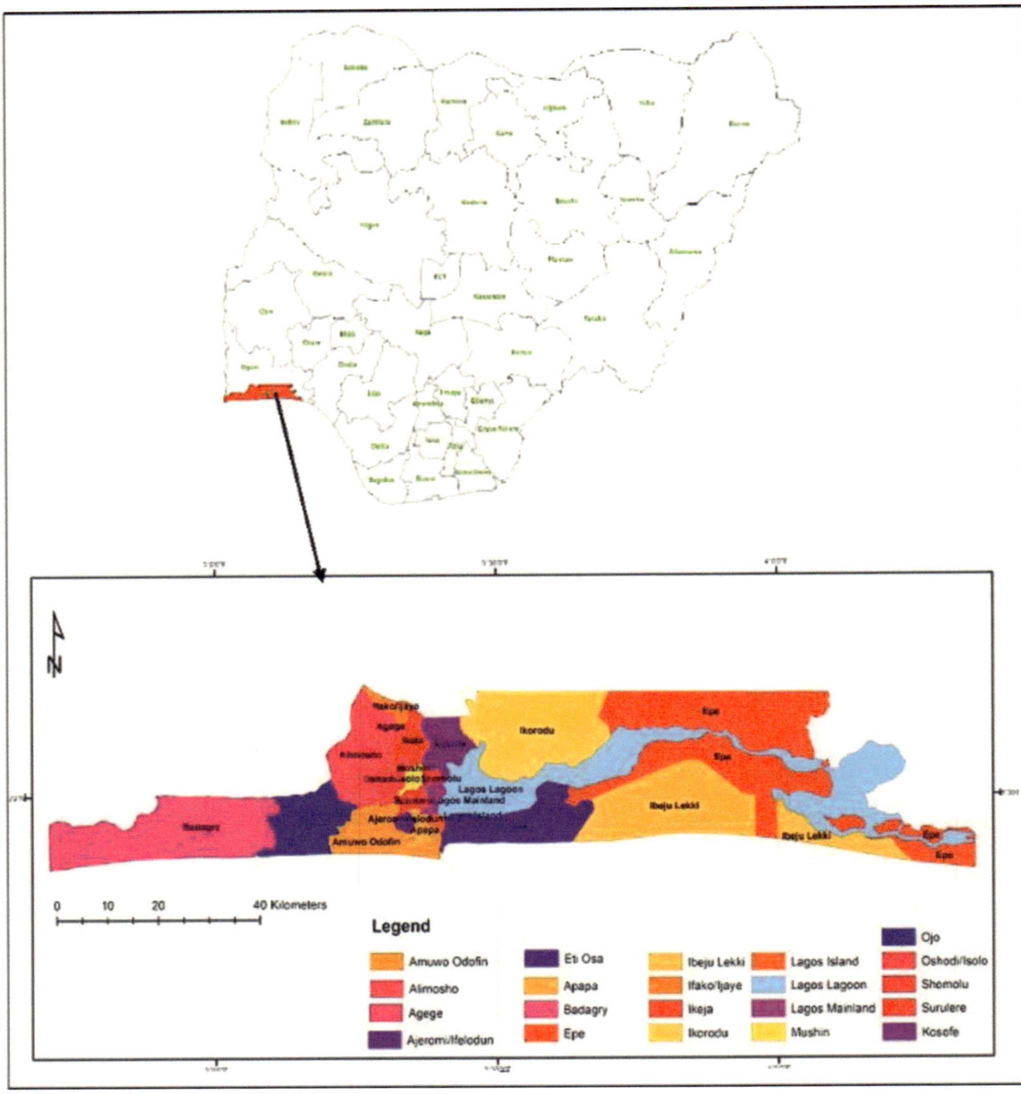

Figure 1. Lagos State map showing local government areas.

2. Methodology

2.1. Sampling and analysis

Samples were collected in two primary schools in each of the 20 local government areas of Lagos State. The schools were selected based on the proximity of the school to major roads and where possible, nearness to industrial activities. A point was picked aimlessly in the play area and entrance and a squared framework of around 2 km side was drawn from the spot. In the vicinity of four and five discrete soil samples were gathered from each of the grid. The number of samples from each site was determined based on the circumstances existing at that site, i.e., high foot traffic (entrance), spots where the children frequently play during recess (playgrounds). A total of 80 topsoil samples were obtained. Plastic parker and brush were used to collect the surface soil. Composite samples were obtained by mixing all the subsamples to form one single sample (about 2 kg of soil), which was used for chemical analysis. Particulates such as grass, leaves, polythene bags, and papers present in soil samples were removed after gentle shaking to remove the soil attached around them. All soil samples were air-dried and sieved to 2 mm using a stainless-steel sieve and stored in plastic vials until analyzed.

In the classrooms, 40 dusts samples were gathered in the nursery classes (kids in the age scope of 2–4 years of age are more inclined to pica propensities). Dusts were gathered from window ledges, bookshelves, corners in the classrooms, and so forth. Around 3–5 tests were gathered and were blended to frame one composite sample. The soil and dust samples were air-dried and sieved to <2 mm using a stainless-steel sieve and stored in plastic vials until analyzed.

2.2. Sample preparation

The samples were air-dried in the laboratory for 24 h (laboratory dust was avoided). Larger grits and dirt were removed from the dried-dust samples before being homogenized with mortar and pestle. The collected samples were packed and kept in a sealed polythene prior analysis.

2.3. Glassware, reagents, and standards

All glassware and plastic products were first washed with high-grade laboratory cleanser, flushed with deionized water, and then 10% nitric acid was added overnight and rinsed with the deionized water. All solutions and dilutions were performed with double deionized water (18.2 M cm) obtained from Simplicity water purification system. All chemicals utilized through the examination were analytical grade chemicals. There was no further cleaning for purification of all reagents and adjustment benchmarks were provided by Merck Pty Ltd. South Africa. The analysis was carried out in triplicate to obtain the mean value.

2.4. Microwave-assisted digestion

Davidson et al. [36] method for determination of pseudototal metal concentration was adopted. This digestion method is enhanced by extracting most of the potentially mobile fractions but leaves the more resistant silicates undissolved. Approximately, 1 g of the sample and 20 mL of acid were heated at temperature of 180°C and pressure of 120 psi for 10 min. Then, the digested samples were filtered through Whatman No. 125 filter paper into 100 mL volumetric flasks. The digestion vessels were then rinsed with distilled water and filtered into the flasks. Each filtrate was made up to the mark with further distilled water, to give a final sample solution containing 20% (v/v) aqua regia.

2.5. Instrumentation calibration and measurement

1000 μg/mL standard stock solutions of Pb, Cr, Cd, and Mn were prepared from 1000 mg/L certified standard solutions (Merck Ltd., South Africa). The stock solutions were acidified with nitric acid and used for calibration. Calibration solutions concentrations of 5, 10, 20, 30, 40, 50, and 60 ppb (μg/mL) were prepared, respectively, for Pb, Cr, and Cd, while 1, 1.5, 2,

2.5, 3, 3.5, and 4 μg/mL (ppb) were prepared for Mn. A Shimadzu AA-6300 Graphite Furnace Atomic Absorption Spectrophotometer (GFA-EX7) fitted with an Autosampler (ASC)-6100 was used for determination of the heavy metals. The GFAAS was equipped with Cr, Cd, Pb, and Mn hollow cathode lamps (Varian cathode lamps and photon cathode lamps) and were employed for the measurement of the absorbance.

2.6. Risk assessment

The measured total heavy metals from the analysis were used to assess the health risk of the metal on children. The mean heavy metals concentrations were used to estimate intake at different pathways using standard USEPA's exposure equations [37, 38]. Children could be exposed to contaminants from soil via three different pathways that include oral intake ($I_{ingestion}$), inhalation intake ($I_{inhalation}$), and through skin exposure (I_{dermal}) [38]. Based on this fact, cancer/noncancer risk assessment in this study was estimated. For intake estimation via each exposure pathways, the following equations were used.

$$\text{Intake}_{\text{ingestion}} = \frac{C \; X \; IngR \; X \; EF \; X \; ED}{BW \; X \; AT} \; X \; 10^{-6} \tag{1}$$

where C—concentration of a contaminant in soil (mg/kg), IngR—ingestion rate of soil (mg/day), EF—exposure frequency (days/year), ED—exposure duration (years), BW—average body weight (kg), and AT—average time (days) = ED*365

$$\text{Intake}_{\text{inhalation}} = \frac{C \; X \; InhR \; X \; EF \; X \; ED}{PEF \; X \; BW \; X \; AT} \tag{2}$$

where InhR—inhalation rate (m^3/day) and PEF—particle emission factor (m^3/kg).

$$\text{Intake}_{\text{dermal}} = \frac{C \; X \; SA \; X \; SAF \; X \; ABS \; X \; EF \; X \; ED}{BW \; X \; AT} \; X \; 10^{-6} \tag{3}$$

where SA—surface area of the skin that contacts the soil (cm^2), SAF—skin adherence factor for soil (mg/cm^2), and ABS—dermal absorption factor (chemical specific) = 0.001 (for all metals).

After the three exposure pathways were calculated, hazard quotient (HQ) and HI based on cancer/non-cancer toxic risk were calculated as follows [38]:

$$\text{HQ} = \frac{Intake}{RfD} \tag{4}$$

$$\text{HI}_{\text{ex}}\text{P} = \sum HQexP \tag{5}$$

where exP are different exposure pathways, respectively. Reference dose (RfD) (mg/kg/day) is an estimated value of the daily exposure, maximum permissible risk, to the human population, including sensitive subgroups (children) during a lifetime. **Tables 1** and **2** show the exposure parameters, reference doses, and cancer slope factors used for the health risk assessment for standard residential exposure scenario through different exposure pathways.

Parameter	Unit	Child	References
Body weight (*BW*)	kg	15	[39]
Exposure frequency (*EF*)	days/year	360	[38]
Exposure duration (*ED*)	years	6	[40]
Ingestion rate (*IR*)	mg/day	200	[41, 42]
Inhalation rate (*IRair*)	m^3/day	7.6	[43]
Skin surface area (*SA*)	cm^2	2800	[40]
Soil adherence factor (*SAF*)	mg/cm^2	0.2	[40]
Dermal Absorption factor (*ABS*)	none	0.001	[38–40, 43, 44]
Particulate emission factor (*PEF*)	m^3/kg	1.36×10^9	[38, 39]
Average time (*AT*)	days		[39]
For carcinogens		365 × 70	[39]
For non-carcinogens		365 × ED	[39]

Table 1. Exposure parameters used for the health risk assessment through different exposure pathways for soil.

Heavy metal	$RfD_{ingestion}$	RfD_{dermal}	$RfD_{Inhalation}$	$exP_{ingestion}$	exP_{dermal}	$exP_{Inhalation}$	Ref.
Pb	3.50E–03	5.25E-04	3.50E-03	8.50E–03	—	4.20E–02	[39, 45]
Cr	3.00E–03	3.00E–03	3.00E–05	5.00E–01	—	4.10E+01	[39, 46]
Cd	1.00E-03	1.00E-03	5.70E-05	6.30E+00	—	6.30E+00	[39, 45]
Mn	1.40E-1	1.40E-1	—	—	—	—	[39, 46]

Table 2. Reference doses (***RfD***) in (mg/kg-day) and cancer slope factors (exP) for the different heavy metals.

3. Results and discussion

3.1. Concentrations of heavy metals in soil from classrooms and playgrounds

Descriptive statistics of the four groups of schools are shown in **Table 3**. Average concentrations of heavy metals in mg/kg from playgrounds and classrooms in selected primary schools in Lagos State were used to calculate average daily intakes for carcinogenic/non-carcinogenic risk assessment. The results presented showed the average concentrations of heavy metals in classrooms and playgrounds varied significantly and decreased in the order from Group 1 > Group 3 > Group 2 > Group 4 for Pb; Group 4 > Group 3 > Group 1 > Group 2 for Cr; Group 2 > Group 1 > Group 3 = Group 4 for Cd and Group 2 > Group 3 > Group 1 = Group 4 for Mn in the samples. This implies that the heavy metals are distributed irrespective of the locality either in high or low-density playgrounds/classrooms. The results show the magnitude of the heavy metals in the classrooms and playgrounds decreases in the following order:

		GROUP 1	GROUP 2	GROUP 3	GROUP 4
Pb (μg/g)	Min	23.04	24.30	24.96	8.33
	Max	79.01	37.05	41.91	41.91
	Mean + SD	35.17 ± 16.54	30.08 ± 5.35	30.75 ± 6.72	24.21 ± 12.05
	Range	23.04–79.01	24.30–37.05	24.96–41.91	8.33–41.91
Cr (μg/g)	Min	3.25	0.05	6.66	11.42
	Max	16.93	14.52	19.15	20.24
	Mean + SD	11.88 ± 4.50	6.08 ± 4.63	14.58 ± 4.33	15.73 ± 2.87
	Range	3.25–16.93	0.05–14.52	6.66–19.15	11.42–20.24
Cd (μg/g)	Min	1.05	0.90	ND	ND
	Max	1.63	3.10	ND	ND
	Mean + SD	1.28 ± 0.31	1.90 ± 1.07	ND	ND
	Range	1.05–1.63	0.90–3.10	ND	ND
Mn (μg/g)	Min	ND	1.18	1.18	ND
	Max	ND	1.87	1.18	ND
	Mean + SD	ND	1.47 ± 0.36	1.18 ± .00	ND
	Range	ND	1.18–1.87	1.18	ND

Group 1—High density playground; Group 2—Low-density playground; Group 3—High density classroom, and Group 4—Low-density classroom, ND—Not detected.

Table 3. Statistical parameter for the distribution of Pb, Cr, Cd, and Mn in classroom dusts and playgrounds from selected schools in Lagos State.

Pb > Cr > Cd > Mn. The high concentrations of Pb and Cr in the samples could be attributed to the paint chippings peeling off to settle on the classrooms floors as dust [30]. A study from Wright et al. [47] show that chipping house paint is a critical determinant of high blood lead levels more than 10 μg/L found in 70% of children aged 6–35 months. It has been accounted for that emulsion and gloss sorts of paints, which are produced and sold in Nigeria contained considerable levels of lead [48].

Higher concentrations of investigated metals could be due to proximity of the schools to traffic roads and some industrial locations. Lagos State has been under high urbanization in the past few decades. In this chapter, there were no specific pollution sources of toxic metals [30]; hence, the toxic metal contamination of the soils from the classrooms and playgrounds was most likely from continuous urbanization and development, which can influence human health in the contaminated area. It is important to emphasize that Lagos State with her increasing population is known to have very intensive and heavy traffic [30]. The high Pb concentration of the soil could be ascribed to vehicular discharges and metal plating and greasing up oils. It could likewise be because of harsh surfaces of the streets which increment the wearing of tires and run-offs from the roadsides [48]. Lead contamination in urban soil

	Pb	Cr	Cd	Mn
DPR(μg/g)	0.05	0.03	0.01	–
FEPA(μg/g)	0.05	0.03	0.01	–
WHO (μg/g)	0.05	0.02	0.005	5

Table 4. DPR [50], FEPA [51], and WHO [52] permissible limits.

has been ascribed to ignition of fuel that contains tetraethyl lead as anti-knock agent [48, 49]. Compared with recommended maximum allowable limits for Nigeria and WHO as shown in **Table 4**, Pb, Cr, and Cd concentrations were found to be higher than the permissible limits, and hence the classrooms and playgrounds soils were contaminated, and the learners were at high risk. These highest levels of Pb and Cr could possibly be the cause of alleged sickness in children suffering from diarrheal diseases and chest pains. The high concentrations of heavy metals could be because of the nearness of the schools and playgrounds to bus stops, auto workshops, nearby industries, gas stations in addition to paints chippings as reported in the city of Ibadan [53].

3.2. Health risk of heavy metals for children

The mean values of the heavy metals were higher than permissible limits by DPR, FEPA, and WHO, which make the soils/dusts contaminated except for Mn (**Table 4**). The obtained results of non-carcinogenic health risk on children, based on metal concentrations in playgrounds and classrooms soils and exposure by three different pathways (ingestion, inhalation, and dermal) are shown in **Table 5**. When HQ and HI values are less than one, there is no obvious risk to the population, but if these values exceed one, there may be concern for potential non-carcinogenic effects [38]. The results showed that the major path of children exposure to playgrounds soils and classroom dust that may have adverse effect on their health by Pb, Cr, Cd, and Mn is ingestion, followed by dermal exposure (**Figure 2**). Contribution of inhalation exposure to the HQ and HI is the smallest (**Figure 2**). This finding supports other studies to establish the ingestion pathway as the major contributor to HI, particularly for children [6, 17, 28, 31].

In this study, the ingestion had HI values less than 1 for all the observed heavy metals in all the groups. The summation of the total HI-value is approximately 0.4, which is less than one and ingestion pathway is a major contributor though there is no health effect (**Table 5**). Although, samples (classroom dusts and playground soils) were contaminated with Pb, Cd, and Cr but there is no obvious non-cancer health risk to the children. The high concentrations of indicated heavy metal pollution may pose a non-cancer health risk to children learning in such schools. This could be attributed to the short-term exposure duration (ED) of 6 years as recommended by USEPA health assessment model deployed in the study. The results also indicate that children via ingestion pathway could be exposed to potential possibility of non-carcinogenic risk followed by the inhalation pathway if considered at long-term exposure duration.

		GROUP 1	GROUP 2	GROUP 3	GROUP 4
Pb	Intake(ingestion)	4.63E-04	3.96E-04	4.04E-04	3.18E-04
	Intake(inhalation)	1.29E-08	1.11E-08	1.13E-08	8.90E-09
	Intake(dermal)	1.30E-06	1.11E-06	1.13E-06	8.91E-07
	Total Intake	**4.64E-04**	**3.97E-04**	**4.06E-04**	**3.19E-04**
	HQ(ingestion)	1.32E-01	1.13E-01	1.16E-01	9.10E-02
	HQ(inhalation)	2.46E-05	2.11E-05	2.15E-05	1.69E-05
	HQ(dermal)	3.70E-04	3.16E-04	3.24E-04	2.55E-04
	Total HQ	**1.33E-01**	**1.13E-01**	**1.16E-01**	**9.12E-02**
	HI(ingestion)	1.12E-03	9.61E-04	9.82E-04	7.73E-04
	HI(inhalation)	1.03E-06	8.84E-07	9.04E-07	7.12E-07
	HI(dermal)	–	–	–	–
	Total HI	**1.12E-03**	**9.62E-04**	**9.83E-04**	**7.74E-04**
Cr	Intake(ingestion)	1.56E-04	8.00E-05	1.92E-04	2.07E-04
	Intake(inhalation)	4.37E-09	2.23E-09	5.36E-09	5.78E-09
	Intake(dermal)	4.37E-07	2.24E-07	5.37E-07	5.79E-07
	Total Intake	**1.57E-04**	**8.02E-05**	**1.92E-04**	**2.07E-04**
	HQ(ingestion)	5.21E-02	2.67E-02	6.39E-02	6.90E-02
	HQ(inhalation)	1.46E-06	7.45E-07	1.79E-06	1.93E-06
	HQ(dermal)	1.46E-02	7.46E-03	1.79E-02	1.93E-02
	Total HQ	**6.67E-02**	**3.41E-02**	**8.18E-02**	**8.83E-02**
	HI(ingestion)	2.60E-02	1.33E-02	3.20E-02	3.45E-02
	HI(inhalation)	5.97E-05	3.05E-05	7.32E-05	7.90E-05
	HI(dermal)	–	–	–	–
	Total HI	**2.61E-02**	**1.34E-02**	**3.20E-02**	**3.46E-02**
Cd	Intake(ingestion)	1.68E-05	2.50E-05	–	–
	Intake(inhalation)	4.70E-10	6.98E-10	–	–
	Intake(dermal)	4.71E-08	7.00E-08	–	–
	Total Intake	**1.69E-05**	**2.51E-05**	**–**	**–**
	HQ(ingestion)	1.68E-02	2.50E-02	–	–
	HQ(inhalation)	8.25E-06	1.22E-05	–	–
	HQ(dermal)	4.71E-05	7.00E-05	–	–
	Total HQ	**1.69E-02**	**2.51E-02**	**–**	**–**
	HI(ingestion)	1.06E-01	1.57E-01	–	–
	HI(inhalation)	5.20E-05	7.72E-05	–	–
	HI(dermal)	–	–	–	–
	Total HI	**1.06E-01**	**1.57E-01**	**–**	**–**

		GROUP 1	GROUP 2	GROUP 3	GROUP 4
Mn	Intake(ingestion)	—	1.93E-05	1.55E-05	—
	Intake(inhalation)	—	5.40E-10	4.34E-10	—
	Intake(dermal)	—	5.41E-08	4.34E-08	—
	Total Intake	**—**	**1.94E-05**	**1.56E-05**	**—**
	HQ(ingestion)	—	1.38E-04	1.11E-04	—
	HQ(inhalation)	—	—	—	—
	HQ(dermal)	—	3.87E-07	3.10E-07	—
	Total HQ	**—**	**1.38E-04**	**1.11E-04**	**—**

Table 5. Noncarcinogenic risk for children in soil/dust from the classrooms and playgrounds.

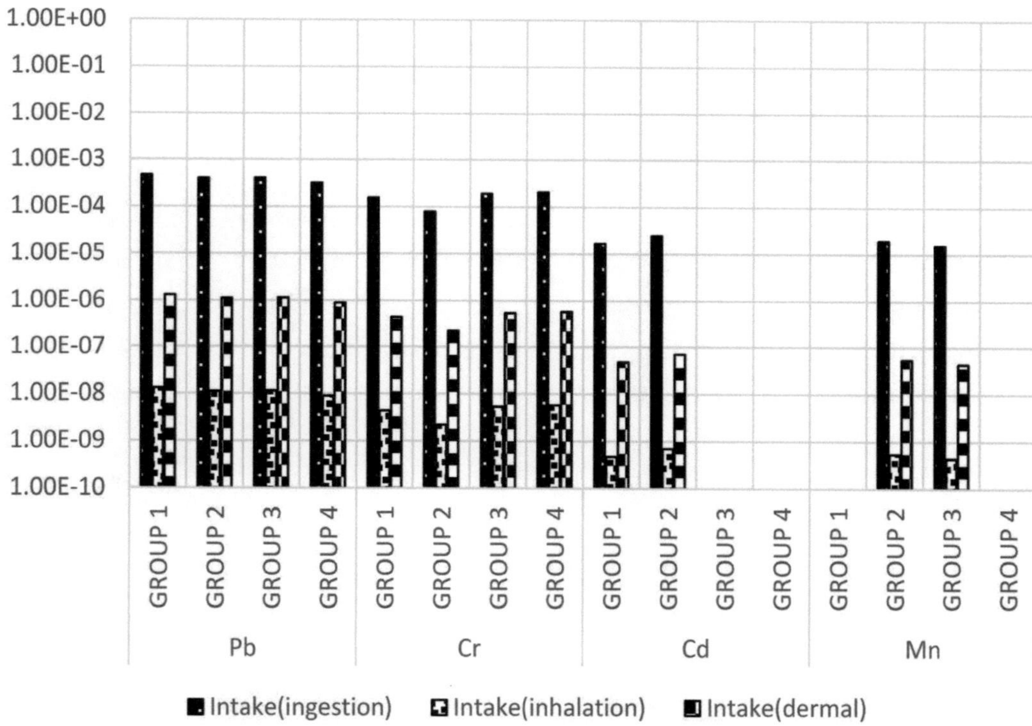

Figure 2. Heavy metal intake (three pathways) values for children in selected school in Lagos State.

Literature shows that the daily ingestion rates of soil by children were calculated to vary between 39 and 270 mg/day [54]. Monitoring of Pb content in soil is of great necessity owing to the adverse effect on the children's central nervous system [55]. Numerous neurological and formative issue might be seen in children' populace because of a long time of exposure and ingestion of specific amount of Pb from contaminated soil; the potential wellbeing dangers incorporate iron deficiency, kidney harm, colic, muscle shortcoming, and cerebrum harm [56]. Ingestion of small Pb from dust may be harmful for blood, development, behavior, and intellectual functioning, as well [56]. Hence, children are more at risk particularly in both high- and low-density classrooms/playgrounds.

The excess lifetime cancer risks for children are calculated from the mean concentrations of individual heavy metals in classrooms and playgrounds for the pathways using Eqs. (4) and (5). **Table 6** presented the calculated carcinogenic risk values for Pb, Cr, Cd, and Mn. Cr and Cd was found to be highest contributor to the cancer risk. The US Environmental Protection Agency considers adequate for administrative purposes a malignancy hazard point of confinement of 1 × 10^{-6} to 10^{-4} [32]. The summation of cancer risk for learners was found to be

		GROUP 1	GROUP 2	GROUP 3	GROUP 4
Pb	Intake(ingestion)	3.96E-05	3.39E-05	3.47E-05	2.73E-05
	Intake(inhalation)	1.11E-09	9.47E-10	9.68E-10	7.63E-10
	Intake(dermal)	1.11E-07	9.49E-08	9.71E-08	7.64E-08
	Total Intake	**3.98E-05**	**3.40E-05**	**3.48E-05**	**2.74E-05**
	HQ(ingestion)	1.13E-02	9.69E-03	9.90E-03	7.80E-03
	HQ(inhalation)	2.11E-06	1.80E-06	1.84E-06	1.45E-06
	HQ(dermal)	3.17E-05	2.71E-05	2.77E-05	2.18E-05
	Total HQ	**1.14E-02**	**9.72E-03**	**9.93E-03**	**7.82E-03**
	HI(ingestion)	9.63E-05	8.23E-05	8.42E-05	6.63E-05
	HI(inhalation)	8.86E-08	7.58E-08	7.75E-08	6.10E-08
	HI(dermal)	—	—	—	—
	Total HI	**9.64E-05**	**8.24E-05**	**8.43E-05**	**6.63E-05**
Cr	Intake(ingestion)	1.34E-05	6.85E-06	1.64E-05	1.77E-05
	Intake(inhalation)	3.74E-10	1.91E-10	4.59E-10	4.95E-10
	Intake(dermal)	3.75E-08	1.92E-08	4.60E-08	4.96E-08
	Total intake	**1.34E-05**	**6.87E-06**	**1.65E-05**	**1.78E-05**
	HQ(ingestion)	4.46E-03	2.28E-03	5.48E-03	5.91E-03
	HQ(inhalation)	1.25E-07	6.38E-08	1.53E-07	1.65E-07
	HQ(dermal)	1.25E-03	6.40E-04	1.53E-03	1.65E-03
	Total HQ	**5.71E-03**	**2.92E-03**	**7.01E-03**	**7.57E-03**
	HI(ingestion)	2.23E-03	1.14E-03	2.74E-03	2.96E-03
	HI(inhalation)	5.11E-06	2.62E-06	6.28E-06	6.77E-06
	HI(dermal)	—	—	—	—
	Total HI	**2.24E-03**	**1.14E-03**	**2.75E-03**	**2.96E-03**

		GROUP 1	GROUP 2	GROUP 3	GROUP 4
Cd	Intake(ingestion)	1.44E-06	2.14E-06	–	–
	Intake(inhalation)	4.03E-11	5.98E-11	–	–
	Intake(dermal)	4.04E-09	6.00E-09	–	–
	Total intake	**1.45E-06**	**2.15E-06**	**–**	**–**
	HQ(ingestion)	1.44E-03	2.14E-03	–	–
	HQ(inhalation)	7.07E-07	1.05E-06	–	–
	HQ(dermal)	4.04E-06	6.00E-06	–	–
	Total HQ	**1.45E-03**	**2.15E-03**	**–**	**–**
	HI(ingestion)	9.09E-03	1.35E-02	–	–
	HI(inhalation)	4.46E-06	6.61E-06	–	–
	HI(dermal)	–	–	–	–
	Total HI	**9.09E-03**	**1.35E-02**	**–**	**–**
Mn	Intake(ingestion)	–	1.66E-06	1.33E-06	–
	Intake(inhalation)	–	4.63E-11	3.72E-11	–
	Intake(dermal)	–	4.64E-09	3.72E-09	–
	Total intake	**–**	**1.66E-06**	**1.33E-06**	**–**
	HQ(ingestion)	–	1.18E-05	9.50E-06	–
	HQ(inhalation)	–	–	–	–
	HQ(dermal)	–	3.31E-08	2.66E-08	–
	Total HQ	**–**	**1.19E-05**	**9.53E-06**	**–**

Table 6. Carcinogenic (three exposure pathways) risk for children in soil and dust from the classrooms and playgrounds.

3.2×10^{-2} (1 in 31 individuals), which is higher than the acceptable values. Be that as it may, ingestion pathway is by all accounts the significant contributor to overabundance lifetime cancer risk, then inhalation pathway. **Tables 5** and **6** show the difference in the exposure duration, which results to the no obvious non-cancer risk (exposure duration of 6 years) and potential cancer risk (exposure duration of 70 years) of the learners in the selected schools. Therefore, there is a need to take necessary action to decontaminate the sites in order to protect children's health. This approach may be helpful in decision-making process in every growing city in the world like Lagos State.

4. Conclusion

Heavy metals concentrations (Pb, Cr, Cd, and Mn) were determined from the playgrounds and classrooms of selected primary schools in Lagos State. The concentrations of the metals detected in classrooms dust and playgrounds soils were higher than recommended permissible limits by DPR, FEPA, and WHO except for Mn. The high concentrations of the heavy metals could be because of the proximity of the schools to bus stops, auto workshops, nearby industries, gas stations in addition to paints chippings. The results showed that the major path of children exposure to playgrounds soil and classroom dust is ingestion, followed by dermal exposure. HI value indicated heavy metal pollution that may pose no obvious non-cancer health risk to children learning in such schools. This is possibly due to the short exposure duration of 6 years. The results also indicate that children via ingestion pathway contributes most to carcinogenic risk followed by the inhalation pathway owing to longer exposure duration of 70 years. The cancer risk for learners was found to be 3.2×10^{-2} (1 in 31 individuals), which is higher than the acceptable values of 1×10^{-6} to 10^{-4}. The results will provide the direct evidence needed by local environmental authorities and school managers to warn learners about the potential health risks caused by heavy metals in playgrounds and classrooms dust.

Author details

Olatunde S. Durowoju[1]*, Joshua N. Edokpayi[1], Oluseun E. Popoola[2] and John O. Odiyo[1]

*Address all correspondence to: durotunde@gmail.com

1 Hydrology and Water Resources Department, University of Venda, Thohoyandou, South Africa

2 Chemical Science Department, Yaba College of Technology, Yaba, Lagos State, Nigeria

References

[1] Population-Lagos State. Lagos State Government. Archived from the original on 18 October 2015. Retrieved 21 December 2017

[2] Nriagu JO. A silent epidemic of environmental metal poisoning. Environmental Pollution. 1988;**50**:139-161

[3] Kreimer A. Environmental management and urban vulnerability. World Bank Discussion Paper 168. World Bank, Washington, DC; 1992

[4] Thornton I. Environmental geochemistry and health in the (1990): A global perspective. Applied Geochemistry (Supplement). 1993;**2**:203-210

[5] Tong ST, Lam KC. Home sweet home? A case study of household dust contamination in Hong Kong. Science of the Total Environment. 2000;**256**(2-3):115-123

[6] Lane TW, Morel FM. A biological function for cadmium in marine diatoms. Proceedings of the National Academy of Sciences of the United States of America. 2009;**9**:462-431

[7] Kabata-Pendias A. Trace Elements in Soil and Plants. 4th ed. Boca Raton, FL, USA: Taylor & Francis; 2011

[8] Ouyang Y, Higman J, Thompson J, Toole OT, Campbell D. Characterization and spatial distribution of heavy metals in sediment from Cedar and Ortega Rivers sub-basin. Journal of Contaminant Hydrology. 2002;**54**:19-35

[9] Knight C, Kaiser GC, Lailor H, Robothum JV. Witter, heavy metals in surface water and stream sediments in Jamaica. Environmental Geochemistry and Health. 1997;**19**:63-66

[10] Loubieres Y, Lassence AD, Bernier M, Baron AV, Schmitt JM, Page B, Jardin F. Acute, fatal, oral chromic acid poisoning. Journal of Toxicology. Clinical Toxicology. 1999;**37**:333-336

[11] Strachan S. Heavy metal. Current Anaesthesia and Critical Care. 2010;**21**:44-48

[12] Dieter HH, Bayer TA, Multhaup G. Environmental copper and manganese in the pathophysiology of neurologic diseases (Alzheimer's disease and Manganism). Acta Hydrochimica et Hydrobiologica. 2005;**33**:72-78

[13] Steenland K, Boffetta P. Lead and cancer in humans: Where are we now? American Journal of Industrial Medicine. 2000;**38**:295-299

[14] Mortada WI, Sobh MA, El-Defrawy MM, Farahat SE. Study of lead exposure from automobile exhaust as a risk for nephrotoxicity among traffic policemen. American Journal of Nephrology. 2001;**21**:274-279

[15] Jarup L. Hazards of heavy metal contamination. British Medical Bulletin. 2003;**68**:167-182

[16] Ogwuegbu MOC, Muhanga W. Investigation of lead concentration in the blood of people in the copper belt province of Zambia. Journal of Environment. 2005;**1**:66-75

[17] Banerjee ADK. Heavy metal levels and solid phase speciation in street dusts of Delhi, India. Environmental Pollution. 2003;**123**:95-105

[18] Davydova S. Heavy metals as toxicants in big cities. Microchemical Journal. 2005;**79**(1-2): 133-316

[19] Bhargava AK, Gupta R, Bhargava S, Paridhi. Effect of automobile exhaust on total N, P and heavy metals of road side sugarcane at district Saharanpur. Advances in Plant Sciences. 2003;**16**:557-560

[20] Ahmed F, Ishiga H. Trace metal concentration in street dusts of Dhaka City, Bangladesh. Atmospheric Environment. 2006;**40**:3835-3844

[21] Lu X, Wang L, Lei K, Huaing J, Zhai Y. Contamination assessment of copper, lead, zinc, manganese and nickel in street dust of Boaji, N.W china. Journal of Hazardous Materials. 2009;**161**:1058-1062

[22] De Miguel E, Llamas JF, Chacon E, Berg T, Larssen S, R›yset O, Vadset M. Origin and patterns of distribution of trace elements in street dust: Unloaded petrol and urban lead. Atmospheric Environment. 1997;**31**:2733-2740

[23] Turer D. Effect of non-vehicular sources on heavy metal concentrations of roadside soils. Water, Air, and Soil Pollution. 2005;**166**:251-264

[24] Shinggu DY, Ogugbuaja VO, Toma I, Barminas JT. Determination of heavy metals in street duct in Yola, Nigeria. African Journal of Pure and Applied Chemistry. 2010;**4**(1):17-21

[25] Abrahams PW. Soils: Their implications to human health. Science of the Total Environment. 2002;**291**(1-3):1-32

[26] Poggio L, Vrscaj B, Schulin R, Hepperle E, Ajmone Marsan F. Metals pollution and human bioaccessibility of topsoils in Grugliasco (Italy). Environmental Pollution. 2009;**157**(2): 680-689

[27] Ljung K, Selinus O, Otabbong E. Metals in soils of children's urban environments in the small northern European city of Uppsala. Science of the Total Environment. 2006; **366**(2-3):749-759

[28] Acosta JA, Cano AF, Arocena JM, Debela F, Martínez-Martínez S. Distribution of metals in soil particle size fractions and its implication to risk assessment of playground in Murcia City (Spain). Geoderma. 2009;**149**(1-2):101-109

[29] Popoola OE, Bamgbose O, Okonkwo OJ, Arowolo TA, Odukoya AT, Popoola AO. Heavy metals content in playground topsoil of some public primary schools in metropolitan Lagos, Nigeria. Research Journal of Environmental and Earth Sciences. 2012;**4**(4):434-439

[30] Okereke CJ, Amadi PU. Accumulation and risk assessment of heavy metal contents in school playgrounds in Port Harcourt Metropolis, Rivers State, Nigeria. Journal of Chemical Health and Safety. 2017. http://dx.doi.org/10.1016/j.jchas.2017.01.002

[31] USEPA. Risk-Based Concentration Table. Philadelphia PA. Washington DC, USA: United States Environmental Protection Agency; 2000

[32] Dooyema CA, Neri A, Lo YC, Durant J, Dargan PI, Swarthout T, Biya O, Gidado SO, Haladu S, Sani-Gwarzo N, Nguku PM, Akpan H, Idris S, Bashir AM, Brown MJ. Outbreak of fatal childhood lead poisoning related to artisanal gold mining in Northwestern Nigeria, 2010. Environmental Health Perspectives. 2012;**120**(4):601-607

[33] Su C, Jiang L, Zhang W. A review of heavy metal contamination in the soil worldwide: Situation impact and remediation techniques. Environmental Skeptics and Critics. 2014;**3**(2):24-38

[34] Wang Z, Chai L, Yang Z, Wang Y, Wang H. Identifying sources and assessing potential risk of heavy metals in soils from direct exposure to children in a mine-impacted city, Changsha, China. Journal of Environmental Quality. 2010;**39**(5):1616-1623

[35] Davidson CM, Duncan AL, Littlejohn D, Ure AM, Garden LM. Critical evaluation of the three-stage BCR sequential extraction procedure to assess the potential mobility and toxicity of heavy metals in industrially-contaminated land. Analytica Chimica Acta. 1998;**363**:45-55

[36] US EPA (United States Environmental Protection Agency). Risk Assessment Guidance for Superfund. Human Health Evaluation Manual. EPA/540/1-89/002, Vol. I. Office of Solid Waste and Emergency Response. 1989. Available from: http://www.epa.gov/superfund/programs/risk/ragsa/index.htm

[37] US DoE. The Risk Assessment Information System (RAIS). Argonne, IL: U.S. Department of Energy's Oak Ridge Operations Office (ORO); 2011

[38] Department of Environmental Affairs. The Framework for the Management of Contaminated Land, South Africa. 2010. Available from: http://sawic.environment.gov.za/documents/562.pdf [Accessed: 22 December 2017]

[39] Zheng N, Liu J, Wang Q, Liang Z. Health risk assessment of heavy metal exposure to street dust in the zinc smelting district, northeast of China. Sci Total Environ 2010;**408**(4):726-33

[40] World Health Organization. Growth reference data for 5-19 years. Geneva: World Health Organization; 2000. Available from: http://www.who.int/growthref/en

[41] US EPA (United States Environmental Protection Agency). Supplemental Guidance for Developing Soil Screening Levels for Super Fund Sites. OSWER 9355.4-24.2001. Office of Solid Waste and Emergency Response. 2001. Available from: http://www.epa.gov/superfund/resources/soil/ssgmarch 01.pdf

[42] Calabrese EJ, Kostecki PT, Gilbert CE. How much dirt do children eat? An emerging environmental health question. Comments Toxicol. 1987;**1**:229-41

[43] US EPA (United States Environmental Protection Agency). Exposure Factors Handbook. EPA/600/P-95/002F. Washington, DC: Environmental Protection Agency, Office of Research and Development; 1997

[44] Luo XS, Ding J, Xu B. Incorporating bioaccessibility into human health risk assessments of heavy metals in urban park soils. Science of the Total Environment. 2012;**424**:88-96

[45] U.S. Environmental Protection Agency. Framework for Determining a Mutagenic Mode of Action for Carcinogenicity: Review Draft. 2007. Available online: http: //www.epa.gov/osa/mmoaframework/pdfs/MMOA-ERD-FINAL-83007.pdf (accessed on 20 December 2017)

[46] Wright NJ, Thacher TD, Pfitzner MA, Fischer PR, Pettifor JM. Causes of lead toxicity in a Nigerian city. Archives of Disease in Childhood. 2005;**90**:262-626

[47] Adebamowo EO, Agbede OA, Sridhar MKC, Adebamowo CA. An evaluation of lead levels in residential paints sold in Nigeria markets. Indoor and Built Environment. 2006;**15**:551-554

[48] Yusuf AA, Arowolo TA, Bamgbose O. Cadmium, copper and nickel levels in vegetables from industrial and residential areas of Lagos. City, Nigeria. Food and Chemical Toxicology. 2003;**41**:375-378

[49] Tuzen M. Determination of heavy metals in soil, mushroom and plant samples by atomic absorption spectrometry. Microchemical Journal. 2003;**74**:289-297

[50] DPR. Environmental Guidelines and Standards of the Petroleum Industry in Nigeria. UK: Ministry of Petroleum Resources Lagos; 1991. pp. 35-75

[51] FEPA. Guideline and Standard for Environmental Pollution Control in Nigeria. Federal Republic of Nigeria; 1991. pp. 61-63

[52] WHO/WMO. Air Monitoring Programme Designed for Urban and Industrial Area Published for Global Environmental Monitoring System by UNEP. Geneva: WHO and WMO; 1971

[53] Onianwa PC, Jaiyeola OM, Egekenze RN. Heavy metals contamination of topsoil in the vicinities of auto-repair workshops, gas stations and motor-parks in a Nigerian City. Geneva, Switzerland: Toxicological and Environmental Chemistry. 2003;**84**(1):33-39

[54] Ljung K, Oomen A, Duits M, Selinus O, Berglund M. Bioaccessibility of metals in urban playground soils. Journal of Environmental Science and Health. 2007;**42**(9):1241-1250

[55] Cicchella D, de Vivo B, Lima A, Albanese S, Mcgill RA, Parrish RR. Heavy metal pollution and Pb isotopes in urban soils of Napoli, Italy. Geochemistry: Exploration, Environment, Analysis. 2008;**8**(1):103-112

[56] Osman K. Health effects of environmental lead exposure in children [Dissertation]. Stockholm: Karolinska Institute, Institute Environmental Medicine; 1998. (Swedish)

Chapter 19

Heavy Metal in Urban Soil: Health Risk Assessment and Management

Fei Li

Additional information is available at the end of the chapter

http://dx.doi.org/10.5772/intechopen.73256

Abstract

With the rapid development of industrialization and urbanization in the developing country, quite a few pollutants produced by anthropological activities enter into urban soil. A large area of urban soil was polluted by different kinds of pollutants. There have been many pollution events like itai-itai disease in the whole world, and many people suffered from urban soil pollution. As the place where people are highly concentrated, urban soil has a great deal with human health. Research has shown that the current soil environmental management and control system may be unable to adapt to changes of the current complex environment. In order to protect urban residents' health, the scientific and reasonable environmental health risk assessment and management system of environmental pollutants in urban soil have strong scientific, realistic, and strategic significances. Soil pollution including its characteristics, resources, and pollutant types were introduced. The definition of urban soil, current situation of urban soil pollutant, and assessment methods of urban soil were also introduced. Also, this chapter introduces main contents of health risk assessment and management on heavy metals in urban soil, and pointed out existing issues and possible research directions in this field, with the aim of offering references.

Keywords: urban soil, soil pollution, heavy metal, health risk assessment, risk management

1. Introduction

World Urbanization Prospects 2014 Revision published by United Nations Population Division showed that from 1950 to 2014, the global urban population increased from 746 million to 3.9 billion, accounting for 54% of the world's total population [1]. Despite the low urbanization rate in Asia, it is still the most populous city in the world due to its large population base, accounting for 53% of the total urban population in the world [2]. Urban is the place where human activities are highly concentrated. The population in urban is dense and unevenly distributed, which consumes a large amount of natural resources and produces a large amount of pollutants [3]. When the pollutants exceed the load of the urban environment, the urban environment would be polluted or even destroyed. With rapid urbanization and industrialization in developing country, relatively short-term and intensive human activities will bring a large number of organic pollutants (such as polycyclic aromatic hydrocarbons and phenols) and inorganic pollutants (such as heavy metals, arsenic compounds, and so on) into the urban environment. These pollutants will be accumulated in the urban environment, and may migrate or transform directly through the soil, water, atmosphere, and other environmental multi-media, directly or indirectly causing risk to human health and ecological security in urban [4–6]. Urban soil is an important part of urban ecological environment, and it is closely related to urban water, atmosphere, groundwater, and other environmental factors [7]. Urban soil is located in the central position of urban ecological environment, which is not only the sink of various pollutants, but also the source of various pollutants; so it is often used as an important indicator of comprehensive environmental quality of urban [8, 9]. Heavy metals are a kind of toxic pollutants, which are difficult to degrade in the environment. Heavy metal pollution in urban soils is difficult to control because of its wide sources, many forms, and complex migration [10]. The three wastes, traffic exhaust, sewage irrigation, improper fertilizer, and pesticide application have become the main sources of heavy metals in urban soils [4, 8, 11]. The land use type, the disturbance time, and degree of human activities are the important factors affecting heavy metal pollution in urban soil [8, 12]. In recent years, with the acceleration of urbanization and the occurrence of environmental public hazards related to heavy metal pollution in urban, the research, and practice of the characteristics, mechanism, assessment, and management of heavy metal pollution in urban soil have become major issues that all countries must be faced with. The research results will be of great significance to the sustainable development of urban economy and society, the maintenance of regional environmental equity, and the construction of ecological civilization in urban.

The Chinese National "12th five-year plan" for Environmental Protection and *National Environment and Health Action Plans (2007–2015)* pointed out that "With the rapid development of industrialization and urbanization in China, problems of environmental pollution affecting people's health and safety have become prominent. Protecting environment and ensuring human health have become the most pressing need of the people". Urban residents may directly or indirectly be exposed to heavy metals in urban soil through the ingestion of soil, skin contact, and inhalation, causing short-term harm or long-term risks [13]. By 2050, the global urban population will increase by 2.5 billion, and the proportion of global urban population will reach 66%, nearly 90% of which is concentrated in Asia and Africa. Countries with the largest increase in urban population will be India, China, and Nigeria in the future. From 2014 to 2050, these three countries will increase by 404, 292, and 212 million of the urban population, totally accounting for 37% of the global new urban population. In developing countries such as China, its urbanization rate has reached 53.73% and the total exceeding standard rate of soil was 16.1% [14]. The research on the relationship between heavy metals in urban soil and urban population health is in line with the actual needs of the country and people. According to the soil census data of 2005–2013 in China, the spatial distribution of the total noncarcinogenic risk factors of heavy metals (including cadmium, mercury, lead, chromium, arsenic, copper, zinc, and nickel) in three pathways of oral intake, skin contact, and respiratory inhalation was obtained, which can be seen from **Figure 1**. For children, 45% of 29 provinces, autonomous regions and municipalities in the mainland of China have a moderate potential noncarcinogenic risk, which did not exceed the acceptable noncarcinogenic risk limits [15]. Moreover, research and practice have shown that the current soil environmental management and control system based on *Soil Environmental Quality Standard (GB15618-1995)* has been unable to adapt to changes of the current complex environment and requirements of "green" socio-economic development. In 2016, the State Council promulgated *Action Plan for Soil Pollution Prevention and Control*. It indicates that the health risk of urban people exposure to heavy metal pollution in urban soil cannot be ignored.

Therefore, hopes are placed on the health risk assessment and management technology, which has been vigorously promoted and developed by the United States Environmental Protection Agency (USEPA) and the International Program on Chemical Safety (IPCS) in the past 20 years. A mature framework for health risk assessment and management of contaminated sites has been developed both at home and abroad. However, there is a big difference in land area, function, and characteristics between soil in contaminated sites and urban soil;

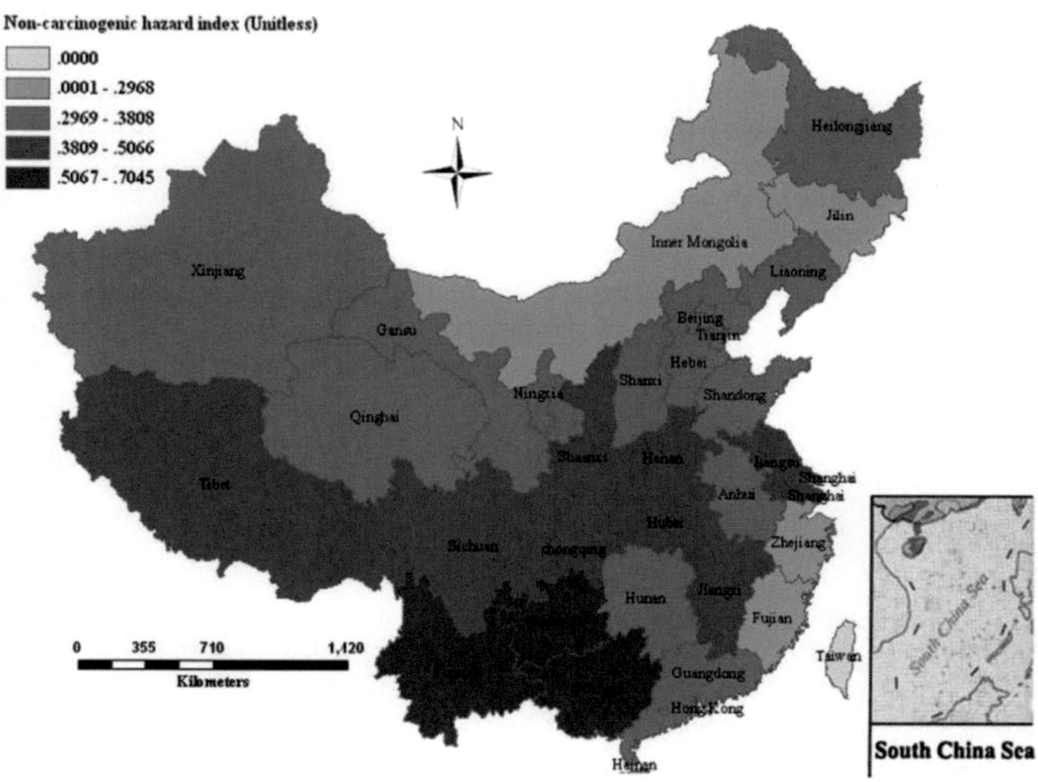

Figure 1. Spatial distribution of total noncarcinogenic risk factors of heavy metals [15].

mechanism of urban soil heavy metal pollution is more complex and involved more factors, so there is still no formation of a scientific and efficient urban soil environmental health risk assessment and management system in China. The World Bank Overview of the Current Situation on Brownfield Remediation and Redevelopment in China points out that at present, China has a large area of Brownfield and a limited amount of restoration funds, and a highly efficient and prioritized Brownfield risk rating system should be established according to the actual situation on Brownfield in China. Considering the latest researches on soil science and the significant differences in land use status, receptor spatial distribution, exposure characteristics, risk assessment and management, and economic input degree between different countries, the key technologies of health risk assessment, risk source analysis and risk quantification management of heavy metal in soil should be further improved, integrated, and developed.

In summary, with the aim of protecting urban residents' health, the scientific and reasonable environmental health risk assessment and management of heavy metal pollution in urban soil have strong scientific, realistic, and strategic significances. It helps to decision-making of regional soil remediation, establishment of regional soil standard value and action value, and establishment and improvement of relevant policies and regulations. Taking scientific health risk assessment and China's national situation as the starting point, how to construct a health risk assessment and management system of environmental pollutants in urban soil with Chinese characteristics and easy to be popularized, has becoming the basic research subject that needs to be explored. The State Council promulgated *Action Plan for Soil Pollution Prevention and Control* in 2016, so the market scale of health risk assessment and remediation of pollutants in urban soil in China will be more than 1000 billion Chinese yuan. It is undoubtedly a huge push to promote the development of related research projects, which obviously makes this study to have strong social and economic prospects.

2. Heavy metal pollution in urban soils

2.1. Introduction of soil pollution

2.1.1. Characteristics of soil pollution

Soil is a historical natural body. It is a loose and heterogeneous aggregate layer with vitality and productivity, which is located on the earth's surface and in the bottom of phreatic area. It is a component of the earth system and a key element of environmental quality control [12]. Soil environment is a complex system, which contains solid phase, liquid phase, and gas phase, and its components change greatly. Moreover, the interaction between the soil environment system and the plant system that grows in soil constitutes the soil-plant ecosystem [12, 16].

However, soil pollution becomes a common thing with the development of industry, and it happens mainly due to human factors. Substances or preparations that are harmful to human beings and other organisms are applied to the soil intentionally or unintentionally, and the content of the new component will be increased or a component becomes significantly higher than its original content, finally results in the existing or potential deterioration of soil environment.

According to soil science, soil pollution has three types of characteristics [12, 16, 17]:

I. Concealed and latent. Water or air pollution is intuitive. It could be easily found by people when water or air pollution is heavy. On the contrary, soil pollution has characters of concealment and latency. It commonly reflected the growth status of crops including grain, vegetables, fruits, and grass and health status of related human or animals, and there is a gradual cumulative process from happening pollution to causing harm.

II. Irreversible and long-term effects. Once the soil is contaminated, it is difficult to recover. The process of soil pollution by heavy metals is irreversible.

III. Consequences are serious. Once the soil is polluted, biodiversity, biological cycles, and water cycles (including water quality and water cycle processes) are bound to be affected accordingly. Besides, because of the concealment and irreversibility of soil pollution, health of animals and human beings often suffer from soil pollution via food chain.

2.1.2. Sources of soil pollution

Soil pollution sources can be divided into natural sources and anthropogenic sources. Natural sources of soil pollution relate to natural disasters such as volcano eruption that will produce harmful substances. These harmful substances will be discharged into the natural environment and result in soil pollution. Anthropogenic sources refer to the pollution produced by anthropogenic activities such as sewage irrigation, utilization of solid waste, use of pesticides and chemical fertilizers, atmospheric deposition and so on, which are the main objects of scientific research.

I. Sewage irrigation. Domestic sewage and industrial wastewater contain nutrient elements that plants growth needed, such as nitrogen, phosphorus and potassium. Therefore, using sewage probably to irrigate farmland can gradually increase production. However, the sewage also contains heavy metals, phenols, cyanide, and many other toxic and harmful substances. If the sewage is used for agricultural irrigation directly without pretreatment, toxic and harmful substances in the sewage will enter into the farmland and pollute the soil.

II. Utilization of solid waste. Industrial waste and municipal solid waste are solid pollutants in soil. For example, a variety of agricultural plastic film is widely used as in greenhouse and plastic mulching. If the management and recovery are not good, a large number of pieces of plastic film will scatter in the field and cause white pollution. Such solid pollutants are not easily evaporated, volatilized, or decomposed by microorganisms, which will remain in the soil for a long time.

III. Use of pesticides and chemical fertilizers. The application of chemical fertilizer is an important measure to increase agricultural yields, but unreasonable use of chemical fertilizer will result in soil pollution. Long-term use of nitrogen fertilizer will destroy soil structure, result in soil compaction, deteriorate biological properties, and finally affect the yields and quality of crops. Part of the pesticide sprayed on the object absorbed by plants escapes into the atmosphere. About half of pesticide for object spraying are scattered on farmland, and this part of pesticide and pesticide directly applied to the farmland constitute the basic sources of pesticides in soil.

IV. Atmospheric deposition. The harmful gases in the atmosphere are mainly poisonous waste gases discharged from industry. Its pollution surface is big, which will cause serious pollution to the soil. Industrial waste gas pollutions include gas pollution and aerosol pollution. Gas pollution is usually caused by sulfur dioxide, fluoride, ozone, nitrogen oxides, and hydrocarbons. The reason for causing aerosol pollution is that solid particles (such as dust, soot, and smoke) and liquid particles (such as mist) enter into soil via deposition and precipitation.

2.1.3. Types of soil pollution

There is no strict classification of soil pollution. Considering from the material properties, soil pollution can be divided into organic pollution, inorganic pollution, microbial pollution, and radioactive pollution.

I. Organic pollution. Natural organic pollutants and synthetic organic pollutants can cause organic pollution. Synthetic organic pollutants include organic wastes and pesticides. Organic pollutants in soil have adverse effects on the growth of crops and the survival of soil organisms. After people contacted with polluted soil, red rash appears on their hands with the feeling of nausea and dizziness. The application of pesticide in agricultural production has received good results, but its residues have contaminated the soil and corresponding food chain.

II. Inorganic pollution. Some inorganic pollutants come into the soil with the natural processes such as crustal change, volcanic eruption and rock weathering, while some enter into the soil with the human production and consumption activities. Human production activities such as mining, smelting, machinery manufacturing, building, and chemical producing produce a great number of inorganic pollutants including harmful heavy metals and their compounds, acids, alkalis, and salts.

III. Microbial pollution. Soil biological pollution means that one or more harmful microbial populations invade the soil from the outside, and multiply in a sharply speed, destroy the original ecological balance, and cause adverse effects on soil ecosystem and human health. The main sources of soil biological pollution are untreated feces, garbage, municipal sewage, and waste of feed yard and slaughterhouse. Soil organisms may do harm to human health, and some plant pathogens that survive in the soil for a long time will seriously harm plants, and resulting in agricultural yield reduction.

IV. Radioactive pollution. Radioactive pollution refers to that radioactive materials emitted by human activities entering into soil makes the radioactivity of soil become higher than its background level. Radioactive pollutants refer to various radionuclides, radioactivity of which has nothing with their chemical states. Radionuclides can pollute soils in many ways. Radioactive wastewater to the ground, radioactive solid waste buried in the underground, and radioactive emission accident in nuclear power plant will cause serious soil pollution in some areas. Deposition of radioactive substances in the atmosphere, application of phosphate fertilizer containing uranium, radium and other radionuclides, and irrigation of water with radioactive river water will also cause soil radioactive pollution. Although the pollution is generally mild, its scope is relatively large. Polluted by radioactive materials, soil can produce alpha, beta, and gamma rays by radioactive decay. These rays can penetrate human tissues, damage cells, cause external radiation damage, and enter the human body through respiratory system or food chain, causing internal radiation damage.

2.2. Overview of urban soil

2.2.1. Definition of urban soil

Bockheim, the first man used the word of "urban soil", believed that urban soil is a kind of urban or suburban soil with thickness of 50 cm formed by the man-made mixture, landfill, or pollution of land [18]. De Kimpe and Morel [19] generalizes urban soil as the soil in urban and suburban areas that are strongly disturbed by human activity. The mainstream explanation of urban soil according to Chinese scholars is that urban soil developed under long-term disturbance or direct assembly of human activities and under the special environmental background of the urban [20]. Comparing with the natural soil and agricultural soil, urban soil not only inherits some characteristics of natural soil, but also has its unique soil environment and forming process. Urban soil has special physical and chemical properties, nutrient cycling and soil biological characteristics. Urban soil is widely distributed in residential areas, urban rivers, roads, parks, suburban areas, stadiums, landfill sites, abandoned factories, and so on. In view of the fact that heavy metals in urban soil under all land use types are likely to be exposed to nearby receptor populations and cause health risks, urban soil defined in this study include urban soil, forest soil, agricultural soil, and potentially contaminated soil [21].

2.2.2. Current situation of urban soil pollution

The earliest cases of heavy metal pollution in foreign was the copper poisoning caused by drainage from the Ashio Copper Mine in Tochigi prefecture, beginning as early as 1878. It was found that the extremely painful itai-itai disease and Minamata disease were caused by mercury and cadmium pollution. Heavy metal pollution caused by the food chain has begun to be much concerned. In the early 1970s, British have carried out studies on heavy metals in soil and dust in London and other big city. It was found that heavy metal pollution in urban soil is closely related to industrial activities and emission of automobile exhaust, and heavy metals in urban surface soil and road dust can be used as indicators of urban air pollution. In 1990s, urban soil research has gradually become new hot topic in the field of soil research all over the world. Heavy metal pollution in urban soil in cities such as major cities in British, major cities in America, Naples and Sicily in Italy, Seville in Spain, Munich in Germany, Hanoi in Vietnam, and Bangkok in Thailand have been studied by many a scholar. Pb, Zn, and Hg were the main pollution factors in Sicily City of Italy, and the enrichment coefficient of Pb was 5–10, and Hg was 35. The correlation between soil heavy metal content and soil properties was determined, and the heavy metal pollution's relationships with urban scale, population density, traffic flow, land use types, and history were discussed. Therefore, the International Union of Soil Sciences established the Soils in Urban and Industrial, Traffic and Mining Areas (SUITMA) working group. Up to now, three international academic conferences had been held, of which heavy metal pollution in urban soils was the main topic.

In 2014, the Ministry of Land and Resources and the Ministry of Environmental Protection jointly released the National Soil Survey Report. The results showed that the soil environment in the whole country was not optimistic and the soil pollution was terribly serious in some areas. The soil environment quality of cultivated land was worth serious consideration while the soil environment problems of abandoned mining industry were prominent [14]. Compared with standard values, the total exceeding rate of soil sampling sites in China was 16.1%, and the proportions of slight pollution, mild pollution, moderate pollution, and severe pollution were 11.2, 2.3, 1.5 and 1.1%, respectively. For exceeding sampling sites, the main type of pollution was nonorganic pollution (82.8%), the second was organic pollution, and the least compound pollution. The problem of heavy metal pollution in urban soil in China is outstanding. Liao et al. [22] studied on distribution characteristics and pollution sources of heavy metals in soils of Jiangsu Province. It was found that contents of Cd and Hg in soil in the South of Jiangsu were generally higher while the soil in Northern Jiangsu was rich of As. Pollution degrees of heavy metals in soil in Suzhou and Wuxi were relatively serious, mainly due to the human activities in the process of industrialization and urbanization. Guo et al. [23] studied on the characteristics of heavy metal pollution in soil under different functional zones of Hohhot and it showed that average contents of Cu and Zn were 2.33 times and 1.85 times than their background values, respectively. Soil heavy metal pollution near the business district and urban roads was more serious, which mainly comes from traffic pollution and living waste dumps.

The population in urban is much more intensive, heavy metals in the soil are easily enter into bodies of human beings via various exposure pathways directly or indirectly. Those heavy metals can interact strongly with proteins and enzymes, and then make proteins and enzymes lose activity. Heavy metals can also be accumulated in organs of human body. Once the accumulated content of heavy metals exceed the limit of the human body can be tolerated, the human body will cause great harm or risk, it will lead to acute poisoning, subacute poisoning, chronic poisoning, and so on [24, 25].

2.3. Assessment methods of heavy metal pollution

2.3.1. Geo-accumulation index

To assess the enrichment degree of heavy metals in the soils, the geo-accumulation index (I_{geo}) was used. I_{geo} can be calculated by the following formula [26–29]:

$$I_{geo} = log_2(k\,C_i/B_i) \tag{1}$$

where C_i is the actually measured concentration of the heavy metal in the samples. k is corrected coefficient, which take account of variation of background value caused by anthropogenic influences or litho logic variations in the soil (in general $k = 1.5$). B_i is the reference value of heavy metal concentration in soil. I_{geo} is classified into seven levels and its corresponding contamination degrees of heavy metal are shown in **Table 1**.

2.3.2. Potential ecological index

In order to evaluate the potential ecological risk caused by heavy metals in soil, the potential ecological risk index (PEI) was utilized. The PEI was established by Hanson to assess the characteristics of heavy metal contaminants on the basis of sedimentary theory [30, 31].

$$RI = \sum_{i=1}^{n} E_r^i, E_r^i = T_r^i \cdot C_r^i = C_D^i \cdot C_B^i \quad (2)$$

where RI is the sum of the potential risk of individual heavy metal, E_r^i is the potential risk of individual heavy metal, T_r^i is the toxic-response factor for a given heavy metal, value of T_r^i for Cd, Ni, Zn, Cu, and Cr are 30, 5, 1, 5, and 2, respectively [30, 32, 33]. C_r^i is the contamination factor, C_D^i is the present concentration of heavy metals in soil, and C_B^i is the reference value of heavy metal concentration in soil. Hanson defined five categories of E_r^i, and four categories of RI, as listed in **Table 2**.

Level	Value	Extent of pollution
I	$I_{geo} \leq 0$	Uncontaminated
II	$0 < I_{geo} \leq 1$	Uncontaminated to moderately contaminated
III	$1 < I_{geo} \leq 2$	Moderately contaminated
IV	$2 < I_{geo} \leq 3$	Moderately contaminated to heavily contaminated
V	$3 < I_{geo} \leq 4$	Heavily contaminated
VI	$4 < I_{geo} \leq 5$	Heavily to extremely contaminated
VII	$I_{geo} > 5$	Extremely contaminated

Table 1. Seven levels and its corresponding contamination degrees.

E_r^i value	Grades of eco-risk of single metal	RI value	Grades of potential eco-risk of the environmental
$E_r^i < 40$	Low risk	$RI < 150$	Low risk
$40 \leq E_r^i < 80$	Moderate risk	$150 \leq RI < 300$	Moderate risk
$80 \leq E_r^i < 160$	Considerable risk	$300 \leq RI < 600$	Considerable risk
$160 \leq E_r^i < 320$	High risk	$RI \geq 600$	Very high risk
$E_r^i \geq 320$	Very high risk		

Table 2. Indices and corresponding grades of potential eco-risk of heavy metals.

3. Health risk assessment of heavy metals in urban soils

3.1. Health risk assessment

Current soil environmental management system can control soil heavy metal pollution to some degree, but it is obviously the lack of consideration on the relationship between soil heavy metal and urban human health. There is no complete set of related laws and regulations, so its efficiency of quantitative control is low. In the past 20 years, driven by Organization for Economic Co-operation and Development (OECD), World Health Organization (WHO), especially the United States Environmental Protection Agency (USEPA), the International Program on Chemical Safety (IPCS), and European Center for Ecotoxicology and Toxicology of Chemicals (ECETOC), the health risk assessment management has developed rapidly [13]. The European Union, the United States, Canada, and Japan have issued relevant laws and regulations, and health risk assessment has become an important basis and scientific reference at national even international level for dealing with the problem of chemical pollution risk management [34]. It provides important references for the establishment of health risk assessment and management system of urban soil.

Health risk refers to the possibility of human exposure to environmental substances causing injury, disease, or death. In 1983, the United States National Academy of Sciences (NAS) published the "Red Book" named *Risk Assessment in the Federal Government: Managing the Process,* which primarily proposed the definition of health risk assessment. It is defined as a characteristic description of the potential adverse health effects of human exposure to environmental hazards. Health risk assessment includes several elements [13, 35, 36]: an assessment of potential adverse health effects based on epidemiological, clinical, toxicological, and environmental studies, extrapolations from assessment results to predict and estimate types and extent of human health effects under certain exposure conditions, determination of the number and characteristics of human exposed to different intensities and times, and summarization of existence and overall level of public health problems. Therefore, risk assessment has the following four steps: hazard identification, dose-response assessment, Exposure assessment, and risk characterization. Then, subsequent risk management is carried out based on risk assessment results. The basic contents of health risk assessment and management are shown as in **Figure 2**.

3.1.1. Hazard identification

Hazard identification is the first step of a human health risk assessment, which aims to identify the character and intensity of risk sources. The NAS defines it as "*Hazard Identification is the process of determining whether exposure to a stressor can cause an increase in the incidence of specific adverse health effects (e.g., cancer, birth defects).*" The types of hazard can be physical, chemical, or biological, and when human has enough exposure dose, it can cause injury, disease, or death. Scientists assess the effects of harmful substances on human and laboratory animal health, understand the characteristics of the substance, identify its potential health

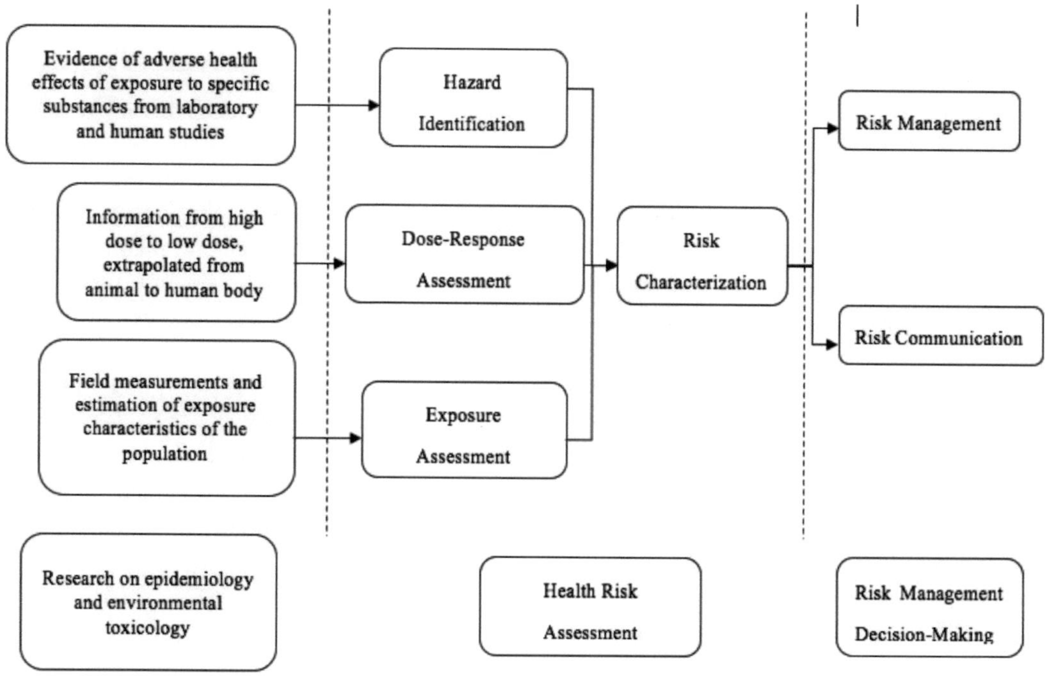

Figure 2. Risk assessment and management of environmental health.

problems, and clearly demonstrate its statistically or biologically significant toxicity [13, 37]. Study on the human directly exposed to hazardous substances can provide the best evidence, but because there are few human clinical trials data, animal experiments, in vitro experiments and activity relationship of chemical structure are often used to estimate the possible health hazards of human exposure to chemicals.

3.1.2. Dose-response assessment

Dose-response assessment is a process of quantitative estimation on the relationship between the exposure level of harmful factors and the incidence of health hazard effect of exposed humans, which is the cornerstone of quantization of health risk assessment [13, 37]. Dose-response assessment belongs to the category of toxicology research, and it has an important creed, that is "Dose makes the poison". For example, the level of clinical dose can determine whether the drug is a therapeutic drug or a potentially lethal drug. Dose-response relationship can be broadly divided into two categories. The first one is the relationship between the dose of exposure to hazardous pollutants and the biological intensity response of individual. The second one is the relationship dose of hazardous pollutants and the proportion of individuals who respond to a particular response in a group like mortality and tumor incidence. Each chemical substance has different dose response relationship according to its toxicity.

3.1.3. Exposure assessment

Exposure assessment refers to quantitative or qualitative methods for estimating or calculating exposure dose, exposure frequency, exposure time, and exposure path [38]. The main exposure pathways include ingestion, derma contact, and inhalation. However, it must be pointed out that no matter how dangerous the pollutant is, "*No exposure, no health risk*" [13]. Exposure assessment often requires exposure factors (including body height, weight, inhalation rate, exposure frequency, and so on) of the local receptor population, and factors related to the transport of pollutants in the environmental media under the local environmental conditions. The United States, Japan, South Korea, and other countries have established a national exposure factors database. Due to China's vast territory, large population and the great differences in life habit, it is necessary to establish our own database of exposure factors gradually. The Ministry of environmental protection published *Exposure Factors Handbook of Chinese Population (adults)* in 2013, which filled the gaps in the national exposure factor manual in China. However, it is clear that the more systematic national exposure factor database is still in the stage of data accumulation, which also brings some uncertainty to the health risk assessment in China [39].

3.1.4. Risk characterization

As the last step of risk assessment, risk characterization based on the data and analysis results of the three steps mentioned above. Finally, the probability of the occurrence of harmful results, the acceptable risk level and the uncertainty of the evaluation results are quantified [13]. Risk characterization is a bridge linking risk assessment and risk management, and risk managers can use risk characterization results to formulate strategies of risk control and pollution remediation. Risk communicators can use risk characterization conclusions to inform the types, sizes, and possibilities of adverse health effects among stakeholders, and disseminate knowledge about risk prevention.

3.1.5. Risk management

Health risk assessment requires collaboration among multiple scientific teams. Similarly, risk management is also a multidisciplinary process. Risk management is a decision-making process base on health risk assessment. It is necessary to balance the political, social, economic and technical information and other related information, to develop, analyze and compare management methods, and ultimately choose appropriate management methods for health risks. That is, to reduce the risk to acceptable levels with acceptable costs [13].

3.2. Researches and issues on health risk assessment and management

3.2.1. Hazard identification

With the development of industrial technology and social economy, the types of materials that people produce and discharge into the environment are increasing rapidly, and the amount of harmful substances that can be analyzed and detected is also increasing. Although people pay more and more attention to the environmental pollution and its health damage, it is impossible to carry out comprehensive management of the pollutants in the environment because of the limit of manpower, material resources, financial resources, and science and technology. At the same time, researches on chemical toxicology have shown that not all of the pollutants have the same harm to human health. Therefore, targeted research and treatment of pollutants with high health hazard effect have gradually become an effective environmental management strategy. Since the mid-twentieth century, many countries, regions and international organizations have developed their own screening methods for pollutants, and compiled their own list of environmental priority pollutants, and applied them to the management of environmental pollutants in practice [13, 40]. The European community, the United States, Japan, Germany, Holland, and other countries have published list of environmental hazardous substances one after another, which plays an important role in promoting environmental protection and governance.

With a short period of about 30 years, China has achieved development, which can catch up with the development of developed countries for 100 years, and achieved brilliant economic achievements. However, due to the extensive development mode and many other reasons, the ecological environment of our country has suffered serious damage, which leads to the environmental problems that should occur at different stages are reflected and erupted in a short time. In view of this, the State Environmental Protection Administration of China presided over research and proposed "Black list of pollution substance for preferential control in water" in the period of seventh five-year. However, over time, the characteristics of China's environmental problems have undergone profound changes, the existing list obviously not able to fully reflect the current situation of environmental risk and research level, and failed to cover the comprehensive environment system of water, soil, and atmosphere [40, 41]. In view of the new changes in the national and world environmental issues, with the reference of the developed countries' catalog making methods, the State Environmental Protection Administration (SEPA) published the *National catalog of Pollutants' Environmental Health Risk* in 2007. It represents the significant progress in the field of hazard identification in China, and provides the basis and method for the decision-making of government, monitoring of environment, formulation of environmental emergency plans, and emergency treatment of environmental pollution accidents.

However, after more than half a century of development, directory management has gradually exposed its limitations. Most of the early catalogs are static, which are balancing results of environmental pollution status, understanding level and management objectives at the beginning of the catalog editing. It cannot reflect the long-term dynamic changes of various factors, making many catalogs cannot effectively support the current environmental management work. Besides, the early catalogs commonly compiled by a single country or region. With the deepening understanding of transboundary migration of pollutants, people have realized that it is necessary to take the joint action of river basin and region to effectively cope with the common environmental problems. Therefore, the hot research spots in the field of hazard identification are mainly [13, 40, 42]: (a) new screening methods of priority pollutants considering pollutants migration and transformation characteristics, (b) construction of dynamic risk list updating mechanism in regional environment, (c) and risk identification of non-occupational long-term exposure with low dose.

3.2.2. *Dose-response assessment*

I. No threshold effect (carcinogenic effect) [13]. In view of humanitarian considerations, most carcinogenic chemicals are based on low dose extrapolation models to assess the risk probability of human exposure. Varieties of evidences have been applied to dose-response assessment of carcinogens, including human epidemiological data, which should be regarded as primary basis. In the absence of appropriate human clinical studies, data on animal species that close to humans should be used, and in the long-term animal studies, the most sensitive and biologically acceptable data should be given to the greatest extent of attention.

II. Threshold effect (noncarcinogenic effects) [13]. The reference dose (RfD) is an important basic factor, which means that when the dose is below this level, the risk of harmful effects will not be expected. At present, the following methods are used to evaluate the dose-response effect of threshold dose: firstly, determine the key toxic effects (the initial deleterious effects at this dose) and the highest dose that does not occur harmful effects (no observed adverse effect level, NOAEL) through reading literatures. Then, the NOAEL is divided by the uncertainty factor to obtain the safety limit value, and the uncertainty factor range from 10 to 1000. The uncertainty factor expresses a variety of internal uncertainties related to the existing data.

The focus of toxicity testing of environmental chemicals is to determine the safety level of human exposure to toxic chemicals. At present, the test has been extended from simple acute and subacute tests to full consideration of various toxicity data, including acute, subacute and chronic toxicity, and some specific toxicities, such as carcinogenicity, mutagenicity, reproductive toxicity, immune toxicity, neurotoxicity, skin toxicity, and other organ tests. In addition to these studies, the toxic dynamics of chemicals and their related mechanisms in the tissues, cells, subcellular, and receptor levels are also expected to further elucidate the toxic mechanisms of environmental chemicals and provide scientific references for potential hazard assessment of human bodies. In recent years, the main research focuses of dose-response assessment are [42, 43]: (a) dose-response relationship of chemical mixture interaction, (b) study on biological dynamics characteristics and establishment of bio dynamic model, (c) study on toxicity mechanism and toxicity kinetics, (d) and study on toxicity mechanism of cellular and subcellular.

3.2.3. *Exposure assessment*

Humans may expose to a variety of pollutants through different kinds of exposure pathways, which can be broadly divided into external exposure and internal exposure [39]. External exposure refers to the concentration of a substance in contact with the receptor, which can be understood as the gastrointestinal epithelium, the lung epithelium during respiration, and the epidermis in contact with the skin [37]. Internal exposure refers to the amount that a substance has been absorbed, that is, the amount that has entered the systemic circulation of the receptor. Bioavailability is defined as the proportion of absorbed in external dose.

Exposure assessment research mainly includes the following aspects [39, 44–47]: (a) analysis and characterization of exposure environment. That is, to describe characteristics of common environmental physics and population. The climate, vegetation, groundwater literature, and surface water in regional need to be determined, and the target receptor population also should be determined, and characteristics of exposure population like the location of the source of pollution and the activity characteristics of the population should be described. Moreover, it is necessary to take into account characteristics of both the current exposure population and future exposure population. (b) Analysis of exposure pathways. According to the location and release of pollution sources, the migration and transformation process of possible chemical substances in multi-media environment and the potential exposure position and activities of people, to analyze and determine exposure points and approaches via each exposure pathway (such as direct skin absorption, ingestion, and so on). (c) Quantification of exposure. To quantify the exposure size, exposure frequency, and exposure duration of each pathway, step of exposure quantification including estimating exposure concentration and calculating intake. (d) Estimation of exposure concentration. To determine concentrations of chemical pollutants in the exposure period, and the exposure concentration is generally estimated by monitoring data or chemical environmental fate model. (e) Calculation of intake. To determine specific chemical exposure dose via each exposure route, and exposure dose is expressed in terms of unit weight per unit time, and the mass of body exposure to chemicals. In recent years, hot research spots in field of exposure assessment mainly include: (a) investigation on regional epidemiology or occupational exposure factors, (b) study on the mechanism and mass model of chemical migration in atmosphere, (c) study on migration mechanism and model of chemicals in water and sediment, (d) study on migration mechanism and model of chemicals in soil, (e) and study on chemo taxis mechanism and corresponding integrated exposure model of chemicals in multi-media environment.

3.2.4. *Risk characterization*

Risk characterization includes qualitative risk characterization and quantitative risk characterization [13]. The qualitative risk characterization use semi quantitative words such as "negligible", "slight", "medium", or "serious" to describe the degree of risk. Quantitative risk characterization can quantify the degree of risk by digital expression, which can more intuitively and effectively describe the degree of risk, and facilitate the screening and ranking of the risk of pollution factors, providing scientific reference for decision makers. The risk characterization methods both at home and abroad are established based on the framework of risk assessment system of USEPA. The quantitative risk characterization can be divided into the carcinogenic risk characterization and noncarcinogenic risk characterization. The conclusion of environmental risk characterization will provide a scientific basis for the formulation of environmental standards or environmental management strategies at the national, local, and organizational levels. And the formulation and implementation of environmental standards is the starting point of environmental administration and an important basis for environmental management.

3.2.5. Risk management

The risk management process is triggered by the concerns of risk of specific uses or special scenarios of chemicals. Environmental risk assessment and risk management are closely related, but their process is different. The characteristics of the risk management decision affect the breadth and depth of risk assessment, while risk assessment provides a scientific basis for risk management. The last game between risk assessment and legal, political, social, and economic and technology based on the present situation and put forward reasonable control measures. Risk management is based on the relationship of risk assessment with law, politics, society, economy, and technology, and puts forward reasonable control measures. Risk management includes the following tasks: classification of risks, determination of risk reduction measures and benefit analysis of risk management, risk reduction, and monitoring and review. In recent years, the new development of risk management research mainly includes [13, 48, 49]: (a) focus on risk reduction and responsibility concerns, that is, the establishment of appropriate laws and regulations, and the responsibility of the original competent authorities transferred to the manufacturing risk industry (manufacturers and importers and its users), (b) risk communication and stakeholder participation. Risk communication is the link between environmental risk assessment and risk management, and the active participation of stakeholders will help to ensure that the assessment results and management practices are better understood and implemented, (c) the integrated framework of information platform for environmental risk assessment and management system. To analyze human health comprehensively, environment, ecological security and socio-economic factors has been one of the main trends in this field, (d) risk cognition. Because the risk (or interest) cognitions of individuals, public, enterprises or employees are different, which will change with time as well, it is of great significance to explore different groups for their own risk assessment or cognitive inertia.

3.2.6. Uncertainty analysis of system

The controversies about risk assessment often revolve around some differences, which aim at the assessment of incomplete and uncertain data and the nature, explanation and demonstration of the methods and assessment models. When science is used for management purposes, decision makers need to know not only the existing scientific knowledge, but also to understand the uncertainty part and blank part of these knowledge, and to distinguish between uncertainty and variability. In view of risk assessment, inaccurate and incorrect parameters, parameter variability, imperfect or unreasonable simplified model theory, and unreasonable establishment of receptor exposure scenarios may lead to uncertainty [13, 34]. In 1999, Cullen and Frey divided the uncertainty in risk assessment into parameter uncertainty, model uncertainty and variability. Because parameter uncertainty is easier to quantify, most of the uncertainty research focus on parameter uncertainty. Zhang Yinghua, Liang Jie, and Babendreier, respectively applied Monte-Carlo algorithm, Bayesian Monte-Carlo neural network method, Monte-Carlo method, and fuzzy mathematics method to effectively control and quantify the parameter uncertainty in assessment to a certain extent, ignoring uncertainty and variability of the model in the evaluation process. Wang Yongjie, Moschander,s and Karuchit found that the influence of model uncertainty and variability on the credibility of assessment results is much higher than that of parameter uncertainty, and made a qualitative/semi-quantitative analysis. Nevertheless, because model uncertainty and variability are not so easy to quantify, related researches tend to move along at a slow.

To sum up, on the basis of risk assessment system in the United States, the European Union, Australia, Canada, Japan and other countries, referring to a large number of literature, the deficiency and difficulty of in the field of soil environmental pollutants health risk research are mainly concentrated in the following aspects:

I. The urban is the center of the crowd gathering, which is easy to occur environmental public hazards. Present researches on the urban soil environment and population health is not deep enough at home and abroad.

II. In the practice of risk assessment and management, how to obtain the regional characteristics of risk assessment and management more efficiently under limited economic budget still needs further exploration.

III. In the field sampling analysis of health risk assessment, it is often used to monitor the distribution of sites, which requires a large amount of human and material input. In view of the significant temporal and spatial variability of soil environment, it is necessary to find out more efficient sampling methods.

IV. A large number of studies show that the chemical fractions of heavy metals in soil can more accurately characterize soil metal bioavailability and mobility, so how to embed heavy metal fraction parameters into the classical model of heavy metal content evaluation needs further study and verification in practice.

V. For health risk assessment, the scientific identification of exposure pathways of different receptor populations is the key to determine the credibility of risk assessment. Most of the existing studies base on questionnaire surveys or theoretical analysis hypotheses. However, these two methods have high cost and high uncertainty, so it is a new research direction to find a low cost and high credibility exposure path analysis and determination method.

VI. In the risk assessment systems implemented in China and abroad, most of them only make the relevant regulations on the uncertainty control of the parameters in the assessment process ignoring model uncertainty and variability. Researches show that influences of model uncertainty and variability on the credibility of assessment results is much higher than that of parameter uncertainty. Because model uncertainty and variability are not so easy to quantify, related researches tend to move along at a slow.

VII. In most of the current studies, the pollutant source analysis is only multivariate statistical analysis of regional pollutant content data. But considering the complexity of urban pollution, single pollution data statistical analysis is limited by the amount and quality of pollution data, which cannot fully reflect the problem obviously. Therefore, it is necessary to take into account ecological environment, human health and social economy and further study the correlation of risk sources, so as to help decision makers make more reasonable and accurate risk management decisions.

VIII. In the process of current health risk assessment, the participation rights and discourse rights of stakeholders are obviously unequal. How to embed the concept of environmental equity into environmental risk assessment system and further develop feasible technical methods has become a new direction of research.

Acknowledgements

This chapter was financially supported by the Humanities and Social Sciences Foundation of Ministry of Education of China (17YJCZH081) and the Science and Technology Research Project of Hubei Provincial Education Department (B2017601).

Author details

Fei Li

Address all correspondence to: lifei@zuel.edu.cn

Research Center for Environment and Health, Zhongnan University of Economics and Law, Wuhan, China

References

[1] Shi JR, Chen KL. Urban environmental safety. Beijing: Chemical Industry Press; 2010

[2] Li QL. Characteristics of Heavy Metals in Soil and Crops in Regional Ecosystems: A Case Study of Chongqing. Beijing: China Environmental Science Press; 2010

[3] Wang Y, Chen YC, Li ZP. Contamination pattern of heavy metals in Chinese urban soils. Environmental Chemistry. 2012;**31**(6):763-770

[4] Cheng HG, Li M, Zhao CD, Li K, Peng M, Qin AH, Cheng XM. Overview of trace metals in the urban soil of 31 metropolises in China. Journal of Geochemical Exploration. 2014;**139**:31-52

[5] Li F, Huang J, Zeng G, Huang XL, Liu WC, Wu HP, Yuan YJ, He XX, Lai MY. Spatial distributions and health risk assessment of heavy metals associated with receptor population density in street dust: A case study of Xiandao District, Middle China. Environmental Science and Pollution Research. 2015;**22**(9):6732-6742

[6] Li F, Huang J, Zeng G, Yuan XZ, Li XD, Liang J, Wang XY, Tang XJ, Bai B. Spatial risk assessment and sources identification of heavy metals in surface sediments from the Dongting Lake, Middle China. Journal of Geochemical Exploration. 2013;**132**:75-83

[7] Qishlaqi A, Moore F, Forghani G. Characterization of metal pollution in soils under two landuse patterns in the Angouran region, NW Iran: A study based on multivariate data analysis. Journal of Hazardous Materials. 2009;**172**(1):374-384

[8] Mihailović A, Budinski-Petković L, Popov S, Ninkov J, Vasin J, Ralevića NM, Vučinić Vasića M. Spatial distribution of metals in urban soil of Novi Sad, Serbia: GIS based approach. Journal of Geochemical Exploration. 2015;**150**:104-114

[9] Wang XT, Chen L, Wang XK, Lei BL, Sun YF, Zhou J, Wua MH. Occurrence, sources and health risk assessment of polycyclic aromatic hydrocarbons in urban (Pudong) and suburban soils from Shanghai in China. Chemosphere. 2015;**119**:1224-1232

[10] Singh R, Gautam N, Mishra A, Gupta R. Heavy metals and living systems: An overview. Indian Journal of Pharmacology. 2011;**43**(3):246-253

[11] Xue JL, Zhi YY, Yang LP, Shi JC, Zeng LZ, Wu LS. Positive matrix factorization as source apportionment of soil lead and cadmium around a battery plant (Changxing County, China). Environmental Science and Pollution Research. 2014;**21**(12):7698-7707

[12] Chen HM. Environmental Soil Science. 2nd ed. Beijing: Science Press; 2010

[13] Leeuwen CJV, Vermeire TG. Risk Assessment of Chemicals: An Introduction. 2nd ed. Heidelberg: Springer Press; 2007

[14] Ministry of Environmental Protection of China (MEPC), Ministry of Land and Resources of China (MLRC). Report on the National General Survey of Soil Contamination. Beijing: MEPC and MLRC; 2014

[15] Chen HY, Teng YG, Lu SJ, Wang YY, Wang JS. Contamination features and health risk of soil heavy metals in China. Science of the Total Environment. 2015;**512-513**:143-153

[16] Wu QT. Environmental Soil Science. Beijing: China Agriculture Press; 2011

[17] Huang CY. Soil Science. Beijing: China Agriculture Press; 2004

[18] Bockheim JG. Nature and properties of highly disturbed urban soils, Philadelphia, Pennsylvania. In: Paper Presented Before Division S-5, Soil Genesis, Morphology and Classification, Annual Meeting of the Soil Science Society of America; 1974. Chicago, America: Soil Science Society; 1974

[19] De Kimpe CR, Morel JL. Urban soil management: A growing concern. Soil Science. 2000;**28**(4):31-40

[20] Zhang GL, Wu YJ, Gong ZT. Urban soils: An ecological protector of urban environment. Chinese Journal of Nature. 2006;**28**(4):205-209

[21] Chu CJ, Ma JH, Zhu YT. Comparison between heavy metal contents in urban soils and their potential ecological hazards in city and town: A case study of Zhengzhou, Zhongmu and Hansi, Henan, China. Chinese Journal of Soil Science. 2010;**41**(2):467-472

[22] Liao QL, Hua M, Jin Y, Huang SS, Zhu BW, Weng ZH, Pan YM. A preliminary study of the distribution and pollution source of heavy metals in soils of Jiangsu Province. Geology in China. 2009;**36**(5):1163-1171

[23] Guo W, Sun WH, Zhao RX, Zhao WJ, Fu RY, Zhang J. Characteristic and evaluation of soil pollution by heavy metal in different functional zones of Hohhot. Environmental Science. 2013;**34**(4):1561-1567

[24] Wu S, Peng S, Zhang X, Wu S, Peng SQ, Zhang XX, Wu DL, Luo W, Zhang TB. Levels and health risk assessments of heavy metals in urban soils in Dongguan, China. Journal of Geochemical Exploration. 2015;**148**:71-78

[25] Xi DL, Sun YS, Liu XY. Environmental Monitoring. 3rd ed. Beijing: China Higher Education Press; 2004

[26] He DM, Wang XF, Chen LJ, Su R. Assessment on heavy metals contaminations of sugarcane soil in Guangxi Province by the geo-accumulation index and potential ecological risk index. Journal of Agricultural Resources and Environment. 2014;**31**(2):126-131

[27] Muller G. Index of geoaccumlation in sediments of the Rhine river. GeoJournal. 1969; **2**(3):108-118

[28] Nezhad MTK, Tabatabaii SM, Gholami A. Geochemical assessment of steel smelter–impacted urban soils, Ahvaz, Iran. Journal of Geochemical Exploration. 2015;**152**:91-109

[29] Karim Z, Qureshi BA, Mumtaz M. Geochemical baseline determination and pollution assessment of heavy metals in urban soils of Karachi, Pakistan. Ecological Indicators. 2015;**48**:358-364

[30] Hakanson L. An ecology risk index for squatic pollution control: A sedimentological approach. Water Research. 1980;**14**(8):975-1001

[31] Islam S, Ahmed K, Mamun HA, Masunaga S. Potential ecological risk of hazardous elements in different land-use urban soils of Bangladesh. Science of the Total Environment. 2015;**512-513**:94-102

[32] Zhang L, Liao Q, Shao S, Zhang N, Shen Q, Liu C. Heavy metal pollution, fractionation, and potential ecological risks in sediments from lake chaohu (eastern china) and the surrounding rivers. International Journal of Environmental Research and Public Health. 2015;**12**(11):14115-14131

[33] Xu ZQ, Ni SJ, Tuo XG, Zhang CJ. Calculation of heavy metals' toxicity coefficient in the evaluation of potential ecological risk index. Environmental Science and Technology. 2008;**31**(2):112-115

[34] Cullen AC, Frey HC. Probabilistic Techniques in Exposure Assessment. New York and London: Plenum Press; 1999

[35] National Research Council. Risk Assessment in the Federal Government: Managing the Process. Washington DC: National Academy Press; 1983

[36] Chen HH, Chen HW, He JT, Li F, Shen ZL, Han B, Sun J. Health-based risk assessment of contaminated sites: Principles and methods. Earth Science Frontiers. 2006;**13**(1):216-223

[37] Zhou ZC. Evaluation of toxic effect threshold and dose response relationship. Journal of Toxicology. 2013;**27**(6):407-408

[38] Qiu FG, Gao ST, Chen Q. Review on research advances in health risk exposure assessment. Journal of Safety and Environment. 2012;**12**(1):126-129

[39] Duan XL. Research Methods of Exposure Factors and Its Application in Environmental Health Risk Assessment. Beijing: Science Press; 2012

[40] Cui XY, editor. Comparative Study on Environmental Health Risk of Chemical Pollutants at Home and Abroad. Beijing: Science Press; 2010

[41] Pei SW, Zhou JL, Liu ZT. Research progress on screening of environment priority pollutants. Journal of Environmental Engineering Technology. 2013;**3**(4):363-368

[42] Mackay D. Multimedia Environmental Models: The Fugacity Approach 2nd ed. UK: Taylor and Francis Group; 2001

[43] Dong WY, Ding ZB, Sun SQ. Algorithm of radiological risk estimation for exposures to depleted uranium based on icrp biokinetic models. Environmental Science and Management. 2008;**33**(2):21-24

[44] Leggett RW. A biokinetic model for zinc for use in radiation protection. Science of the Total Environment. 2012;**420**:1-12

[45] Wethasinghe C, Yuen STS, Kaluarachchi JJ, Hughes R. Uncertainty in biokinetic parameters on bioremediation: Health risks and economic implications. Environment International. 2006;**32**(2):312-323

[46] Duan XL, Zhang K, Qian Y, Li Q, Wang ZS, Zhang JL. The latest developments of human exposure assessment. Academic Symposium and Conference Papers of Chinese Toxicology Society Management Toxicology Specialized Committee; Beijing. 2009

[47] Jin YL, Zhang YF, Min J, Sun JF, Liu P. The application of individual margin of exposure in the health risk assessment of dietary exposure to lead. Chinese Journal of Health Statistics. 2014;**31**(6):943-945

[48] Wang JN, Cao GZ, Cao D, Yu F, Bi J. A framework analysis on national environmental risk management system for China. China Environmental Science. 2013;**33**(1):186-191

[49] Zhang HY, Ge Y, Li FY, Yan J, Bi J. A review of psychometric paradigm in environmental risk perception. Journal of Natural Disasters. 2010;**19**(1):78-82

Chapter 20

Heavy Metal Pollution of Ecosystem in an Industrialized and Urbanized Region of the Republic of Azerbaijan

Fagan Aliyev, Hadiya Khalilova and Farhad Aliyev

Additional information is available at the end of the chapter

http://dx.doi.org/10.5772/intechopen.74600

Abstract

Along with other oil-producing regions of the world, environmental pollution is characteristic also for the Absheron Peninsula of Azerbaijan, which has long been subject to the impact of various anthropogenic factors. A large amount of such toxic substances as heavy metals, hydrocarbons and surface-active agents, and so on were released into soil and water basins leading to the change of natural ecosystem throughout the region. A primary goal of this chapter was to study the level of heavy metal pollution in the Absheron industrial region, while evaluating the potential ecological risk posed by each toxic metal including Hg, Cd, As, Cr, Pb, Cu, Ni, and Zn. Analysis of the calculated values of pollution index (PI), enrichment factor (EF), geoaccumulation index (I_{geo}), and ecological risk factor (E_i) indicates the contribution of anthropogenic sources to heavy metal accumulation in soils and sediments of the study area.

Keywords: Absheron Peninsula, heavy metals, ecological risk, soil, sediment

1. Introduction

The Republic of Azerbaijan is located at 38°25′–41°55′ North Latitude and 44°50′–50°51′ East Longitude. It is the largest country among other South Caucasus states according to its territory, number of population, rich fuel energy, and other resources. It occupies an area of 866,000 km^2. The total length of the country's border comes to 2646 km including 390 km with Russian Federation, 480 km with Georgia, 1007 km with Armenia, 756 km with Islamic Republic of Iran, and 13 km with Turkey. The Caspian Sea forms its eastern border.

The Republic of Azerbaijan is characterized by special wonderful nature, climate condition, and physicogeographical features. Its territory is rich in valuable natural resources including oil-gas fields, ore deposits, and mineral waters. The industry is specialized in the development of hydrocarbon and other mineral deposits, oil refining, petrochemical, and chemical industries. Cattle breeding, grain farming, and gardening as well as cotton, tea, and vegetable growing are the main agricultural occupations [1].

In recent years, Azerbaijan has achieved significant economic growth due to its oil and gas resources. A reliable oil-industrial infrastructure has been established in the republic. Along with the strengthening of the social and economic situation in the country, these achievements also contribute to the development of non-oil sector.

Azerbaijan also serves as an important gateway for oil and gas transportation. Construction of the Baku-Tbilisi-Ceyhan (BTC) oil pipeline in 2005 also contributed to the country's economic growth to some extent. A significant increase in the gross domestic production has been observed during the last decades. Most of the national revenue comes from the oil industry.

At the same time, Azerbaijan has serious environmental problems due to the intensive development of the region's hydrocarbon resources, increasing the amount of consumed fossil fuels and greenhouse gas emissions. The major environmental challenges are as follows [1, 2]:

- Pollution of water resources with wastewater, including transboundary pollution
- Inadequate supply of quality water to human settlements, wastage in delivery, and shortage of sewer lines (insufficient quantity of sewerage systems)
- Air pollution by industrial plants and vehicles
- Degradation of fertile soil lands (contamination, erosion, salinity, etc.)
- Improper disposal of solid industrial and municipal wastes including hazardous wastes
- Biodiversity decline including the depletion of forests, flora, and fauna

It is well known that pollution of the environment with toxic compounds causes serious problems due to their negative impact on all the components of ecosystem especially on human health. Unfavorable consequences of ecosystem pollution are appearing more acutely in urbanized territories characterized with a high density of population and industrial enterprises. The Absheron peninsula is the most urbanized territory of the republic of Azerbaijan, where two large industrial cities—Baku and Sumgait—and about 60% of onshore oil production are located.

The Absheron peninsula has about 200,000-ha territory on the coastal line of the Caspian Sea in the southwestern edge of the Great Caucasus Mountain. The region is characterized by a dry semi-desert climate. Strong northern winds are the main distinguishing feature of the peninsula. The average annual air temperature is +14.2°C and the average annual rainfall is about 200–300 mm. The soil cover of the territory mainly consists of gray-brown and coastal sandy soils. The Absheron peninsula is also the location of a great number of oil and gas fields. The region is specialized on the production, initial processing and transportation of crude oil and gas, oil machine building, and fishery. Much of the peninsula is crossed by oil and gas pipelines. Despite the fact that the history of oil production in Absheron dates to ancient times, the industrial development of oil-gas fields has started from the middle of the nineteenth century. About 70% of the republic's industrial potential is concentrated in this territory. Like other industrialized and urbanized regions of the world, the Absheron peninsula could not avoid environmental pollution. There are a number of technogenic-polluted aerials. Oil industry is the main pollution source of the environment. Since the end of the nineteenth century, the peninsula has been subject to intense industrialization and urbanization. Accelerated development of hydrocarbon producing and refining industries has radically changed the historically formed geochemical balance of the environment due to the removal of different chemical elements to the Earth's surface. The territory has over 200 basins of natural and artificial origin. A huge amount of toxic hydrocarbons and heavy metals accumulate in soil and deep sediments, thus leading to the change of the qualities of natural ecosystem. According to its ecological status and climatic landscape character, the Absheron region is categorized to be a territory with the most acute environmental situation in the republic [3, 4].

Currently, environmental pollution is a serious concern worldwide. Oil and oil products contribute significantly to the pollution of environment in oil-bearing regions. The exploration of hydrocarbon deposits, as well as extraction, transportation, storage, and use of oil and gas causes pollution of the environment with hazardous substances. Crude oil includes hydrocarbons such as alkanes, cycloalkanes, unsaturated aliphatic hydrocarbons, and aromatic hydrocarbons, as well as heterocyclic compounds containing nitrogen, oxygen and sulfur, metals, and natural radionuclides.

Past studies have indicated that there are high concentrations of contaminants in soil and water systems of the Absheron peninsula [5–7]. Sometimes, the degree of oil contamination in soil varies from 20 to 30% and more. A large part of the peninsula is occupied by destroyed oil-polluted and bituminous lands. Contaminants accumulated in soils migrate into lakes, reservoirs, surface, and groundwaters throughout the site. Migration of pollutants into deep sections of soil and groundwaters causes serious danger to both local and regional viewpoints.

Today, the pressure on the environment in the peninsula is increasing continuously due to the ever-increasing population, plus industrial and urban growth.

1.1. Heavy metals and their environmental effects

Heavy metals are the most hazardous environmental pollutants due to their toxicity and accumulation ability. According to their classification, heavy metals have a density of more than 5 g/cm^3 (about 6.0 g/cm^3 or more, which is much higher than the average particle density of soils—2.65 g/cm^3) and occur naturally in rocks but concentrations are frequently elevated as a result of contamination. Rare and noble metals are not included in heavy metals category. According to the biological classification of chemical elements, heavy metals belong to a group of micro- and ultra-microelements. In the vast majority of researchers, Pb, Cu, Zn, Ni, Cd, Co, Sb, Sn, Bi, and Hg are considered heavy metals [8–10].

Heavy metals are characterized by two main features—toxic effects on living organisms in relatively low concentrations and bioaccumulative abilities. Being rich in heavy metal soils, plants and bottom sediments become toxic over time, presenting a serious danger to all living things.

Both the accumulation of heavy metals in the environment and their negative impact on the ecosystem are widely studied by scientists from all over the world. The work of many scientists was devoted to the research of heavy metals in biosphere [11–14].

Compared to other ecosystem components, soil accumulates the highest concentration of heavy metals. Being in interaction with other components of ecosystem such as atmosphere, hydrosphere, and plants, soil contributes significantly to the ingress of heavy metals into the human organism. Penetrating into soil, they are accumulated in different parts of agricultural products through root system, and in aquatic organisms and bottom sediments when washed with surface waters. One of the major global problems of the present times—acid rain—is one of the factors increasing heavy metals' level in ecosystem. Washing soil rocks with acid rain increases the amount of metals in lakes and other water basins. Migration of metals in the ecosystem promotes their accumulation in the human body (**Figure 1**).

Heavy metals can be of both native and anthropogenic origins. Their natural sources are rocks and soils, but mainly in the forms inaccessible to wildlife, such as constituents of rocks and soil minerals. When the metal pollution is caused by anthropogenic sources, this can seriously influence all the ecosystem components. Waste disposal, transport emissions, oil-gas and mining industries, atmospheric depositions, and land application of fertilizers are anthropogenic sources of heavy metals [10, 15, 16].

Most oils have certain content of heavy metals. Starting from gasoline, all fractions of oil include metals. The crude oil extracted from the hydrocarbon deposits of Azerbaijan is dominated by iron (up to 0.74%), chrome, and nickel (often more than $n \times 10^{-2}$ ppm). Asphaltene fraction of some oil fields of Absheron peninsula is rich in such microelements as Co, Br, Ag, Au, La, Sb, and Sc (**Table 1**). As a result of neutron activation analysis, $1.2 \times 10^{-3}\%$ Rb, $6.3\% \times 10^{-5}\%$ Cs, and $2.1\% \times 10^{-5\%}$ Eu were found in the crude oil of the Balakhani oil field [17]. Usually, a majority of heavy metals are found in the asphaltene-tar fractions of oil. They are also found in the composition of oil cuttings and drilling fluids.

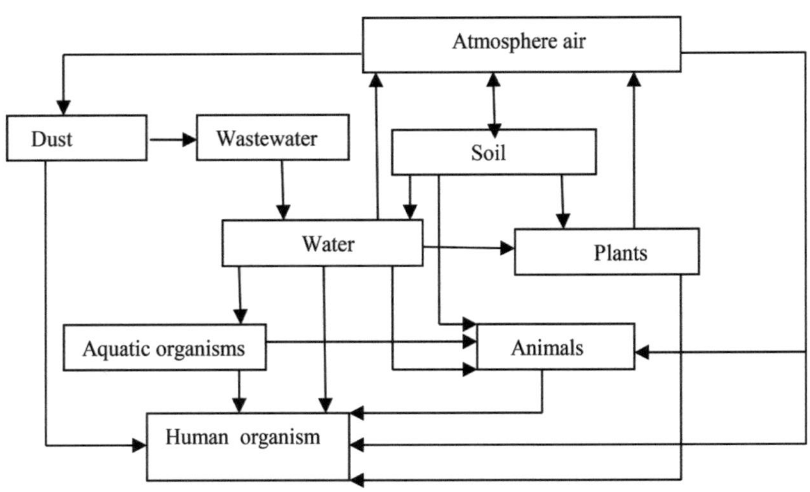

Figure 1. Scheme of heavy metals penetrating into the human organism.

Oil fields	Metals (%)														
	Fe	Ni	Cr	Zn	Ba	V	Co	Hg	Ag	Au	La	Sb	Sc	Hf	Se
	$\times 10^{-4}$			$\times 10^{-5}$				$\times 10^{-6}$					$\times 10^{-7}$		
	Asphaltenes														
1.	600	57	96	750	53	151	110	nd*	289	140	87	300	1500	28	47
2.	138	58	174	270	265	260	64	160	nd	8	28	19	20	26	32
3.	400	63	68	110	540	142	35	570	nd	6	nd	4	13	14	nd
	Tars														
4.	170	8	20	80	5	37	2	14	10	150	4	2	10	23	75
5.	34	23	12	3	nd	62	10	nd	2.4	6	nd	nd	2	52	60
6.	110	19	9	8	99	47	4	2250	14	4	nd	9	11	127	46
7.	180	65	27	14	8	53	6	nd	nd	1	nd	1	nd	18	nd

*nd, compound not detected or below instrumental direction limits.

Table 1. Metal content of the Absheron oil fields.

According to the researches of specialists, oil and gas complex creates pedochemical factors of having an impact due to the existing natural technogenic hydrogeological conditions resulting in over-moistening and extent of contamination. There is a potential hazard of washing out of oil and oil products into rivers, bays, ground, and surface waters. Abnormal redistribution of chemical elements forming the secondary lithochemical areas has been recorded in several oil fields of the Absheron peninsula. Thus, some mobile elements including Pb, Zn, Cu, Ni, and Cr are accumulated in soils jointly and form paragenic association. They are able to migrate and accumulate in the appropriate barriers. Their concentrations exceed the normal values typical of the natural landscapes in the Absheron peninsula [5]. There is a tendency of the growth of heavy metals concentration both in the bottom sediments and in some hydrobionts in the Absheron territory. As a result of contamination by wastewater from various anthropogenic sources including offshore oil fields, heavy metals penetrate into the coastal line, natural lakes, and other reservoirs.

Usually, the most commonly encountered metals in polluted areas are lead (Pb), chromium (Cr), mercury (Hg), arsenic (As), zinc (Zn), cadmium (Cd), copper (Cu), and nickel (Ni). It is well known that heavy metals are nondegradable and persistent pollutants and accumulate in the environment for a long time. In soil, they could be adsorbed and gathered in different parts of plants through root systems. It has been revealed that heavy metal pollution causes adverse effects on the quality and yield of agricultural plants, resulting in the changes of the number, composition, and activity of microorganisms. Practice shows that the geochemical cycle of heavy metals under anthropogenic impacts represents serious, at times unpredictable, environmental effects. As mentioned earlier, due to their migration and accumulation in the environment, most heavy metals can easily enter the food chain and create serious threat to human health. Many heavy metals have a strong affinity for sulfur and disturb the enzyme function forming bonds with sulfur groups. They can get into a food chain and accumulate in an organism reaching toxic levels. These may cause massive mortality of various animal species [18].

In order to find a proper solution to the problem, we studied the level of heavy metal pollution in ecosystem while evaluating the contribution of anthropogenic factors and the potential ecological risk posed by individual elements.

The main purpose of this work was to investigate the heavy metal pollution of

- soils from two types of land uses in the study site, including the areas of oil wells and transport road;
- ground and groundwaters in a zone of influence of the oil-industrial waste;
- waters and sediments of two natural water basins—the Boyuk shore and Bulbula Lakes—that had undergone a long-term impact of anthropogenic sources.

The studies have revealed increased levels of some heavy metals in soils, ground, and surface water basins in the territories subjected to the impact of various anthropogenic sources.

2. The objects and methods of research

The objects of the present research were the territories of oil production enterprises, transport roads, as well as ground and groundwaters around an aeration station and surface water basins in industrial territories (**Figure 2**). The main sources of ecosystem pollution in the study site are the industrial discharges formed during oil fields development and production activities. Along with the produced waters, these discharges also contain drilling solutions and various chemicals used in the drilling and production operations. Thus, the territory is subject to the impact of oil, oil products, heavy metals, and other technogenic waste. The main pollution sources of ground and groundwaters in the territory are the substances characterized with a high migration ability such as crude oil and heavy metals. It should be noted that the atmosphere emissions of oil-gas refineries and petrochemical industries are among the factors contributing to the ecological situation in the study area. In order to evaluate the impact of transport emissions on the heavy metal pollution, soil samples were collected at a 10-m distance of Baku-Mardakan highway.

One of the major objects of the research—Boyuk-shore lake—is located at the center of the Absheron peninsula at an absolute height of 12 m. It is the largest and most polluted lake of the peninsula with a 45-million-m^3 volume, a 15-km^2 surface area, and a 4.2-m maximum depth. Being an enclosed water body, the lake feeds underground and surface waters of water catchment. Starting from 1930s, a huge amount of oil-industrial wastewater has been discharged to this basin (**Picture 1**). According to the information of Ministry of Ecology and Natural Resources, the total daily volume of wastewater released from industrial and communal-domestic objects to the Boyuk-shore lake amounted to 18,000 m^3 [19].

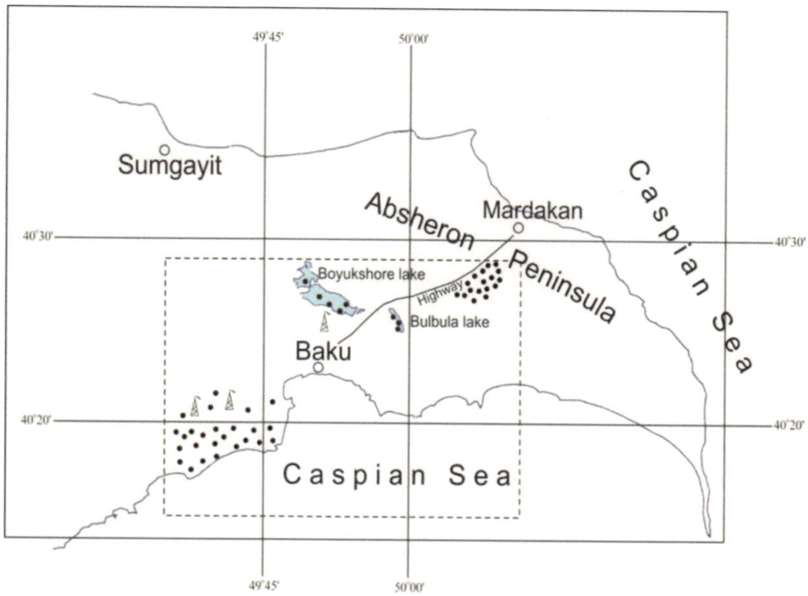

Figure 2. The map of the study area.

Picture 1. Oil-polluted area in the northern part of Boyuk-shore Lake (taken in 2017).

The Bulbula Lake is situated at the center of the Absheron peninsula in the northeastern part of Baku city at 8 m below the ocean level. It is an enclosed saline basin with a 2-km^2 surface area and a 2.05 million m^3 volume. Its maximal depth is 3–4 m. Till the end of the nineteenth century, the water regime of this lake was dependent on seasonal fallings and inflows from adjacent areas. Starting from that time, a large volume of oil-industrial wastewater have been discharged into this basin that resulted in dynamics of its level and flooding the residential buildings, social objects, and infrastructures in the territory (**Picture 2**). In order to lower the lake level, a pumping station with a 60,000-m^3 daily capacity was built for pumping about 30,000-m^3 water to Hovsan treatment plant. The lake is surrounded with five population settlements occupying an area of 122.5 ha with 3690-m length and 320–430 m width and is therefore permanently subject to the impact of domestic waste.

Analysis of water, sediment, and soil samples was carried out according to standard methods [20, 21]. Inductive Coupled Plasma-Optical Emission Spectrometry (ICP-OES) device was used in the analysis of heavy metals (As, Ba, Cd, Cr, Cu, Fe, Mn, Pb, and Zn), except Hg that was analyzed by Cold Vapor Atomic Fluorescence (CVAF) apparatus.

For the first time, the overall pollution status of the study area was evaluated with the calculation of pollution indices of heavy metals.

Various methods have been used to evaluate the heavy metal pollution. The calculation of single pollution index (PI_i), enrichment factor (EF), geoaccumulation index (Igeo), Nemerow pollution index (PI_N), and potential ecological risk index (RI) is widely applied for the assessment of heavy metals' level in soils and sediments [19, 22].

The pollution indexes used during our researches to assess the heavy metal contamination in the study sites are as follows:

2.1. Pollution index

The pollution level of each heavy metal was evaluated using pollution index (PI_i), calculated as the ratio between the metal concentration (C_i) in the sample and its reference value—the national criteria of metal (S_i):

Picture 2. Wastewater flows to Bulbula Lake from its northern side (taken in 2015).

$$PI_i = \frac{C_i}{S_i} \tag{1}$$

The maximum permissible concentrations (MPCs) of pollutants established by Azerbaijan legislation were taken as S_i values.

2.2. Enrichment factor

Enrichment factor (EF) was initially calculated to identify the origin of elements in atmosphere, precipitation, or seawater, and it was further extended to the study of soils, aquatic sediments, and other components of ecosystem [23]. In this study, the enrichment factors of heavy metals in soils and sediments are calculated to assess the contribution of anthropogenic sources to the natural levels of heavy metals in the Absheron soils and lake sediments. The following formula was used to calculate EF:

$$EF = \frac{C_i/C_r}{B_i/B_r} \tag{2}$$

where C_i and C_r are the concentrations of the target metal and the reference metal in the sample, while B_i and B_r are the background concentrations (BCs) of the target metal and the reference metal for the natural soils of the Absheron peninsula. Immobile elements such as Al, Fe, Ti, or Mn have been used as the reference metals for EF calculation. EFs for all the heavy metals were calculated using Mn as the reference metal in our study.

2.3. Index of geoaccumulation (I_{geo})

A geoaccumulation index was originally defined by Muller [24] for the evaluation of metal contamination in aquatic sediments, but it was also applied in assessing the metal contamination of soils. The formula used for the calculation of geoaccumulation index soils and sediments is.

$$I_{geo} = Log_2\left(\frac{C_n}{1.5\,B_n}\right) \tag{3}$$

where C_n is the measured content of element (n), B_n is the BC of element n, and 1.5 is the background matrix correction factor due to lithogenic effects (the constant 1.5 is introduced to minimize the effect of possible variations in the background values which may be attributed to lithologic variations in soils).

2.4. Ecological risk factor

The potential ecological risk of heavy metal pollution in the soils and sediments of water basins in the study area was assessed using the ecological risk index (RI) [25]. RI is the comprehensive potential ecological index, which is the sum of individual heavy metals—E_i. It represents the sensitivity of the biological community to the toxic substance and illustrates the potential ecological risk caused by the overall contamination. The RI was calculated as the sum of risk factors of heavy metals:

Metals	Hg	Cd	As	Pb	Cu	Cr	Ni	Zn
MPC (mg/kg)	2.1*	1.0	2.0	32.0	3.0	6.0	4.0	23.0
BC (mg/kg)	0.4	3.0	15	20	100	40	20	70
T_i	40	30	10	5	5	2	5	1

*The value of 2.1 is acceptable only for Hg pollution of soils in nonresidential areas.

Table 2. MPC, BC, and toxic-response factors of heavy metals.

$$RI = \sum E_i \tag{4}$$

where E_i is the single risk factor for heavy metal i and is defined as.

$$E_i = T_i f_i = T_i \frac{C_i}{B_i} \tag{5}$$

where T_i is the toxic-response factor for heavy metal i. The ratio f_i is the metal pollution factor calculated from the measured concentration C_i and the background concentration B_i of metal in soils. MPC and BC, and the T_i values defined by Hakanson for the measured heavy metals that were used during the calculation of pollution indices are given in **Table 2**.

The calculated values of PI_i, E_i, EF, and I_{geo} given in subsequent tables are the means of these indices from at least 10 samples.

3. Results and discussion

Initial studies focused on the assessment of heavy metal pollution in the territory of oil-gas production enterprises. The study was carried out several years ago. Ten soil samples were analyzed for metals including potentially toxic elements—As, Cd, Cr, Hg, Pb, Cu, Zn, and Ba. The values derived from the analyses were compared with both the background concentrations and the maximum permissible concentrations of these elements established by the Cabinet of Ministers of the Republic of Azerbaijan. The results are presented in **Table 3**.

The values in **Table 3** show that the concentrations of the majority of heavy metals, including As, Ba, Cr, Cu, Pb, and Zn, exceed their permissible levels established in the republic. The contents of most toxic metals like As, Pb, and Zn are found to be from 3.4 to 8.2, 14.3 to 42.2 and 11.5 to 105 mg/kg, respectively. Their concentrations were higher than the background levels in some sampling areas. Meanwhile, no exceeding of maximum permissible and background levels on Hg, Cd, and Mn was observed in this area during our researches.

According to their impact on the ecosystem, many of these metals belong to the following categories, with exception of Fe, which is low toxic:

1. As, Hg, Pb, Cd, and Zn—super dangerous
2. Cr and Cu—dangerous
3. Ba and Mn—moderately dangerous

Samples	Heavy metals (mg/kg)									
	As	Ba	Cd	Cr	Cu	Fe	Mn	Hg	Pb	Zn
1	5.8	730	0.28	20.4	67.1	16,500	380	0.04	27.2	60.7
2	7.6	214	0.10	27.1	30.7	30,600	494	0.03	25.3	32.5
3	5.4	1290	0.41	21.6	44.4	28,600	691	0.11	37.3	105
4	8.2	335	0.09	27.7	26.6	15,500	541	0.07	27.2	54.9
5	7.6	330	0.12	14.4	20.8	16,200	294	0.02	26.2	17.1
6	3.4	360	0.15	18.9	57.2	7900	188	0.04	37.1	37.5
7	3.9	45	0.02	6.5	5.0	4500	130	0.01	25.1	11.5
8	4.4	760	0.23	17.7	38.4	13,900	435	0.19	29.8	32.4
9	7.5	570	0.37	16.9	60.0	13,500	474	0.07	42.2	76.2
10	7.8	354	0.08	28.4	29.9	15,600	485	0.03	14.3	52.1
MPC (mg/kg)	2.0	200–350	1.0	6.0	3.0		1500	2.1	6.0	23
BC (mg/kg)	15		3	40	100		250	0.4	20	70

Table 3. The content of heavy metals in oil-contaminated soils.

It is necessary to have knowledge about sediment pollution situation to evaluate the level of heavy metals' impact on water biological system. With this purpose, sediment samples from 12 distinct water bodies (including natural and artificial basins) in the study site were analyzed for heavy metal contamination. It should be noted that along with oil-gas production enterprises, there are some other industrial objects in this area. The results of analyses are presented in **Table 4**. The table indicates that significantly high quantities of heavy metals are accumulated in the bottom sediments of surface waters.

As can be seen from **Table 4**, in many samples, the concentrations of toxic heavy metals, especially Cd, Cr, Cu, and Zn, were found to be greater than their permissible levels accepted in the republic—1, 6, 3, and 23 mg/kg, respectively. In several stations, the contents of most toxic metals like Cd and Zn ranged from 17.8 to 33.0 mg/kg and from 28.3 to 86.8 mg/kg, respectively. The highest concentrations of Cd were recorded in the study sites located closer to cement manufacture and nonferrous metal-processing plants. In a large majority of samples, the values of Cd were higher than its background level (3 mg/kg) to a considerable extent. The content of Cr was significantly higher than its MPC at most of sampling stations. However, no exceeding of the MPC on Pb (32 mg/kg) was detected in the samples during the study period.

Further studies were carried out to assess the potential ecological risk caused by the heavy metal contamination in the Absheron industrial zone through the calculation of single pollution index (PI_i), enrichment factor (EF), geoaccumulation index (Igeo), and potential ecological risk index (RI). The studies focused on the evaluation of the contamination levels of various ecosystem components with toxic metals. At first, the contamination of ground and groundwaters was studied in one of the polluted areas of the Absheron peninsula, in the territory of Hovsan Aerator Station.

Samples	Heavy metals (mg/kg)								
	Ni	Co	Pb	Mn	Cr	Zn	Cu	Cd	V
1	44.7	18.4	18.9	136.4	11.9	20.4	14.9	33.0	3.3
2	3.0	9.6	7.0	155.0	16.1	20.1	12.9	ND	3.0
3	36.4	7.6	6.9	213.4	11.3	16.9	58.9	17.8	1.8
4	15.2	10.9	6.9	162.0	14.9	21.0	14.8	ND	3.2
5	13.2	9.8	6.3	159.1	18.4	20.3	13.6	ND	2.1
6	50.1	19.9	12.1	636.0	23.9	30.0	13.1	29.8	6.9
7	44.6	9.8	9.6	351.0	16.9	86.8	83.1	28.4	4.0
8	52.1	25.0	10.9	546.0	20.6	32.3	11.0	29.4	8.9
9	49.6	24.6	11.3	513.2	23.9	29.6	10.6	1.0	9.3
10	46.8	21.0	11.6	596.0	25.1	28.3	12.9	31.0	5.7
11	15.8	12.1	7.0	169.3	15.9	21.9	16.1	ND	3.4
12	13.4	11.3	6.8	156.8	13.8	19.8	15.4	ND	2.4

Table 4. The content of heavy metals in bottom sediments of surface waters within the Absheron industrial region.

For the evaluation of the pollution level and the potential ecological risk caused by heavy metal pollution in soils, enrichment factor and ecological risk index were calculated for the most important heavy metals with regard to potential hazard such as Cd, Cr, Pb, Zn, and Ni, with contamination values exceeding MPC.

The data derived from heavy metal analysis of ground and groundwaters of the study site are presented in **Tables 5** and **6**. The results indicated that in some groundwater samples, the concentrations of Cd and Zn were found to be higher than their MPC established by the national legislation. No exceeding of MPC was recorded in the values of Pb, Cr, Ni, and Mn during our observations. Overall, the concentrations of the majority of the analyzed heavy

Metals	MPC, mg/l	Sampling stations								
		1	2	3	4	5	6	7	8	9
Cd	0.001	0.010	0.008	0.009	0.009	0.005	0.008	0.014	0.012	0.006
Zn	0.5	0.09	0.09	0.012	0.015	0.014	0.116	0.248	0.190	0.120
Pb	0.03	0.009	0.013	0.006	0.014	0.007	0.012	0.019	0.015	0.008
Cr	0.1	0.007	0.008	0.005	0.012	0.005	0.025	0.040	0.011	0.007
Ni	0.1	0.005	0.018	0.014	0.012	0.006	0.008	0.021	0.010	0.005
Mn	0.05	0.013	0.014	0.016	0.008	0.015	0.027	0.014	0.010	0.008

Table 5. The content of heavy metals in groundwaters (mg/l).

Metals	Sampling stations								
	1	2	3	4	5	6	7	8	9
Cd	3.9	1.4	4.4	6.4	2.2	5.6	3.5	3.3	2.5
Zn	99	140	130	120	110	77	110	80	130
Pb	7	9	8	14	16	9	5	12	14
Cr	30	92	68	55	70	46	130	57	59
Ni	20	30	40	36	25	21	35	53	18
Mn	360	230	270	320	220	320	400	280	260

Table 6. The content of heavy metals in grounds (mg/kg).

metals in groundwaters were at the level of national and international standards. According to the international water-quality standards, the minimal risk concentration of Zn, Cu, Ni, and Pb should be 20, 10, 2, and 10 μq/l, respectively [26, 27].

Analysis of the data presented in **Table 6** shows unequal distribution of heavy metals in the polluted grounds of the territory. The variability in the concentrations of metals may be associated with the origins of their pollution sources. The maximum heavy metal pollution levels observed were 6.4 mg/kg for Cd, 140 mg/kg for Zn, 130 mg/kg for Cr, and 53 mg/kg for Ni. With exception of Mn and Pb, the average concentrations of Cd, Zn, Cr, and Ni in the ground samples were found to be greater than their MPC about 3, 5, 10, and 9 times, respectively.

Figures 3 and **4** show the calculated EF and E_i values of Cd, Zn, Pb, Cr, and Ni. The results of calculation showed that the values of EF of Zn and Cr are 2.18 and 2.50 in the ground from station 2, indicating moderate enrichment by these elements, whereas the EF values of Cd, Pb and Ni were 0.51, 0.49, and 0.71, respectively, exhibiting depletion to mineral enrichment (<2). EF values of all the studied metals in both stations 4 and 8 recorded the values lower than 2, indicating that the grounds of these areas are in the state of depletion to mineral enrichment with respect to these metals. The data presented in **Figure 1** show the same EF values for Pb at all three stations, which can be associated with the composition of the parent materials of soils. A relatively high EF value of Cd (1.67) was observed at sampling station 4.

Comparative analysis of the calculated values of E_i index with the data given in **Table 7** shows that among other elements, Cr and Cd demonstrated a higher ecological risk during the observations, indicating that the site was impacted by the anthropogenic sources of these metals. The calculated E_i values were 46 for Cr at station 2, and 64 for Cd at station 4 exhibiting a moderate potential ecological risk. According to the results of calculations, the area of station 8 can be characterized by a relatively lower enrichment and a potential ecological risk with regard to heavy metal pollution. As can be seen from **Figure 4**, the potential environmental risk of metal in a greater extent depends on its toxicity to the ecosystem, rather than the content of the element in the soil. At station 8, the concentration of Cd exceeded MPC by 3.3 times, and the concentration of Zn exceeded MPC by 3.5 times, whereas according to the calculation results, the values of their ecological risk index E_i were 33 and 1.2, respectively.

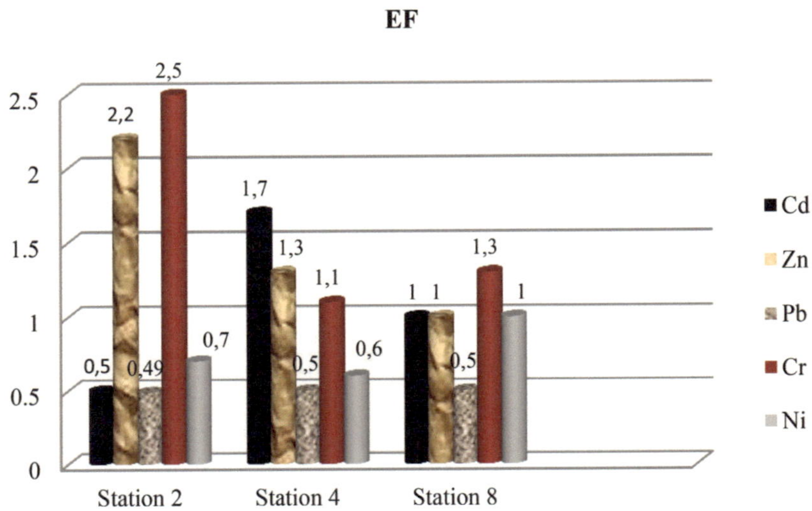

Figure 3. Change of the values of enrichment factor of heavy metals in the grounds of three study stations.

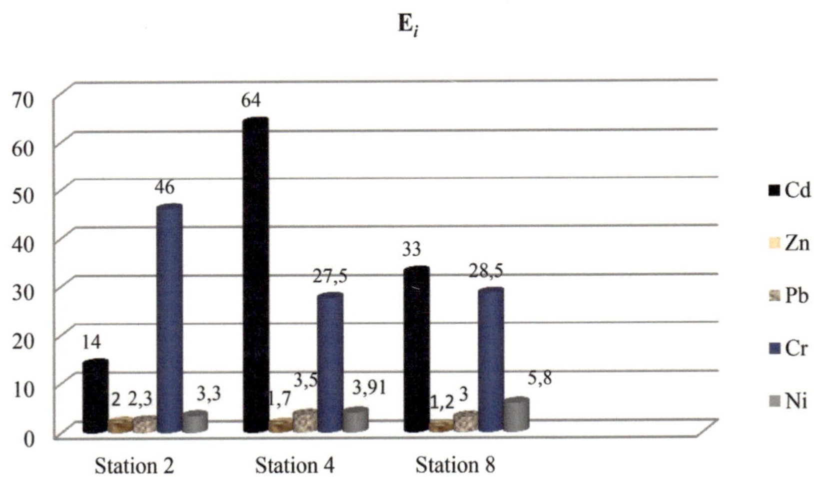

Figure 4. Change of the values of potential ecological risk index of heavy metals in the grounds of three study stations.

Part of the research was carried out to assess the current level of heavy metal pollution of soils and surface water systems in the oil-industrial zone of the Absheron peninsula. The researches focused on the heavy metal pollution of soils from two types of land uses including two natural lakes—Boyuk shore and Bulbula, which have been subject to a long-term impact of various anthropogenic sources. The heavy metal pollution of the study site was assessed through the calculation of the single pollution index (PI_i), geoaccumulation index (I_{geo}), enrichment factor (EF), and ecological risk factor (E_i).

Comparison of the results given in **Table 8** with the pollution classes in **Table 7** shows that oil field area can be categorized as unpolluted by Pb, Cd, Zn, Cu, and Hg, and moderately

Class	I_{geo}		EF		E_i	
1	<0	Unpolluted/slightly polluted	<2	Depletion to mineral enrichment	<40	Low potential ecological risk
2	0–1	Moderately polluted	2–5	Moderate enrichment	40–80	Moderate potential ecological risk
3	1–3	From moderately to strongly polluted	5–20	Significant enrichment	80–160	Considerable potential ecological risk
4	3–5	Strongly polluted	20–40	Very high enrichment	160–320	High potential ecological risk
5	>5	Extremely polluted	>40	Extremely high enrichment	320	Very high potential ecological risk

Table 7. Classification of heavy metal pollution levels in soils and sediments based on I_{geo}, EF, and E_i values.

to strongly polluted by As and Cr, which mean I_{geo} values are higher than 1 to some extent. The mean values of EF are 0.0367 for Cd and 0.8900 for Pb, indicating that the territory can be characterized by the first classification level—depletion to mineral enrichment. According to the mean E*i* values of individual metals within the range of 0.68 (Zn)–7.28 (Pb), the soils of oil field area demonstrate a low potential ecological risk regarding heavy metal contamination.

Figure 5 presents the percentage of individual heavy metals in potential ecological risk index (RI) for the soils of oil field area. It is seen from this figure that the soils from this area had the highest E_i percentage for Pb, Hg, and As compared to other heavy metals.

In order to study the impact of transport emissions on heavy metal contamination of soils, the content of heavy metals was measured in soil samples collected in the vicinity of two main roads. The results are given in **Table 9**.

Elements	Concentration (mg/kg)			PI_i	E_i	EF	$I_{geo.}$
	Min.	Max	Mean	Mean	Mean	Mean	Mean
As	3.4	8.2	4.16	2.8	4.16	0.2683	1.4623
Cd	0.02	0.41	0.18	0.18	1.85	0.0367	−4.5658
Cr	6.5	28.4	19.9	3.36	0.99	0.3000	1.7297
Cu	5.0	67.1	37.1	12.7	1.89	0.2250	−1.9808
Hg	0.01	0.19	0.06	0.03	6.10	0.0912	−3.3219
Pb	14.3	42.2	29.2	0.9	7.28	0.8900	−0.0389
Zn	11.5	105	47.9	2.3	0.68	0.4107	−1.1292
Mn	130	691	410				

Table 8. The content and pollution indices of heavy metals in soils in oil field area.

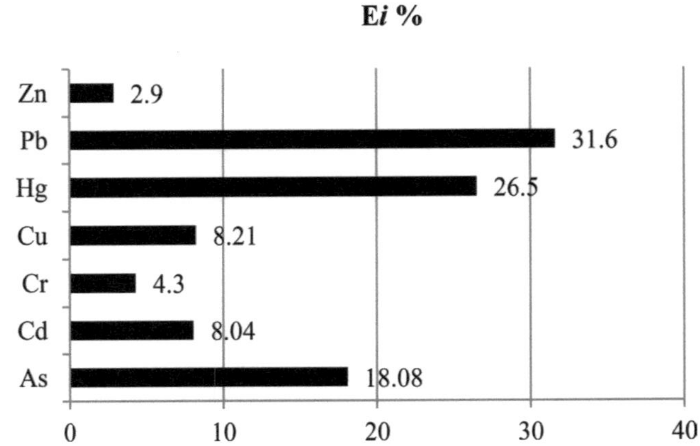

Figure 5. Percentage of heavy metals in potential ecological risk (RI) for the soils of oil fields area.

The calculated values of pollution indices for the soils near to Baku-Mardakan highway that was polluted predominantly by transport emissions (**Table 10**) varied from their values for oil-polluted site. The highest pollution indices in this area were recorded for Pb. The mean values of I_{geo}, EF, and E_i of this metal were 4.1429, 2.1250, and 132.5, respectively, demonstrating strongly polluted, moderate enrichment, and considerable ecological risk levels. The calculated I_{geo} values indicate that the site can be categorized as strongly polluted with respect to Pb and Cr and from moderately to strongly polluted area with respect to Zn and Cu. Relatively low pollution levels were detected for Hg and As in the soils of this area.

Figure 6 indicates that an increasingly high E_i percentage was recorded for Pb, Cd, and Cr in the soils from the area of transport road. Combustion of petroleum products by transport facilities is a major source of environment pollution by these metals.

Initial studies on the ecological situation of the abovementioned lakes were conducted to evaluate the level of contamination of lake waters with oil products and heavy metals. During the studies, the highest levels of heavy metal and oil pollution were recorded in the samples collected from industrial wastewater flow to Boyuk shore Lake (**Table 11**). Among toxic heavy metals, Zn, Pb, and Cd each had the highest concentrations exceeding the MPC by 70, 19, and 44 times, respectively. Increasingly high concentrations of petroleum compounds were observed in the waters of both lakes exceeding the MPC by 10-fold. Elevated levels of Cu contamination were observed at all sampling stations.

The contents and calculated results of pollution indices of heavy metals in the sediments of Boyuk shore and Bulbula Lakes are presented in **Tables 12** and **13**. Minor concentrations of As

Roads	**Concentration of metals, mg/kg**									
	Pb	**Cd**	**Ni**	**Zn**	**Mn**	**Hg**	**Cr**	**As**	**Co**	**Cu**
Baku-Sumqait	470	4.5	375	430	1950	0.019	860	4.08	52	115
Baku-Mardakan	565	5.8	440	515	2175	0.023	909	5.29	57	150

Table 9. The content of metals along transport roads.

Elements	Concentration (mg/kg)			PIi	Ei	EF	$I_{geo.}$
	Min.	Max	Mean	Mean	Mean	Mean	Mean
As	1.25	6.17	5.30	2.65	3.5	0.0266	−2.0858
Cd	1.30	5.80	4.50	4.5	45.0	0.1416	0
Cr	215	926	860	143	43.0	1.5000	3.8413
Cu	150	740	575	191	28.7	0.5500	1.9386
Hg	0.007	0.034	0.028	0.013	2.80	0.0219	−7.7433
Pb	75	588	530	16	132.5	2.1250	4.1429
Zn	126	517	435	19	6.20	0.5357	2.0506
Mn	1755	3390	2170				

Table 10. Concentration and mean values of pollution indices of heavy metals in soils in the area of Baku-Mardakan highway.

were detected in the sediments of both lakes during the studies. Consequently, the pollution indices were calculated for six metals including Cd, Cr, Cu, Hg, Pb, and Zn.

It is seen from **Table 12** that the average PI_i values of Cd, Cr, Cu, and Zn in the Boyuk shore Lake sediments showed the mean concentrations to be significantly higher than the MPC of these metals. The maximum concentrations of Cd, Pb, and Zn each were 5, 47.2, and 510 mg/kg, respectively. Laboratory analyses showed that the concentrations of the majority of heavy metals in the sediments of Bulbula Lake were lower than those in Boyuk shore Lake, exclusively Cu which had the highest mean PI_i value of 7.5. The variations in the heavy metals pollution levels can be explained with the origins of their sources.

The results of calculations suggest that in both lakes, the mean values of $I_{geo.}$ and EF for the studied metals represent unpolluted condition, while there are some elevated levels of Cd, Pb, and Zn, especially in the sediments of Boyuk shore Lake. The results showed that the maximum EF value in the sediments of Boyuk shore Lake observed in Zn is greater than 3, indicating that the lake was severely impacted by anthropogenic sources of this metal.

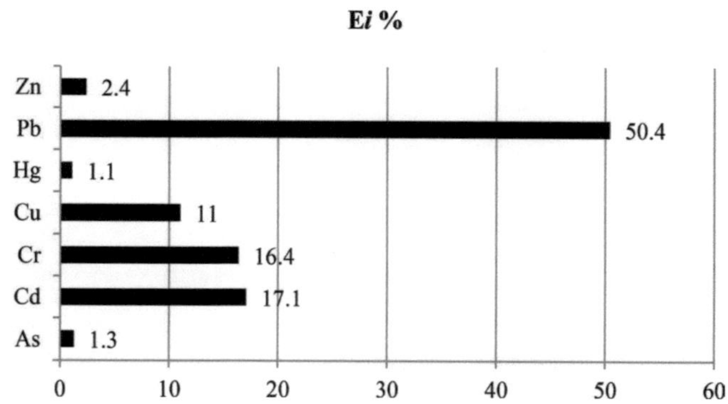

Figure 6. Percentage of heavy metals in potential ecological risk (RI) for the soils of the area close to transport road.

Pollutants (mg/l)	MPC (mg/l)	Boyuk shore Lake	Bulbula Lake
Petroleum compounds	0.05	2.6	2.05
Zn	0.5	35.06	ND
Cu	0.01	1.72	0.37
Ni	0.1	6.07	0.006
Pb	0.03	0.56	0.01
Cd	0.001	0.044	<0.001

Table 11. The content of petroleum compounds and heavy metals in two natural lakes of the Absheron Peninsula.

Elements	Concentration (mg/kg)			PI_i	E_i	EF	$I_{geo.}$
	Min.	Max	Mean	Mean	Mean	Mean	Mean
Cd	1.0	5,0	1.7	1.7	17.0	0.3916	−1.4043
Cr	15.6	59.0	25.1	4.2	1.21	0.4562	−1.2572
Cu	10.6	24.4	14.7	4.9	0.74	0.1125	−3.3510
Hg	0.004	0.063	0.0085	0.04	0.85	0.1500	−6.1413
Pb	16.7	47.2	28.5	0.9	7.12	1.0040	−0.0740
Zn	20.3	510	86.8	3.8	1.24	0.8714	−0.2746
Mn	213.4	546	355				

Table 12. The content and mean values of pollution indices of heavy metals in the sediments of Boyuk shore Lake.

Elements	Concentration (mg/kg)			PI_i	E_i	EF	$I_{geo.}$
	Min.	Max	Mean	Mean	Mean	Mean	Mean
Cd	<0.5	1.4	0.6	0.7	6.00	0.2150	−2.9068
Cr	11.3	28.9	17.7	2.9	0.88	0.2937	−1.7612
Cu	13.1	35.0	22.6	7.5	1.13	0.1475	−2.7305
Hg	0.002	0.029	0.006	0.003	0.60	0.0993	−6.6438
Pb	6.9	30.2	18.4	0.6	4.60	0.6101	−0.7052
Zn	6.6	36.8	20.3	0.9	0.29	0.1928	−2.3708
Mn	179.4	636.0	376.2				

Table 13. The content and mean values of pollution indices of heavy metals in the sediments of Bulbula Lake.

Table 13 shows that the calculated EF values for the studied metals in the sediments of Bulbula Lake are lower than 1, indicating depletion to mineral enrichment. The means of $I_{geo.}$ for all metals were negative in the Bulbula Lake sediments, based on which the lake is categorized with "unpolluted/slightly polluted" class.

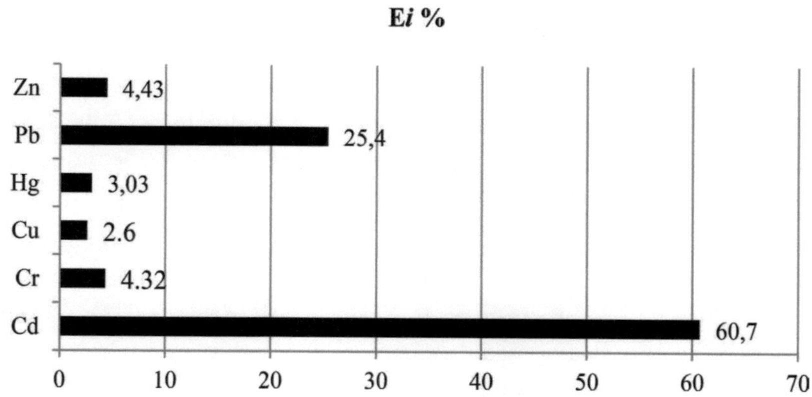

Figure 7. Percentage of heavy metals in potential ecological risk (RI) for the sediments of Boyuk shore Lake.

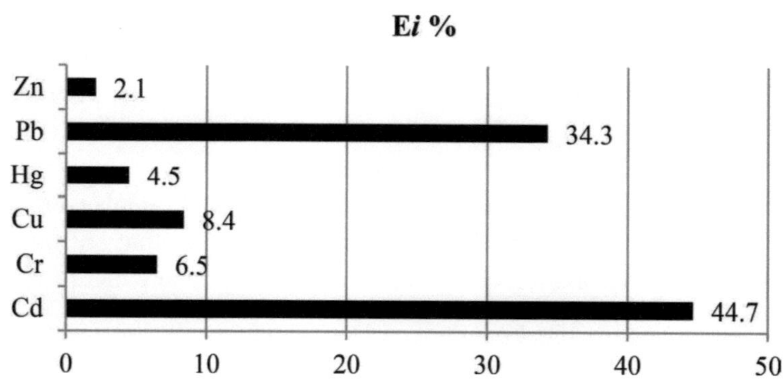

Figure 8. Percentage of heavy metals in potential ecological risk (RI) for the sediments of Bulbula Lake.

The calculated values of E_i indicate that in both lakes, compared to other metals, relatively higher potential risks are related to Pb and Cd (**Figures 7** and **8**). The highest risk was revealed for Cd in Boyuk shore Lake sediments.

4. Conclusions

In summary, the levels of heavy metal contamination of the main ecosystem components including soils, grounds, surface, and groundwaters were assessed in the oil-industrial region of the Absheron peninsula. In order to evaluate the contribution of anthropogenic sources to heavy metal contamination, pollution indexes were calculated for most toxic metals—As, Cd, Cr, Hg, Pb, Cu, Zn, and Ni.

Based on the results of researches, it can be concluded that the long-term impact of anthropogenic discharges has led to the pollution of all ecosystem components in the study site. The majority of the studied metals exceeded their permissible levels. The laboratory analysis revealed high levels of toxic heavy metals in grounds and groundwaters in the territory of aerator station. The studies showed that the concentrations of Cr, Cd, Zn, and Ni in the grounds of this site exceeded their MAC by several times. In order to evaluate the metal pollution status, the EF and E_i indexes of metals were calculated for three areas. Despite the fact that in the majority of samples, the concentrations of heavy metals were higher than their permissible levels, the calculated EF and E_i values suggest that the grounds in the site can be categorized as depletion to mineral enrichment area representing a moderate potential ecological risk with respect to heavy metals. Thus, the grounds in the territory of the aerator station cannot be characterized by a serious potential risk from heavy metals.

Based on the results of the comparative studies carried out in the oil-industrial area and the Baku-Mardakan transport road, we can say that the long-term impact of anthropogenic sources has led to the pollution of ecosystem by heavy metals in this area. The researches have revealed increased levels of heavy metal contamination. High concentrations of Pb, Cr, Zn, and Cu were detected in soils exceeding their permissible limits by several magnitudes that pose serious threat. Both oil industry and transport are the main sources of soil pollution in this site, where transport emission is the dominant factor.

Calculations showed that there is some correlation between the values of EF and $I_{geo.}$ for Boyuk shore and Bulbula Lakes, respectively. The accumulation of metals in the sediments of these lakes was not insignificant, indicating the condition of depletion to mineral enrichment. Analysis of the calculated EF, I_{geo}, and E_i values confirmed that in both lakes, the highest and lowest values of the mentioned indexes were recorded almost in the same metals that can be explained by their lithogenic sources. Higher quantities of heavy metals were detected in the samples from Boyuk shore Lake. This is the evidence of the fact that anthropogenic factors strongly influence the contamination of the lake sediments by heavy metals.

In recent years, according to the Presidential Decree, comprehensive works were implemented on the rehabilitation of nine natural lakes of the Absheron peninsula [28]. The works implemented on the rehabilitation and protection of the Boyuk shore Lake were divided into two phases. The first phase covered a 283-ha lake area and an area of 500 m from the lake. At this stage, the division of the lake into two parts was intended for the eastern part of the lake, adjacent to the Baku Olympic Stadium. The first phase was completed in 2015. However, it is important to have a detailed knowledge on the ecological situation of these lakes and surrounding areas to evaluate the extent of the works to be undertaken for further rehabilitation.

Overall, the potential ecological risk indices indicated a low potential risk (<40) from individual heavy metals, with exception of Pb posing a considerable potential risk level in the territory of Baku-Mardakan highway.

The calculated values suggest that the study area cannot be characterized by a high potential risk from heavy metals but increasing industrial development and urban growth can lead to a higher potential risk in the future years.

In order to prevent pollution and protect the ecosystem from the future disasters, it is possible to recommend

1. development and implementation of complex measures based on a scientifically proven approach to protect the ecosystem from the impact of toxic compounds including heavy metals;

http://dx.doi.org/10.5772/intechopen.74600

2. application of ICT systems for the development of an electronic database of the ecosystem pollution including the maps of heavy metals distribution indicating the anthropogenic sources of individual elements;
3. development of a well-rounded program to investigate the heavy metals impact on soil and water ecosystems taking into account their geological transformations, migration, and accumulation conditions including the acidity and alkalinity of soils, oxidation-reductive mode, the content of humus and the presence of readily soluble salts, and so on;
4. development and application of environmental-friendly technologies aimed at the minimization of industrial discharges and heavy metal pollution at all operational stages;
5. development and implementation of an integrated program for the elimination of transport emissions in the Absheron region, mainly in the city of Baku.

Author details

Fagan Aliyev[1], Hadiya Khalilova[2*] and Farhad Aliyev[3]

*Address all correspondence to: khalilova@rambler.ru

1 Azerbaijan Architecture and Construction University, Baku, Azerbaijan

2 Institute of Physics, Azerbaijan National Academy of Sciences, Baku, Azerbaijan

3 International Ecoenergy Academy, Baku, Azerbaijan

References

[1] Aliyev F, Khalilova H. The anthropogenic impact on surface water resources in Azerbaijan. Energy and Environment. 2014;**25**(2):343-356

[2] Second National Communication to the United Nations Framework Convention on Climate Change. Report of the Ministry of Ecology and Natural Sources of the Republic of Azerbaijan, Baku. 2010. 83 p

[3] Alekperov A. Absheron: Problems of Hydrogeology and Geoecology. Baku: State Book Chamber of Azerbaijan; 2000. 484 p. (in Rus.)

[4] Talibov A. Cartographic Analysis of Landscape-Ecological Conditions of the Absheron Peninsula. Baku: Chashioglu Publishing; 2004. 191 p. (in Rus.)

[5] Isayev S, Babayev F, Ragimzadeh A, Sultanov R. Ecologic-Geochemical Assessment of Changes in Biosphere of the Absheron Peninsula. Baku: MBM Publishing House; 2007. 470 p. (in Rus.)

[6] Khalilova H, Mamedov V. Assessing heavy metal pollution of sediments of the Boyuk-Shore lake in the Absheron industrial region of Azerbaijan. In: Proceedings of the 18th

Conference "Engineering geology and geoecology. Fundamental problems and applied tasks" dedicated to 25th anniversary of foundation of Institute of Geoecology of Russian Academy of Sciences named after E. M. Sergeyev; 25-26 March 2006; Moscow. pp. 429-433 (in Rus.)

[7] Aliyev M. Ecological Monitoring of the Caspian Sea Water Area. Baku: Tahsil Publishing; 2009. 290 p. (in Azerbaijani)

[8] Reymers N. Ecology (Theory, Rules, Laws, Principles and Hypotheses). Moscow: Nauka; 1994. 367 p. (in Rus.)

[9] Puschenreiter M, Horak F, Hartt W. Low cost agricultural measures to reduce heavy metal transfer into food chain, a review. Plant, Soil and Environment. 2005;**51**:1-11

[10] Teplaya G. Heavy metals as a factor environmental pollution, a review. Bulletin of Environmental Education. 2013;**1**(23):182-192 (in Rus.)

[11] Smith C, Hopmans P, Cook F. Accumulation of Cr, Pb, cu, Ni, Zn and cd in soil following irrigation with treated urban effluent in Australia. Environmental Pollution. 1996;**94**:317-323

[12] Mammadov V. Lakes of Absheron Peninsula. National Atlas of the Republic of Azerbaijan. Cartography Factory: Baku; 2014. 187 p

[13] Dobrovolskiy V. Textbook for the Students of Higher Education Institutions. Moscow: "Academy" Publishing House; 2003. 400 p. (in Rus.)

[14] Osunkiyesi A, Taiwo A, Olawunmi O, et al. Environmental impact of heavy metal contaminants and micro nutrients in soil samples of metal dumpsites in Abeokuta, Ogunstate, Nigeria. IOSR Journal of Applied Chemistry. 2008;**7**(5):52-55

[15] Samedov P, Bababekova L, Aliyeva V, Mamedzade V. Biological Characteristics of Technogenic Contaminated Soils. Baku: Elm Publishing; 2011. 104 p. (in Rus.)

[16] Urushadze T, Ghambashidze G, Blum W, Mentler A. Soil contamination with heavy metals in Imereti region (Georgia). Bulletin of the Georgian National Academy of Sciences. 2007;**175**(1):122-130

[17] Mirbabayev M. Microelement composition of oils according to neutron activation analysis. Chemistry and Technology of Fuels and Oils. 1997;**5**:46-47 (in Rus)

[18] Qiu H. Migration mechanism of organic pollutants in national water-body sediments. Geography and Geology. 2011;**1**:239-246

[19] Khalilova H, Mammadov V. Assessing the anthropogenic impact on heavy metal pollution in soils and sediments in urban areas of Azerbaijan's oil industrial region. Polish Journal of Environmental Studies. 2016;**25**(1):159-166. DOI: 10.152.44/pjoes/60723

[20] Standard Methods for Examination of Water and Wastewater. 19th ed. Washington, DC: American Public Health Association/American Water Works Association/Water Environment Federation; 1995

[21] Test Methods for Evaluating Solid Wastes. USEPA; 1986

[22] Hu Y, Liu X, Bai J, et al. Assessing heavy metal pollution in the surface soils of a region that had undergone three decades of intense industrialization and urbanization. Environmental Science and Pollution Research. 2013;**20**:6150-6159. DOI: 10.1007/s11356-013-1668-z

[23] Liu J, Zhuo Z, Sun S, et al. Concentrations of heavy metals in six municipal sludges in Guangzhou and their potential ecological risk assessment for agricultural land use. Polish Journal of Environmental Studies. 2015;**24**(1):165-174. DOI: 10.15244/pjoes/28348

[24] Muller G. Schwermetallein den sediments des Rheins-VeraderungenSeitt. Umschan. 1979;**79**:778-783

[25] Hakanson L. An ecological risk index for aquatic pollution control. A sedimentological approach. Water. Research. 1980;**14**(8):975-1001

[26] Uzoekwe S, Achudume A. Pollution status and effect of crude oil spillage in Ughoton stream ecosystem in Niger Delta. Journal of Ecology and the Natural Environment. 2011;**3**(15):469-473. DOI: 10.5897/JENE11.071

[27] Water Quality Criteria (WQC). A Report of the Committee on Water Quality Criteria. Washington, DC: NAS; 1972. 593 p

[28] Report of Ministry of Emergency Situations on the Management of Lakes, Reservoirs, Surface and Underground Waters in the Absheron Peninsula. Baku; 2015. 263 s

Chapter 21

Heavy Metal Pollution as a Biodiversity Threat

Efraín Tovar-Sánchez, Isela Hernández-Plata,
Miguel Santoyo Martínez,
Leticia Valencia-Cuevas and Patricia Mussali Galante

Additional information is available at the end of the chapter

http://dx.doi.org/10.5772/intechopen.74052

Abstract

Heavy metals exert their toxic effects through different mechanisms. Lately, increasing attention has been focused on understanding the long-term ecological effects of chronically exposed populations and communities and their consequences to the ecosystem. The long-term exposure to heavy metals in the environment represents a threat to wild populations, affecting communities and putting ecosystem integrity at risk. Therefore, this type of exposure represents a threat to biodiversity. In the field, metal exposure is generally characterized by low doses and chronic exposures. This type of exposure exerts alterations across levels of biological organization. Distribution and abundance of populations, the community structure and the ecosystem dynamics may be altered. This chapter will focus on how chronically metal exposures in the field affect negatively populations and communities becoming a threat to biodiversity. Also, attention is put on the tools that enable to characterize and analyze the detrimental effects of heavy metal exposure on wild populations. Hence, the use and development of biomarkers in ecotoxicology will be discussed.

Keywords: biological organization levels, biomarkers, ecosystem health, ecotoxicology, biodiversity

1. Introduction

Heavy metals can be emitted into the environment by natural sources and anthropogenic activities, being the anthropogenic activities the main causes of emission. Among these, mining operations represent the greatest threat to ecosystem integrity due to the persistence of heavy metals in the environment, which persist for hundreds of years after the cessation of mining operations [1]. In environmental exposures, these toxicants exert their effects trough different mechanisms, being chronic exposures at low doses of complex metal mixtures the responsible for the effects observed in wild animal populations and communities, with implications at the ecosystem level [2]. Therefore, this type of exposure represents a threat to biodiversity.

Exposed individuals integrate exposure to contaminants in their environment and respond in some measurable and predictable way, being these responses observed across different levels of biological organization [3]. Hence, to better understand the ecological consequences of metal exposure, the use of biological markers or biomarkers is necessary. Biomarkers are tools that enable the analysis of the extent of exposure and the effects of environmental chemical contamination [4]. These measures offer valuable predictors of ecologically relevant effects. However, in ecotoxicology, where exposed populations, communities and the consequences at the ecosystem level are the point of interest, the use of biomarkers is not an easy task, since the responses to toxic chemical stress become less specific and many variables interfere with physiological responses. In this context, Bickham and coworkers explain that although the damage from xenobiotic exposure is at the cellular or genetic levels, effects can be observed at higher levels of biological organization (emergent effects) [5]. It is important to take into consideration that the biomarker response must be tightly and regularly connected to responses at these higher levels, particularly if the biomarkers are to be used as effect indicators [6]. Also, biomarkers of exposure (external dose, internal dose; bioaccumulation levels), biomarkers of biological effective dose (DNA adducts) and biomarkers of effect (DNA breaks) must be used to analyze the relationship between the cellular and genetic effects with ecological responses.

At the population level, some considerations must be taken into account when analyzing the emergent effects. Some of these effects are: Shifts in sex proportions, age structure alterations, low reproductive success, inbreeding, genetic structure and diversity alterations, low fitness and population declines [7]. However, these effects are not specific to environmental metal exposures. Hence, differences in biomarker response among populations of a species may be taken carefully by analyzing geographical influences, habitat influence, population vulnerability (specific to the population in question) and exposure history [6]. In the last decade, one of the emergent effects that has been evaluated in environmentally exposed populations are shifts in their genetic pool, which were defined by Mussali-Galante and collaborators as permanent biomarkers [2].

At the community level: Shifts in diversity and species richness, changes in dominant species, changes in species composition and biodiversity loss may be some of the emergent effects. However, due to the complexity of species interactions, such effects cannot be accurately predicted from effects at the population level, as was recognized many years ago [8–10].

Studies assessing community level responses to environmental metal stress are mostly conducted in aquatic ecosystems using invertebrate and fish communities. Among the few studies conducted in terrestrial ecosystems, insect communities are the point of interest [11, 12]. In these type of studies bioaccumulation levels are analyzed in different invertebrate groups and the relationship between bioaccumulation levels and community effects (mainly species richness and composition) is examined.

At the ecosystem level, biomagnification (bioaccumulation within successive trophic levels) has been well documented for some metals. Trophic chain effects have been observed where individuals that feed lower on the food chain generally are exposed to lower metal concentrations. In these type of studies, the primary producers (plants) represent an important step in metal transfer since they constitute the foundation of the food chain. Hence, certain metals can be transported from plants to higher strata of the food chain, representing a threat to biodiversity and to ecosystem integrity [13].

2. Heavy metal effects on terrestrial wild animal populations and communities

Heavy metal (HM) exposure affect the health and survival of the individuals, resulting in negative impacts in the subsequent levels of biological organization, like populations and communities. The first step in ecotoxicological studies is to determine HM concentrations in soil and in the organism in question and then, to analyze the effects at the population level using emergent properties like male/female ratio, age class, reproductive success, inbreeding, genetic diversity and fitness. At the community level, species richness and diversity, dominant species, and species composition [7] are some of the characteristics that have been analyzed. In general, the Shannon-Wiener diversity index is one of the most used parameters in ecotoxicological studies. For a better understanding of the effects of HM on terrestrial ecosystems, the study of the functioning of detritivore soil communities [14–16] has been incorporated, where parameters such as biomass, soil organic matter content, microbial respiration, microbial biomass carbon, and the phosphatase activity have been analyzed.

2.1. Heavy metal accumulation in terrestrial invertebrates

Heavy metals may enter the trophic chain trough primary producers and invertebrates that live in soils [17]. Invertebrate are widely used in ecotoxicological studies due to their easy capture, wide distribution, high abundance, their key ecological roles such as soil decomposers, constitute the first step in trophic chains, low mobility, and are in close contact with soils [18–20]. For example, Gramigni et al. detected a relationship between HM (Zn, Ni, Mn, Cd and Pb) in soils and their bioaccumulation in ants *Crematogaster scutellaris* [21].

In invertebrate communities HM bioaccumulation has been observed in target organs. For example: Wilczek and Babczyńska found that spiders (*P. amentata, L. triangularis, M. segmentata, A. diadematus, A. marmoreus*) had higher bioaccumulation levels of Cu, Zn and Cd in their hepatopancreas than in the gonads [22]. Also, Gramigni et al. documented that ants (*Crematogaster scutellaris*) bioaccumulate Mn and Zn in their intestines, being Zn accumulated specifically in Malpighi tubules and low Zn concentrations were find in fat tissue [21]. In contrast, Ni, Pb, and Cd did not bioaccumulate specifically in target organs.

Some of the most invertebrate Phyla used in ecotoxicological studies are Arthropoda and Annelida [18]. Bioaccumulation patterns depend on the species or taxonomic group in question (Class, Order, Family, Genera). For example, Wilczek and Babczyńska studied different spider species (*Pardosa amentata, Linyphia triangularis, Metellina segmentata, Araneus diadematus, Araneus marmoreus*) [22]. They found that the spider *P. amentata* bioaccumulates more Zn in the hepatopancreas than other spider species that inhabit in the same polluted site. Also, HM bioaccumulation differences have been observed in different taxonomic classes. It was evidenced that bioaccumulation of Cr, Cu, Ni, Pb and Zn in the order Araneida (Class Arachnida) were higher than those detected in for the order Coleoptera (Class Insecta) [17]. Moreover, they found an effect of HM bioaccumulation depending on the analyzed order, being the order Araneida who presented a positive correlation between Zn bioaccumulation in spiders and Zn concentrations in soils. In contrast, in the order Coleoptera a positive correlation was register between Cd and Pb bioaccumulation and soil concentrations, these differences may be attributed to feeding preferences between both orders (predators and herbivores).

2.2. Heavy metal effects on populations and communities of terrestrial invertebrates

It has been documented that HM bioaccumulation in organisms may modify their body size. For example: Jones and Hopkin studied woodlice populations (*Porcellio scaber*) in polluted environments (Zn, Cd, Pb and Cu) [23]. The author reported that HM bioaccumulation had an effect on head size independently of the gender (male vs. female), being bigger in individuals from contaminated sites. They propose that environmental stressors generate costs in individuals because of the detoxification process, a fact that results in negative effects on their health and survival rates. Hence, HM exposure may have consequences in higher levels of biological organization such as populations and communities.

In ecotoxicology, gradient studies are necessary. They offer the visualization of gradual changes in HM soil concentrations in polluted sites and their relationship with population distribution and abundance and with some community structure parameters. Spurgeon and Hopkin found a negative correlation between the distance from the pollution source and HM concentrations (Pb, Cd, Zn and Cu) [24]. Also, a negative correlation was registered between HM concentration and the absolute abundance of six earthworm species (Phyllum Annelida: *L. rubellus*, L. cataneus, *L. terrestris, A. rosea, A. caliginosa, A. chlorotica*). Particularly, the species *L. rubellus, L. castaneus* and *L. terrestris* were present in most sites, whereas the species *A. rosea, A. caliginosa* and *A. chlorotica* were absent at the nearest sites to the pollution source. These differences were attributed to differences in calcium metabolism between earthworm species. Also, earthworm communities (*L. terrestres, L. rubellus, L. castaneus, A. chlorotica, A. rosea, A. caliginosa, A. longa, E. tetraedra, M. minúscula, O. cyaneum*) in a polluted site (Pb, Cd, Zn and Cu) were studied [25]. They found that the relative abundance and the Shannon-Weiner diversity index were lower on the nearest sites to the pollution source and higher in the farthest sites from the pollution source. However, the dominance index (Berger-Parker) was higher in sites near the pollution source because *L. rubellus, L. castaneus* and *L. terrestres* were the dominant species in those sites. The study by Jung et al. demonstrated that although the Shannon-Wiener diversity index values were similar between contaminated and less contaminated sites (Pb, Cd) for six different spider families, the abundance of the Linyphiidae family was correlated with metal concentrations in soils and enabled the discrimination between contaminated and less contaminated sites [12]. In contrast, others analyzed the community of the Phyla Annelida and Arthopoda, they found that absolute and relative abundances of all the organisms was lower in the most polluted sites [26]. Specifically, the earthworm species *A. caliginosa* had higher densities in non-polluted sites and was absent in polluted sites, on the contrary, the larvae of the coleopteran Hoplinae predominated in highly Zn polluted sites.

Also, there are studies that have shown that HM exposure does not affect some taxonomic groups. For example, Zaitsev and van Straalen studied the mite community (Phyllum Arthropoda) from contaminated soils (Pb, Zn, Cu, Fe, Cd) [19]. They evidenced that although a metal contamination gradient was found in the soils, this gradient was not detected in bioaccumulation of HM in mites, and no effects were found in the community structure. Likewise, Migliorini et al. did not find differences between the abundance of some arthropod groups (Collembola, Protura and Diplura), in contaminated sites by Pb and Sb, while other groups were absent (Symplyla) [20]. Hence, HM exposure affects differently the community structure of different invertebrate groups. Through the study of the functioning of detritivore soil communities [14–16] some parameters like biomass, soil organic matter content, microbial respiration, microbial biomass carbon, and the phosphatase activity are analyzed as biomarkers for HM effects at the community level. Hobbelen and colleagues studied millipedes, isopods, and earthworms in contaminated zones (Zn, Cu, Cd), where no correlation was found between community structure (richness and density) and soil metal content [14]. On the contrary, Zn concentration correlated positively with biomass of the earthworm *Lumbricus rubellus*. On the other hand, the soil organic matter content explained the variation in species density, showing that HM concentration in soils is not the only variable that influences the community structure.

The aforementioned studies evidenced that community structure and function of terrestrial invertebrates, facilitates the evaluation of HM impact on the first trophic chain levels, as well as their incorporation and biomagnification patterns. Therefore, studies assessing HM bioaccumulation in other trophic levels like terrestrial vertebrates complete the knowledge of the effects of HM in the ecosystem health.

2.3. Heavy metal bioaccumulation effects on health of terrestrial vertebrates

In wild vertebrates, information regarding HM bioaccumulation and their effects on target organs is vast. Some examples of wild vertebrate species used in ecotoxicological studies are: Brown bears (*Ursus arctos*), Gray wolfs (*Canis lupus*), Eurasian badgers (*Meles meles*) and Pine martens (*Martes martes*) [27], bank vole (*Clethrionomys glareolus*), yellow-necked mouse (*Apodemus flavicollis*) [28–30], wood mice (*Apodemus sylvaticus*) [31, 32], tuco-tuco (*Ctenomys torquatus*) [33], greater white-toothed (*Crosidura russula*) [34], *Peromyscus melanophrys*, pygmy mice (*Baiomys musculus*) [7, 35].

In general, studies on HM bioaccumulation on wild life, have detected an effect of the study species and the target organ on bioaccumulation patterns. For example, Bilandžić and collaborators analyzed HM bioaccumulation on wild carnivores, the authors report that the highest Cd concentration was present in kidney and liver of the Eurasian badger (*M. meles*) [27]. While Cu concentration in liver decreased among studied species showing the next pattern: Eurasian badger > Brown bear > Pine marten > Eurasian lynx > Gray wolf. The Eurasian badger registered the highest concentrations in muscles (As, Cu, Pb), liver (As, Cd, Cu, Pb) and kidneys (Cd, Pb) and the Pine marten accumulated the highest concentrations in kidneys (As, Cu, Hg). However, in the scientific literature, contrasting results about the effects of HM exposure on the health of the individuals are present. In this context, Levengood and Heske showed that white-footed mice that inhabit in a Cd and Zn polluted site, registered the highest Cd, Cu and Zn concentrations in liver, in comparison to unexposed individuals [36]. In spite of the bioaccumulation levels observed, they did not detect changes in the health of the individuals (reproductive and fitness parameters).

In contrast, other studies have detected histological changes in exposed individuals to HM. Damek-Poprawa and Sawicka-Kapusta found that yellow-necked mouse individuals that live in a polluted site, bioaccumulate more Pb and Cd that unexposed individuals [28]. In particular, individuals bioaccumulate Pb in their femur and Cd in kidneys. Also, histopathological studies showed that exposed individuals presented multiple organ alterations in liver, kidneys and testicles. Similar results were found for wood mice and the greater white-toothed living in landfill zones [31].

Studies assessing the effects of HM bioaccumulation on population and community parameters are scarce, a fact that may be attributed to sampling technique, which is influenced by the size and mobility of the individuals and trapping success (e.g. site perturbance, water availability, predator activity, migration index, etc.), among others. Therefore, in order to infer the population health status, some studies have considered the gender, age, (age class), reproductive condition, litter size (number of embryos, placental scars; embryos/scars per female) embryos weight, trap success, and condition index. In this regard, Santolo documented that male and female individuals of the deer mice exposed to Se from a contaminated site, had a lower condition Index than those from the reference site [37]. In addition, the ratio of males to females age class and reproductive condition were similar between individuals from both sites. Except from individuals from the polluted site that their reproductive condition was lower. This last result suggest that Se exposure affects negatively rodent populations, among other factors.

In addition, in terrestrial vertebrate communities, changes or alterations in community parameters may be due to the competitive selection of the most tolerant species. Moreover, some species may be opportunistic and HM residues may serve a protection mechanism against their predators [38].

Other parameters used as population level biomarkers are: residual index "RI" (linear regression between body weight and body length without tail) and kidney size proportion. RI is used as "energy reserve" measure. Individuals with positive RI values are considered as better fitted individuals, and the increase in kidney relative weight suggest the presence of a stressor [38].

2.4. Heavy metal effects on genetic diversity of exposed populations

Although HM exposure has immediate effects at the molecular and cellular levels, they may extend to higher levels of biological organization, like the genetic structure and diversity of the exposed populations [3]. Chronic exposures at low doses is one of the factor implicated in changes in the genetic pool of the populations, especially if chemical agents are capable of inducing DNA damage, such as HM. In general, there are four mechanisms by which HM exert their effects on the genetic diversity of exposed populations: (1) Some HM are genotoxic, mutagenic and alter DNA repair processes, increasing the mutational load of the individuals; (2) HM exposure favors the presence of tolerant genotypes and the elimination of intolerant ones, changing the genetic composition of the exposed population; (3) HM may cause bottlenecks and (4) alter migration patterns, increasing or decreasing genetic flow between populations [39–41].

Exposed population to HM pollution may have two types of response on genetic diversity levels: (a) increase in genetic diversity levels as a consequence of induced mutations by genotoxins or (b) decrease in genetic diversity levels as a result of bottlenecks [7]. In both cases, these responses are the consequence of the adaptation of the population to polluted environments [3, 41–44].

In general, studies where 11 mammal species were analyzed for the effects of HM exposure (being Cd, Zn, Cu and Pb the most common) on genetic diversity, the pattern found was that 45.4% of the analyzed species displayed a decrease in genetic diversity levels in comparison to non-exposed populations. Some examples are: *Peromyscus melanophrys*: As, Pb, Cd, Cu, Zn [7], *Cognettia sphagnetorum*: Cu [45]; *Talitrus saltator*: Cd, Hg, Cu, [46]; *Pachygrapsus marmoratus*: As, Pb, Cd, Co [47]; *Ficedula hypoleuca*: Cd, Zn, Pb, Cu, Ni, Al, As, Cr, Se [48]. In contrast, an increase in genetic diversity levels in 36.4% of the analyzed species was found, like *Lumbricus rubellus*: Cd, Zn, Cu, Pb [49]; *Cepaea nemoralis*: Cd, Cr, Cu, Ni, Pb, Zn [50]; *Parus major*: Cd, Zn, Cu, Pb, Ni, Al, As, Cr, Sn [48] and *Larus argentatus* (steel), [51]. Meanwhile, the remaining 18.2% of the studied species did not register changes in genetic diversity levels, such as: *Apodemos sylvaticus*: Cd, Zn, Cu, Pb, Ni, Al, Ag, As, Co, Mn, Fe [42] and *Succinea putris*: Cd, Cr, Cu, Ni, Pb, Zn [50].

Changes in genetic diversity levels as a consequence of metal pollution may serve as a biomarker of permanent effects. Mussali-Galante et al. defined permanent biomarkers as changes in genetic structure and diversity due to metal pollution that cannot be the same as they were before the exposure [2].

Finally, polluted environments may be considered as unique systems because their different origin, pollution type and degree are specific for each site. Additionally, edaphic characteristics, weather and vegetation type differ between sites. Hence, it is expected that there will be differences among exposed sites that may or may not alter populations and communities; therefore, the use of bioindicator o sentinel species becomes important. Species that represent biological diversity in terms of feeding preferences, life cycles, trophic chain position, etc. are the point of interest. This last initiative permits to identify susceptible species to environmental stressors such as HM. Basu et al. suggested that sentinel species should have: wide geographical distribution, high abundance, capacity to bioaccumulate HM, easy capture and sampling, low mobility and well known biology [52]. In terrestrial environments, small mammals are commonly used because of their similar physiological systems to humans.

3. Heavy metal effects on ecosystems

Ecosystems are open thermodynamic systems of matter and energy effluxes, which maintain stable from the balance of their biotic and abiotic components [53]. Ecosystem stability may be altered because of the incorporation of HM, derived from mining activities [54].

HM incorporation in the ecosystem depends mainly on their bioavailability and thereafter, through their incorporation into the trophic chain, reaching their highest concentrations in the last levels of the chain, a process known as "biomagnification" [55, 56]. Under this situation, ecosystems may be or may not be affected by HM, depending on the magnitude and exposure time, or if one of the functions that maintain the ecosystem integrity is compromised (e.g. nutrient cycles, energy efflux) due to biodiversity loss [57].

Many ecotoxicological studies that assess HM effects on terrestrial ecosystems have focused on the analysis of HM concentrations in soils, their bioavailability and bioaccumulation, but few have analyzed their biomagnification trough the trophic chain and their effects on ecosystem integrity [54].

3.1. Heavy metal incorporation into the ecosystems

The first step of HM incorporation into the ecosystems is because their bioavailability potential and soil mobility, where metallic cations adhere to negative charged particles like clay and organic matter, when metals separate from these soil particles, they enter the soluble soil fraction, being bioavailable and having the potential to bioaccumulate in different organisms [56]. Microorganisms, plants and invertebrate species have mechanisms to incorporate trace metals for their development and survival (e.g. Cu, Ni, Fe, Co, Mn and Mg), however, these can be toxic in higher concentrations. Also, these same mechanisms facilitate the entrance of non-trace metals (As, Cd, Hg and Pb) in the organisms, which are highly toxic at low concentrations [58].

Microorganism are vital elements of soils, they participate in nutrient and inorganic element recycling, like minerals and trace metals, for plants that constitute the first trophic level in terrestrial ecosystems [59]. However, HM pollution may affect microorganism communities, generating changes in their structure and biodiversity, which in turn, has consequences on the soil processes in which they participate [60, 61]. For example, development alterations and in biochemical processes of microorganism have been reported [61–63]. Such alterations affect organic matter decomposition process, reducing nutrient accumulation and availability for plants and compromising matter and energy fluxes at the base of the trophic chains [61]. On the other hand, soil invertebrates can bioaccumulate HM because of their feeding preferences, like crustaceans, snails, and earthworms that inhabit leaf litter and feed on organic matter with high HM concentrations [64, 65]. In fact, it has been proven that these invertebrate groups had the highest HM concentrations in comparison to beetles or butterflies, or even higher concentrations than some vertebrate groups [18, 64].

3.2. Heavy metal transfer along food chain

In particular, it has been suggested that HM hyperaccumulation by plants as a defense mechanism against herbivores, may transcend to higher trophic levels. For example, the plant *Streptanthus polygaloides* (Brassicaceae) is a Ni hyperaccumulator species, that is consumed by the herbivore *Melanotrichus boydi* (Hemiptera) who bioaccumulates Ni in its body as a defense mechanism against the predator species *Misumena vatia* (Araneae) [66].

HM transfer along the trophic chains varies depending on the type of metal, the trophic level in question and the number and type of species that integrate it. For example, [18] report that

Cd, Cu, Pb and Zn concentrations among invertebrate groups registered the next pattern: Isopoda > Lumbricidae > Coleoptera which is attributed to their different feeding patterns [57]. Another interesting example is that Cd is more mobile towards herbivores and their predators, while Zn is less efficient in its transfer to higher trophic levels [67, 68].

HM transfer along the trophic chain has been reported, for example for the Ni hyperaccumulator plant *Alyssum pintodalsilvae* (Brassicaceae), that transfers Ni to grasshoppers (herbivore) and spiders (carnivorous insect), having the spiders higher Ni concentrations [69]. Similar results were reported by Boyd and Wall [66]. These studies demonstrate that HM can be transferred among invertebrate species, mobilizing metals from one trophic level to another, reaching animals such as small mammals [70–72]. These last studies evaluated HM concentration in small mammals, reporting higher HM levels in carnivorous or omnivorous mammals in comparison to those that feed only by plants. So, HM transfer along the trophic chain not only depends on the magnitude of exposure, but on the species type, season, gender, age and metal type [73].

Additionally, HM bioaccumulation in plants may also affect interactions with their pollinators, since HM can transfer to nectar, a fact that alters pollinators feeding patterns, suggesting that metals and metalloids such as Se found in pollen and nectar affect negatively the pollinators, which results in changes in plant communities due to the nonappearance of pollinators on such plants [74–77].

Most of the studies about metal transfer along trophic chains in terrestrial ecosystems focus on at least three trophic chains levels. In contrast, a study by Hsu and collaborators in more than three trophic levels, they report high levels of Cd, Hg, Pb and Sn and their biomagnification in all analyzed levels, being the snails and the earthworms the groups who registered the highest metal concentrations [64].

All the aforementioned studies have evaluated HM transfer in small trophic chains in terrestrial ecosystems, the majority of them analyze three trophic levels. These studies are of great importance because the information generated helps to understand the general patterns of HM transfer along trophic chains, especially for the most common metals like Cd, Cu, Pb and Zn. Also, these studies highlight that HM transfer, assimilation and excretion in organisms along the trophic chain, can have extended effects, at the individual level (altering their health and physiology) at the population level (modifying population dynamics, abundance, distribution and their genetic pool) at the community level (altering species richness and diversity), affecting then, the ecosystem dynamics. Therefore, is very important to conduct studies where more than three trophic chain levels are analyzed and to integrate new biomarkers (e.g. stable isotope techniques; which enable to follow HM transfer along trophic chains by knowing the extent of the pollutant flux in the chain [78]).

4. Conclusions and perspectives

Chronic environmental metal exposures exert their negative effects on individuals health, having consequences at the population and community levels, putting ecosystem integrity at risk. However, the recognition and use of biomarkers in ecotoxicology has been a difficult task, due to the unspecific responses and multiple variables that affect physiological responses to toxic stress. Therefore, it becomes necessary that ecotoxicological studies include: HM concentrations in soils, bioaccumulation parameters in vertebrate and invertebrate species, the relationship between these biomarkers with morphological, anatomic and physiological alterations that may alter population parameters. In particular, the use of bioindicator or sentinel species is necessary in order to evidence the consequences of HM exposure in wild populations.

Terrestrial invertebrates have been used as an ideal system to evaluate community responses to environmental chemical stress, due their easy capture, wide distribution, great abundance, low mobility and close contact to HM from soils. Especially, earthworms and arthropods are the most studied organisms. On the contrary, the studies that evaluate HM effects on vertebrate community structure are scarce, probably due to their body size, mobility and sampling difficulties. However, when working with vertebrates, an excellent alternative has been the study of small mammal species that serve as good bioindicators and the results may be easily compared to humans. Also, a methodological strategy in many studies has been the use of pollution gradients in order to visualize slight changes in HM concentrations along the soil gradient and to relate these changes to some community structure parameters. At the moment, we can conclude that HM affect differently the community structure and the community functioning of the different animal groups studied so far.

At the community level, the search for new biomarkers continues. In this context, abundance changes in different guilds that conform the community may also be used as a biomarker since changes in abundance or guild disappearance in exposed communities may serve as an ecological response to chemical stress.

At the ecosystem level, ecotoxicological studies are very limited. Trophic chain alterations, biomagnification and modifications in nutrient and energy cycles have been reported. Studies generally asses HM transfer along three trophic levels, such studies have concluded that metal flux depends on the biology of the species, on the trophic position in the chain and on the metal type or metal mixture in question. Mainly, HM transfer from plants to invertebrate herbivores (insects) and from insects to other invertebrates (spiders) or predator vertebrates (small mammals) has been the point of interest. The information from these studies has gained attention, especially because human beings represent the last level of the trophic chain, such as in the case of agroecosystems. It is desirable to use as biomarkers in ecosystem studies, measures of stable carbon and nitrogen isotopes for evaluation of HM transfer along terrestrial trophic chains.

Finally, it necessary that future efforts integrate different biological and ecological responses across all levels of biological organization as a result of biomarker approaches. Moreover, study designs should be more rigorous, including multispecies and multibiomarkers that permit the evaluation of HM exposure in a more realistic way, which in turn will allow to predict, understand and resolve in a better way HM pollution problems worldwide.

Acknowledgements

We thank the Consejo Nacional de Ciencia y Tecnología (CONACyT) for the scholarships to IHP and to MSM.

Conflict of interest

The authors declare that there is no conflict of interest.

Author details

Efraín Tovar-Sánchez[1], Isela Hernández-Plata[2], Miguel Santoyo Martínez[1], Leticia Valencia-Cuevas[1] and Patricia Mussali Galante[2]*

*Address all correspondence to: patricia.mussali@uaem.mx

1 Laboratory of Molecular Markers, Center for Research on Biodiversity and Conservation, Autonomous University of Morelos State, Cuernavaca, Morelos, México

2 Laboratory of Environmental Research, Center for Biotechnology Research, Autonomous University of Morelos State, Cuernavaca, Morelos, México

References

[1] Duruibe JO, Ogwuegbu MOC, Egwurugwu JN. Heavy metal pollution and human biotoxic effects. International Journal of Physical Sciences. 2007;**2**:112-118

[2] Mussali-Galante P, Tovar-Sánchez E, Valverde M, Rojas E. Genetic structure and diversity of animal populations exposed to metal pollution. Reviews of Environmental Contamination and Toxicology. 2014;**227**:79-106. DOI: 10.1007/978-3-319-01327-5_3

[3] Bickham J, Sandhu S, Hebert P, Chikhi L, Athwal R. Effects of chemical contaminants on genetic diversity in natural populations: Implications for biomonitoring and ecotoxicology. Mutation Research. 2000;**463**:33-51. DOI: 10.1016/S1383-5742(00)00004-1

[4] Handy R, Galloway T, Depledege M. A proposal for the use of biomarkers for the assessment of chronic pollution and in regulatory toxicology. Ecotoxicology. 2003;**12**:331-343. DOI: 10.1023/A:1022527432252

[5] Bickham J, Smolen M. Somatic and heritable effects of environmental genotoxins and the emergence of evolutionary toxicology. Environmental Health Perspectives. 1994; **102**:25-28

[6] Forbes VE, Annemette P, Lis B. The use and misuse of biomarkers in ecotoxicology. Environmental Toxicology and Chemistry. 2006;**25**:272-280. DOI: 10.1897/05-257R.1

[7] Mussali-Galante P, Tovar-Sánchez E, Valverde M, Rojas E. Biomarkers of exposure for assessing environmental metal pollution: From molecules to ecosystems. Revista Internacional de Contaminación Ambiental. 2013;**29**:117-140

[8] Forbes T, Forbes V. A critique of the use of distribution-based extrapolation models in ecotoxicology. Functional Ecology. 1993;**7**:249-254. DOI: 10.2307/2390202

[9] Hopkin S. Ecological implications of 95% protection levels for metals in soil. Oikos. 1993;**66**:137-141. DOI: 10.2307/3545206

[10] Lagadic L, Caquet T, Ramade F. The role of biomarkers in environmental assessment. Invertebrate populations and communities. Ecotoxicology. 1994;**3**:193-208. DOI: 10.1007/BF00117084

[11] Nahmani J, Lavelle P. Effects of heavy metal pollution on soil macrofauna in a grassland of northern France. Ecotoxicology and Environmental Safety. 2012;**86**:188-197. DOI: 10.1016/j.ecoenv.2012.09.013

[12] Jung MP, Kim ST, Kim H, Lee J. Species diversity and community structure of ground-dwelling spiders in unpolluted and moderately heavy metal-polluted habitats. Water, Air, and Soil Pollution. 2008;**195**:15-22. DOI: 10.1007/s11270-008-9723-y

[13] Peralta-Videa J, Laura Lopez M, Narayan M, Saupe G, Gardea-Torresdey J. The biochemistry of environmental heavy metal uptake by plants: Implications for the food chain. International Journal of Biochemistry & Cell Biology. 2009;**41**:1665-1677. DOI: 10.1016/j.biocel.2009.03.005

[14] Hobbelen HF, van de Brinkb PJ, Hobbelena JF, van Gestela CA. Effects of heavy metals on the structure and functioning of detritivore communities in a contaminated flood-plain area. Soil Biology and Biochemistry. 2006;**38**:1596-1607. DOI: 10.1016/j.soilbio.2005.11.010

[15] van Gestel CA, Koolhaas JE, Hamers T, van Hoppe M, van Roovert M, Korsman C, Reinecke SA. Effects of metal pollution on earthworm communities in a contaminated floodplain area: Linking biomarker, community and functional responses. Environmental Pollution. 2009;**157**:895-903. DOI: 10.1016/j.envpol.2008.11.002

[16] Niemeyer JC, Nogueira MA, Carvalho GM, Cohin-De-Pinho SJ, Outeiro US, Rodrigues GG, da Silva EM, Sousa JP. Functional and structural parameters to assess the ecological status of a metal contaminated area in the tropics. Ecotoxicology and Environmental Safety. 2012;**86**:188-197. DOI: 10.1016/j.ecoenv.2012.09.013

[17] Schipper AM, Wijnhoven S, Leuven RS, Ragas AM, Hendriks AJ. Spatial distribution and internal metal concentrations of terrestrial arthropods in a moderately contaminated lowland floodplain along the Rhine River. Environmental Pollution. 2008;**151**:17-26. DOI: 10.1016/j.envpol.2007.03.007

[18] Heikens A, Peijnenburg WJ, Hendriks AJ. Bioaccumulation of heavy metals in terrestrial invertebrates. Environmental Pollution. 2001;**113**:385-393. DOI: 10.1016/S0269-7491(00)00179-2

[19] Zaitsev AS, van Straalen NM. Species diversity and metal accumulation in oribatid mites (Acari, Oribatida) of forests affected by a metallurgical plant. Pedobiologia. 2001;**45**:467-479. DOI: 10.1078/0031-4056-00100

[20] Migliorini M, Pigino G, Bianchi N, Bernini F, Leonzio C. The effects of heavy metal contamination on the soil arthropod community of a shooting range. Environmental Pollution. 2004;**129**:331-340. DOI: 10.1016/j.envpol.2003.09.025

[21] Gramigni E, Calusi N, Gelli L, Giuntini M, Massi G, Delfino G, Chelazzi D, Barachi D, Frizzi F, Santini G. Ants as bioaccumulators of metals from soils: Body content and tissue-specific distribution of metals in the ant *Crematogaster scutellaris*. European Journal of Soil Biology. 2013;**58**:24-31. DOI: 10.1016/j.ejsobi.2013.05.006

[22] Wilczek G, Babczyńska A. Heavy metals in the gonads and hepatopancreas of spiders (Araneae) from variously polluted areas. Ekológia (Bratislava). 2000;**19**:283-292

[23] Jones DT, Hopkin SP. Reduced survival and body size in the terrestrial isopod *Porcellio scaber* from a metal-polluted environment. Environmental Pollution. 1998;**99**:215-223. DOI: 10.1016/S0269-7491(97)00188-7

[24] Spurgeon DJ, Hopkin SS. The effects of metal contamination on earthworm populations around a smelting works: Quantifying species effects. Applied Soil Ecology. 1996;**4**:147-160. DOI: 10.1016/0929-1393(96)00109-6

[25] Spurgeon DJ, Hopkin SP. Seasonal variation in the abundance, biomass and biodiversity of earthworms in soils contaminated with metal emissions from a primary smelting works. Journal of Applied Ecology. 1999;**36**:173-183. DOI: 10.1046/j.1365-2664.1999.00389.x

[26] Nahmani J, Lavelle P. Effects of heavy metal pollution on soil macrofauna in a grassland of northern France. European Journal of Soil Biology. 2002;**38**:297-300. DOI: 10.1016/S1164-5563(02)01169-X

[27] Bilandžić N, Dežđek D, Sedak M, Đokić M, Simić B, Rudan N, Brstilo M, Lisicin T. Trace elements in tissues of wild carnivores and omnivores in Croatia. Bulletin of Environmental Contamination and Toxicology. 2012;**88**:94-99. DOI: 10.1007/s00128-011-0449-y

[28] Damek-Poprawa M, Sawicka-Kapusta K. Damage to the liver, kidney, and testis with reference to burden of heavy metals in yellow-necked mice from areas around steelworks and zinc smelters in Poland. Toxicology. 2003;**186**:1-10. DOI: 10.1016/S0300-483X(02)00595-4

[29] Metcheva R, Teodorova S, Topashka-Ancheva M. A comparative analysis of the heavy metal loading of small mammals in different regions of Bulgaria I: Monitoring points and bioaccumulation features. Ecotoxicology and Environmental Safety. 2003;**54**:176-187. DOI: 10.1016/S0147-6513(02)00052-0

[30] Al Sayegh PS, Kopušar N, Kryštufek B. Small mammals as biomonitors of metal pollution: A case study in Slovenia. Environmental Monitoring and Assessment. 2014;**186**:4261-4274. DOI: 10.1007/s10661-014-3696-7

[31] Sánchez-Chardi A, Peñarroja-Matutano C, Borrás M, Nadal J. Bioaccumulation of metals and effects of a landfill in small mammals. Part III: Structural alterations. Environmental Research. 2009;**109**:960-967. DOI: 10.1016/j.envres.2009.08.004

[32] Tête N, Durfort M, Rieffel D, Scheifler R, Sánchez-Chardi A. Histopathology related to cadmium and lead bioaccumulation in chronically exposed wood mice, *Apodemus sylvaticus*, around a former smelter. Science of the Total Environment. 2014;**481**:167-177. DOI: 10.1016/j.scitotenv.2014.02.029

[33] da Silva J, de Freitas TR, Heuser V, Marinho JR, Bittencourt F, Cerski CT, Kliemann LM, Erdtmann B. Effects of chronic exposure to coal in wild rodents (*Ctenomys torquatus*) evaluated by multiple methods and tissues. Mutation Research. 2000;**470**:39-51. DOI: 10.1016/S1383-5718(00)00094-2

[34] Sánchez-Chardi A, Nadal J. Bioaccumulation of metals and effects of landfill pollution in small mammals. Part I. The greater white-toothed shrew, *Crocidura russula*. Chemosphere. 2007;**68**:703-711. DOI: 10.1016/j.chemosphere.2007.01.042

[35] Tovar-Sánchez E, Cervantes T, Martínez C, Rojas E, Valverde M, Ortiz-Hernández L, Mussali-Galante P. Comparison of two wild rodent species as sentinels of environmental contamination by mine tailings. Environmental Science and Pollution Research. 2012;**19**:1677-1686. DOI: 10.1007/s11356-011-0680-4

[36] Levengood JM, Heske EJ. Heavy metal exposure, reproductive activity, and demographic patterns in white-footed mice (*Peromyscus leucopus*) inhabiting a contaminated floodplain wetland. Science of the Total Environment. 2008;**389**:320-328. DOI: 10.1016/j.scitotenv.2007.08.050

[37] Santolo GM. Small mammals collected from a site with elevated selenium concentrations and three reference sites. Archives of Environmental Contamination and Toxicology. 2009;**57**:741-754. DOI: 10.1007/s00244-009-9322-y

[38] Sánchez-Chardi A, Peñarroja-Matutano C, Ribeiro CA, Nadal J. Bioaccumulation of metals and effects of a landfill in small mammals. Part II. The wood mouse, *Apodemus sylvaticus*. Chemosphere. 2007;**70**:101-109. DOI: 10.1016/j.chemosphere.2007a.06.047

[39] Van Straalen N. Genetic biodiversity in toxicant-stressed populations. Progress in Environmental Science. 1999;**1**:195-201

[40] Van Straalen N, Timmermans M. Genetic variation in toxicant stressed populations: An evaluation of the "genetic erosion" hypothesis. Human and Ecological Risk Assessment. 2002;**8**:983-1002. DOI: 10.1080/1080-700291905783

[41] Maes GE, Raeymaekers JAM, Pampoulie C, Seynaeve A, Goemans G, Belpaire C, Volckaert FAM. The catadromous European eel *Anguilla anguilla* (L.) as a model for freshwater evolutionary ecotoxicology: Relationship between heavy metal bioaccumulation, condition and genetic variability. Aquatic Toxicology. 2005;**73**:99-114. DOI: 10.1016/j.aquatox.2005.01.010

[42] Berckmoes V, Scheirs J, Jordaens K, Blust R, Backeljau T, Verhagen R. Effects of environmental pollution on microsatellite DNA diversity in wood mouse (*Apodemus Sylvaticus*) populations. Environmental Toxicology and Chemistry. 2005;**24**:2898-2907. DOI: 10.1897/04-483R.1

[43] Gardeström J, Dahl U, Kotsalainen O, Maxson A, Elfwing T, Grahn M, Bengtsson B, Breitholtz M. Evidence of population genetic effects of long-term exposure to contaminated sediments: A multiendpoint study with copepods. Aquatic Toxicology. 2008;**86**:426-436. DOI: 10.1016/j.aquatox.2007.12.003. Get rights and content

[44] Durrant J, Stevens JR, Hogstrand C, Burya NR. The effect of metal pollution on the population genetic structure of brown trout (*Salmo trutta* L.) residing in the river Hayle, Cornwall, UK. Environmental Pollution. 2011;**59**:3595-3603. DOI: 10.1016/j.envpol.2011.08.005

[45] Haimi J, Knott K, Selonen S, Laurikainen M. Has long-term metal exposure induced changes in life history traits and genetic diversity of the enchytraeid worm *Cognettia sphagnetorum*? (Vejd). Environmental Pollution. 2006;**140**:463-470. DOI: 10.1016/j.envpol.2005.08.009

[46] Ungherese G, Mengoni A, Somigli S, Baroni D, Focardi S, Ugolini A. Relationship between heavy metals pollution and genetic diversity in Mediterranean population of the sandhopper *Talitrus saltator* (Montagu) (Crustaceae, Amphipoda). Environmental Pollution. 2010;**158**:1638-1643

[47] Fratini S, Zane L, Ragionieri L, Vannini M, Cannicc S. Relationship between heavy metal accumulation and genetic variability decrease in the intertidial crab *Pachygrapsus marmoratus* (Decapoda; Grapsidae). Estuarine, Coastal and Shelf Science. 2008;**79**:679-686. DOI: 10.1016/j.ecss.2008.06.009

[48] Eeva T, Belskii E, Kuranov B. Environmental pollution affects genetic diversity in wild bird populations. Mutation Research. 2006;**608**:8-15. DOI: 10.1016/j.mrgentox.2006.04.021

[49] Peles JD, Towler WI, Guttman SI. Population genetic structure of earthworms (*Lumbricus rubelluz*) in soils contaminated by heavy metals. Ecotoxicology. 2003;**12**:379-386. DOI: 10.1023/A:1026269804938

[50] Jordaens K, De Wolf H, Van Houtte N, Vandecasteele B, Backeljau T. Genetic variation in two land snails *Cepaea nemoralis* and *Succinea putris* (Gastropoda), from sites differing in heavy metal content. Genetica. 2006;**128**:227-239. DOI: 10.1007/s10709-005-5705-9

[51] Yauk C, Fox G, McCarry B, Quinn J. Induced minisatellite germline mutations in herring gulls (*Larus argentatus*) living near steel mills. Mutation Research. 2000;**452**:211-218. DOI: 10.1016/S0027-5107(00)00093-2

[52] Basu N, Scheuhammer AM, Bursian SJ, Elliott J, Rouvinen-Watt K, Chan HM. Mink as a sentinel species in environmental health. Environmental Research. 2007;**103**:130-144. DOI: 10.1016/j.envres.2006.04.005

[53] Maintaining NR. Ecological integrity of landscape and eco-region. In: Noss RF, editor. Ecological Integrity: Integrating Environmental, Conservation and Health. Washington, DC: Island Press; 2001. pp. 191-208

[54] Gall JE, Boyd RS, Rajakaruna N. Transfer of heavy metals through terrestrial food webs: A review. Environmental Monitoring and Assessment. 2015;**187**:201

[55] Hossain MA, Piyatida P, Teixeria da Silva JA, Fujita M. Molecular mechanism of heavy metal toxicity and tolerance in plants: central role of glutathione in detoxification of

reactive oxygen species and methylglyoxal and in heavy metal chelation. Journal of Botany. 2012;**2012**:37. DOI: 10.1155/2012/872875

[56] Kim RY, Yoon JK, Kim TS, Yang JE, Owens G, Kim KR. Bioavailability of heavy metals in soils: Definitions and practical implementation—A critical review. Environmental Geochemistry and Health. 2015;**37**:1041. DOI: 10.1007/s10653-015-9695-y

[57] Mann RM, Vijver MG, Peijnenburg WJGM. Metals and metalloids in terrestrial systems: Bioaccumulation, biomagnification and subsequent adverse effects. In: Sánchez-Bayo PJ, van den Brink PJ, Mann RM, editors. Ecological Impacts of Toxic Chemicals. Sharjah: Bentham; 2011. pp. 43-62

[58] Gadd GM. Metals, minerals and microbes: Geomicrobiology and bioremediation. Microbiology. 2010;**156**:609-643. DOI: 10.1099/mic.0.037143-0

[59] Harris J. Soil microbial communities and restoration ecology: Facilitators or followers? Science. 2009;**325**:573-574. DOI: 10.1126/science.1172975

[60] Giller KE, Witter E, McGrath SP. Heavy metals and soil microbes. Soil Biology and Biochemistry. 2009;**41**:2031-2037. DOI: 10.1016/j.soilbio.2009.04.026

[61] Boshoff M, De Jonge M, Dardenne F, Blust R, Bervoets L. The impact of metal pollution on soil faunal and microbial activity in two grassland ecosystems. Environmental Research. 2014;**134**:169-180. DOI: 10.1016/j.envres.2014.06.024

[62] Smejkalova M, Mikanova O, Boruvka L. Effects of heavy metal concentrations on biological activity of soil micro-organisms. Plant, Soil and Environment. 2003;**49**:321-326. DOI: 10.1.1.557.375

[63] Abdousalam A. Effect of heavy metals on soil microbial processes and population. Egyptian Academic Journal of Biological Sciences. 2010;**2**:9-14

[64] Hsu MJ, Selvaraj K, Agoramoorthy G. Taiwan's industrial heavy metal pollution threatens terrestrial biota. Environmental Pollution. 2006;**43**:327-334. DOI: 10.1016/j.envpol.2005.11.023

[65] Rogival D, Scheirs J, Blust R. Transfer and accumulation of metals in a soil–diet–wood mouse food chain along a metal pollution gradient. Environmental Pollution. 2007;**145**:516-528. DOI: 10.1016/j.envpol.2006.04.019

[66] Boyd RS, Wall MA. Responses of generalist predators fed high-Ni *Melanotrichus boydi* (Heteroptera: Miridae): Elemental defense against the third trophic level. The American Midland Naturalist. 2001;**146**:186-198. DOI: 10.1674/0003-0031(2001)146[0186:ROGPFH]2.0.CO;2

[67] Green ID, Diaz A, Tibbett M. Factors affecting the concentration in seven-spotted ladybirds (*Coccinella septempunctata* L.) of Cd and Zn transferred through the food chain. Environmental Pollution. 2010;**158**:135-141. DOI: 10.1016/j.envpol.2009.07.032

[68] Dar MI, Khan FA, Green ID, Naikoo MI. The transfer and fate of Pb from sewage sludge amended soil in a multi-trophic food chain: A comparison with the labile elements

Cd and Zn. Environmental Science and Pollution Research. 2015;**22**:16133-16142. DOI: 10.1007/s11356-015-4836-5

[69] Peterson LR, Trivett V, Baker AJM, Aguiar C, Pollard AJ. Spread of metals through an invertebrate food chain as influenced by a plant that hyperaccumulates nickel. Chemoecology. 2003;**13**:103-108. DOI: 10.1007/s00049-003-0234-4

[70] Hamers T, van den Berg JHJ, van Gestel CAM, van Schooten FJ, Murk AJ. Risk assessment of metals and organic pollutants for herbivorous and carnivorous small mammal food chains in a polluted floodplain (Biesbosch, The Netherlands). Environmental Pollution. 2006;**144**:581-595. DOI: 10.1016/j.envpol.2006.01.020

[71] Veltman K, Huijbregts MAJ, Hamers T, Wijnhoven S, Hendriks AJ. Cadmium accumulation in herbivorous and carnivorous small mammals: Meta-analysis of field data and validation of the bioaccumulation model optimal modeling for ecotoxicological applications. Environmental Toxicology and Chemistry. 2007;**26**:1488-1496. DOI: 10.1897/06-518R.1

[72] Wijnhoven S, Leuven RSEW, Van der Velde G, Jungheim G, Koelemij EI, de Vries FT, Eijsackers HJP, Smits AJM. Heavy-metal concentrations in small mammals from a diffusely polluted floodplain: Importance of species- and location-specific characteristics. Archives of Environmental Contamination and Toxicology. 2007;**52**:603-613

[73] Van den Brink N, Lammertsma D, Dimmers W, Boerwinkel MC, Van der Hout A. Effects of soil properties on food web accumulation of heavy metals to the wood mouse (*Apodemus sylvaticus*). Environmental Pollution. 2010;**158**:245-251. DOI: 10.1016/j.envpol.2009.07.013

[74] Hladun KR, Parker DR, Trumble JT. Selenium accumulation in the floral tissues of two Brassicaceae species and its impact on floral traits and plant performance. Environmental and Experimental Botany. 2011;**74**:90-97. DOI: 10.1016/j.envexpbot.2011.05.003

[75] Quinn CF, Prins CN, Freeman JL, Gross AM, Hantzis LJ, Reynolds RJB, Yang S, Covey PA, Bañuelos GS, Pickering IJ, Fakra SC, Marcus MA, Arathi HS, Pilon-Smits EAH. Selenium accumulation in flowers and its effects on pollination. The New Phytologist. 2011;**192**:727-737. DOI: 10.1111/j.1469-8137.2011.03832.x

[76] Meindl GA, Ashman TL. The effects of aluminum and nickel in nectar on the foraging behavior of bumblebees. Environmental Pollution. 2013;**177**:78-81. DOI: 10.1016/j.envpol.2013.02.017

[77] Meindl GA, Bain DJ, Ashman TL. Nickel accumulation in leaves, floral organs and rewards varies by serpentine soil affinity. AoB Plants. 2014;**6**:1-9. DOI: 10.1093/aobpla/plu036

[78] Croteau MN, Luoma SN, Stewart AR. Trophic transfer of metals along freshwater food webs: Evidence of cadmium biomagnification in nature. Limnology and Oceanography. 2005;**50**:1511-1519. DOI: 10.4319/lo.2005.50.5.151